Natural Resources: Ecology, Pollution and Management

Natural Resources: Ecology, Pollution and Management

Editor: Stacy Keach

Callisto Reference
www.callistoreference.com

Callisto Reference,
118-35 Queens Blvd., Suite 400,
Forest Hills, NY 11375, USA

Visit us on the World Wide Web at:
www.callistoreference.com

ISBN: 978-1-63239-824-6 (Hardback)

The publisher's policy is to use permanent paper from mills that operate a sustainable forestry policy. Furthermore, the publisher ensures that the text paper and cover boards used have met acceptable environmental accreditation standards.

Printed in China.

Cataloging-in-publication Data

Natural resources : ecology, pollution and management / edited by Stacy Keach.
p. cm.
Includes bibliographical references and index.
ISBN 978-1-63239-824-6
1. Natural resources--Environmental aspects. 2. Ecology. 3. Pollution--Environmental aspects. 4. Environmental management. 5. Climatic changes. 6. Atmospheric ozone--Physiological effect. 7. Plant-atmosphere relationships.
I. Keach, Stacy.
HC85 .N38 2017
333.7--dc23

Table of Contents

Preface

There has been a long standing debate on unequal allocation of resources owing to the increasing scarcity and shortages. This book covers in detail some existent theories and innovative concepts revolving around natural resources, its challenges and its different aspects like ecology, pollution and management. It aims to provide comprehensive details on the multiple techniques of resource management and ecosystem preservation. The text has addressed the burning issue of the depletion of natural resources. It aims to serve as a resource guide for students and experts alike and contribute to the growth of the discipline. As this field is emerging and rapid pace, the contents of this book will help the readers understand the modern concepts and applications of the subject by presenting researches and case studies contributed by experts from across the globe.

This book was inspired by the evolution of our times; to answer the curiosity of inquisitive minds. Many developments have occurred across the globe in the recent past which has transformed the progress in the field.

This book was developed from a mere concept to drafts to chapters and finally compiled together as a complete text to benefit the readers across all nations. To ensure the quality of the content we instilled two significant steps in our procedure. The first was to appoint an editorial team that would verify the data and statistics provided in the book and also select the most appropriate and valuable contributions from the plentiful contributions we received from authors worldwide. The next step was to appoint an expert of the topic as the Editor-in-Chief, who would head the project and finally make the necessary amendments and modifications to make the text reader-friendly. I was then commissioned to examine all the material to present the topics in the most comprehensible and productive format.

I would like to take this opportunity to thank all the contributing authors who were supportive enough to contribute their time and knowledge to this project. I also wish to convey my regards to my family who have been extremely supportive during the entire project.

Editor

1

Estimating Soil Organic Carbon Stocks and Spatial Patterns with Statistical and GIS-Based Methods

Junjun Zhi[1], Changwei Jing[2], Shengpan Lin[1], Cao Zhang[1], Qiankun Liu[1], Stephen D. DeGloria[3], Jiaping Wu[2]*

1 College of Environmental and Resource Sciences, Zhejiang University, Hangzhou, China, **2** Ocean College, Zhejiang University, Hangzhou, China, **3** Department of Crop and Soil Sciences, Cornell University, Ithaca, New York, United States of America

Abstract

Accurately quantifying soil organic carbon (SOC) is considered fundamental to studying soil quality, modeling the global carbon cycle, and assessing global climate change. This study evaluated the uncertainties caused by up-scaling of soil properties from the county scale to the provincial scale and from lower-level classification of Soil Species to Soil Group, using four methods: the mean, median, Soil Profile Statistics (SPS), and pedological professional knowledge based (PKB) methods. For the SPS method, SOC stock is calculated at the county scale by multiplying the mean SOC density value of each soil type in a county by its corresponding area. For the mean or median method, SOC density value of each soil type is calculated using provincial arithmetic mean or median. For the PKB method, SOC density value of each soil type is calculated at the county scale considering soil parent materials and spatial locations of all soil profiles. A newly constructed 1:50,000 soil survey geographic database of Zhejiang Province, China, was used for evaluation. Results indicated that with soil classification levels up-scaling from Soil Species to Soil Group, the variation of estimated SOC stocks among different soil classification levels was obviously lower than that among different methods. The difference in the estimated SOC stocks among the four methods was lowest at the Soil Species level. The differences in SOC stocks among the mean, median, and PKB methods for different Soil Groups resulted from the differences in the procedure of aggregating soil profile properties to represent the attributes of one soil type. Compared with the other three estimation methods (i.e., the SPS, mean and median methods), the PKB method holds significant promise for characterizing spatial differences in SOC distribution because spatial locations of all soil profiles are considered during the aggregation procedure.

Editor: Ben Bond-Lamberty, DOE Pacific Northwest National Laboratory, United States of America

Funding: This study was supported by the National Natural Science Foundation of China (No. 30771253) and the Key Project of the Science and Technology Department of Zhejiang Province (No. 2006C22026). The funders had no role in study design, data collection and analysis, decision to publish, or preparation of the manuscript.

Competing Interests: The authors have declared that no competing interests exist.

* E-mail: jw67@zju.edu.cn

Introduction

Soil organic carbon (SOC), which plays a critical role in the global carbon cycle, comprises a major part of the terrestrial carbon reservoir [1–3]. In terrestrial ecosystems, SOC stock is almost three times the size of carbon storage in the vegetation of terrestrial ecosystems [4], and approximately twice as large as carbon storage in the atmosphere [5]. Because of the important role of SOC and its large quantity stored in terrestrial ecosystems, a slight change in SOC stock may influence global climate [6–8]. Accurately quantifying SOC stock has become a focus of present research on global climate change, and is considered essential for studying soil quality, modeling the global carbon cycle, and assessing global climate change [8–12].

Different SOC stock estimates, however, can vary greatly at both global and regional scales [13–17]. For global scales, Bohn [18] estimated the total SOC stock was 3,000 Pg (1 Pg = 10^{15} g), whereas Bolin [19] estimated only 710 Pg, over a four-fold difference. In China, estimated SOC stocks for terrestrial ecosystems range from 50 Pg [20] to 185.7 Pg [21], also approximately a four-fold difference. The methodology for estimating SOC stocks, such as soil profile-based [22–24] and model-based (e.g., the CENTURY [25] and DeNitrification-DeComposition models [26]), is one of the main factors attributed to the wide range of differences in SOC stock estimation from different studies. The soil profile-based methodology calculates SOC stock using soil profiles and their corresponding areas obtained from soil survey products (the soil type method), vegetation type maps (the vegetation type method), or life zone maps (the life zone method), among which the soil type method is the most widely used [9,18–19,27–28]. According to the sources of soil area, the soil type method contains the Soil Profile Statistics (SPS) method and the GIS-based Soil Type (GST) method [2]. The SPS method calculates SOC stock by multiplying the SOC density value of a soil type by its corresponding field survey area recorded in soil survey reports (e.g., Soil Species of China [29]). The GIS-based Soil Type method calculates areas of various soil types accurately based on digital soil map and can provide information on the spatial distribution of SOC stocks.

In China, different scale (i.e., county, provincial, and national) soil survey products (e.g., soil survey reports, soil maps) of the Second National Soil Survey of China are the most important data sources for SOC stock estimations [15,20–21,30–33]. For the

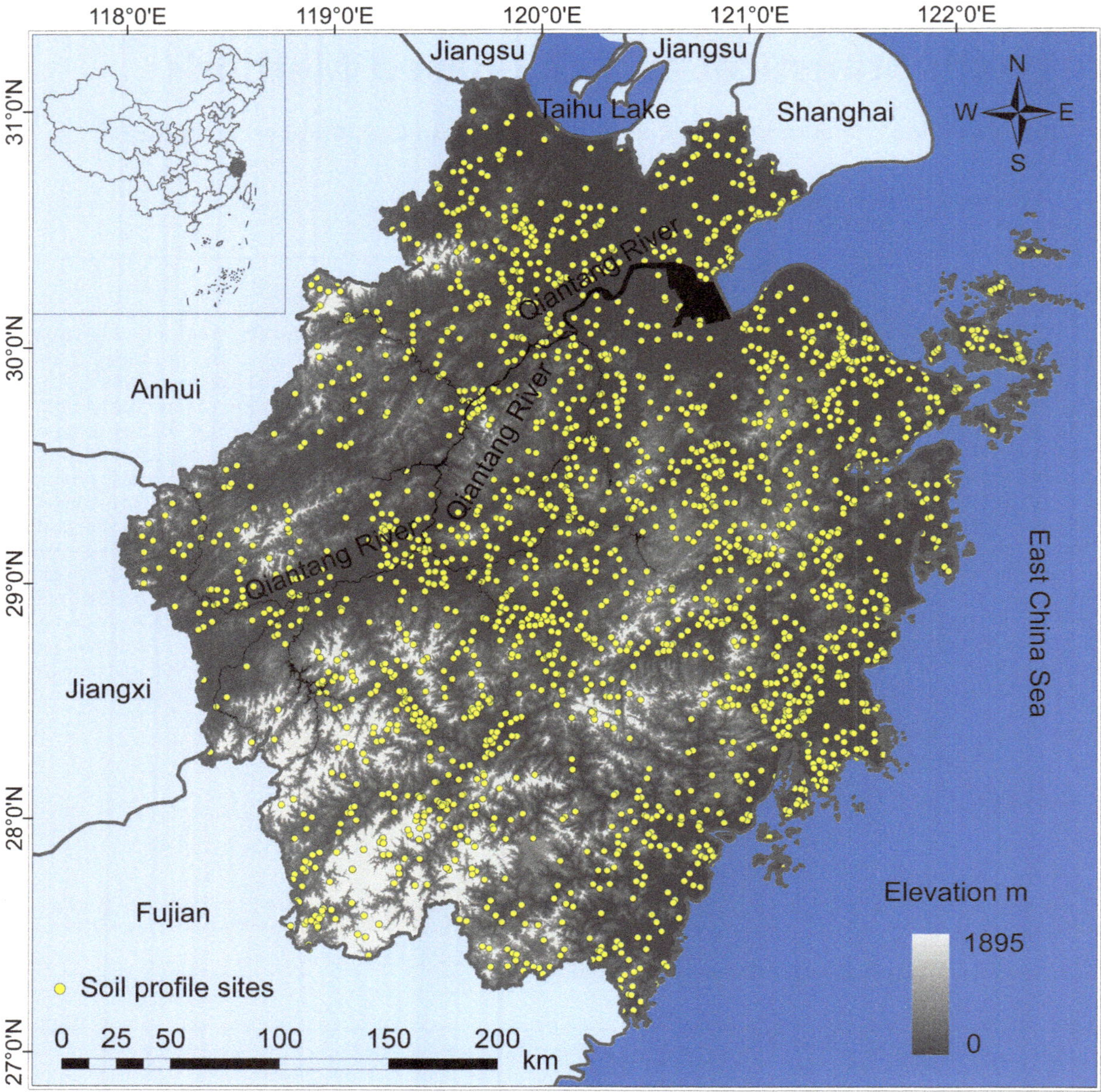

Figure 1. Study area and distribution of soil profile sites.

Second National Soil Survey of China, soil profile properties were aggregated sequentially in the county, provincial, and national scales. Different number of sampling profiles can be used to represent the attributes of one soil type at different scales and can cause uncertainties in SOC stock estimation. At the provincial scale, calculating the mean and median of SOC density values of multiple soil profiles of the same soil type name to represent the SOC density value of that soil type are the two most commonly used GIS-based Soil Type methods [32–34]. At the county scale, the SPS method [20,30–32] and the pedological professional knowledge based (PKB) method [2,15,34–36] are the two most commonly used methods. Compared with the median or mean method, the PKB method aggregates soil profile properties downscaling from the provincial scale to more detail soil map units, which links a SOC density value of each soil profile to a digital soil map according to the identity or similarity in soil parent materials and spatial locations of all soil profiles at the county scale. However, the uncertainties of soil profile properties caused by up-scaling from the county scale to the provincial scale among the four methods remain unknown.

Soil map up-scaling includes soil classification level up-scaling and resolution up-scaling. The former aggregates Soil Species to higher soil classification levels (e.g., Soil Group, Subgroup, Soil Family); the latter aggregates higher resolution soil maps to lower resolution ones, such as from 1:50,000-county-scale map to 1:1,000,000-country-scale map. Both the soil classification level up-scaling and resolution up-scaling can cause uncertainties in SOC stock estimation. Zhao et al. [34] tested the effects of soil map resolution up-scaling from 1:500,000 to 1:10,000,000 using the three GIS-based Soil Type methods (i.e., the mean, median, and PKB methods) on SOC stock estimation for Hebei Province, China. However, the effects of these three GIS-based Soil Type

Table 1. Methods used to estimate SOC (soil organic carbon) stocks in Zhejiang Province, China.

Method	SOC density value	Scale	Note
SPS[a]	Mean	County; soil species	(1) One soil species has multiple areas, which were surveyed county by county. (2) Mean SOC density value was calculated from one or multiple profiles within the county. (3) 2154 soil profiles were used.
Mean or median	Mean or median	Province; soil species	(1) One soil species has one area calculated from the digital soil map. (2) Mean or median SOC density value was calculated from one or multiple profiles within the Province. (3) 2154 soil profiles were used.
PKB[a]	Mean	County; soil map unit	(1) Soil map units were derived from the digital soil map county by county. (2) One soil map unit in one county may have one or multiple areas calculated from the digital soil map. (3) Mean SOC density value was calculated from one or multiple profiles located within one polygon; polygons belong to one soil map unit in one county may assigned different SOC density values. (4) 2154 soil profiles were used.

[a]SPS, Soil Profile Statistics; PKB, pedological professional knowledge based.

methods on SOC stock estimation at soil map resolution larger than 1:500,000 remain unknown. Moreover, the uncertainties in SOC stock estimation caused by soil classification level up-scaling using the SPS, mean, median, and PKB methods remain unknown.

The goal of this study is to evaluate the uncertainties of changes in scale among the four estimation methods (i.e., the SPS, mean, median, and PKB methods) by using a newly completed 1:50,000 soil survey geographic database of Zhejiang Province, China. Specifically, the main objectives of this study were to: (1) evaluate the uncertainties caused by up-scaling of soil profile properties from the county scale to the provincial scale and soil classification levels from Soil Species to Soil Group among the four methods; and (2) quantify spatial differences in SOC stock estimation among the three GIS-based Soil Type methods.

Materials and Methods

Study Area

Zhejiang Province is located between 27°06′ and 31°03′ N latitude and 118°01′ and 123°10′ E longitude in southeastern China (Figure 1). The Province covers a land area of $1.02*10^5$ km^2. Hills and mountains are dominant terrain, occupying 70.4% of the total land area. The plains and basins account for 23.2%, and the remaining 6.4% is comprised of lakes and rivers. Provincial topography is characterized by high mountains in the southwest and low plains in the northeast. The elevation ranges from 0 to 1895 m with an average of 296 m above sea level.

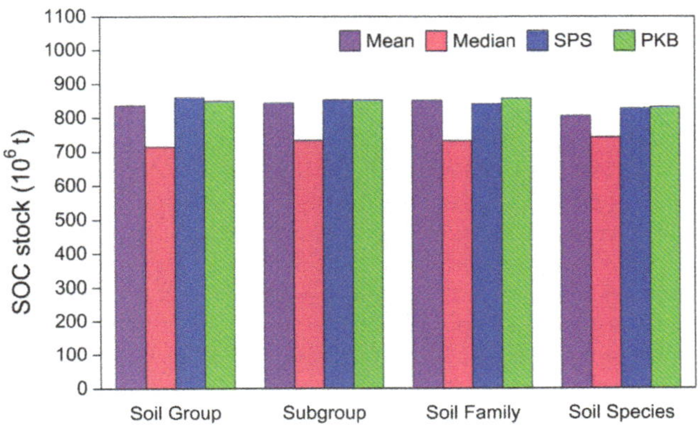

Figure 2. The estimates of SOC (soil organic carbon) stocks for soil classification levels up-scaling from Soil Species to Soil Group using the mean, median, SPS (soil profile statistic), and PKB (pedological professional knowledge based) methods.

With a subtropical monsoon climate, Zhejiang Province has an annual average temperature from 15 to 18°C and an annual precipitation of 1200 to 1800 mm. As substantial differences exist in climate, geomorphology, geohydrology, land use, and parent material throughout the Province, its vegetation and soils and their spatial distribution patterns vary greatly. According to Soils of Zhejiang Province [37], Red soils are the dominant Soil Group, accounting for approximately 39% of the total area of Zhejiang soils.

Data Sources

A recently completed 1:50,000 soil survey geographic database of Zhejiang Province [38] was used in the study, which is the most detailed soil survey data at the provincial scale in China to date. The database includes Soil Spatial Data and Soil Attributes Data. The Soil Spatial Data include a 1:50,000 digital soil map of the Province and a digital map of 2154 geo-referenced soil sampling profile sites, both of which were derived by digitizing and re-compiling field soil survey maps at 1:50,000 from 76 counties in Zhejiang. The Soil Attributes Data contain the properties of the 2154 geo-referenced soil sampling profiles, which were taken from original county soil survey reports. For each profile, there are one to seven soil layers; for each layer, there are up to 104 soil physical and chemical properties, including geographic location, depth, bulk density, organic matter content, and gravel content, etc.

The soil survey geographic database was used to estimate SOC stocks with the mean, median, and PKB methods. This information from the Second National Soil Survey of China conducted in early 1980's was the most comprehensive and detailed inventory of soil characteristics in Zhejiang Province. The Genetic Soil Classification System of Zhejiang Province was used during the field soil survey. The soils were classified using a six-level hierarchical scheme: Order, Suborder, Group, Subgroup, Family, and Species. The Soil Species is the basic classification level, and the Soil Group is the most commonly used classification level in China [2]. According to the Genetic Soil Classification

Table 2. Descriptive statistics of profile soil organic carbon densities by Soil Group.

Soil Group	Soil Order of U.S. Taxonomy[a]	N[b]	Range	Min[b]	Max[b]	Mean	Median	SD[b]	CV/%[b]	Skew[b]	Kurt[b]
Red soils	Alfisols, ultisols, inceptisols	372	32.98	0.87	33.85	6.97	6.15	3.93	15.5	2.29	9.48
Yellow soils	Alfisols, inceptisols	126	91.26	1.93	93.19	16.56	13.39	14.61	213.3	3.25	13.16
Purple soils	Inceptisols, entisols	84	12.10	0.29	12.38	5.19	4.84	2.36	5.6	0.70	0.28
Limestone soils	Mollisols, inceptisols	22	20.52	1.95	22.47	9.14	9.68	5.04	25.4	0.51	0.92
Skel soils	Inceptisols, entisols	114	36.71	0.10	36.81	4.89	3.30	5.27	27.8	3.56	16.24
Red clay soils	Inceptisols, alfisols	4	2.22	4.80	7.02	5.48	5.04	1.04	1.1	1.92	3.75
Mountain meadow soils	Histosols, inceptisols	4	241.54	37.98	279.52	104.32	49.89	116.98	13680.0	1.98	3.94
Fluvio-aquic soils	Inceptisols, entisols	189	20.34	0.94	21.28	6.38	5.78	3.48	12.1	0.96	1.66
Coastal saline soils	Inceptisols	64	14.25	0.92	15.17	7.37	7.50	3.19	10.2	−0.05	−0.55
Paddy soils	Anthrosols	1175	143.90	1.97	145.87	9.82	8.54	7.13	50.9	8.10	124.50
All profiles		2154	279.42	0.10	279.52	9.06	7.62	9.41	88.6	14.17	346.41

[a]Reference conversion between Soil Group of the Genetic Soil Classification System of Zhejiang Province and Soil Order of the U.S. Taxonomy.
[b]N, Min, Max, SD, CV, Skew, Kurt are the abbreviations of the number of soil profiles occurring in a Soil Group, minimum, maximum, standard deviation, coefficient of variation, skewness and kurtosis, respectively.

System of Zhejiang Province, the collected 2154 soil profiles originated from 10 Soil Groups, 21 Subgroups, 99 Soil Families and 277 Soil Species.

Estimation of Organic Carbon Density for Soil Profiles

For each of the 2 154 soil profiles, SOC density was calculated with the formula [34,39]:

$$\mathrm{SOCD}_D = \sum_{i=1}^{n} [(1-\theta_i\%) \times \rho_i \times C_i \times T_i/100] \qquad (1)$$

where SOCD_D (kg m^{-2}) is the SOC density of a soil profile within a depth of D (cm), n is the number of soil layers in the soil survey, $\theta_i\%$ represents the volumetric percentage of gravel (>2 mm) content, ρ_i is the soil bulk density (g/cm^3), C_i is the organic carbon content (C g/kg), and T_i represents the thickness (cm) of the layer i. Density of SOC was estimated to a maximum depth of 100 cm to facilitate comparison among data sets. For profiles with actual depths greater than or equal to 100 cm, but less than 100 cm was observed, data for the unobserved profile sections were derived from the mean values of all the corresponding soil profiles of the same Soil Species or Soil Family [15,33,40]. Organic carbon content is calculated by multiplying soil organic matter content by 0.58 (the Bemmelen index), which is based on the assumption that soil organic matter contains approximately 58% organic carbon [41].

Estimation of Regional Soil Organic Carbon Stocks

The SOC stocks of the Province were estimated by using the SPS, mean, median, and PKB methods (Table 1).

For the SPS method, the mean SOC density value of each soil type in one county was calculated from one or multiple profiles belonging to that soil type; then the SOC stock value for that soil type was calculated by multiplying its SOC density value by its corresponding area; finally, the SOC stock values of all soil types were summed up as the total SOC stock for the Province. The area calculated from the 1:50,000 digital soil map was used in this study to facilitate comparison among different methods.

For the mean or median method, the arithmetic mean or median of SOC density values of all soil profiles of the same soil type name in the Province was calculated first as the SOC density value of that soil type; secondly, SOC stocks for polygons on the 1:50,000 digital soil map were calculated by multiplying the mean or median values by corresponding areas of the soil types and then were summed up as the total stock for the Province.

For the PKB method, the SOC density value of each of the 2154 soil profiles was linked to corresponding polygons on the 1:50,000 digital soil map according to the same soil type name, the soil parent materials, and the spatial locations by county. When two or more soil profiles of the same soil type name were located in one polygon, the mean SOC density value of these soil profiles was calculated and used for the linkage. The SOC stock values for all polygons were calculated and then were summed up as the total stock for the Province.

The estimation of SOC stocks using the three GIS-based Soil Type methods were performed using ESRI software ArcGIS 10.0 (Redlands, CA) with water and urban areas excluded from the calculation.

Table 3. Estimates of SOC (soil organic carbon) stocks and SOC density values of various Soil Groups using the mean, median, SPS (soil profile statistic), and PKB (pedological professional knowledge based) methods.

Soil Group	Area[a]	PKB		Mean			Median			SPS		
	(km^2)	SOC density ($kg\ m^{-2}$)	SOC stock (10^6 t)	SOC density ($kg\ m^{-2}$)	SOC stock (10^6 t)	%[b]	SOC density ($kg\ m^{-2}$)	SOC stock (10^6 t)	%[b]	SOC density ($kg\ m^{-2}$)	SOC stock (10^6 t)	%[b]
Red soils	39681.3	6.53	259.10	6.73	266.96	3.0	6.24	247.71	−4.4	6.50	258.11	−0.4
Yellow soils	10013.5	16.92	169.45	14.00	140.15	−17.3	12.24	122.54	−27.7	16.72	167.44	−1.2
Purple soils	3597.7	5.73	20.60	5.00	18.01	−12.6	4.77	17.15	−16.7	5.65	20.34	−1.2
Limestone soils	1571.0	10.18	16.00	8.56	13.45	−15.9	9.43	14.82	−7.4	10.47	16.45	2.8
Skel soils	13736.8	5.11	70.20	5.26	72.30	3.0	4.29	58.99	−16.0	5.09	69.95	−0.4
Red clay soils	29.1	5.45	0.16	5.45	0.16	0.0	5.45	0.16	0.0	5.45	0.16	0.0
Mountain meadow soils	3.3	45.30	0.15	104.32	0.34	130.3	49.89	0.16	10.1	45.30	0.15	0.0
Fluvio-aquic soils	4318.3	7.43	32.10	7.33	31.66	−1.4	7.17	30.98	−3.5	7.44	32.13	0.1
Coastal saline soils	2793.3	7.79	21.75	7.58	21.19	−2.6	7.50	20.95	−3.7	7.80	21.78	0.1
Paddy soils	24995.8	9.68	241.98	9.67	241.66	−0.1	9.20	229.86	−5.0	9.64	241.06	−0.4
Total soils	100740.1	8.25	831.49	8.00	805.88	−3.1	7.38	743.32	−10.6	8.21	827.57	−0.5

[a]Area used for the four methods in this study was from the 1:50,000 digital soil map.
[b]Percentage of difference of estimated SOC stocks between the mean, median, or SPS method and the PKB method.

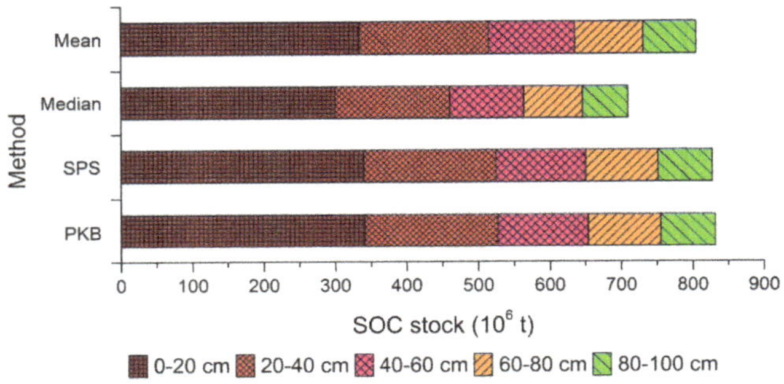

Figure 3. Estimated SOC stocks at different soil depths using the mean, median, SPS (soil profile statistic), and PKB (pedological professional knowledge based) methods.

Results

Effect of Soil Classification Levels Up-Scaling from Soil Species to Soil Group on SOC Estimation Using Different Methods

The SOC stocks for soil classification levels up-scaling from Soil Species to Soil Group using the mean, median, SPS, and PKB methods were calculated (Figure 2). Among the four methods, the estimated SOC stocks for the study area using the median method were always the lowest regardless of soil classification level. Because the SOC densities of the 2154 soil profiles showed a positively skewed distribution with skewness of 14.17 (Table 2), the estimated SOC stocks using the mean method were much higher than those using the median method at all four soil classification levels.

With soil classification levels up-scaling from Soil Species to Soil Group, the estimated SOC stocks presented small variations with coefficient of variations of 2.1%, 1.4%, 1.4%, and 1.2% for the mean, median, SPS, and PKB methods, respectively. The coefficient of variations of estimated SOC stocks among different methods presented an increasing trend, with 4.4%, 6.2%, 6.3%, and 7.1% for soil classification levels of Soil Species, Soil Family, Subgroup, and Soil Group, respectively. The estimated SOC stocks using the mean, median, and SPS methods at the Soil Species level were 3.1%, 10.6%, and 0.5% lower than that obtained by the PKB method, respectively. The variation of estimated SOC stocks among different methods was obviously larger than that among different soil classification levels.

Effect of Different Methods on the Estimates of SOC Stocks for Various Soil Groups

Soil Group is the most stable and consistent soil classification level commonly used in China [2]. In this study, the SOC stocks and SOC densities for various Soil Groups estimated by the mean, median, SPS, and PKB methods are presented in Table 3. For the Soil Groups with SOC density values of soil profiles presented positively skewed distribution, the estimated SOC stocks using the median method were usually lower than that using the mean and PKB methods (Table 2, Table 3). The estimated SOC stocks for various Soil Groups using the SPS method were most similar to that obtained by the PKB method because both methods aggregate soil profile properties at the county scale. The estimated SOC stocks for various Soil Groups using the mean and median methods showed large differences compared with the PKB method because both the mean and median methods aggregating soil profile properties at the provincial scale that the SOC density values of all soil profiles belonging to a soil type in the Province were aggregated to one value.

The most pronounced differences of the estimated SOC stocks for the ten Soil Groups (i.e., Red soils, Yellow soils, Purple soils, Limestone soils, Skel soils, Red clay soils, Mountain meadow soils, Fluvio-aquic soils, Coastal saline soils and Paddy soils) among the three GIS-based Soil Type methods occurred on Yellow soils and Mountain meadow soils, which presented high SOC density values with high coefficient of variances. Thus, the process of linking the

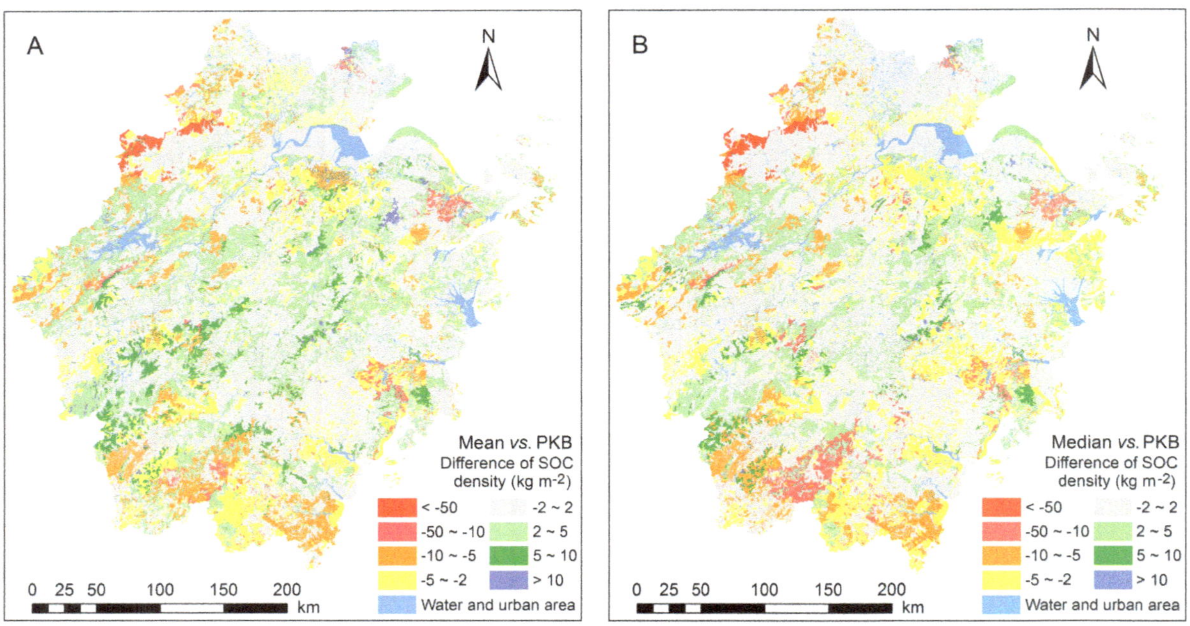

Figure 4. Spatial differences of estimated SOC (soil organic carbon) densities between the mean and PKB (pedological professional knowledge based) methods (A), and between the median and PKB methods (B).

aggregated SOC density values of soil profiles belonging to these two Soil Groups to corresponding polygons on the 1:50,000 digital soil map would significantly influence the estimates of SOC stocks. One important factor leading to the significantly high variation of SOC in Mountain meadow soils was the limited number of soil profiles ($n=4$) in the soil survey area.

Effect of Different Methods on the SOC Stock Estimation at Different Soil Depths

For the four methods, the trend of estimated SOC stocks at each of the five soil depths (0–20, 20–40, 40–60, 60–80, and 80–100 cm) is consistent with the soil depth of 0–100 cm, followed the order: PKB>SPS>mean>median (Figure 3). At each of the five soil depths, the estimated SOC stock using the SPS method presented the smallest difference while the median method presented the largest difference compared with the PKB method. Decreasing SOC stock with increasing soil depth was evident independent of estimation method, and these decreasing trends are consistent among all four methods. At the depth of 0–20 cm, SOC stocks estimated by the mean, median, SPS, and PKB methods account for 41.5%, 42.3%, 41.1%, and 41.1%, of total SOC stocks, respectively, while the SOC stocks at the depth of 80–100 cm estimated by the four methods account for only 9.2%, 8.9%, 9.1%, and 9.1%, of the total SOC stocks, respectively.

Comparison of the Differences in the Spatial Patterns of SOC Distribution among the Three GIS-based Soil Type Methods

Estimated SOC densities using three GIS-based Soil Type methods showed clearly spatial differences (Figure 4). The largest differences were as high as -229.63 kg m^{-2} between the median and PKB methods and -175.20 kg m^{-2} between the mean and PKB methods. Polygons mostly belonging to the Soil Groups of Mountain meadow soils and Yellow soils showed substantial differences (<-50 kg m^{-2}) in SOC densities among the three GIS-based Soil Type methods. For the polygons with large spatial differences in SOC densities ranging from -50 kg m^{-2} to -5 kg m^{-2} and >5 kg m^{-2}, the locations were usually in mountain and hilly areas with dominate Soil Groups of Yellow soils, Red soils, and Skew soils, with the remainder in plain areas with dominate Soil Group of Paddy soils. The total areas for spatial differences in SOC densities with absolute value larger than 5 kg m^{-2} between the mean and PKB methods and the median and PKB methods were estimated to be 12883.4 km^2 and 11428.0 km^2, accounting for 12.8% and 11.3% of the total area of Zhejiang soils, respectively. The total areas for spatial differences in SOC densities with absolute value lower than 2 kg m^{-2} between the mean and PKB methods and the median and PKB methods were estimated to be 55316.1 km^2 and 59147.7 km^2, accounting for 54.9% and 58.7% of the total area of Zhejiang soils, respectively.

Discussion

With soil classification levels up-scaling from Soil Species to Soil Group, the estimated SOC stock values using the SPS method were most similar to that obtained by the PKB method. The difference of the estimated SOC stocks between the SPS method and the PKB method occurred in the cases where two or more soil profiles belonging to one soil species existed in one county. However, for most cases, there was only one soil profile belonging to one soil species in one county, the SOC density value of that soil profile was used to represent the SOC density value of that soil species in that county, which resulted in the same estimated SOC stock value of that soil species in that county using the SPS and PKB methods. Therefore, more number of soil profiles is needed to further compare the difference between these two methods.

The estimated SOC stock values using the mean method were similar to that using the PKB method. When two or more soil profiles located in the same polygon within a county, the mean SOC density value of these soil profiles was calculated and used to link with that polygon, so the PKB method approaches the mean method in linking SOC density values with polygons to some extent [34], especially for higher soil classification levels (e.g., Soil Group, Subgroup, Soil Family).

With soil profile properties up-scaling from the county scale (the SPS and PKB methods) to the provincial scale (the mean and median methods), the SOC density values of all soil profiles belonging to a soil type in the Province were aggregated to one value. The aggregating procedure for the mean or median method could cause large uncertainties because information about how much area one profile can represent was missing, thus different polygons belonging to the same map unit on the digital soil map were linked with the same SOC density value. For the PKB method, different polygons belonging to the same map unit may be assigned different SOC density values depending on their locations, thus has an obvious advantage in the demonstration of spatial differences in SOC distribution. The difference in aggregating procedure led to Soil Groups (e.g., Mountain meadow soils, Yellow soils) with soil profiles of high variations in SOC density values presented significant spatial differences among the three GIS-based Soil Type methods.

Conclusions

The up-scaling of soil classification levels from Soil Species to Soil Group has small effect on SOC stock estimation. However, obvious differences occurred with different estimation methods, especially for Soil Groups with high variations in SOC densities and spatial distribution. The PKB method, which links soil profile properties to spatial databases by county, produced more stable results for soil types, soil parent materials, and spatial locations of all soil profiles under consideration. Thus this method has an obvious advantage in the demonstration of differences in spatial patterns of SOC distribution than both the mean and median methods. We recommend the PKB method as a prior option rather than the mean, median, and SPS methods for SOC stock estimation in China, especially when 1:50,000 soil survey geographic database (soil map) is available. It potentially reduces uncertainties related to up-scaling soil profile properties.

Acknowledgments

We are very grateful to the Academic Editor and two reviewers for their useful comments and suggestions on the original manuscript.

Author Contributions

Conceived and designed the experiments: JJZ JPW. Performed the experiments: JJZ CWJ CZ QKL. Analyzed the data: JJZ CZ QKL. Contributed reagents/materials/analysis tools: JJZ. Wrote the paper: JJZ. Revised the manuscript: SPL JPW SDD.

References

1. Davidson EA, Janssens IA (2006) Temperature sensitivity of soil carbon decomposition and feedbacks to climate change. Nature 440: 165–173.
2. Zhang Y, Zhao YC, Shi XZ, Lu XX, Yu DS, et al. (2008) Variation of soil organic carbon estimates in mountain regions: a case study from Southwest China. Geoderma 146: 449–456.
3. Zhou Y, Pei Z, Su J, Zhang J, Zheng Y, et al. (2012) Comparing Soil Organic Carbon Dynamics in Perennial Grasses and Shrubs in a Saline-Alkaline Arid Region, Northwestern China. PLoS ONE 7 (8): e42927. doi:10.1371/journal.-pone.0042927
4. Schlesinger WH (1990) Evidence from chronosequence studies for a low carbon-storage potential of soils. Nature 348: 232–234.
5. Eswaran H, Van Den Berg E, Reich PF (1993) Organic carbon in soils of the world. Soil Science Society of America Journal 57: 192–194.
6. Lal R, Kimble JM, Follett RF, Cole CV (1998) The Potential of U.S. Cropland to Sequester Carbon and Mitigate the Greenhouse Effect. Chelsea: Ann Arbor Press.
7. Uri ND (2000) Conservation practices in U.S. agriculture and their implication for global climate change. Science of the Total Environment 256: 23–38.
8. Li MM, Zhang XC, Pang GW, Han FF (2013) The estimation of soil organic carbon distribution and storage in a small catchment area of the Loess Plateau. Catena 101: 11–16.
9. Batjes NH (1996) Total carbon and nitrogen in the soils of the world. European Journal of Soil Science 47: 151–163.
10. Tan G, Shibasaki R (2003) Global estimation of crop productivity and the impacts of global warming by GIS and EPIC integration. Ecological Modelling 168: 357–370.
11. Janzen HH (2004) Carbon cycling in earth systems-a soil science perspective. Agriculture, Ecosystems & Environment 104: 399–417.
12. Morisada K, Ono K, Kanomata H (2004) Organic carbon stock in forest soils in Japan. Geoderma 119: 21–32.
13. Wang SQ, Huang M, Shao XM, Robert AM, Li KR, et al. (2004) Vertical distribution of soil organic carbon in China. Environmental Management 33: 200–209.
14. Leifeld J, Bassin S, Fuhrer J (2005) Carbon stocks in Swiss agricultural soils predicted by land-use, soil characteristics, and altitude. Agriculture, Ecosystems & Environment 105: 255–266.
15. Yu DS, Shi XZ, Wang HJ, Sun WX, Chen JM, et al. (2007) Regional patterns of soil organic carbon stocks in China. Journal of Environmental Management 85: 680–689.
16. Arnold RW (1995) Role of soil survey in obtaining a global carbon budget. In: Lal R, Kimble J, Levine E, Stewart BA, editors. Soils and global change. Boca Raton: CRC Press. 257–263.
17. Rapalee G, Trumbore SE, Davidson EA, Harden JW, Veldhuis H (1998) Soil carbon stocks and their rates of accumulation and loss in a boreal forest landscape. Global Biogeochemical Cycles 12: 687–701.
18. Bohn HL (1976) Estimate of organic carbon in world soils. Soil Science Society of America Journal 40: 468–470.
19. Bolin B (1977) Change of land biota and their importance for the carbon cycle. Science 196: 613–615.
20. Pan GX (1999) Study on carbon reservoir in soils of China. Bulletin of Science and Technology 15: 330–332.
21. Fang JY, Liu GH, Xu SL (1996) Carbon sinks of terrestrial ecosystems in China. In: Wang RS, editor. Study on Key Issues of Modern Ecology. Beijing: Press of Chinese Sciences and Technologies, 251–267 p.
22. Homann PS, Sollins P, Fiorella M, Thorson M, Kern JS (1998) Regional soil organic carbon storage estimates for western Oregon. Soil Science Society of America Journal 62: 789–796.
23. Batjes NH (2000) Effects of mapped variation in soil conditions on estimates of soil carbon and nitrogen stocks for South America. Geoderma 97: 135–144.
24. Wu J, Ransom MD, Kluitenberg GJ, Nellis MD, Seyler HL (2001) Land-use management using a soil survey geographic database for Finney County, Kansas. Soil Science Society of America Journal 65: 169–177.
25. Alvaro-Fuentes J, Plaza-Bonilla D, Arrue JL, Lampurlanes J, Cantero-Martinez C (2014) Soil organic carbon storage in a no-tillage chronosequence under Mediterranean conditions. Plant and Soil 376: 31–41.
26. Xu SX, Zhao YC, Shi XZ, Yu DS, Li CS, et al. (2013) Map scale effects of soil databases on modeling organic carbon dynamics for paddy soils of China. Catena 104: 67–76.
27. Batjes NH (2006) Soil carbon stocks of Jordan and projected changes upon improved management of croplands. Geoderma 132: 361–371.
28. Liebens J, VanMolle M (2003) Influence of estimation procedure on soil organic carbon stock assessment in Flanders, Belgium. Soil Use and Management 19: 364–371.
29. National Soil Survey Office (1993) Soil Species of China, Volumes 1–6. Beijing: China Agriculture Press.
30. Wang SQ, Zhou CH (1999) Estimating soil carbon reservoir of terrestrial ecosystem in China. Geographical Research 18: 349–356.
31. Jin F, Yang H, Cai ZC, Zhao QG (2001) Calculation of density and reserve of organic carbon in soils. Acta Pedologica Sinica 38: 522–528.
32. Yu JJ, Yang F, Wu KN, Li L, Lü QL (2008) Soil organic carbon storage and its spatial distribution in Henan Province. Chinese Journal of Applied Ecology 19: 1058–1063.
33. Zhao YC, Shi XZ, Yu DS, Pagella TF, Sun WX, et al. (2005) Soil organic carbon density in Hebei Province, China: estimates and uncertainty. Pedosphere 15: 293–300.
34. Zhao YC, Shi XZ, Yu DS, Sun WX (2006) Map scale effect on soil organic carbon stock estimation in north China. Soil Science Society of America Journal 70: 1377–1386.
35. Shi XZ, Yu DS, Warner ED, Pan XZ, Petersen GW, et al. (2004) Soil database of 1:1000000 digital soil survey and reference system of the Chinese Genetic Soil Classification System. Soil Survey Horizons 45: 111–148.
36. Liu QH, Shi XZ, Weindorf DC, Yu DS, Zhao YC, et al (2006) Soil organic carbon storage of paddy soils in China using the 1:1000000 soil database and their implications for C sequestration. Global Biogeochemical Cycles 20: GB3024.
37. Zhejiang Soil Survey Office (1994) Soils of Zhejiang Province. Hangzhou: Zhejiang Science and Technology Press.
38. Wu JP, Hu YL, Zhi JJ, Jing CW, Chen HJ, et al. (2013) A 1:50000 soil database of Zhejiang Province, China. Acta Pedologica Sinica 50: 30–40.
39. Kazuhito M, Ono K, Kanomata H (2004) Organic carbon stock in forest soils in Japan. Geoderma 119: 21–32.
40. Sun WX, Shi XZ, Yu DS (2003) Distribution pattern and density calculation of soil organic carbon in profile. Soils 35: 236–241.
41. Nelson DW, Sommers LE (1982) Total carbon, organic carbon and organic matter. In: Page AL, Miller RH, Keeney DR, editors. Methods of Soil Analysis. Part 2: Chemical and Microbiological Properties. Wisconsin: American Society of Agronomy. 539–579.

2

Effect of Acute Ozone Induced Airway Inflammation on Human Sympathetic Nerve Traffic: A Randomized, Placebo Controlled, Crossover Study

Jens Tank[1]*, Heike Biller[2], Karsten Heusser[1], Olaf Holz[2], André Diedrich[3,4], Theodor Framke[5], Armin Koch[5], Anika Grosshennig[5], Wolfgang Koch[2], Norbert Krug[2], Jens Jordan[1], Jens M. Hohlfeld[2]

1 Institute of Clinical Pharmacology, Hannover Medical School, Hannover, Germany, **2** Fraunhofer Institute for Toxicology and Experimental Medicine (ITEM), Hannover, Germany, **3** Division of Clinical Pharmacology, Department of Medicine, Autonomic Dysfunction Center, Vanderbilt University School of Medicine, Nashville, Tennessee, United States of America, **4** Department of Biomedical Engineering, Vanderbilt University School of Engineering, Nashville, Tennessee, United States of America, **5** Institute for Biometry, Hannover Medical School, Hannover, Germany

Abstract

Background: Ozone concentrations in ambient air are related to cardiopulmonary perturbations in the aging population. Increased central sympathetic nerve activity induced by local airway inflammation may be one possible mechanism.

Methodology/Principal Findings: To elucidate this issue further, we performed a randomized, double-blind, cross-over study, including 14 healthy subjects (3 females, age 22–47 years), who underwent a 3 h exposure with intermittent exercise to either ozone (250 ppb) or clean air. Induced sputum was collected 3 h after exposure. Nineteen to 22 hours after exposure, we recorded ECG, finger blood pressure, brachial blood pressure, respiration, cardiac output, and muscle sympathetic nerve activity (MSNA) at rest, during deep breathing, maximum-inspiratory breath hold, and a Valsalva maneuver. While the ozone exposure induced the expected airway inflammation, as indicated by a significant increase in sputum neutrophils, we did not detect a significant estimated treatment effect adjusted for period on cardiovascular measurements. Resting heart rate (clean air: 59±2, ozone 60±2 bpm), blood pressure (clean air: 121±3/71±2 mmHg; ozone: 121±2/71±2 mmHg), cardiac output (clean air: 7.42±0.29 mmHg; ozone: 7.98±0.60 l/min), and plasma norepinephrine levels (clean air: 213±21 pg/ml; ozone: 202±16 pg/ml), were similar on both study days. No difference of resting MSNA was observed between ozone and air exposure (air: 23±2, ozone: 23±2 bursts/min). Maximum MSNA obtained at the end of apnea (air: 44±4, ozone: 48±4 bursts/min) and during the phase II of the Valsalva maneuver (air: 64±5, ozone: 57±6 bursts/min) was similar.

Conclusions/Significance: Our study suggests that acute ozone-induced airway inflammation does not increase resting sympathetic nerve traffic in healthy subjects, an observation that is relevant for environmental health. However, we can not exclude that chronic airway inflammation may contribute to sympathetic activation.

Editor: Murielle Bochud, University Institute of Social and Preventive Medicine, Switzerland

Funding: This work was supported by the Deutsche Forschungsgemeinschaft, Project BI 1403/1-1. The funders had no role in study design, data collection and analysis, decision to publish, or preparation of the manuscript.

Competing Interests: The authors have declared that no competing interests exist.

* E-mail: tank.jens@mh-hannover.de

Introduction

Based on large epidemiological studies, the World Health Organization estimated that air pollution is the 13th leading cause of mortality worldwide.[1] A large proportion of the excess mortality can be attributed to cardiovascular deaths.[2] Long term exposure to fine particles and to ozone were associated with an increased cardiopulmonary mortality.[3] Short term exposure to air pollutants is sufficient to elicit changes in cardiovascular and pulmonary function in healthy subjects[4] and in patients[5]. The rapid onset implicates a neural mechanism. Indeed, autonomic nervous system imbalance with raised sympathetic and attenuated parasympathetic activity may contribute to cardiovascular morbidity and mortality in this setting.[6] Increased ozone and fine particle exposure were associated with attenuated heart rate variability (HRV) and excessive oxidative stress and inflammatory biomarkers in venous blood samples.[7,8,9] Ozone exposure reduced HRV in asthma patients [10] but not in otherwise healthy older subjects.[11] Field studies cannot dissect the individual contribution of pollutants on cardiovascular control. The mechanisms through which ozone and fine particle influence autonomic function may differ.[12,10,9] Moreover, all previous studies relied on indirect methodologies, such as HRV analysis, to assess autonomic cardiovascular regulation. None employed direct measurements of muscle sympathetic nerve activity (MSNA) or detailed plasma catecholamine analysis. In animals, experimental inflammation in the kidney[13] or in gastrointestinal organs[14] increases central sympathetic activity through afferent neural pathways. Ozone induced airway inflammation [15] could elicit a similar response. A similar mechanism could also contribute to sympathetic activation in patients with chronic obstructive pulmonary disease[16] or asthma.[17] We challenged healthy volunteers with ozone in a double blind, randomized, and cross-over fashion to test the hypothesis that neutrophilic airway inflammation induced by acute ozone exposure is accompanied by increased muscle sympathetic nerve activity.

Methods

Participants

Healthy ozone responsive women and men aged 22 to 47 years with forced expiratory volume in the first second (FEV_1) >80% were eligible for our study. Subjects had to respond to ozone during a screening challenge with 250 ppb ozone over three hours. We defined a response as ≥10% increase in sputum neutrophils 6 hours after the start of ozone exposure. Subjects with respiratory tract infection in four weeks before screening or a positive skin prick test to common aeroallergens were excluded. Subjects regularly taking prescription or over the counter medication except acetaminophen for pain relief, oral contraceptives, hormonal replacement therapy, or vitamin supplements were also excluded.

Ethics statement

The ethics committee of the Hannover Medical School, Carl Neuberg Str. 1, 30625 Hannover, Germany, approved the study and all patients gave written informed consent.

Protocol

We exposed healthy ozone responsive subjects in a randomized, double-blinded, cross-over fashion for 3 hours to ozone (250 ppb) or clean air. In the Fraunhofer ozone exposure chamber ($2.7 \times 2.3 \times 2.5$ m^3), air temperature and relative humidity were kept in a range of 20–25°C and 40–60%, respectively. A high purity ozone mass flux was generated using commercial generator (COM_ADM, ANSEROS GmbH, Tübingen, Germany). Ozone concentration in the chamber was continuously monitored by two independent analyzers (Ozomat MP, ANSEROS, Tübingen, and 400A, MLU-Messtechnik für Luft und Umwelt GmbH, Essen). During exposure, subjects conducted intermittent bicycle ergometer training for 15 minutes at intensities increasing ventilation to 20 $l/min/m^2$ alternating with 15 minutes rest. Sessions were conducted at least 2 weeks apart.

We obtained induced sputum 6 hours after the start of challenge. Subjects inhaled ultrasonically nebulized pyrogen-free hypertonic saline through a mouthpiece while wearing a nose clip. After 5 minutes inhalation, we asked subjects to rinse the mouth and blow the nose to minimize sputum contamination. Then, subjects expectorated into a sterile container. Saline concentration was increased stepwise from 3% to 4% and 5% with expectoration after each step. Sputum was immediately processed. In short, sputum plugs were selected from the whole expectorate (including saliva), and transferred to a pre-weighted cup to determine its weight. To homogenize the sputum, 4 volumes of freshly prepared 0.1% sputolysin (dithiothrietol) was added and incubated for 15 min on a bench rocker. The dispersed sputum was filtered through a 70 µm cell strainer and the cell number was counted by hemacytometer. The differential cell count was assessed on cytospin slides by counting at least 400 non-squamous cells.[18,19]

Blood samples were obtained from a cubital vein before as well as 5, 7, and 24 hours after start of exposure challenge. Spirometry was performed according to ATS standards before as well as 3, 6, and 24 hours after start of challenge. Cardiovascular and microneurographic measurements were obtained approximately 19 to 22 hours after end of exposure.

Cardiovascular and sympathetic measurements

We conducted our measurements after an overnight fast in the morning hours. During testing, subjects remained in the supine position. Electrocardiogram, beat-by-beat blood pressure (Finapres, Ohmeda, Englewood, CA, U.S.A.), and brachial blood pressure (Dinamap, Critikon, Tampa, FL, U.S.A.) were determined. We inserted one antecubital venous catheter for blood sampling. Muscle sympathetic nerve activity was recorded from the right peroneal nerve (Nerve Traffic Analyzer 662C-3, Biomedical Engineering Department, University of Iowa, Iowa City, IA, USA) as described previously.[20] Briefly, a tungsten electrode with uninsulated tip (diameter 1–5 µm, shaft diameter 200 µm, 2 Mega Ohm, Frederick Haer and Co, Bowdoinham, MA) was inserted into the muscle nerve fascicles of the peroneal nerve at the fibular head. The raw nerve signal was band-pass filtered (700–2000 Hz), amplified (100×999.9), rectified and integrated (time constant of 0·1 s) to obtain mean voltage neurograms (IMSNA) using the nerve traffic analysis system (662C-3, University of Iowa). Satisfactory recordings of MSNA were defined by (1) heart pulse synchronicity; (2) facilitation during Valsalva straining and suppression during the hypertensive overshoot phase after release; (3) increases in response to breath-holding; and (4) no change during tactile or auditory stimulation.

MSNA bursts were identified by an automated detection algorithm with artifact elimination, dynamic noise level detection, and signal-to-noise estimation in the integrated signal. Bursts were accepted if the signal-to-noise ratio was greater than 2:1 and synchronization to a previous cardiac event was found in an interval between 1.2–1.6 seconds. All detections were visually verified.

After a resting period of at least 30 minutes, we obtained baseline recordings over 15 minutes. Then, we obtained blood samples for plasma catecholamine determination with high pressure liquid chromatography. Cardiac output measurements were obtained using an inert gas rebreathing method (Innocor, Innovision A/S, Odense, Denmark).

After baseline measurements had been obtained, we determined cardiovascular responses to fixed breathing at a rate of six breaths per minute in order to reach the highest heart rate variability values in individual subjects.[21] Then, subjects were asked to perform a Valsalva maneuver, by exhaling into a mouthpiece fixed to a mercury manometer with a pressure of 40 mmHg column for 15 sec. Finally, subjects conducted a maximum voluntary inspiratory breath hold maneuver.

Signals were digitized at a sampling rate of 500 Hz and 16 bit resolution using hardware based intelligent oversampling method (DI720USB, DATAQ Instruments, Akron, OH), and then processed with user software written in PV-Wave (Visual Numerics Inc., Houston, TX). We determined the following MSNA parameters from the integrated nerve signal: the burst frequency, i.e. the number of MSNA bursts per minute (bursts/min), the burst incidence, i.e. the number of bursts per 100 heart beats (bursts/100 heart beats), as well as the total activity, i.e. the area under the bursts per minute as arbitrary units per minute (au/min). [20]

Beat-to-beat values of detected R–R intervals and BP values were interpolated, low-pass filtered (cutoff 0.5 Hz) and re-sampled at 4 Hz. Data segments of 300 s were used for spectral analysis. Linear trends were removed and power spectral density was estimated with the FFT-based Welch algorithm using segments of 256 data points with 50% overlapping and Hanning window. The power in the frequency range of low frequencies (LF: 0.04 to 0.15 Hz) and high frequencies (HF: 0.15 to 0.40 Hz) was calculated following Task Force recommendations.[22]

Baroreflex gain was defined as the mean magnitude value of the transfer function between systolic blood pressure and R-R intervals in the low-frequency (BRSLF) and high-frequency (BRSHF) band with negative phase and squared coherence value greater than 0.5. We calculated for our settings (300 sec at 4 Hz, 256 data points segment length, Hanning window) coherence values of 0.33 for $p<0.01$ and 0.229 for $p<0.05$. Thus, the coherence limit of 0.5 applied by us and others is suitable for the calculation of baroreflex sensitivity.[23] Spontaneous baroreflex slope was calculated as the slope of the linear regression line between the systolic BP and the subsequent R-R intervals using sequences defined as an episode of at least three heart beats with more than 0.5 mm Hg systolic BP changes and 5 msec R-R interval changes. For sequences with rising (BRSup) and falling BP (BRSdown) the averaged value of slopes with a correlation coefficient greater than 0.85 was calculated.

Statistical analysis

All data are expressed as mean ± SEM. All parameters were evaluated by a standard cross over analysis with a two-sided t-test adjusting for a potential period effect. Continuous variables were checked for the normal distribution assumption using the D'Agostino and Pearson omnibus normality test. In case of a non Gaussian distribution of the variable the results were confirmed with a non parametric test (Wilcoxon signed rank test) using GraphPad-Prism 5.

For description of the adjusted period effect, means, standard deviations, minimum, maximum and medians for each period and sequence were analyzed. 95% confidence intervals and p-values were calculated. R 2.11.0 has been used for calculations (including packages lattice 0.18–8 and reshape 2.11.0). For the power/sample size calculation, R 2.11.0 and nQuery 7 have been used. A value for $p<0.05$ was considered significant. Based on the assumption of a relevant treatment difference of 5 bursts/min a sample size of 11 subjects was calculated for the parameter burst frequency at baseline to achieve a power of 80% to detect a difference in measurements after ozone or clean air (two sided t-test, type I error of 5%).

Results

Cohort characteristics and disposition

We included fifteen healthy, ozone-responsive subjects in our study (12 men, 3 women, age 22 to 47 years (34 +/− 10 years), body weight 59 to 104 kg (80 +/− 11 kg). One subjects had to be withdrawn due to an upper respiratory tract infection. Two successful nerve recordings of comparable quality could be obtained in 11 subjects. Characteristics of the subjects are shown in table 1.

Ozone-induced airway inflammation

The percentage of neutrophils in induced sputum obtained 6 hours after the start of challenge ($n=14$) was increased after ozone compared to clean air (estimated treatment effect adjusted for period: 16% (difference of percentages), 95% CI: 4.06% to 28.22%, $p<0.05$), indicating the presence of neutrophilic airway inflammation (table 2). FEV_1 at baseline was 4.4 +/− 0.6 L before the clean air exposure and 4.4 +/− 0.7 L before the ozone exposure. Forced vital capacity (FVC) at baseline was 5.5 +/− 0.8 L before the clean air exposure and 5.5 +/− 0.7 L before the ozone exposure. Immediately after leaving the challenge chamber, FEV_1 and FVC were significantly decreased (FEV_1: estimated treatment effect adjusted for period: −0.37 L, 95% CI: −0.58 L to −0.15 L, $p<0.005$; FVC: estimated treatment effect: −0.32 L, 95% CI: −0.48 L to −0.16 L, $p<0.005$). However, 24 hours after the start of challenge, the effect of ozone exposure on lung function parameters was no longer observed (FEV_1: estimated treatment effect adjusted for period: 0.007 L, 95% CI: −0.07 L to −0.085 L, $p=0.844$; FVC: estimated treatment effect: 0.042 L, 95% CI: −0.047 L to −0.131 L, $p=0.321$). Airway inflammation was paralleled by signs of systemic inflammation. The percentage of neutrophils in peripheral blood was increased 5 hours after start of ozone challenge compared with clean air (estimated treatment effect adjusted for period: +10.24% (difference of percentages), 95% CI: 5.95% to 14.52%, $p<0.0005$). At 24 hours after challenge, all changes in the peripheral white blood cell count had recovered to baseline (estimated treatment effect adjusted for period difference of percentages of neutrophils in peripheral blood: −0.964%, 95% CI: −7.993% to 6.064%, $p<0.768$).

Hemodynamics and sympathetic regulation

Resting HR was 59±2 bpm after exposure to clean air and 60±2 after ozone exposure. Resting systolic blood pressure (clean air: 121±3 mmHg; ozone: 121±2 mmHg), diastolic blood pressure (clean air: 71±2 mmHg; ozone: 71±2 mmHg), cardiac output (clean air: 7.42±0.29 mmHg; ozone: 7.98±0.60 l/min), plasma norepinephrine levels (clean air: 213±21 pg/ml; ozone: 202±16 pg/ml), and plasma epinephrine levels (clean air: 23±3 pg/ml; ozone: 23±3 pg/ml) were similar on both study days. No significant estimated treatment effect adjusted for period was detected for HR, BP, CO and plasma catecholamines.

Figure 1 illustrates sympathetic nerve recordings on both study days for two subjects with different sequence. Muscle sympathetic nerve activity was similar after exposure to ozone and to clean air. Figure 2 shows individual values of muscle sympathetic nerve

Table 1. Characteristics of the study population.

	Sequence 1 (air -> ozone)	Sequence 2 (ozone -> air)	Total
Number	6	8	14
Male (%)	4 (66.7%)	7 (87.5%)	11 (78.6%)
Age [years]	33.2 (±10.34)	34.1 (±9.5)	33.7 (±9.5)
Weight [kg]	76.7 (±10.7)	83.12 (±11.9)	80.4 (±11.4)
Height [cm]	179.8 (±8.4)	185.6 (±5.7)	183.1 (±7.3)
BMI [kg/m^2]	23.6 (±1.8)	24.0 (±2.3)	23.9 (±2.0)

Shown are either absolute frequencies and percentages or means and standard deviations.

Table 2. Differential cell count in induced sputum at baseline and the exposure effect adjusted for period 6 hours after the start of exposure to ozone compared to clean air.

Parameter	Baseline	Treatment effect	95%CI		p-value
Neutros/g	42.2±10.3	91.82±39.7	5,23	178,4	0,0394
Neutros perc	42.4±5.4	16.14±5.54	4,06	28,22	0,0131
Total cells	171.3±46.3	98.38±82.5	−81,4	278,2	0,2563
Total cells/g	84.6±11.4	92.41±47.7	−11,4	196,3	0,0764
Macros perc	25.0±4.0	−11±4.61	−21,1	−0,95	0,0345
Eos perc	0.02±0.02	−0.12±0.06	−0,25	0,01	0,0693
Lymphos perc	1.04±0.19	−0.48±0.44	−0,47	1,43	0,2935
Monos perc	0.79±0.18	−0±0.51	−1,11	1,12	0,9946
Epis perc	30.8±4.6	−5.5±3.12	−12,3	1,3	0,1034
Squamous cells	26.7±4.0	−5.52±4.46	−15,2	4,19	0,2392
Cell viability	58.1±5.5	−5.44±3.73	−13,6	2,7	0,1711

Mean ± SEM, n = 14, CI confidence limits, g = gram, perc = per cent, cells/g = cells/gram, neutros = neutrophil granulocytes, macros = macrophages, eos = eosinophil granulocytes, lymphos = lymphocytes, epis = epithelial cells.

activity measured after clean air and ozone exposure. Table 3 presents the mean values of sympathetic nerve activity as burst frequency, burst incidence and area under the burst measured on both study days at rest, during deep breathing, at the end of maximum inspiratory apnea and during late phase II of the Valsalva maneuver. No significant estimated treatment effect adjusted for period was detected. MSNA at the end of apnea quantified as burst area tended to be increased with ozone.

Heart rate variability and baroreflex sensitivity

Table 4 presents the mean values of heart rate variability, blood pressure variability, and baroreflex sensitivity measured on both study days at rest and during deep breathing. All calculated parameters were similar on both study days. No significant estimated treatment effect adjusted for period was detected.

Discussion

The main finding of our study is that neutrophilic airway inflammation induced by acute ozone exposure was not accompanied by changes in muscle sympathetic nerve activity in healthy younger subjects. We combined the microneurography

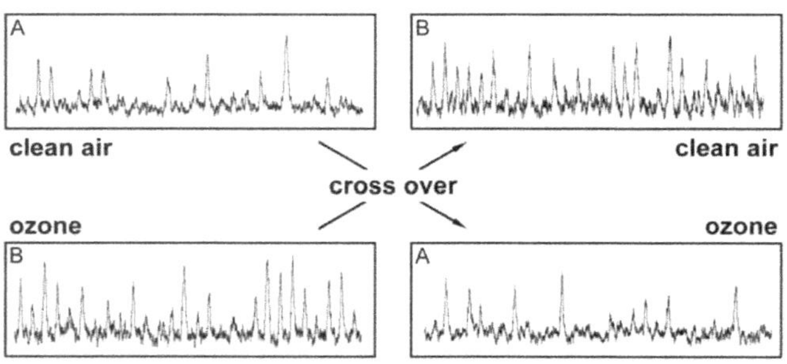

Figure 1. Examples of integrated muscle sympathetic nerve activity recorded at about 19–22 hours after exposure to clean air or to ozone in two subjects.

technique with hemodynamic measurements, plasma catecholamine determination, cardiovascular autonomic function testing, and heart rate as well as blood pressure variability analysis, which provides more comprehensive insight in autonomic regulation than either method alone. None of these measurements changed with ozone.

Ozone exposure is an established model to induce neutrophilic airway inflammation. The increase in sputum neutrophil counts observed after ozone challenge in our study is comparable with earlier results.[15,24] Thus, 250 ppb ozone exposure over three hours in combination with intermittent exercise leads to consistent increases in neutrophilic airway inflammation. We assessed airway inflammation six hours after the start of ozone exposure. Given recent recommendations not to perform more than one sputum induction within 48 hours, we did not obtain additional samples at a later time point.[25] However, other studies applying similar methodology showed that neutrophilic airway inflammation persists for at least 24 hours.[18,26] In contrast, blood neutrophil counts rapidly normalize after ozone exposure. Thus, our study allowed to study influences of local inflammation in the lung without a major systemic inflammatory response. The issue is important because systemic inflammation could mask sympathetic activation induced by airway inflammation.[27] For ethical reasons we did not obtain airway biopsies for histological analysis. Therefore, we do not have direct evidence that increased neutrophil counts in sputum are associated with inflammation of deeper airway wall portions. However, in previous studies, bronchial biopsies taken 6 hours[28,29] or 18 hours [30,31] after an acute ozone exposure in healthy subjects showed increased percentage of neutrophils and total protein concentration in the bronchial fraction. The observation is consistent with acute inflammatory cell influx across airway walls. In one study, inflammatory markers in biopsies did not change with ozone exposure.[29] However, epithelial shedding and substance P release from subepithelial sensory nerves was demonstrated suggesting that mucosal nerve endings are affected by ozone exposure.[32] Furthermore, experiments in anesthetized and artificially ventilated dogs showed ozone responsive bronchial C fibers and rapidly adapting receptors [33]. Chest discomfort, which commonly occurs with deep ozone inhalation, suggest a similar mechanism in human beings. Therefore, we suggest that the design of our study is suitable to assess acute influences of superficial airway inflammation on cardiovascular autonomic control.

In contrast to many other cardiovascular autonomic measurements, resting MSNA exhibits surprisingly little intraindividual variability in healthy subjects.[34] The technique has been proven useful studying interactions between respiratory tract and the cardiovascular system. For example, sympathetic vasomotor tone has been shown to be tightly regulated by chemoreflex mechanisms.[35] Furthermore, MSNA is substantially increased in patients with obstructive sleep apnea [36,37] and in patients with chronic obstructive pulmonary disease[38]. Sympathetic activity increases profoundly during prolonged breath-holding in some cases about tenfold above baseline.[39] Finally, using the microneurography technique, active smoking and passive smoking were shown to raise sympathetic activity.[40,41]

Given the excellent reproducibility of the microneurography technique, we are confident that our study excludes a major change in muscle sympathetic nerve activity elicited by ozone-induced neutrophilic airway inflammation in young healthy subjects. In an earlier study, sympathetic vasomotor tone was 34 bursts/min in control subjects and 61 bursts/min in patients with severe chronic obstructive pulmonary disease.[38] Our study was

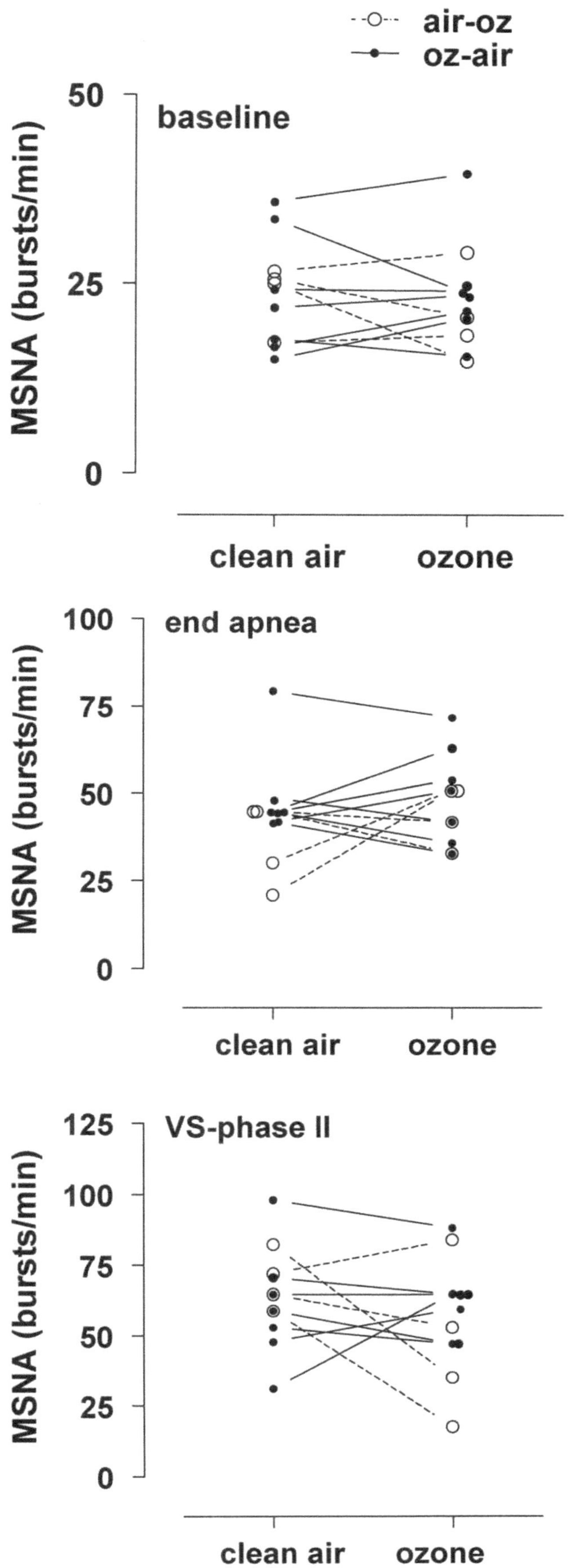

Figure 2. Individual values of muscle sympathetic nerve activity (burst frequency in bursts/min) measured after clean air and ozone exposure at rest (top), at the end of maximum inspiratory apnea (middle) and during late phase II of the Valsalva maneuver (bottom) for subjects exposed first to clean air (empty circles, n = 4) and for subjects exposed first to ozone (filled circles, n = 7).

powered to detect 5 bursts/min differences in sympathetic vasomotor tone. Sympathetic reflexes regulating the cardiovascular system often interact with each other as exemplified by baroreflex/chemoreflex interactions.[42] Possibly, sympathetic activation through another reflex pathway could be modulated by neutrophilic airway inflammation. In the event, ozone exposure did not change the sympathetic response to respiratory stimuli like deep breathing, maximum endinspiratory apnea, or the Valsalva-Maneuver. Thus, our study challenges the idea that neutrophilic airway inflammation is sufficient to drive an increase in muscle sympathetic nerve activity in the absence of abnormalities in blood gases or pulmonary hemodynamics. Provided that the acute stimulus was sufficiently strong and properly targeted, our study suggests that airway inflammation does not elicit a major change in muscle sympathetic nerve activity. However, we cannot exclude that more chronic changes in airway inflammation contribute to sympathetic activation in chronic obstructive lung disease or in asthma. As our subjects were healthy, we cannot exclude the possibility that ozone exposure produces or exacerbates sympathetic activation in COPD or asthmatic patients.

Sympathetic nerve activity during apnea quantified as area under the burst tended to be increased with ozone exposure. The

Table 3. Muscle sympathetic nerve activity (burst frequency, burst incidence and area under the burst) measured on both study days at rest, during deep breathing (6 breaths per minute), at the end of the maximum voluntary inspiratory breath hold maneuver (end apnea) and during late phase II of the Valsalva maneuver.

parameter	clean air	ozone	difference	p-value
baseline				
MSNA (bursts/min)	23±2	23±2	−0.68	0.7161
MSNA (bursts/100 beats)	42±5	39±5	−2.69	0.4886
MSNA (area/min)	1.23±0.30	1.04±0.15	−0.14	0.5861
deep breathing				
MSNA (bursts/min)	20±2	22±2	1.45	0.4119
MSNA (bursts/100 beats)	33±4	35±4	1.84	0.5322
MSNA (area/min)	1.20±0.28	1.16±0.19	−0.04	0.8802
end apnea				
MSNA (bursts/min)	44±4	48±4	5.94	0.1614
MSNA (bursts/100 beats)	68±5	71±4	6.16	0.3127
MSNA (area/min)	3.79±0.45	5.05±0.75	1.5	0.0862
Valsalva late phase II				
MSNA (bursts/min)	64±5	57±6	−10.19	0.1496
MSNA (bursts/100 beats)	75±4	68±6	−10.28	0.1293
MSNA (area/min)	3.94±0.36	4.83±0.96	0.74	0.4950

mean values ± SEM (n = 11), MSNA = muscle sympathetic nerve activity.

Table 4. Heart rate variability (HRV), blood pressure variability (BPV) and baroreflex sensitivity (BRS) data measured on both study days at rest and during deep breathing.

parameter	clean air	ozone	difference	p-value
baseline				
HRV and BPV				
TP (ms²)	2722±424	3433±688	874	0.2212
LF (ms²)	1099±241	1363±346	333	0.3132
HF (ms²)	540±80	906±287	398	0.0925
LF/HF	2.15±0.34	2.05±0.43	−0.03	0.9343
LF-SBP (mmHg²)	8±1	9±2	1.24	0.4831
BRS (ms/mmHg)				
BRSup	22±2	26±3	3.89	0.1911
BRSdown	21±3	23±3	1.91	0.5779
BRSLF	13±2	13±2	0.17	0.9306
BRSHF	27±3	30±4	3.25	0.4400
deep breathing				
HRV and BPV				
TP (ms²)	11254±2101	12692±2541	1658	0.2840
LF (ms²)	8815±1653	10652±2188	2035	0.1720
LF-SBP (mmHg²)	21±3	25±5	4.51	0.2051
BRS (ms/mmHg)				
BRSup	35±3	39±4	3.88	0.3087
BRSdown	21±2	19±2	−3.19	0.1657
BRSLF	16±1	16±2	0.20	0.8649

mean values ± SEM, TP= total power, LF = low frequency, HF = high frequency, LF/HF = LF to HF ratio, LF-SBP = Low frequency power of systolic blood pressure variability, BRSup = baroreflex sensitivity calculated by the sequence technique for increasing blood pressure, BRSdown = baroreflex sensitivity calculated by the sequence technique for decreasing blood pressure, BRSLF= baroreflex sensitivity calculated by cross spectral analysis in the low frequency band, BRSHF= baroreflex sensitivity calculated by cross spectral analysis in the high frequency band.

observation might suggest that airway inflammation exacerbates sympathetic activation when an additional stimulus is present.

Our observations on experimental ozone exposure are relevant for environmental health, which largely relies on epidemiological observations. Autonomic nervous system imbalance with increased sympathetic and decreased parasympathetic cardiovascular control may be mechanistic link between air pollution and excess cardiovascular mortality.[1] The rapid onset of cardiovascular events after exposure supports this idea.[43] Indeed, in an epidemiological study in 2681 men and women, resting heart rate increased slightly during an air pollution episode.[44] Fine particle exposure during the previous day reduced heart rate variability in another study.[11] Reduced heart rate variability was also observed with increased exposure to ozone.[45] However, the relationship may be attenuated after statistical adjustment for fine particle exposure.[46] Finally, experimental exposure to an ozone/fine particle mixture raised diastolic blood pressure.[47] Changes in autonomic regulation with exposure to air pollution may result from direct effects on rapidly adapting receptors or C-fibers in the lung or oxidative stress and release of inflammatory cytokines in the lung or elsewhere in the body.[43] Our study suggests that persistent airway inflammation induced by acute exposure to ozone without additional pollutants is unlikely to induce an acute change in cardiovascular autonomic regulation. The idea is supported by the observation that environmental ozone exposure is associated with increased risk for death from respiratory causes whereas cardiovascular risk is related to fine particle exposure.[3]

Conclusion

We applied ozone exposure as a model for neutrophilic airway inflammation, which is a common feature observed in chronic obstructive lung disease. Given the design of our study, we were only able to test the effect of an acute airway inflammation, therefore we cannot exclude that chronic airway inflammation has the potential to affect autonomic cardiovascular regulation. Nevertheless, our study suggests that acute airway inflammation is not sufficient to drive sympathetic activation, an observation that is also relevant for environmental health. Over the last decades, pulmonary medicine and cardiovascular medicine developed exciting new methodologies. Our study underscores the need to join forces for patient oriented research projects on mechanisms mediating cardiovascular disease in pulmonary patients.

Acknowledgements

We would like to express our gratefulness to the study nurses of the department of clinical airway research for patient recruitment and management. The authors thank Rudolf Jörres for advice with ozone challenge and induced sputum methodology.

Author Contributions

Conceived and designed the experiments: JMH JJ NK WK KH JT. Performed the experiments: JT HB KH OH. Analyzed the data: JT AD TF AK AG HB. Contributed reagents/materials/analysis tools: AD. Wrote the paper: JT JJ JH OH.

References

1. World Health Organisation World Health Report: reducing risks and promoting healthy life. (2002) Geneva: World Health Organisation.
2. Pope CA, III, Burnett RT, Thurston GD, Thun MJ, Calle EE, et al. (2004) Cardiovascular mortality and long-term exposure to particulate air pollution: epidemiological evidence of general pathophysiological pathways of disease. Circulation 109(1): 71–77.
3. Jerrett M, Burnett RT, Pope CA, III, Ito K, Thurston G, et al. (2009) Long-term ozone exposure and mortality. N Engl J Med 360(11): 1085–1095.
4. Kim CS, Alexis NE, Rappold AG, Kehrl H, Hazucha MJ, et al. (2011) Lung Function and Inflammatory Responses in Healthy Young Adults Exposed to 0.06 ppm Ozone for 6.6 Hours. Am J Respir Crit Care Med, [Epub ahead of print].
5. Liu L, Poon R, Chen L, Frescura AM, Montuschi P, et al. (2009) Acute effects of air pollution on pulmonary function, airway inflammation, and oxidative stress in asthmatic children. Environ Health Perspect 117(4): 668–674.
6. Thayer JF, Yamamoto SS, Brosschot JF (2010) The relationship of autonomic imbalance, heart rate variability and cardiovascular disease risk factors. Int J Cardiol 141(2): 122–131.
7. Chuang KJ, Chan CC, Su TC, Lee CT, Tang CS (2007) The effect of urban air pollution on inflammation, oxidative stress, coagulation, and autonomic dysfunction in young adults. Am J Respir Crit Care Med 176(4): 370–376.
8. Park SK, O'Neill MS, Vokonas PS, Sparrow D, Schwartz J (2005) Effects of air pollution on heart rate variability: the VA normative aging study. Environ Health Perspect 113(3): 304–309.
9. Gold DR, Litonjua A, Schwartz J, Lovett E, Larson A, et al. (2000) Ambient pollution and heart rate variability. Circulation 101(11): 1267–1273.
10. Power KL, Balmes J, Solomon C (2008) Controlled exposure to combined particles and ozone decreases heart rate variability. J Occup Environ Med 50(11): 1253–1260.
11. Luttmann-Gibson H, Suh HH, Coull BA, Dockery DW, Sarnat SE, et al. (2006) Short-term effects of air pollution on heart rate variability in senior adults in Steubenville, Ohio. J Occup Environ Med 48(8): 780–788.
12. Fakhri AA, Ilic LM, Wellenius GA, Urch B, Silverman F, et al. (2009) Autonomic effects of controlled fine particulate exposure in young healthy adults: effect modification by ozone. Environ Health Perspect 117(8): 1287–1292.

13. Ye S, Gamburd M, Mozayeni P, Koss M, Campese VM (1998) A limited renal injury may cause a permanent form of neurogenic hypertension. Am J Hypertens;116 Pt 1): 723–728.
14. Ordway GA, Boheler KR, Longhurst JC (1988) Stimulating intestinal afferents reflexly activates cardiovascular system in cats. Am J Physiol 254(2 Pt 2): H354–H360.
15. Holz O, Tal-Singer R, Kanniess F, Simpson KJ, Gibson A, et al. (2005) Validation of the human ozone challenge model as a tool for assessing anti-inflammatory drugs in early development. J Clin Pharmacol 45(5): 498–503.
16. Bernardi L, Casucci G, Haider T, Brandstatter E, Pocecco E, et al. (2008) Autonomic and cerebrovascular abnormalities in mild COPD are worsened by chronic smoking. Eur Respir J 32(6): 1458–1465.
17. Kumar M, Verma NS, Tiwari S, Pandey US (2005) Sympathetic hyperactivity in patients of bronchial asthma. Indian J Physiol Pharmacol 49(1): 89–94.
18. Holz O, Jorres RA, Timm P, Mucke M, Richter K, et al. (1999) Ozone-induced airway inflammatory changes differ between individuals and are reproducible. Am J Respir Crit Care Med 159(3): 776–784.
19. Holz O, Richter K, Jorres RA, Speckin P, Mucke M, et al. (1998) Changes in sputum composition between two inductions performed on consecutive days. Thorax 53(2): 83–86.
20. Tank J, Schroeder C, Diedrich A, Szczech E, Haertter S, et al. (2003) Selective impairment in sympathetic vasomotor control with norepinephrine transporter inhibition. Circulation 107(23): 2949–2954.
21. Hirsch JA, Bishop B (1981) Respiratory sinus arrhythmia in humans: how breathing pattern modulates heart rate. Am J Physiol 241(4): H620–H629.
22. Heart rate variability. Standards of measurement, physiological interpretation, and clinical use. Task Force of the European Society of Cardiology and the North American Society of Pacing and Electrophysiology. (1996) Eur Heart J 17(3): 354–381.
23. Immink RV, van den Born BJ, van Montfrans GA, Koopmans RP, Karemaker JM, et al. (2004) Impaired cerebral autoregulation in patients with malignant hypertension. Circulation 110(15): 2241–2245.
24. Holz O, Mucke M, Paasch K, Bohme S, Timm P, et al. (2002) Repeated ozone exposures enhance bronchial allergen responses in subjects with rhinitis or asthma. Clin Exp Allergy 32(5): 681–689.
25. Djukanovic R, Sterk PJ, Fahy JV, Hargreave FE (2002) Standardised methodology of sputum induction and processing. Eur Respir J Suppl 37: 1s–2s.
26. Mudway IS, Kelly FJ (2004) An investigation of inhaled ozone dose and the magnitude of airway inflammation in healthy adults. Am J Respir Crit Care Med 169(10): 1089–1095.
27. Sayk F, Vietheer A, Schaaf B, Wellhoener P, Weitz G, et al. (2008) Endotoxemia causes central downregulation of sympathetic vasomotor tone in healthy humans. Am J Physiol Regul Integr Comp Physiol 295(3): R891–R898.
28. Stenfors N, Pourazar J, Blomberg A, Krishna MT, Mudway I, et al. (2002) Effect of ozone on bronchial mucosal inflammation in asthmatic and healthy subjects. Respir Med 96(5): 352–358.
29. Krishna MT, Madden J, Teran LM, Biscione GL, Lau LC, et al. (1998) Effects of 0.2 ppm ozone on biomarkers of inflammation in bronchoalveolar lavage fluid and bronchial mucosa of healthy subjects. Eur Respir J 11(6): 1294–1300.
30. Balmes JR, Aris RM, Chen LL, Scannell C, Tager IB, et al. (1997) Effects of ozone on normal and potentially sensitive human subjects. Part I: Airway inflammation and responsiveness to ozone in normal and asthmatic subjects. Res Rep Health Eff Inst;78): 1–37.
31. Aris RM, Christian D, Hearne PQ, Kerr K, Finkbeiner WE, et al. (1993) Ozone-induced airway inflammation in human subjects as determined by airway lavage and biopsy. Am Rev Respir Dis 148(5): 1363–1372.
32. Krishna MT, Springall D, Meng QH, Withers N, Macleod D, et al. (1997) Effects of ozone on epithelium and sensory nerves in the bronchial mucosa of healthy humans. Am J Respir Crit Care Med 156(3 Pt 1): 943–950.
33. Coleridge JC, Coleridge HM, Schelegle ES, Green JF (1993) Acute inhalation of ozone stimulates bronchial C-fibers and rapidly adapting receptors in dogs. J Appl Physiol 74(5): 2345–2352.
34. Fagius J, Wallin BG (1993) Long-term variability and reproducibility of resting human muscle nerve sympathetic activity at rest, as reassessed after a decade. Clin Auton Res 3(3): 201–205.
35. Somers VK, Mark AL, Zavala DC, Abboud FM (1989) Contrasting effects of hypoxia and hypercapnia on ventilation and sympathetic activity in humans. J Appl Physiol 67(5): 2101–2106.
36. Somers VK, Dyken ME, Clary MP, Abboud FM (1995) Sympathetic neural mechanisms in obstructive sleep apnea. J Clin Invest 96(4): 1897–1904.
37. Grassi G, Facchini A, Trevano FQ, Dell'Oro R, Arenare F, et al. (2005) Obstructive sleep apnea-dependent and -independent adrenergic activation in obesity. Hypertension 46(2): 321–325.
38. Heindl S, Lehnert M, Criee CP, Hasenfuss G, Andreas S (2001) Marked sympathetic activation in patients with chronic respiratory failure. Am J Respir Crit Care Med 164(4): 597–601.
39. Heusser K, Dzamonja G, Tank J, Palada I, Valic Z, et al. (2009) Cardiovascular regulation during apnea in elite divers. Hypertension 53(4): 719–724.
40. Hausberg M, Mark AL, Winniford MD, Brown RE, Somers VK (1997) Sympathetic and vascular effects of short-term passive smoke exposure in healthy nonsmokers. Circulation 96(1): 282–287.
41. Narkiewicz K, van de Borne PJ, Hausberg M, Cooley RL, Winniford MD, et al. (1998) Cigarette smoking increases sympathetic outflow in humans. Circulation 98(6): 528–534.
42. Somers VK, Mark AL, Abboud FM (1991) Interaction of baroreceptor and chemoreceptor reflex control of sympathetic nerve activity in normal humans. J Clin.Invest 87(6): 1953–1957.
43. Brook RD (2008) Cardiovascular effects of air pollution. Clin Sci (Lond) 115(6): 175–187.
44. Peters A, Perz S, Doring A, Stieber J, Koenig W, et al. (1999) Increases in heart rate during an air pollution episode. Am J Epidemiol 150(10): 1094–1098.
45. Park SK, O'Neill MS, Vokonas PS, Sparrow D, Schwartz J (2005) Effects of air pollution on heart rate variability: the VA normative aging study. Environ Health Perspect 113(3): 304–309.
46. Holguin F, Tellez-Rojo MM, Hernandez M, Cortez M, Chow JC, et al. (2003) Air pollution and heart rate variability among the elderly in Mexico City. Epidemiology 14(5): 521–527.
47. Urch B, Silverman F, Corey P, Brook JR, Lukic KZ, et al. (2005) Acute blood pressure responses in healthy adults during controlled air pollution exposures. Environ Health Perspect 113(8): 1052–1055.

3

Regulatory Response to Carbon Starvation in *Caulobacter crescentus*

Leticia Britos[1], Eduardo Abeliuk[1,2], Thomas Taverner[3], Mary Lipton[3], Harley McAdams[1], Lucy Shapiro[1]*

1 Department of Developmental Biology, Stanford University School of Medicine, Stanford, California, United States of America, 2 Department of Electrical Engineering, Stanford University, Stanford, California, United States of America, 3 Environmental Molecular Sciences Laboratory, Pacific Northwest National Laboratory, Richland, Washington, United States of America

Abstract

Bacteria adapt to shifts from rapid to slow growth, and have developed strategies for long-term survival during prolonged starvation and stress conditions. We report the regulatory response of *C. crescentus* to carbon starvation, based on combined high-throughput proteome and transcriptome analyses. Our results identify cell cycle changes in gene expression in response to carbon starvation that involve the prominent role of the FixK FNR/CAP family transcription factor and the CtrA cell cycle regulator. Notably, the SigT ECF sigma factor mediates the carbon starvation-induced degradation of CtrA, while activating a core set of general starvation-stress genes that respond to carbon starvation, osmotic stress, and exposure to heavy metals. Comparison of the response of swarmer cells and stalked cells to carbon starvation revealed four groups of genes that exhibit different expression profiles. Also, cell pole morphogenesis and initiation of chromosome replication normally occurring at the swarmer-to-stalked cell transition are uncoupled in carbon-starved cells.

Editor: Roy Martin Roop II, East Carolina University School of Medicine, United States of America

Funding: This work was supported by National Institutes of Health (NIH - http://nih.gov/) grant GM32506 to LS and by Department of Energy (DOE - http://www.energy.gov/) grant DE-FG02ER64136 to LS and HM. Proteomic analyses were performed in the Environmental Molecular Sciences Laboratory, a Department of Energy/Office of Biological and Environmental Research (DOE/BER) national scientific user facility on the Pacific Northwest National Laboratory (PNNL) campus in Richland, Washington. PNNL is a multiprogram national laboratory operated by Battelle for the DOE under Contract DE-AC05-76RL01830. The funders had no role in study design, data collection and analysis, decision to publish, or preparation of the manuscript.

Competing Interests: The authors have declared that no competing interests exist.

* E-mail: shapiro@cmgm.stanford.edu

Introduction

Starvation for nutrient and energy sources are common stresses confronted by bacteria in natural environments. Bacteria have limited energy reserves, so they need robust mechanisms to quickly shift between rapid and slow growth, as well as a strategy for long-term survival during periods of prolonged starvation. The response to starvation comprises an initial stage of scavenging and metabolic adaptation. If the missing essential nutrients are not replenished, there is a second stage of physiological adaptation, which includes the inhibition of growth and cell division, in order to retain viability.

We have used *Caulobacter crescentus*, a gram-negative oligotrophic bacterium [1], as a model system to study the response to carbon deprivation. *C. crescentus* has a dimorphic life cycle. Each asymmetric division yields a chemotactically-competent flagellated cell (swarmer cell) and a sessile cell with a polar stalk (stalked cell). The core genetic network that drives cell cycle progression and cell division in *C. crescentus* is well characterized [2,3,4,5].

In *Escherichia coli* and *Bacillus subtilis*, the RpoS (σ^S) and SigB (σ^B) sigma factors are the master regulators of the general starvation-stress response [6,7]. *C. crescentus* lacks orthologs of the *rpoS* and *sigB* genes, as do all other α-proteobacteria [8,9]. An equivalent master regulator has not been identified. We have recently identified two factors involved in the adaptation of *C. crescentus* to carbon starvation: the SpoT ppGpp synthetase/hydrolase, which contributes to the regulation of the DnaA protein stability and initiation of DNA replication [10], and the CrfA small noncoding regulatory RNA controlling the mRNA stability of 27 transcripts [11].

A comparison of gene expression profiles of *C. crescentus* growing under carbon- and nitrogen-limited continuous flow cultures was recently reported [12]. In these experiments, slow cell growth was supported by a constant supply of nutrients at low concentration, and the gene expression profiling captured the metabolic adaptations to the relative levels of carbon and nitrogen. This study identified genes differentially induced in carbon- versus nitrogen-limited conditions, among which are those predicted to require the alternative sigma factor RpoN for complete induction [12]. Here, we have studied the cells' response to an abrupt loss of carbon source resulting in the inhibition of mass accumulation and cell cycle progression. We performed a global analysis of the differences between growing and carbon-starved *C. crescentus* cultures, through combined high-throughput proteome and transcriptome assays. By analyzing the response at the protein level, we took into account the fact that *C. crescentus* uses a complex array of regulatory strategies that include targeted proteolytic events [13]. We identified genetic regulatory pathways that mediate the transduction of carbon starvation signals and found that the SigT ECF sigma factor controls a core set of genes that are activated by carbon depletion, osmotic stress and exposure to heavy metals. We identified gene clusters that are differentially expressed upon carbon starvation at specific stages of the cell cycle.

Results

The proteome and transcriptome profile of *C. crescentus* cells starved for carbon

Transfer of *C. crescentus* cultures in exponential growth to media lacking a carbon source results in immediate growth arrest. Under prolonged starvation, there is a pronounced loss in viability [10]. In order to identify proteins that participate in the carbon starvation response, we incubated cultures in minimal M2 medium in the absence of glucose, and the respective non-starved controls in the presence of 0.2% glucose, for 30 and 60 minutes, as described in the Methods section. Samples were analyzed by liquid chromatography coupled to tandem mass spectrometry. We identified 2471 distinct proteins across all conditions tested, accounting for 65.6% of 3767 *C. crescentus* predicted protein-coding genes. This is high coverage for a prokaryotic proteome, lower only than that reported for *Mycoplasma mobile* (88.6%), which has a genome of one-sixth the size of that of *C. crescentus* (approx 0.78 Mb) (Table S1). Previous electrophoresis-based proteome studies of *C. crescentus* identified 81 cell cycle-regulated proteins [14], 39 stalk-specific proteins [15], and 86 membrane-associated proteins [16,17].

The levels of 513 proteins were found to change reproducibly in cells starved for carbon for 30 and/or 60 min (Table S2). Figure 1 shows the upregulated and downregulated proteins by functional categories, according to COG classifications [18]. Those categories with a statistically significant enrichment in either upregulated or downregulated proteins are indicated (see Methods for details on the enrichment analysis). Functional categories comprising transport and metabolism of amino acids (COG E), nucleotides (COG F), carbohydrates (COG G) and cell motility (COG N) were significantly enriched in proteins whose levels decreased upon carbon starvation. On the other hand, the categories comprising energy production and conversion (COG C), inorganic ion transport and metabolism (COG P) and defense mechanisms (COG V) were significantly enriched in proteins whose levels increased upon carbon starvation. Of the 40 COG P proteins that are upregulated upon carbon starvation, 21 belong to the TonB-dependent receptor family, which are highly abundant in the *C. crescentus* genome, including those controlled by the CrfA small noncoding RNA described by Landt *et. al.* [11]. These observations are compatible with metabolic adaptation of the cell to carbon starvation.

The proteome profile of carbon-starved cells also provided insights into the physiological adaptation associated with stasis. Among the proteins that change significantly, there was notable downregulation of three essential cell cycle regulators (CtrA, DnaA and GcrA), and two proteins (PopZ and ParA) that mediate chromosome segregation [13,19,20]. The decrease in the levels of DnaA, an essential protein that serves both as an activator of replication initiation and global transcription factor [21], is consistent with the previously reported starvation-induced ClpP-mediated proteolysis of DnaA [22]. The decrease in DnaA levels prevents the initiation of DNA replication. The CtrA response regulator inhibits DNA replication in the swarmer cell, and acts as a transcription factor regulating the expression of nearly 100 genes. It was previously shown that CtrA is proteolyzed in swarmer cells starved for carbon, but with different kinetics than that observed under normal growth conditions [10,22]. Given the number of CtrA-regulated genes that change upon carbon starvation, it is clear that, at least in part, an important role of CtrA in the starvation response relies on its function as transcription factor. Additionally, the level of the FtsZ cell division protein, that is essential for cytokinesis, is shown to decrease significantly upon carbon starvation; FtsZ was previously shown to decrease during stationary phase in *C. crescentus* [23].

Total RNA was extracted from samples parallel to those used for the proteome analysis. As described in Methods, cDNA was synthesized, fragmented, labeled and hybridized to the CauloHI1 chip. A minimum two-fold change requirement between the starvation and non-starvation conditions, with a false discovery rate (FDR) below 1%, resulted in upregulation of 607 of the 3767 *C. crescentus* protein coding genes after 30 min of carbon starvation, and 700 genes after 60 min (553 of these were significantly upregulated at both time points); 725 and 618 genes were downregulated after 30 and 60 min of carbon starvation, respectively; 603 genes were equally downregulated for both time points. All genes that changed significantly are listed in Table S3. 16.3% of the 753 genes that were significantly upregulated after either 30 or 60 min of carbon starvation, and 19.9% of the 739 downregulated genes were classified as cell cycle regulated transcripts by Laub *et al.* [24] (Table S3).

Carbon starvation transcriptional regulators

Cell cycle transcription profiles were previously used to identify predicted cell cycle regulons and their conserved promoter motifs. Fourteen conserved promoter motifs were identified, seven of which shared significant similarity with binding motifs for previously characterized regulators [25]. We analyzed these sets of genes to determine if any of them were significantly enriched in genes whose expression was up- or down-regulated upon carbon starvation. As shown in Figure 2A, out of 14 gene sets, each of which shared a distinct promoter motif, six were significantly enriched in genes upregulated upon carbon starvation (sets corresponding to motifs cc_1, cc_2, cc_3, cc_4, cc_7, cc_8), while two were enriched in genes downregulated upon carbon starvation (corresponding to motifs cc_6 and cc_13) (Table S4). Motif cc_6 is the cognate binding motif for the RpoD sigma factor, which drives the expression of biosynthetic and housekeeping genes throughout the cell cycle [26] (Figure 2B); as expected, the great majority of the genes with cc_6 motif (27 out of 34) were down-regulated upon carbon starvation in our experiments.

Carbon starvation FixK regulon. Genes with motifs cc_1 to cc_4 are expressed in the swarmer cell stage (Figure 2B). A subset of these genes are likely activated by the stress suffered by cells during the synchronization process, and not swarmer cell specific genes, as they were activated equally by carbon starvation in stalked cells. Motif cc_3 corresponds to the DNA binding motif of the FixK transcription factor. Microarray analysis of a *C. crescentus fixK* knockout strain identified downstream targets whose expression changed in a FixK-dependent manner under hypoxia conditions [27]. We found that five out of eight genes with motif cc_3 were upregulated upon 30 min of carbon starvation. Expression of three of those genes was FixK-dependent under hypoxia conditions [27]. Eleven additional FixK-dependent genes identified by Crosson *et al.*, which lack the cc_3 motif, showed a significant change upon carbon starvation, and are either indirect FixK targets, or are transcribed as part of an operon directly regulated by FixK.

Carbon starvation SigT regulon. Motif cc_7 is the binding motif for the SigT alternative sigma factor, and genes in this cluster have a distinct peak of expression at the swarmer-to-stalked cell transition (Figure 2B). SigT, which belongs to the ECF (extra-cytoplasmic function) sigma factor family, is a regulator of the osmotic and oxidative stress responses in *C. crescentus* [28]. Of 26 genes with motif cc_7, 7 were upregulated after 30 min of carbon starvation, and an additional 7 genes were upregulated after 60 min.

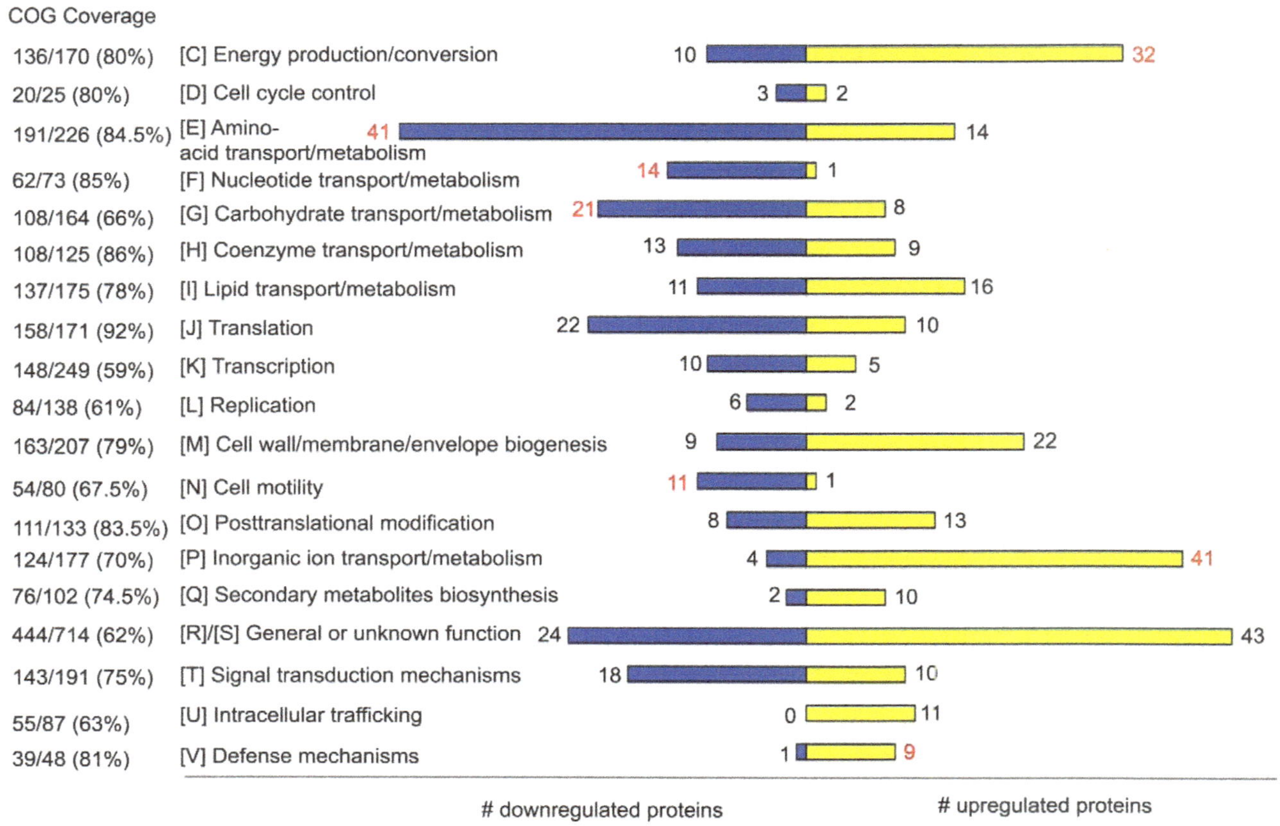

Figure 1. Proteome profile of *C. crescentus* subjected to carbon starvation. Proteins that change significantly upon 30 and/or 60 minutes of carbon starvation classified by NCBI Clusters of Orthologous Genes (COG). Coverage for all categories of the COG scheme (except for A and B for which no proteins were detected), is indicated as the ratio of the number of detected proteins in each category over the total number of proteins assigned to that category (the corresponding percentage value is indicated in parentheses). Yellow bars represent upregulated proteins and blue bars, downregulated proteins in each COG category. Values that meet a statistically significant enrichment, assuming a hypergeometric distribution, are denoted in red. The proteins within each category are listed in Table S2, along with 43 additional proteins that are not included in the COG classification.

In order to characterize the carbon starvation SigT regulon, we performed microarray experiments comparing gene expression profiles of a *C. crescentus* wild-type strain (LS101) and a *sigT* null mutant strain (LS3554), in M2 cultures incubated in the absence of glucose for 15 min. Genes showing a greater than two-fold difference in expression between the mutant and the wild type strain under starvation conditions, and meeting statistical significance as defined in the Methods section, were considered to be SigT-dependent upon carbon starvation (Table 1). Thirteen of the 27 SigT-dependent carbon starvation genes have the previously characterized [25,28] SigT binding motif in their promoters and are presumably direct targets of the regulator. Four additional targets are in close proximity to genes that possess the motif, suggesting they are part of directly regulated operons. Sixteen of the 27 genes whose expression is significantly reduced in the *sigT* mutant, and the two genes with increased expression in the mutant strain, do not show SigT-dependency under osmotic stress, as reported by Alvarez-Martinez *et al.* [28].

The main functional groups represented in the putative SigT carbon starvation regulon are signal transduction and gene regulation (7 genes), and transport and metabolism (5 genes). The gene encoding the SigU ECF sigma factor has a SigT binding motif in its promoter, suggesting it is directly regulated by SigT. SigU is activated by SigT during cell cycle progression [18], upon osmotic stress [25,28] and upon carbon starvation (this work).

In order to determine if any of the SigT-dependent genes lacking the conserved SigT promoter binding motif were regulated by a pathway involving SigU, we used microarrays to compare the carbon starvation-induced expression levels in a wild type strain (LS101) and a *sigU* null mutant strain (LS3547). Only one gene, CC_3466, encoding a hypothetical protein, appeared to be SigU-dependent under carbon starvation. This gene, which was also found to be SigT-dependent, does not have the conserved SigT motif in its promoter, consistent with its SigT dependence being, directly or indirectly through the function of the SigU transcription factor. The CC3466 transcript is cell cycle regulated [24], with a distinct peak at the swarmer-to-stalked cell transition. This pattern was also observed in the cell cycle microarray results of McGrath *et al.* [25]. CC_3466 is not reported to be SigT-dependent upon osmotic stress [28]. On the other hand, its expression is induced by exposure to chromate and dichromate heavy metals [29], incubation in minimal media vs rich media [30], as well as carbon limitation (compared to nitrogen limitation) [12]. The predicted amino acid sequence of CC_3466 (102 amino acids with a predicted molecular weight of 11 KDa) yielded very few low scoring BLASTP hits outside of the *Caulobacter* genus. In

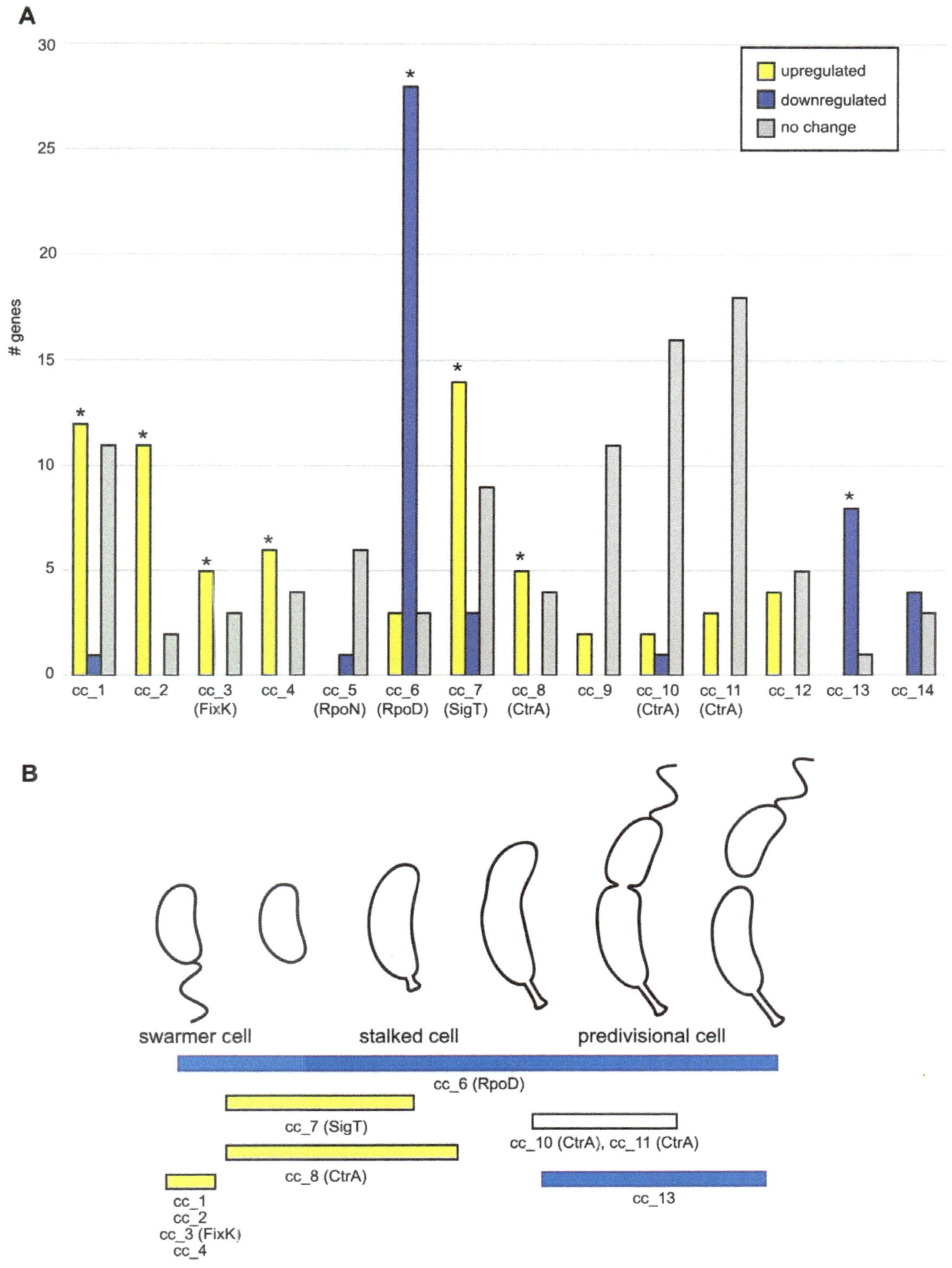

Figure 2. Putative regulators of the carbon starvation response. A. Previously identified clusters of *C. crescentus* genes that share conserved promoter motifs were analyzed for enrichment in genes that change significantly upon carbon starvation. For each gene set, cc_1 through cc_14 (identified by a shared promoter motif), the number of carbon starvation upregulated genes is indicated in yellow, downregulated genes in blue and genes that don't change in gray. Asterisks denote statistically significant enrichment. Gene numbers correspond to 30 min of carbon starvation, except for motif cc_7, for which gene numbers correspond to 60 min of carbon starvation. Identical results were obtained in terms of significant enrichment for the 30 and 60 min carbon starvation data for most gene sets, except for cc_8, which was only enriched in starvation upregulated genes at 30 min, and cc_7, which was only enriched at 60 min. **B**. Cell cycle patterns of expression, as previously determined [25], of carbon-starvation enriched gene sets. Yellow indicates enrichment in genes up-regulated upon 30 min of carbon starvation, while blue corresponds to enrichment in genes down-regulated upon 30 min of carbon starvation.

Table 1. SigT-dependent carbon starvation regulon.

Gene	Annotation	COG	Fold reduction in Δ*sigT* mutant	SigT motif?	Sig-T dependent in osmotic stress?	Up-regulated by exposure to multiple heavy metals?
CC_0284	Two-component receiver domain protein lovR	K	2.65	Yes	Yes	No
CC_1356	Transcriptional regulator	K	3.29	Yes	Yes	No
CC_2883	RNA polymerase ECF-type sigma factor sigU	K	15.93	Yes	Yes	Yes
CC_3477	Hybrid sigma factor/two-component receiver protein phyR	K	3.10	Yes	No	No
CC_0285	Photosensory histidine protein kinase lovK	T	3.94	No (*)	No	No
CC_2330	HTH transcriptional regulator	K	2.25	No	No	No
CC_3258	chemotaxis receiver domain protein cheYIII	K	2.05	No	No	Yes
CC_2549	Putative heme-binding protein	NC	3.97	Yes	Yes	Yes
CC_1048	Acylamino-acid-releasing enzyme	E	0.42	No	No	No
CC_2318	ABC transporter permease protein	Q	2.15	No	No	No
CC_3001	TonB-dependent receptor	P	2.08	No	No	No
CC_3572	Carbonic anhydrase	P	0.43	No	No	No
CC_0280	Conserved cell surface protein	NC	20.99	Yes	Yes	Yes
CC_0673	Hypothetical protein	NC	3.59	Yes	No	Yes
CC_0717	Hypothetical membrane associated protein	S	2.01	Yes	No	No
CC_1179	hypothetical protein	NC	3.14	Yes	Yes	Yes
CC_0163	Conserved hypothetical protein	S	5.60	No (*)	Yes	No
CC_1178	Hypothetical cytosolic protein	R	2.10	No (*)	Yes	Yes
CC_0557	Hypothetical protein	NC	2.78	No	No	No
CC_1532	Conserved hypothetical protein	S	23.41	No	Yes	Yes
CC_3466	Hypothetical protein	NC	7.67	No	No	Yes
CC_3473	Entericidin B homolog	NC	2.57	Yes	No	Yes
CC_1682	Transglycosylase associated protein	S	2.57	Yes	No	Yes
CC_0201	Outer membrane protein	M	2.11	Yes	Yes	Yes
CC_0428	Methyl-accepting chemotaxis protein	N	2.82	Yes	No	No
CC_0164	Chain length regulator (capsular polysaccharide biosynthesis)	D	2.18	No (*)	No	No
CC_0556	Catalase	NC	2.00	No	No	No

Genes with reduced or increased expression in the Δ*sigT* strain upon 15 min of carbon starvation, with respect to the NA1000 wild type strain under the same conditions. SigT-dependency under osmotic stress is according to [28]. Upregulation by exposure to multiple heavy metals corresponds to genes upregulated by two or more metals according to [29]. Genes annotated as No(*) in the 'sigT motif' column correspond to cases where the sigT motif is found upstream of an adjacent gene as part of a putative operon (CC_0284 and CC_0285; CC_1179 and CC_1178; CC_0162, CC_0163 and CC_0164). COG groups: K = Transcription; T = Signal transduction mechanisms; E = Amino acid transport and metabolism; Q = Secondary metabolites biosynthesis; P = Inorganic ion transport and metabolism; S = Function unknown; R = General function prediction only; M = Cell wall/membrane/envelope biogenesis; N = Cell motility; D = Cell cycle control; NC = Not classified.

our proteome analysis, peptide levels for CC_3466 increased significantly after 60 min of carbon starvation.

Carbon starvation CtrA regulon. The gene cluster sharing motif cc_8 was shown to be enriched in genes that were upregulated upon carbon starvation: five out of nine genes with this motif increased significantly upon 30 min of carbon starvation. Motifs cc_08, cc_10 and cc_11 share similarity with the binding motif for the CtrA cell cycle master regulator. While genes with motifs cc_10 and cc_11 showed an expression peak at the predivisional stage, genes with motif cc_8 showed an earlier peak, at the swarmer to stalk cell transition [25]. The CtrA regulon has been characterized by chIP-chip analysis, identifying genomic regions where CtrA binds, and microarray analysis, identifying genes with affected expression in a CtrA temperature sensitive mutant strain [31]. We cross-correlated these results with our carbon starvation microarray dataset and found that the expression of a significant number of both cell-cycle regulated and non cell-cycle regulated CtrA targets was altered upon 30 min of carbon starvation (Table 2). Among these targets was the CC_2644 gene, whose transcript was one of the most strongly upregulated transcripts upon carbon starvation. This gene, which remains uncharacterized in *C. crescentus*, encodes a protein belonging to the PhoH family. The *E. coli phoH* gene, which defines the family, was shown to be induced by phosphate starvation and to have ATPase activity [32].

Table 2. CtrA targets that change significantly upon carbon starvation.

Gene	Annotation	Transcript levels carbon starvation/control	Cell cycle regulated?
CC_2165	division plane positioning ATPase *mipZ*	5.88	Yes
CC_2540	Cell division protein *ftsZ*	2.15	Yes
CC_2063	Flagellar basal-body rod protein *flgF*	2.31	Yes
CC_2316	Transcriptional regulator	2.18	Yes
CC_3219	Two-component sensor histidine kinase	2.54	Yes
CC_2644	*phoH* protein	44.56	Yes
CC_1396	Lactate 2-monooxygenase	6.28	Yes
CC_1101	Protoporphyrinogen oxidase	2.37	Yes
CC_1307	Aspartyl protease *perP*	3.05	Yes
CC_1211	Hypothetical protein	2.01	Yes
CC_3291	Hypothetical protein	9.90	Yes
CC_1072	Ribonuclease H	3.35	Yes
CC_3218	Hypothetical protein	2.12	Yes
CC_1872	peptidoglycan-specific endopeptidase, M23 family	0.30	Yes
CC_0484	peptidyl-tRNA hydrolase	0.41	Yes
CC_1923	SSU ribosomal protein S2P	0.05	Yes
CC_3202	SSU ribosomal protein S12P	0.04	Yes
CC_2166	pantoate-beta-alanine ligase	0.37	Yes
CC_0050	S-adenosylmethionine synthetase	0.02	Yes
CC_0350	hypothetical protein with pentapeptide repeats	0.48	Yes
CC_3454	Acyl-CoA dehydrogenase	4.47	No
CC_0651	Conserved hypothetical protein	2.21	No
CC_0696	GumN superfamily protein	4.72	No
CC_1106	Ice nucleation protein	2.30	No
CC_2869	Hypothetical protein	2.32	No
CC_2966	3-oxoacyl-[acyl-carrier protein] reductase	3.45	No
CC_1892	aspartyl-tRNA synthetase	0.09	No
CC_2451	DNA topoisomerase I	0.11	No
CC_3155	chemotaxis receiver domain protein cheYIII	0.39	No
CC_2241	heat shock protein *hsp33*	0.30	No
CC_1034	GTP-binding protein *lepA*	0.30	No
CC_0249	SCO1/SenC family protein	0.46	No

Genes that change significantly upon 30 min of carbon starvation, and that are directly regulated by CtrA [31], and are either cell cycle regulated and non cell cycle regulated [24].

We performed the same enrichment analysis with the motifs identified by McGrath *et al.* for genes responding to heavy metal stress [25]. Sets sharing motifs m_2 (corresponding to the RpoD binding motif), m_3 (similar to motif cc_13) and m_4 were enriched in genes downregulated upon 30 min of carbon starvation, while sets sharing motifs m_5 (similar to motif cc_1) and m_13 were enriched in genes upregulated upon 30 min of carbon starvation. Since none of the additional sets had a known cognate regulator, this analysis did not identify further pathways related to the carbon starvation response.

Cell-stage specific response to carbon starvation

We explored the response to carbon starvation separately in swarmer and stalked cell populations. We isolated swarmer cells and incubated them in M2 medium lacking glucose for 15 min before collecting samples (Figure 3A). To study the stalked cell response to carbon starvation, a population of swarmer cells was allowed to proceed through the cell cycle for 60 min and transition into stalked cells in complete medium, before removing the carbon source and incubating for 15 min in M2 medium. RNA was extracted from starved swarmer cells and starved stalked cells, and hybridized to the CauloHI1 chip. To account for changes in transcript levels normally occurring as the cell cycle progresses, we used the normalized values corresponding to 15 min and 75 min of the cell cycle microarray profiles obtained from the data of McGrath *et al.* [25], as the non-starved controls for swarmer and stalked cells, respectively. The log-2 values of the expression ratios of the starved and non-starved cells for both stages were used to determine the cell-stage specific responsiveness of each gene, and clustered accordingly (Figure 3B and Table S5). Six clusters were identified. Most of the genes showed a similar expression change in swarmer and stalked cells: out of 667 genes, 165 were upregulated (cluster 3), and 342 genes were downregulated (cluster 6), in both stages. The remaining clusters displayed a stage-differential response to carbon starvation: 27 genes did not change significantly in swarmer cells, but were down-regulated preferentially in stalked cells (cluster 1); 24 genes were upregulated in swarmer cells, but did not change significantly in stalked cells (cluster 2); 71 genes did not change significantly in swarmer cells, but were upregulated in stalked cells (cluster 4); 37 genes were down-regulated in swarmer cells, but did not change significantly in stalked cells (cluster 5). Using MEME motif finder [33], we identified five novel conserved DNA motifs in the upstream regions of genes corresponding to three of the clusters (motif a in cluster 1; b, c and d in cluster 3; and e in cluster 6).

Adaptive changes of swarmer cells subjected to carbon starvation

In *Caulobacter*, the chromosomal replication origin is positioned at the cell pole, and upon initiation of replication, a copy of the origin sequence is moved to the opposite cell pole. Therefore, we can follow the onset of DNA replication by following the cellular position of the origin bound to the ParB segregation factor. In order to connect the stage-specific response to carbon starvation observed at the molecular level with cell cycle and differentiation events, we obtained transmission electron microscope images of swarmer cells immediately following their isolation, and then after subjecting them to carbon starvation. We examined the replication and segregation of the chromosomal origin (*ori*) locus, under the same conditions, using epifluorescence imaging of a strain carrying a fluorescently-tagged version of the ParB centromere-binding protein. 65% of the isolated swarmer cells developed short incipient stalks when incubated in the absence of glucose for 2 hs (Figure 4A). After 8 hs of starvation, these stalks failed to elongate. Under the same conditions, only about a tenth of the population had duplicated the chromosomal *ori* locus (Figure 4B), whereas after 30 min in complete media, 57% of the swarmers showed fully or partially segregated origins (Figure 4C). Thus, for a significant percentage of the population, carbon starvation uncoupled the initiation of stalk biogenesis and the G1-to-S transition, events that occur coincidently when nutrients are sufficient.

SigT-dependent degradation of CtrA in swarmer cells upon carbon starvation

The CtrA cell cycle master regulator is a critical element of the core machinery that regulates cell cycle progression. CtrA binds to the chromosomal origin of replication and blocks replication initiation in the swarmer cell [34]. Upon differentiation of the swarmer cell to the stalked cell, CtrA is cleared from the cell by proteolysis, allowing the initiation of DNA replication. CtrA is re-synthesized following replication initiation. Our proteome studies showed a decrease in the protein levels of the CtrA master regulator in a mixed population of *Caulobacter* cells subjected to carbon starvation: CtrA levels in cells incubated in the absence of glucose for 30 min were half of that of cells incubated in the presence of glucose (Table 2). We observed a similar decrease in the relative levels of CtrA by immunoblot assays of carbon starved mixed cell populations (not shown) and swarmer cells, as previously reported [10]. Swarmer cells incubated for 150 min in the absence of glucose showed low levels of the CtrA protein, while those incubated for 150 min in the presence of glucose had progressed through the cell cycle and reaccumulated CtrA. In contrast, when the same assay was performed with swarmer cells carrying a *sigT* deletion, a less significant decrease in CtrA levels was observed in the absence of glucose, indicating that SigT contributes to the clearance of CtrA protein in carbon starved swarmer cells (Figure 5A and D). We measured the activity of the CtrA promoter under carbon starvation in a wild type and *sigT* deletion background, using a transcriptional fusion of the CtrA promoter region to *lacZ*. We found only a minimal difference in the transcriptional activity of the reporter in both backgrounds (Figure 5B). Further experiments will be carried out to determine if SigT affects the starvation-induced clearance of CtrA at the post-transcriptional level.

Since SigT is a transcriptional regulator, the simplest explanation for these results is that a downstream target whose expression is SigT-dependent is responsible for the effect on CtrA protein stability. Consequently, we compared the starvation and non-starvation levels of CtrA (as described for Figure 5A) in different deletion mutant backgrounds for selected SigT carbon starvation targets identified in our microarray analysis. We also assayed mutants carrying deletions in genes that are colocated with *sigT* in the chromosome (Figure 5C) and are presumed regulators of its activity. As shown in Figure 5D, deletions in both ECF Sigma factor *sigU* and the conserved hypothetical protein CC_1532 showed levels of CtrA stabilization that are similar to those observed for the *sigT* deletion strain, suggesting they participate in the same pathway that leads to CtrA degradation under starvation. This is also the case for the genes encoding the histidine kinase HK4, and the response regulator PhyR (CC_3473), that was recently postulated to function as an anti-anti-sigma factor for SigT (or homologous sigma factor) in *Methylobacterium extorquens* [35], *Sinorhizobium meliloti* [36] and *C. crescentus* [37]. In this assay, we did not test *nepR*, the postulated anti-sigma factor for SigT [37].

Discussion

When faced with nutrient limitation, bacterial cells must deploy an array of scavenging systems, adapt their metabolic fluxes to compensate for missing compounds, and limit energy-consuming growth and cell division processes. *C. crescentus* is an oligotroph whose physiology is adapted for survival in environments characterized by low and fluctuating nutrient levels [1]. Here we have examined the response of *C. crescentus* to the sudden onset of carbon starvation.

Carbon starvation uncouples swarmer cell differentiation and the G1-to-S phase transition

In the presence of adequate nutrient levels, the swarmer-to-stalked-cell transition involves the loss of the polar flagellum, the biogenesis of a stalk in its place, and the concurrent initiation of chromosome replication. It has been shown however, that these processes can be uncoupled. When the DnaA activator of DNA replication is depleted [21], or when CtrA inactivation is prevented [38], swarmer cells undergo polar morphological changes, but fail to initiate chromosome replication. Meanwhile, swarmer cells carrying a deletion in the gene encoding the PleD response regulator replicate their chromosome in the absence of polar morphogenesis [39]. We show here that wild type swarmer cells subjected to abrupt carbon depletion were capable of initiating, but not completing, stalk development (see Figure 4A), while the replication and segregation of the chromosomal origin was inhibited (see Figure 4B, [10]). Because the incipient stalks could not be detected by light microscopy, it was previously postulated that carbon starvation inhibits swarmer differentiation. Furthermore, in these experiments cell populations were starved for carbon prior to isolation of swarmer cells [10]. However, electron microscope images (see Figure 4A) revealed the presence

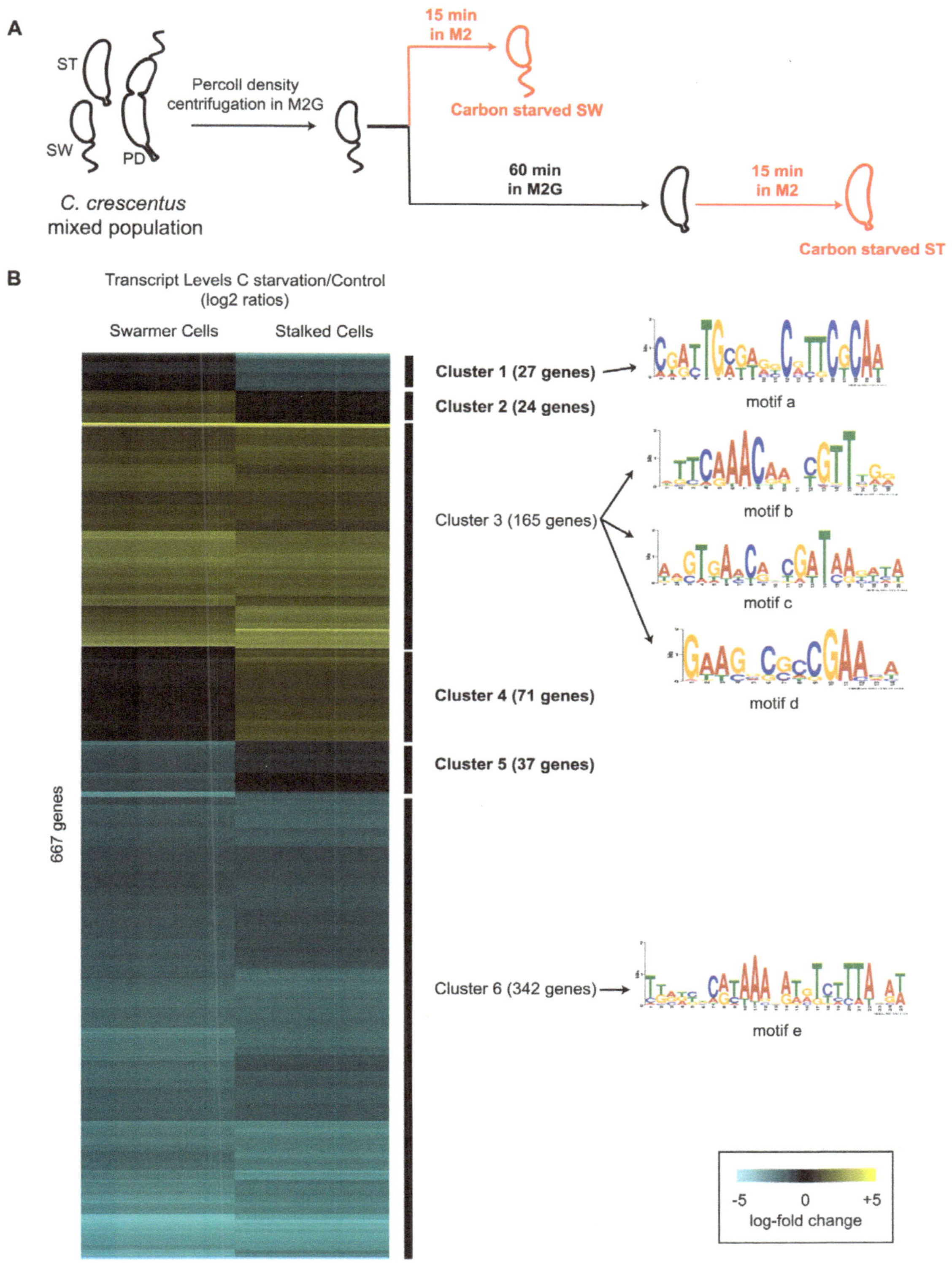

Figure 3. Cell stage-specific carbon starvation response. A. Diagram of the experimental design to explore the cell-stage specific response to carbon starvation. Isolated swarmer (SW) cells were subjected to glucose starvation for 15 min. To assess the response at the stalked cell stage (ST), swarmer cells were allowed to differentiate into stalked cells in complete minimal media for 60 min, and then subjected to glucose starvation for 15 minutes. At the indicated times, cell samples were collected and RNA extracted and transcribed to cDNA to hybridize onto *Caulobacter* microarray chips. PD = predivisional cell. M2 and M2G media are described in the Methods section. **B.** Hierarchical clustering of the transcriptional response to carbon starvation in swarmer cells and stalked cells. The values plotted are the log2-fold change ratios of the cells subjected to 15 minutes of carbon starvation and the non-starved controls. The promoter regions of the genes in each cluster (from −200 to +50 with respect to the translational start site) were used as input in the search for shared motifs using MEME. The five motifs with significant E-values and information content are shown.

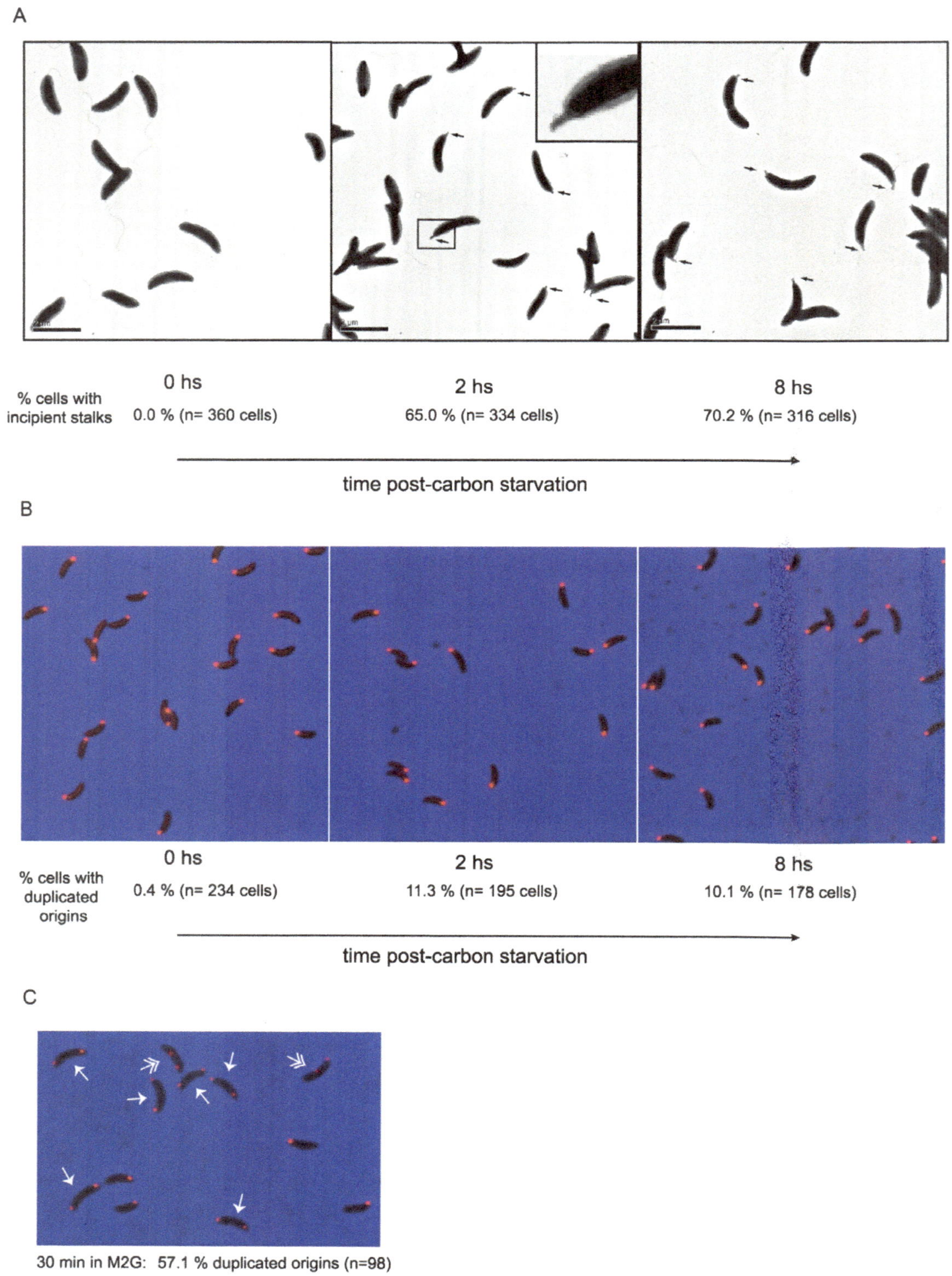

Figure 4. Adaptive changes of swarmer cells upon carbon starvation. A. Polar morphogenesis of wild type swarmer cells subjected to carbon starvation. Swarmer cells were incubated in M2 medium in the absence of glucose for 0, 2 and 8 hs, and visualized by electron microscopy, as described in Methods. The number of cells with incipient stalks (indicated by arrows and detailed in the inset in middle panel) was tallied for several fields and the corresponding percentage is indicated for each time point. **B.** Replication initiation in carbon starved swarmer cells was observed in a strain in which the *parB* gene was replaced with an *ecfp-parB* fusion, treated as described in A. At the indicated times, a sample was removed from the cultures, transferred unto an agarose pad an imaged. The number of cells with duplicated ECFP-ParB foci was counted for each time point and the percentage is indicated. **C.** Swarmer cells isolated for the experiment described in B were incubated in complete M2G media and imaged after 30 min. The percentage of cells with duplicated ECFP-ParB foci is indicated. Arrows indicate cells that have completed origin segregation, as evidence by ECFP-ParB foci in opposite poles, while double arrows indicate cells in the process of segregation.

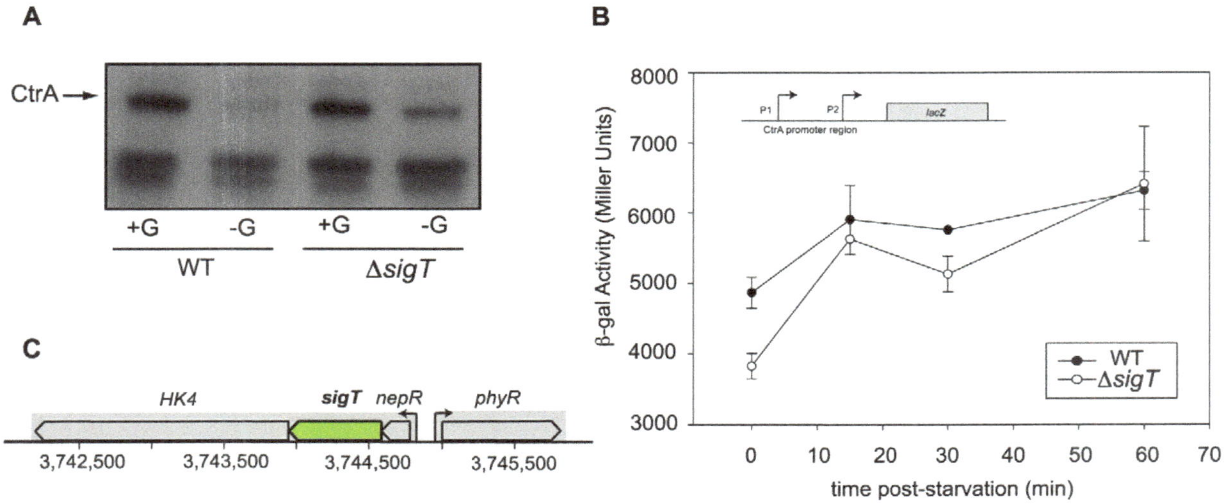

D

Strain	Ratio CtrA Protein -G/+G
WT	**0.26 ± 0.07**
Δ*sigT*	0.57 ± 0.18
Δ*HK4*	0.61 ± 0.09
Δ*phyR*	0.58 ± 0.18
Δ*sigU*	0.46 ± 0.01
Δ*CC_1532*	0.49 ± 0.14

Figure 5. SigT-dependent degradation of CtrA upon carbon starvation. A. Levels of CtrA protein in swarmer cells in the presence and absence of a carbon source, in wild type cells and cells carrying a *sigT* deletion. Isolated swarmer cells from both genetic backgrounds were incubated in M2 medium in the absence or presence of 0.2% glucose. After 150 min, samples from these cultures were subjected to immunoblot analysis with an anti-CtrA polyclonal antibody. The band corresponding to CtrA is indicated with an arrow. **B.** A construct with the complete CtrA promoter region fused to a promoterless *lacZ* reporter in pRKlac290 was introduced into wild-type and *sigT* deletion strains. ß-galactosidase activity was measured in swarmer cells starved for carbon for up to 60 min. **C.** Genomic context of *sigT*. The coordinates correspond to the *C. crescentus* NA1000 genome. CC_3474 (HK = Histidine Kinase) corresponds to CCNA_3588 in NA1000; *sigT* corresponds to CC_3475 (CCNA_3589); *nepR* corresponds to CC_3476 (CCNA_3590); *phyR* corresponds to CC_3477 (CCNA_3591). **D.** Relative levels of the CtrA protein in the presence and absence of carbon, assayed as described in A, are shown for different mutant backgrounds, including deletions of genes in the genomic region of *sigT*, and of SigT-dependent genes (see Table 1). +G corresponds to M2 medium in the presence of 0.2% glucose; -G corresponds to M2 in the absence of glucose. The sampling times are 150 min after starvation, as in A. The error for the ratios corresponds to the standard deviation of the mean for at least two experiments.

of incipient stalks under these conditions. Thus, the point of commitment to initiate stalk morphogenesis appears to precede that of initiation of chromosomal replication. In the population of swarmer cells obtained by the synchronization procedure, the younger cells that are most recently derived from cell division might block both processes, accounting for the 30% of cells that did not develop an incipient stalk by 8 hs of carbon starvation. On the other hand, the majority of swarmer cells in the population that had already committed to polar morphogenesis, blocked the initiation of chromosome replication. Since the stalk is associated with the ability to scavenge for nutrients [15], the initiation of stalk development upon sustained carbon starvation might prime the cell for quick resumption of cell cycle progression once nutrients become available.

Transcript and protein changes in the response to carbon starvation

Transcript profiles as a function of the cell cycle have revealed just-in-time transcriptional activation of distinct functional modules in *C. crescentus* [24]. However, multiple layers of post-transcriptional regulation are known to be involved in the complex orchestration of cell cycle progression and polar differentiation in this bacterium (see [13,40] for reviews). For this reason, our global analysis included not only the transcriptome but also the proteome profile of *C. crescentus* cells subjected to carbon starvation.

Table 3 shows previously characterized proteins of interest from selected functional groups, that change significantly upon 30 and 60 min of carbon starvation. The levels of FtsZ, essential for cytokinesis in *C. crescentus* [41], and FtsX, a predicted ABC transporter that is needed for cell division in *E. coli* [42], decrease significantly upon carbon starvation. This is consistent with the inhibition of cell division that occurs upon carbon starvation.

Proteases have traditionally been associated with the response to environmental stress, as cells need to re-engineer the cellular landscape, recycle damaged and unwanted proteins, and selectively target key regulators [43]. The levels of two major proteolysis-related factors increased significantly upon carbon starvation, and could be fulfilling this role in *C. crescentus*: the FtsH

Table 3. Proteins of interest in selected functional categories that change upon carbon starvation.

Functional category	Upregulated proteins	Downregulated proteins
Cell wall/membrane/envelope biogenesis (COG M) and cell division	TolA, TolB, TolR, TolQ, BamA, BamB, BamE, BamD, DipM, MreC	FtsZ, FtsX
Cell cycle control		CtrA, DnaA, GcrA
Post-translational modification (COG O)	ClpA, FtsH, SppA	DnaJ, Hsp33
Signal transduction and gene regulation	DivL (HK), FixL (HK), CckA (HK-RR)	FlbD (RR), PleD (RR), DivK (RR), McpA, CheAI, CheYI, CheD
Chromosome structure and dynamics	Smc	PopZ, ParA

The classification is based upon COG (NCBI) and the literature. Abbreviations: HK = histidine kinase; RR = response regulator; HK-RR = hybrid histidine kinase-response regulator.

protease, previously characterized as a component of *C. crescentus'* general stress response [44], and the ClpA chaperone, which normally works in concert with the ClpP protease, and has not been thus far associated with the stress response.

The level of Smc (CC_0373), a nucleoid-associated protein required for chromosome structure maintenance and segregation in *C. crescentus* [45], increased significantly upon carbon starvation. This raises the possibility that the nucleoid of growth-arrested cells adopts a different compaction state than that in exponentially growing cells, possibly contributing to the inhibition of DNA replication and segregation, as well as an increased tolerance to stress-related damaging agents. *E. coli's* nucleoid protein composition has been shown to be growth phase-dependent [46] and a factor in long term survival [47].

Respiratory metabolism generates reactive oxygen species that may damage membranes, DNA and proteins. As long as the environment promotes growth and continued de novo synthesis, the oxidized macromolecules are rapidly diluted, but this is not the case during growth-arrest of metabolically active cells. Correlation of our carbon starvation datasets with microarray analysis of *C. crescentus'* response to metal stress [29], revealed that of 222 genes that were induced by two or more of the four heavy metals tested, 38 were upregulated in our carbon starvation proteomics samples, and 103 were induced after 30 minutes of carbon starvation at the transcript level (Table S6). These results suggest that a subset of the genes that respond to carbon starvation are part of a general stress response. In *E. coli* and *B. subtilis*, growth arrest caused by starvation has also been shown to elicit increased synthesis of proteins normally induced by oxidative stress, and starved cells display cross-resistance to these stresses [7,48,49].

The correlation of the set of transcripts and the set of proteins that change significantly upon starvation revealed interesting insights regarding the levels of regulation that might be operating in the carbon starvation response (Figure 6 and Table S7). Considering the 1364 genes for which proteome data were obtained and there was a significant change upon carbon starvation at either the transcript or protein level, or both, in more than half of the cases, the change was only observed at the transcript level (28.5% of genes were up-regulated and 34.1% were downregulated). Some of these might reflect an inherent greater sensitivity of the microarray technique used for detection of transcript changes, than that of the mass spectrometry method used to probe changes in protein levels. 14.8% of the genes showed changes in the same direction at both the transcript and protein level, consistent with transcriptional regulation. For a similar proportion of genes (13.6%), no change was observed at the transcript level, but the corresponding protein was significantly up- or down-regulated. This group contains candidates for post-transcriptional regulation, either at the translation or protein stability levels. For 9.0% of the genes, observed changes at transcript and protein levels occurred in opposite directions. Overall, our observations are in line with a complex interplay of regulatory mechanisms operating in bacteria, and the lack of correlation between mRNA and protein levels observed in *E. coli* [50,51].

Regulatory networks controlling *C. crescentus'* response to carbon starvation

Analysis of transcriptional profiles allowed us to identify regulators and regulatory modules involved in the response to carbon starvation and, more specifically, connections with the regulatory network that controls cell cycle progression. The enrichment analysis based on previously characterized clusters of co-expressed cell cycle genes that share conserved promoter motifs [25], indicated that the RpoD housekeeping sigma factor, the FixK transcription factor, the SigT ECF sigma factor, and the CtrA cell cycle master regulator play significant roles in the response of *C. crescentus* to carbon starvation. The diagram shown in Figure 7 integrates our transcriptome and proteome datasets of the response to carbon starvation, with regulatory pathways derived from microarray data published for CtrA [31], FixK [27], and LexA [52] (See Tables 2 and S8). The pathways for SigT and SigU under starvation conditions were inferred from data presented in this paper (See Table 1). The proposed transcriptional regulatory interactions were assumed to be direct when the cognate DNA binding motif for the regulator was located in the promoter region of the target gene.

RpoD. The transcript levels of housekeeping sigma factor RpoD and 30 genes identified as RpoD targets by the presence of its cognate DNA motif (cc_6), changed significantly upon carbon starvation, the great majority of them being downregulated. Five of the proteins predicted to be encoded by these genes also changed significantly upon carbon starvation (Figure 7 and Table S8). A decrease in RpoD protein levels was not observed under these conditions, pointing to an alternative mechanism of regulation of its activity that would explain the down-regulation of its target genes, such as the competition from alternative sigma factors induced upon starvation, in binding to the core RNA polymerase.

FixK. The FixL-FixJ-FixK pathway was shown to be a major component of *C. crescentus'* response to hypoxia by Crosson *et al.* [27]. Our analysis of the response to carbon starvation revealed that the protein levels of sensor histidine kinase FixL increased, while 16 of the genes directly (bearing the motif cc_3 in their

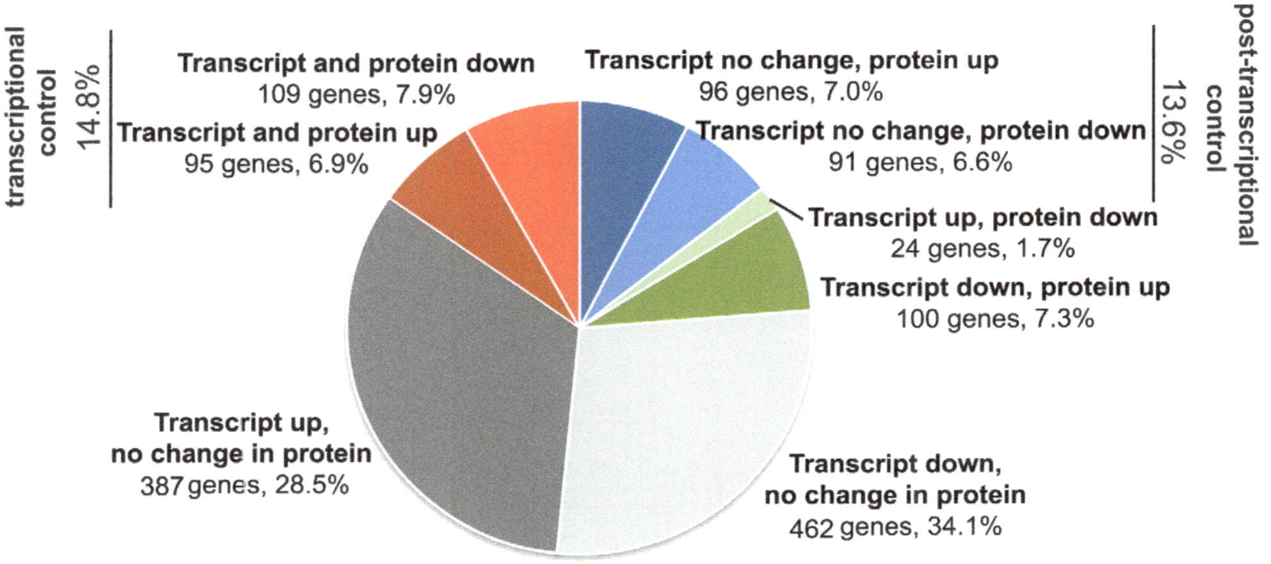

Figure 6. Distribution of carbon starvation-induced changes at the transcript and protein levels. For the 1364 genes for which we had both microarrays and proteomics valid data, and a significant change had been observed for at least one of the datasets, the intersection and exclusion sets are represented. Table S7 lists the genes that belong to each group in the distribution.

promoter regions) or indirectly regulated by transcriptional regulator FixK, changed significantly upon carbon starvation (Figure 7). Two proteins encoded by these FixK-regulated genes were observed to be up-regulated under the same conditions. Interestingly, only half of the genes that have the cc_3 motif were identified as part of the hypoxia-induced FixK regulon by Crosson *et al.*, suggesting that FixK might activate different sets of genes in response to different environmental stimuli or that additional regulatory pathways or factors are involved in these responses.

CtrA. Out of 55 cell cycle-regulated CtrA target genes, we found that 13 were upregulated and 7 were downregulated after 30 minutes of carbon starvation. Out of 32 non-cell cycle-regulated CtrA target genes, 6 were upregulated and 6 were downregulated after 30 minutes of carbon starvation (Table 2 and Figure 7). Our proteome analysis showed a significant decrease in the protein level of CtrA in response to carbon starvation, consistent with previous analyses [10,22]. Of the CtrA-regulated cell cycle transcripts [25] that were upregulated upon carbon starvation (Table 2), 11 out of 13 peaked at the stalked or predivisional stage, while the levels of 10 of these 13 transcripts were reported to increase in a strain carrying a temperature sensitive allele of CtrA ($CtrA^{ts}$) incubated at restrictive temperature [31]. These observations are compatible with these genes being de-repressed as CtrA protein levels drop in swarmer cells starved for carbon. Conversely, 5 out of the 7 CtrA cell cycle target genes that are down-regulated in response to carbon starvation, show lower transcript levels at the swarmer stage of the cell cycle, with 2 of them displaying increased levels in the CtrA thermosensitive mutant at the restrictive temperature. These genes are most likely deactivated in carbon starved swarmer cells as CtrA levels drop. The remaining 12 CtrA target genes, whose transcript levels remain constant during cell cycle progression in complete media, point to a novel role for CtrA in activating and repressing genes involved in the response to starvation and other stress signals, in addition to those involved in cell cycle progression.

LexA. Upon carbon starvation, the protein levels of LexA –a CtrA target- were observed to decrease to 60% of the non-starvation levels (although the p-value (0.09) didn't meet the cutoff established for significance (0.05)). The SOS response, the prototypical response to DNA damage in prokaryotes, is controlled by the opposing activities of the LexA repressor and the RecA activator proteins [53]. Genes in the SOS regulon, repressed by LexA under basal conditions, are activated in response to single-stranded DNA regions, often as result of DNA replication inhibition [54]. Activation of the SOS response in *E. coli* leads to blocked FtsZ ring formation and cell division, via the SulA protein [55]. It is possible that a LexA target interacts with FtsZ in *C. crescentus*, to mediate the cell division arrest observed upon starvation [10]. Out of the 44 previously characterized direct LexA target genes [52], two were downregulated and nine were upregulated upon 30 min of carbon starvation, consistent with the decrease in LexA protein levels; one of the upregulated genes, encoding the CC_2589 hypothetical protein, was also upregulated at the protein level.

SigT is a regulator of the starvation-stress response in *C. crescentus* and is involved in the starvation-induced degradation of the CtrA master regulator

Alternative sigma factors play key roles in various stress responses and morphological differentiation in bacteria. Upon activation by environmental or internal signals, alternative sigma factors direct RNA polymerase promoter specificity to activate different regulons. In most gram-negative bacteria, the transcriptional response to environmental stresses is dominated by alternative sigma factor RpoS (see [56] for a review). In gram-positive bacteria, that role falls upon the alternative sigma factor SigB (see [57] for a review). However, the α-proteobacteria lack homologues of either of these sigma factors [8,58].

Three of the 13 *C. crescentus* ECF alternative sigma factors have been implicated in the response to specific stress conditions: SigF mediates the response to oxidative stress in stationary phase [59], SigE mediates the response to cadmium, organic hydroperoxide, singlet oxygen and UV [60], and SigT mediates the response to osmotic and oxidative stress [28]. Our results support a broader

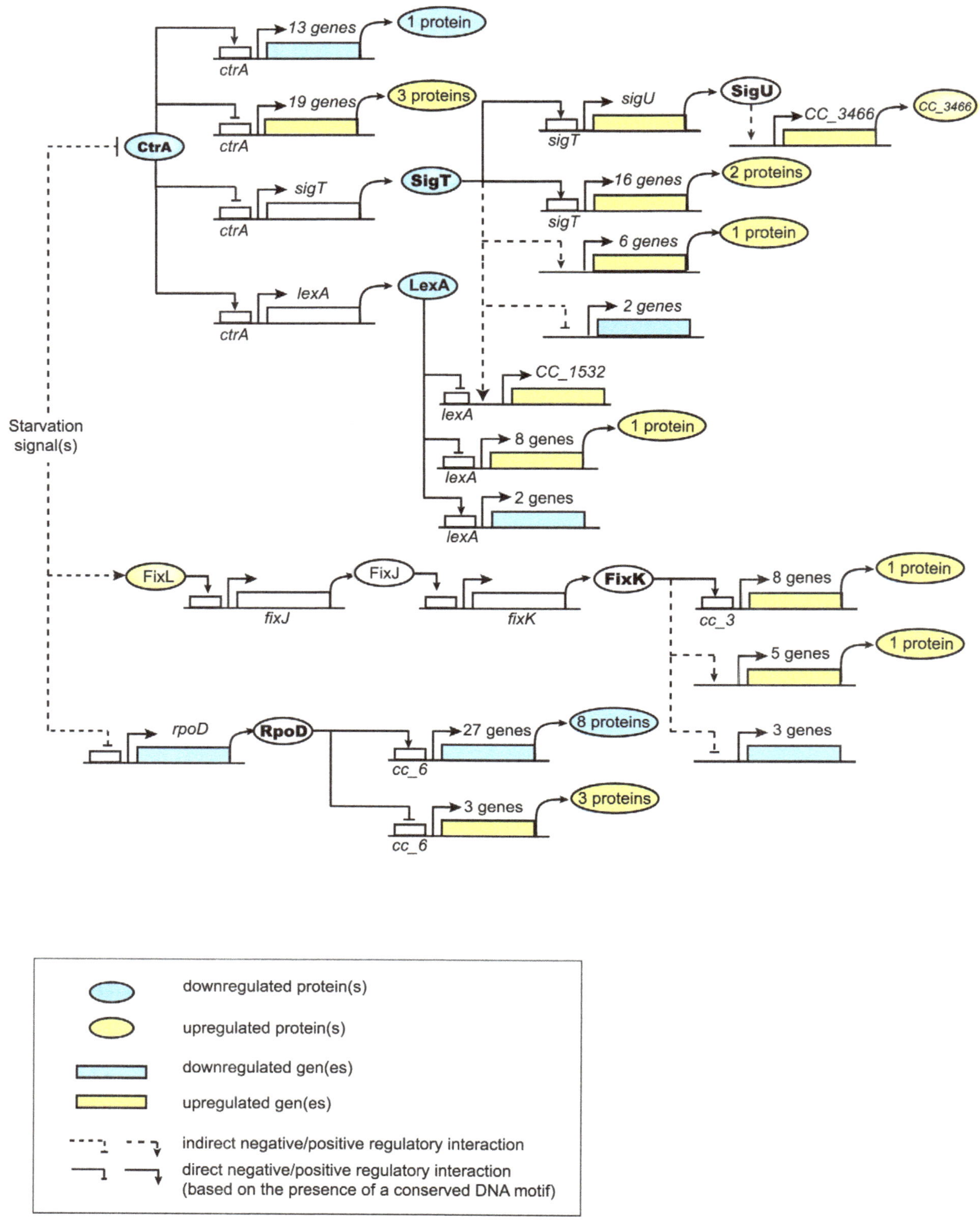

Figure 7. Carbon starvation regulatory pathways. Diagram of regulatory pathways involved in the response to carbon starvation, derived from the analysis of proteomic and gene expression profile changes. Ovals represent proteins and rectangles represent genes. Genes and proteins whose levels increase significantly upon starvation are represented in yellow, and downregulated genes and proteins in blue. For example, we indicate that of the 13 genes controlled by CtrA whose transcripts were downregulated in the absence of carbon (blue rectangle), we were able to detect the down-regulation of one of the corresponding protein products (blue oval). The same schematic representation was used for all genes shown. The genes and proteins that change significantly upon carbon starvation are listed in Table S8. Putative direct regulatory interactions are based on the presence of conserved promoter sequences, while an indirect regulation is proposed if there is no evidence of DNA binding or of the presence of a

conserved promoter element. For the transcript changes, the 30 min time point was considered, while both the 30 and 60 min time points were considered for the protein changes. The CtrA regulon comprises direct CtrA targets, as previously determined [31], that changed upon 30 min of carbon starvation determined by microarray analysis (listed in Table 2), as well as proteins encoded by direct CtrA targets that changed in the proteomic analysis at 30 and/or 60 min of carbon starvation (while most of those proteins were seen to change at the transcript level as well, SigU and LexA were observed to change only at the protein level). CtrA activation or repression was inferred by the direction of the change observed upon starvation: those genes with increased transcript levels upon starvation were inferred to be negatively regulated by CtrA (and consequently derepressed as CtrA protein levels decrease upon starvation); genes with decreased transcript levels were inferred to be positively regulated by CtrA. The SigT regulon includes the genes that showed a significant difference in transcript levels changes upon 15 min of carbon starvation in the *sigT* deletion strain with respect to wild type (listed in Table 1). Genes with a SigT binding motif in their promoters are represented as direct targets, while those lacking the motif, as indirect targets. The levels of proteins encoded by a subset of these genes changed significantly upon 30 and/or 60 min of carbon starvation. The SigU-dependent gene, CC_3466, showed reduced transcript levels under carbon starvation in a *sigU* deletion strain with respect to wild type. The LexA (SOS) regulon comprises those genes belonging to the SOS regulon –as previously determined [52] - that changed significantly after 30 min of carbon starvation. The topology of the FixL-FixJ-FixK pathway is as determined by Crosson *et al.* [27]. The FixK direct targets are those that have the cc_3 motif in their promoter, while the indirect targets were shown to be FixK-dependent by Crosson *et al.*, but lack the cc_3 motif. The RpoD regulon comprises the genes with the cc_6 promoter motif whose transcripts levels were upregulated or downregulated upon 30 min of carbon starvation.

role for SigT. The 27 genes that respond to carbon starvation in a SigT-dependent manner comprise the carbon starvation SigT regulon (Table 1 and Figure 7). Some of these genes were also found to change at the protein level, including CC_3466 (Table 1, Table S2), which appears to be regulated through the SigU sigma factor. Moreover, there is an overlap of the SigT carbon starvation regulon with both the SigT osmotic stress regulon [28] (10 genes in common between the two sets), and the set of genes that are activated by exposure to several heavy metals [29] (12 genes in common). These observations suggest that a core set of SigT regulated genes belongs to a general starvation-stress response. Members of SigT's subfamily of ECF sigma factors from *Sinorhizobium meliloti* and *Methylobacterium extorquens* have been implicated in the regulation of the general stress response [35,61]. Interestingly, some of the transcripts encoding regulatory proteins that were shown to be SigT-dependent upon carbon starvation -namely CC_1356, CC_0284 and CC_1178- required RpoN for induction in carbon versus nitrogen-limited conditions [12]. Two of these genes, CC_1356 and CC_0284 were reported to be SigT-dependent under conditions of osmotic stress [28].

The *sigT* gene belongs to the CtrA cell cycle regulon. CtrA is a negative regulator of *sigT* expression, and *sigT* transcripts peak when CtrA is cleared from the cell at the swarmer to stalked cell transition [31]. The transcript levels of *sigT* were not seen to change significantly upon carbon starvation at the sampled times, in either mixed population or isolated swarmer and stalked cells, while the SigT protein level was found to decrease after 60 minutes of carbon starvation. It is likely that SigT is regulated post-translationally. SigT belongs to the ECF sub-family (ECF15 or EcfG-like) that is characterized by a conserved genomic context, which includes the genes encoding the HK4 histidine kinase and the PhyR response regulator (see Fig. 5C), putative candidates to modulate the activity of the sigma factor [62]. A recent report postulates that *C. crescentus*' *sigT* is regulated by the anti-sigma factor NepR (CC_3476, cotranscribed with *sigT*), and the gene encoding PhyR, acting as an anti-anti-sigma factor, that is transcribed divergently from *sigT* [37].

We propose that a SigT-dependent pathway is involved in the degradation of CtrA upon carbon starvation, and that this pathway includes the response regulator PhyR and the HK4 histidine kinase. The ability of the anti-anti-sigma factor PhyR to sequester the anti-sigma factor NepR, releasing this SigT to activate transcription, has been shown to be dependent on PhyR phosphorylation state [37]. It is possible that the HK4 histidine kinase, which possesses a predicted signal peptide that would target it to the membrane, is the protein responsible for sensing environmental cues and inducing a response in the CtrA starvation degradation pathway, via PhyR (Figure 8).

Materials and Methods

Bacterial strains and growth conditions

All strains were derived from wild type *C. crescentus* (NCBI Taxonomy ID: 155892) strain CB15N (NA1000) [63]. PYE medium (0.2% Bacto Peptone (Difco), 0.1% yeast extract (Difco), 1 mM $MgSO_4$, and 0.5 mM $CaCl_2$) was used to grow strains for cloning purposes. In all other cases, strains were grown in M2 minimal medium (6.1 mM Na_2HPO_4, 3.9 mM KH_2PO_4, 9.3 mM NH_4Cl, 0.5 mM $MgSO_4$, 10 μM $FeSO_4$ (EDTA chelate; Sigma), 0.5 mM CaCl2) with 0.2% glucose as the sole carbon source (referred to as M2G). All *Caulobacter* strains were grown at 28°C and were not allowed to reach an OD_{600} higher than 0.4 at any time, to minimize differences due to physiological adaptation of cultures to stationary phase conditions. *Escherichia coli* strain TOP10 (Invitrogen) was used for cloning following standard procedures.

Construction of gene deletion strains

In frame deletions of *sigT* (CC_3475), *sigU* (CC_2883), *phyR* (CC_3477), *HK4* (CC_3474), CC_3466 and CC_1532 were obtained by a two-step *sacB* counterselection procedure, as previously described [64].

Culture synchronization to obtain isolated swarmer and stalked cells

Swarmer cells were isolated from mixed population cultures using a modified version of the percoll density centrifugation protocol [65]. Isolated colonies from a PYE plate were used to inoculate an M2G culture, grown to an optical density at 600 nm of 0.3. Cultures were cooled on ice, pelleted at 9,000×g at 4°C, and resuspended in 300 μl of ice-cold M2G. After adjusting the volume to 1 ml, one volume of ice-cold Percoll was added and

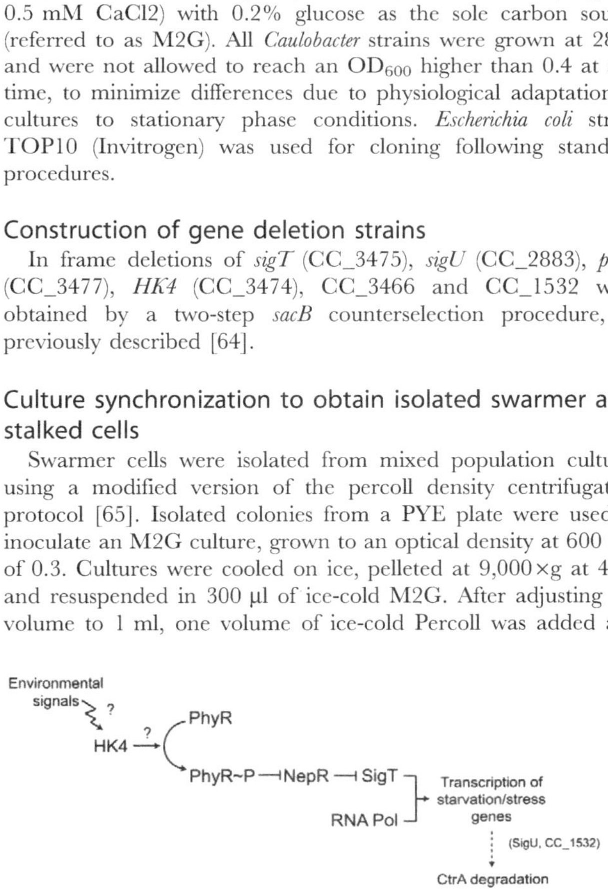

Figure 8. Diagram of the putative HK4-SigT signal transduction pathway. Shown is an inferred pathway based on our results, suggesting that SigT, HK4 (CC_3474), PhyR, SigU and CC_1532 contribute to the degradation of CtrA upon carbon starvation.

mixed thoroughly. The suspension was centrifuged at 11,000×g for 20 min at 4°C, to separate the swarmer cells (lower band) from the rest of the population (top band). To obtain the stalked cells samples for the experiment in Figure 3, the synchronization procedure was performed as described, swarmers were resuspended in M2G, and allowed to differentiate for 60 min at 28°C. At that point, loss of motility of the majority of the population was confirmed by microscopy.

Carbon starvation of wild-type swarmers and stalked cells, and of mixed populations of wild-type, Δ*sigT* and Δ*sigU* strains

Cells (swarmers or mixed populations, as indicated in each case) were washed twice in either ice-cold M2G (control samples) or ice-cold M2 (carbon starvation samples), and resuspended in 2 ml of pre-warmed (28°C) M2G or M2 media. Optical density was adjusted to ~0.3–0.4. After 15 min of incubation at 28°C, cells were pelleted, frozen in liquid nitrogen, and transferred to −80°C. RNA isolation (see below) was always performed within 24 hs of obtaining the samples.

Carbon starvation of *C. crescentus* mixed population for proteomics and microarrays assays

Cells were grown overnight in small cultures in M2G medium (never reaching $OD_{600\ nm}$ greater than 0.4). These cultures were used to inoculate 1200 ml cultures, which were incubated at 28°C until they had reached an $OD_{600\ nm}$ of 0.3–0.4. Cells were collected by centrifugation (5 min at 9,000×*g*) and washed twice with a large volume of ice-cold M2 medium, resuspending by gentle pipetting. Control cells were mock-washed with ice-cold M2 with 0.2% glucose. Washed cells were resuspended in pre-warmed M2 or M2G media and incubated at 28°C in a water bath with shaking. Samples were collected at 30 and 60 minutes, frozen in liquid nitrogen and stored at −80°C.

Liquid chromatography-mass spectrometry analysis

Cells were pelleted, frozen in liquid nitrogen and stored at −80°C for no more than 2 days. Lysis and protein treatment was done as described previously [66]. Cell were lysed by bead beating. Proteins were denatured, reduced, alkylated, concentrated by solid-phase extraction, and digested with trypsin. For each injection, 1 µL of each sample of resuspended peptides was injected onto a reversed-phase column using an Isco LC system (Teledyne Isco) and eluted into a Thermo LTQ mass spectrometer (ThermoFinnigan, Inc., San Jose, CA). The mass spectrometer was operated in a data-dependent scan mode as previously described [67]. Raw data were analyzed with the Sequest program with an in silico database obtained from simulated tryptic digestion of the *C. crescentus* genome. Peptides from this analysis meeting our previously-used scoring standards [67] were used for abundance measurements. Peptide abundances were estimated from ion chromatograms using Viper [68]. Data were extracted and tabulated using SQL queries generated by a custom query from in-house databases. Data analysis, following the general normalization and rollup protocol described previously [69], was performed using the in-house quantitative proteomics tool DAnTE [70] and the statistical software package R (r-project.org). Recorded intensities were converted to natural logarithms throughout. Datasets consisted of the raw peptide intensities, reported as ion counts, for all experiments. All data sets were aligned together as a single batch using MultiAlign, and centered to the most complete data set using the *C. crescentus* accurate mass tag (AMT) database [68]. For subsequent work, peptide intensities were linear regressed to the median value of each experimental replicate data set. Normalized peptide intensities were rolled up into effective intensities for each protein in each experimental replicate using the RRollup algorithm [70]. For the binary comparisons of interest, the mean differential expression was calculated for each protein and the statistical significance of the differential expression was established using a two-sided t-test. Significance was established at $p<0.05$. To take into account very low abundance proteins, the counts of observed peptides under different treatment conditions were also compared using a standard Fisher's Exact Test on the 5 most abundant peptides. Proteins that showed significantly different peptide counts ($p<0.05$) were pooled together with the proteins that showed a significant change by intensity analysis, to report the final list of significant changes.

Transmission electron microscopy

Cells were collected by centrifugation, fixed for 15 min at room temperature in 4% glutaraldehyde in 100 mM cacodylate-HCl buffer pH 7.4, washed, resuspended in cacodylate buffer, and preserved at 4°C until imaging. Fixed cells were spotted onto glow discharged formvar-carbon coated 300 mesh copper grids (Electron Microscopy Sciences), and allowed to settled for 1 min. Grids were then washed with two drops of mQ water, stained for 15 seconds with 1% uranyl acetate, washed again with two drops of mQ water and air dried. Grids were imaged at 80 kV on a JEOL TEM1230 system. Images were captured with a Gatan 967 slow-scan, cooled CCD camera, using the associated Gatan software. The numbers of cells with or without incipient stalks were counted manually from the exported images.

Light and fluorescent microscopy

Swarmer cells from a strain carrying a *ecfp-parB* fusion in place of the *parB* gene in the chromosome (MT190) [71] were isolated, washed as described above to remove glucose, and immobilized onto freshly prepared 1% agarose-M2 pads onto microscopy slides, at the indicated times. Microscopy was performed on a DM6000B upright microscope (Leica) fitted with a 100× 1.46 NA HCX Plan APO oil immersion objective (Leica) and a Hamamatsu C9100 EM CCD camera. Phase contrast and fluorescent images (CFP channel, Ex 438/24 nm, Em 483/32 nm) were taken at 40 and 100 ms exposure times, respectively. Images were acquired using KAMS-acquire, a custom software program developed in-house [72], to control the microscope and camera. We used Photoshop CS4 (Adobe) to make false color merges of phase and fluorescent images. In order to tally cells with duplicated and non-duplicated origins, we processed KAMS-acquire files with a Matlab script developed in-house to obtain ordered images of individual cells (S. Hong, unpublished), which were then counted manually.

Immunoblot analysis

Cell samples normalized by OD_{600} were lysed by boiling in 2× sample buffer (4% Sodium Dodecyl Sulphate (SDS), 20% glycerol, 0.01% Bromophenol Blue, 0.125 M Tris-HCl pH 6.8), and loaded in 8–16% Precise polyacrylamide gradient gels (Pierce), followed by electrophoretic transfer to a PVDF membrane (Millipore). Immunoblotting was done using anti-CtrA polyclonal serum (1:10,000), and horseradish-peroxidase conjugated goat anti-rabbit IgG (1:20,000). Signal was detected with chemiluminescence reagent (Perkin-Elmer) and BioMax MR film. The developed film was scanned, processed with Photoshop CS4 (Adobe), and band intensities were determined using ImageQuant (Molecular Dynamics).

β-galactosidase assays

The β-galactosidase activity of a wild type and a Δ*sigT* strain carrying the plasmid pCtrA290 [73], with the CtrA promoter region fused to *lacZ*, was assayed after removal of glucose from the cultures in M2 medium, using o-nitrophenyl-β-D-galactoside (ONPG) [74].

RNA Extraction, cDNA synthesis and processing, and microarrays hybridization

Total RNA was extracted from cells using the Purelink Mini Total RNA Purification System (Invitrogen), according to the manufacturer's instructions for on-column DNAse treatment to remove contaminating DNA. A maximum of 2 ml of culture of $OD_{600\ nm}$ between 0.3 and 0.4 was used per ml of Trizol. Integrity of the RNA was confirmed using the RNA 6000 Nano kit (Agilent) on a Agilent 2100 Bioanalyzer, and its concentration was calculated from $OD_{260\ nm}$ measurements on a nanodrop. cDNA was synthesized using Super Script II (Invitrogen) with random hexamers, according to the manufacturer's instructions. After reverse transcription, RNA was removed by 30 min NaOH incubation at 65°C and cDNA was purified using MinElute columns (Qiagen). Purified cDNA was fragmented by a 5 min incubation with DNAseI (Invitrogen; 0.6 U/μg cDNA) in a thermocycler, followed by 15 min enzyme inactivation at 98°C. DNAseI activity was previously titrated, and the same batch of enzyme and thermocycler were used with the same settings across all experiments to obtain reproducible fragmentation. RNA 6000 Nano kit (Agilent) on a Agilent 2100 Bioanalyzer was used to control for homogeneous fragmentation (50–200 bp range) of samples. Fragmented cDNA was biotinylated at the 3′ termini with GeneChip DNA Labeling Reagent (Affymetrix) and hybridized onto the CauloHI1 chip (Affymetrix). Hybridization was performed at Stanford's Protein and Nucleic Acid facility, as previously described [25]. All data are MIAME compliant and raw data have been deposited in NCBI Gene Expression Omnibus (www.ncbi.nlm.nih.gov/geo/, with series accession number GSE25999).

Gene annotation

The initial analysis of transcripts and proteins took as a reference the coordinates, ORF prediction and annotation for the *C. crescentus* reference strain CB15 (GeneBank accession AE005673, RefSeq NC_002696). Upon release of the genomic sequence of the *C. crescentus* laboratory strain CB15N (also known as NA1000; GeneBank accession CP001340, RefSeq NC_011916), which includes an updated ORF prediction and gene annotation [75], all results were mapped to this new annotation. Supplementary tables include gene IDs corresponding to both genomes nomenclatures. To generate and process diverse gene lists from the microarrays and proteomics results, a Pathways and Genome Database (PGDB) was built using the PathoLogic and Pathway/Genomes Editor software, in the Pathway Tools platform [76]. The *C. crescentus* NA1000 PGDB was manually curated to incorporate information for diverse published datasets, and is available upon request.

Statistical analysis of microarray data from carbon starved *C. crescentus* mixed population

The RMA statistical algorithm [77], available under the Bioconductor software package of R, was used for background noise removal, normalization and summarization of microarray data corresponding to two independent experiments for each condition (cells starved for carbon for 30 and 60 min, and the respective controls). All data are MIAME compliant, and raw and normalized data files were submitted to NCBI Gene Expression Omnibus (accession number GSE25996). A Significance Analysis of Microarrays (SAM) [78] was applied to the dataset, with the following parameters: unpaired, logged, median centered, T statistic. A 2-fold minimum change was selected as cutoff, and a delta value yielding a false discovery rate lower than 1%.

Clustering of transcript changes in swarmer and stalked cells, and search for conserved promoter motifs

Microarray data were normalized and summarized as described in the previous section. All data are MIAME compliant and raw and normalized data has been deposited in NCBI Gene Expression Omnibus (www.ncbi.nlm.nih.gov/geo/, with accession number GSE25997). For each one of the 3767 genes analyzed, we created a profile vector composed of two values: i) log-ratio between expression values measured for swarmer cells after 15 minutes of carbon starvation and the non-starved control, ii) log-ratio between expression values for stalked cells after 15 minutes of carbon starvation and non-starved control. The non-starved controls values were obtained by linearly interpolating the data from [25], at the corresponding time-points (15 minutes into the cell cycle, as a control for swarmer cells, and 75 minutes for stalked cells). We defined a gene as not responding to carbon starvation if the fold-change in expression was less than 2×. Similarly, we defined a gene as responding to carbon starvation if the fold-change in expression was greater than 3×. We only considered genes that, for both carbon starved swarmer and stalked cells, either responded to carbon starvation or did not (no change: $-1<|\text{log-ratio}|<1$; change: $|\text{log-ratio}|>1.585$). We then clustered the profiles of the genes using a bottom-up hierarchical clustering approach (Bioconductor R package), using the Pearson correlation distance and complete linkage for measuring inter-cluster distances. To search for conserved motifs, we extracted 250 bp of sequence (from −200 to +50 with respect to the translational start site) for the genes in each cluster, and used MEME [79], with the following parameters: distribution of motif occurrences: zero or one per sequence; minimum motif width: 6; maximum motif width: 25. We selected the motifs with an E-value< = 0.2 and average Relative Entropy> = 1bit/bp.

Functional categories enrichment of gene and protein sets

In order to identify functional categories that were significantly over or under-represented within the set of transcripts and proteins that changed upon carbon starvation (30 minutes for transcripts and 30 and/or 60 minutes for proteins), we calculated p-values for each COG-category based on hyper-geometric distributions (p-value< = 1%). Similarly, in order to identify known motifs, associated with cell-cycle and metal stress [25], that were significantly over or under-represented in the upstream regions of the set of genes whose expression changed significantly upon 30 minutes of carbon starvation, we calculated p-values based on hyper-geometric distributions (p-value< = 5%).

Microarray analysis of Δ*sigT* and Δ*sigU* strains upon starvation

For each microarray experiment, expression signals for 3767 genes were analyzed. The RMA statistical algorithm [77], available under the Bioconductor software package of R, was used for background noise removal, normalization and summarization of the microarray data. All data are MIAME compliant, and raw and normalized data files were submitted to NCBI Gene

Expression Omnibus (accession number GSE25998). For each gene i, we calculated the difference between sample means and a two-sample t statistic. For each gene and each condition ($\Delta sigT$, $\Delta sigU$ and wild-type strains starved for glucose for 15 minutes), we then calculated a p-value derived from the gene-specific t-statistic, as well as the empirical distribution derived from the t-statistics of all genes. A gene was considered to be differentially expressed, if two conditions were met: the difference between the mutant and WT was greater than two-fold, and the p-value derived from the corresponding t-statistics was smaller than 5%.

Supporting Information

Table S1 Proteome coverage of bacterial species. The highest proteomic coverages for bacterial organisms are shown with the corresponding reference. The proteomic coverage is expressed in the percentage of annotated genes for which the predicted encoded protein has been detected, as reported by the cited publications.

Table S2 List of proteins that change significantly upon carbon starvation. Gene ID (for strains CB15 and CB15N), annotation, COG category (if applicable), fold change and corresponding p-value are shown for the proteins that were found to change reproducibly in cells starved for carbon for 30 and/or 60 min.

Table S3 List of transcripts that change significantly upon carbon starvation. Gene ID (for strains CB15 and CB15N), annotation, COG category (if applicable), average fold change and corresponding standard deviation are shown for the transcripts that were found to change reproducibly in cells starved for carbon for 30 and/or 60 min. For each transcript it is also indicated whether it was previously classified as cell cycle regulated [24].

Table S4 List of genes that change significantly upon carbon starvation in the sets represented in Figure 2. Gene ID (for strains CB15 and CB15N), annotation, and fold change between starved and non-starved cells at 30 min (except for motif cc_7, for which the values correspond to 60 min) are shown.

Table S5 List of genes in the clusters represented in Figure 3B. Gene ID (for strains CB15 and CB15N), log-2 ratio of transcript levels of starved and non-starved swarmer (SW) and stalked (ST) cells after 15 min, annotation, and COG category (if applicable), are shown for all genes clustered in Figure 3B.

Table S6 List of proteins and transcripts that are upregulated upon carbon starvation and heavy metal stress. Genes that were induced by two or more heavy metals [29], and were upregulated in carbon starvation at the level of protein or transcript.

Table S7 Lists of genes represented in Figure 6. This spreadsheet contains eight separate tabs with the lists of genes corresponding to the categories in the chart in Figure 6. Each list was obtained from the intersections of the lists of genes and proteins determined to be up- and down-regulated according to the criteria described in Methods, for those 1378 genes for which there was valid data at both the transcript and the protein level, and a significant change in at least one of the datasets (microarrays or proteomics).

Table S8 Lists of genes represented in Figure 7. Genes in the FixK carbon starvation regulon were obtained from the intersection of up- and down-regulated genes after 30 min of carbon starvation with genes with motif cc_3 according to McGrath *et. al.* [25], and FixK-dependent genes in hypoxia according to Crosson *et. al.* [27]. Genes in the LexA carbon starvation regulon were obtained from the intersection of up- (1) and down-regulated (2) genes after 30 min of carbon starvation with genes in the SOS regulon [52]. Genes in the RpoD carbon starvation regulon were obtained from the intersection of up- (1) and down-regulated (2) genes after 30 min of carbon starvation with genes with motif cc_6 according to McGrath *et. al.* [25]. Genes in the SigT carbon starvation regulon are listed in Table 1. The proteins encoded by SigT targets from Table 1 that are upregulated after 30 or 60 min of carbon starvation are listed here. The CtrA regulon was determined by intersecting the list of CtrA targets [24] (Table 2) with the list of genes up- and down-regulated after 30 min of carbon starvation (Table S3), and the list of proteins that are up- or down-regulated after 30 and/or 60 min of carbon starvation (Table S2). Only the genes for which the transcript and protein changed upon starvation (sets 1 and 2) were included in Figure 7.

Acknowledgments

The authors would like to thank Peter Karp, Alexander Shearer and Suzanne Paley (Bioinformatics Research Group, SRI International, Menlo Park, California, United States), for assistance with development of the *C. crescentus* NA1000 Pathways and Genome Database; John Perrino (Cell Sciences Imaging Facility, Stanford University, Stanford, California, United States), for training in the use of the transmission electron microscope; Elizabeth Zuo and Natalia Kosovilka (Protein and Nucleic Acid Facility, Stanford University, Stanford, California, United States), for microarray samples hybridization; John Coller and Elena Seraia (Stanford Functional Genomics Facility, Stanford University, Stanford, California, United States), for assistance with Bioanalyzer assays; Mike Fero, for development of KAMS image analysis software; Sun-Hae Hong, for image analysis scripts; and Paola Mera (Shapiro Laboratory, Stanford University, Stanford, California, United States), for critical reading of the manuscript.

Author Contributions

Conceived and designed the experiments: LB HHM LS. Performed the experiments: LB TT. Analyzed the data: LB EA TT ML. Wrote the paper: LB LS.

References

1. Poindexter JS (1964) Biological Properties and Classification of the Caulobacter Group. Bacteriol Rev 28: 231–295.
2. Brown PJ, Hardy GG, Trimble MJ, Brun YV (2009) Complex regulatory pathways coordinate cell-cycle progression and development in Caulobacter crescentus. Adv Microb Physiol 54: 1–101.
3. Goley ED, Iniesta AA, Shapiro L (2007) Cell cycle regulation in Caulobacter: location, location, location. J Cell Sci 120: 3501–3507.
4. Holtzendorff J, Reinhardt J, Viollier PH (2006) Cell cycle control by oscillating regulatory proteins in Caulobacter crescentus. Bioessays 28: 355–361.
5. Jacobs-Wagner C (2004) Regulatory proteins with a sense of direction: cell cycle signalling network in Caulobacter. Mol Microbiol 51: 7–13.
6. Lange R, Hengge-Aronis R (1991) Identification of a central regulator of stationary-phase gene expression in Escherichia coli. Mol Microbiol 5: 49–59.

7. Volker U, Maul B, Hecker M (1999) Expression of the sigmaB-dependent general stress regulon confers multiple stress resistance in Bacillus subtilis. J Bacteriol 181: 3942–3948.
8. Chiang SM, Schellhorn HE (2010) Evolution of the RpoS regulon: origin of RpoS and the conservation of RpoS-dependent regulation in bacteria. J Mol Evol 70: 557–571.
9. Hecker M, Pane-Farre J, Volker U (2007) SigB-dependent general stress response in Bacillus subtilis and related gram-positive bacteria. Annu Rev Microbiol 61: 215–236.
10. Lesley JA, Shapiro L (2008) SpoT regulates DnaA stability and initiation of DNA replication in carbon-starved Caulobacter crescentus. J Bacteriol 190: 6867–6880.
11. Landt SG, Lesley JA, Britos L, Shapiro L (2010) CrfA, a small noncoding RNA regulator of adaptation to carbon starvation in Caulobacter crescentus. J Bacteriol 192: 4763–4775.
12. England JC, Perchuk BS, Laub MT, Gober JW (2010) Global regulation of gene expression and cell differentiation in Caulobacter crescentus in response to nutrient availability. J Bacteriol 192: 819–833.
13. McAdams HH, Shapiro L (2009) System-level design of bacterial cell cycle control. FEBS Lett 583: 3984–3991.
14. Grunenfelder B, Rummel G, Vohradsky J, Roder D, Langen H, et al. (2001) Proteomic analysis of the bacterial cell cycle. Proc Natl Acad Sci U S A 98: 4681–4686.
15. Ireland MM, Karty JA, Quardokus EM, Reilly JP, Brun YV (2002) Proteomic analysis of the Caulobacter crescentus stalk indicates competence for nutrient uptake. Mol Microbiol 45: 1029–1041.
16. Molloy MP, Phadke ND, Chen H, Tyldesley R, Garfin DE, et al. (2002) Profiling the alkaline membrane proteome of Caulobacter crescentus with two-dimensional electrophoresis and mass spectrometry. Proteomics 2: 899–910.
17. Phadke ND, Molloy MP, Steinhoff SA, Ulintz PJ, Andrews PC, et al. (2001) Analysis of the outer membrane proteome of Caulobacter crescentus by two-dimensional electrophoresis and mass spectrometry. Proteomics 1: 705–720.
18. Tatusov RL, Natale DA, Garkavtsev IV, Tatusova TA, Shankavaram UT, et al. (2001) The COG database: new developments in phylogenetic classification of proteins from complete genomes. Nucleic Acids Res 29: 22–28.
19. Bowman GR, Comolli LR, Gaietta GM, Fero M, Hong SH, et al. (2010) Caulobacter PopZ forms a polar subdomain dictating sequential changes in pole composition and function. Mol Microbiol 76: 173–189.
20. Figge RM, Easter J, Gober JW (2003) Productive interaction between the chromosome partitioning proteins, ParA and ParB, is required for the progression of the cell cycle in Caulobacter crescentus. Mol Microbiol 47: 1225–1237.
21. Hottes AK, Shapiro L, McAdams HH (2005) DnaA coordinates replication initiation and cell cycle transcription in Caulobacter crescentus. Mol Microbiol 58: 1340–1353.
22. Gorbatyuk B, Marczynski GT (2005) Regulated degradation of chromosome replication proteins DnaA and CtrA in Caulobacter crescentus. Mol Microbiol 55: 1233–1245.
23. Wortinger MA, Quardokus EM, Brun YV (1998) Morphological adaptation and inhibition of cell division during stationary phase in Caulobacter crescentus. Mol Microbiol 29: 963–973.
24. Laub MT, McAdams HH, Feldblyum T, Fraser CM, Shapiro L (2000) Global analysis of the genetic network controlling a bacterial cell cycle. Science 290: 2144–2148.
25. McGrath PT, Lee H, Zhang L, Iniesta AA, Hottes AK, et al. (2007) High-throughput identification of transcription start sites, conserved promoter motifs and predicted regulons. Nat Biotechnol 25: 584–592.
26. Malakooti J, Wang SP, Ely B (1995) A consensus promoter sequence for Caulobacter crescentus genes involved in biosynthetic and housekeeping functions. J Bacteriol 177: 4372–4376.
27. Crosson S, McGrath PT, Stephens C, McAdams HH, Shapiro L (2005) Conserved modular design of an oxygen sensory/signaling network with species-specific output. Proc Natl Acad Sci U S A 102: 8018–8023.
28. Alvarez-Martinez CE, Lourenco RF, Baldini RL, Laub MT, Gomes SL (2007) The ECF sigma factor sigma(T) is involved in osmotic and oxidative stress responses in Caulobacter crescentus. Mol Microbiol 66: 1240–1255.
29. Hu P, Brodie EL, Suzuki Y, McAdams HH, Andersen GL (2005) Whole-genome transcriptional analysis of heavy metal stresses in Caulobacter crescentus. J Bacteriol 187: 8437–8449.
30. Hottes AK, Meewan M, Yang D, Arana N, Romero P, et al. (2004) Transcriptional profiling of Caulobacter crescentus during growth on complex and minimal media. J Bacteriol 186: 1448–1461.
31. Laub MT, Chen SL, Shapiro L, McAdams HH (2002) Genes directly controlled by CtrA, a master regulator of the Caulobacter cell cycle. Proc Natl Acad Sci U S A 99: 4632–4637.
32. Kim SK, Makino K, Amemura M, Shinagawa H, Nakata A (1993) Molecular analysis of the phoH gene, belonging to the phosphate regulon in Escherichia coli. J Bacteriol 175: 1316–1324.
33. Bailey TL, Elkan C (1994) Fitting a mixture model by expectation maximization to discover motifs in biopolymers. Proc Int Conf Intell Syst Mol Biol 2: 28–36.
34. Quon KC, Yang B, Domian IJ, Shapiro L, Marczynski GT (1998) Negative control of bacterial DNA replication by a cell cycle regulatory protein that binds at the chromosome origin. Proc Natl Acad Sci U S A 95: 120–125.
35. Francez-Charlot A, Frunzke J, Reichen C, Ebneter JZ, Gourion B, et al. (2009) Sigma factor mimicry involved in regulation of general stress response. Proc Natl Acad Sci U S A 106: 3467–3472.
36. Bastiat B, Sauviac L, Bruand C (2010) Dual control of Sinorhizobium meliloti RpoE2 sigma factor activity by two PhyR-type two-component response regulators. J Bacteriol 192: 2255–2265.
37. Herrou J, Foreman R, Fiebig A, Crosson S (2010) A structural model of anti-anti-sigma inhibition by a two-component receiver domain: the PhyR stress response regulator. Mol Microbiol 78: 290–304.
38. Hung DY, Shapiro L (2002) A signal transduction protein cues proteolytic events critical to Caulobacter cell cycle progression. Proc Natl Acad Sci U S A 99: 13160–13165.
39. Aldridge P, Jenal U (1999) Cell cycle-dependent degradation of a flagellar motor component requires a novel-type response regulator. Mol Microbiol 32: 379–391.
40. Curtis PD, Brun YV (2010) Getting in the loop: regulation of development in Caulobacter crescentus. Microbiol Mol Biol Rev 74: 13–41.
41. Quardokus E, Din N, Brun YV (1996) Cell cycle regulation and cell type-specific localization of the FtsZ division initiation protein in Caulobacter. Proc Natl Acad Sci U S A 93: 6314–6319.
42. Schmidt KL, Peterson ND, Kustusch RJ, Wissel MC, Graham B, et al. (2004) A predicted ABC transporter, FtsEX, is needed for cell division in Escherichia coli. J Bacteriol 186: 785–793.
43. Gottesman S (2003) Proteolysis in bacterial regulatory circuits. Annu Rev Cell Dev Biol 19: 565–587.
44. Fischer B, Rummel G, Aldridge P, Jenal U (2002) The FtsH protease is involved in development, stress response and heat shock control in Caulobacter crescentus. Mol Microbiol 44: 461–478.
45. Jensen RB, Shapiro L (1999) The Caulobacter crescentus smc gene is required for cell cycle progression and chromosome segregation. Proc Natl Acad Sci U S A 96: 10661–10666.
46. Ali Azam T, Iwata A, Nishimura A, Ueda S, Ishihama A (1999) Growth phase-dependent variation in protein composition of the Escherichia coli nucleoid. J Bacteriol 181: 6361–6370.
47. Claret L, Rouviere-Yaniv J (1997) Variation in HU composition during growth of Escherichia coli: the heterodimer is required for long term survival. J Mol Biol 273: 93–104.
48. Jenkins DE, Chaisson SA, Matin A (1990) Starvation-induced cross protection against osmotic challenge in Escherichia coli. J Bacteriol 172: 2779–2781.
49. Jenkins DE, Auger EA, Matin A (1991) Role of RpoH, a heat shock regulator protein, in Escherichia coli carbon starvation protein synthesis and survival. J Bacteriol 173: 1992–1996.
50. Taniguchi Y, Choi PJ, Li GW, Chen H, Babu M, et al. (2010) Quantifying E. coli proteome and transcriptome with single-molecule sensitivity in single cells. Science 329: 533–538.
51. Lee PS, Shaw LB, Choe LH, Mehra A, Hatzimanikatis V, et al. (2003) Insights into the relation between mrna and protein expression patterns: II. Experimental observations in Escherichia coli. Biotechnol Bioeng 84: 834–841.
52. da Rocha RP, Paquola AC, Marques Mdo V, Menck CF, Galhardo RS (2008) Characterization of the SOS regulon of Caulobacter crescentus. J Bacteriol 190: 1209–1218.
53. Friedberg EC, Walker GC, Siede W (1995) DNA repair and mutagenesis. Washington, D.C.: ASM Press.
54. Sassanfar M, Roberts JW (1990) Nature of the SOS-inducing signal in Escherichia coli. The involvement of DNA replication. J Mol Biol 212: 79–96.
55. Bi E, Lutkenhaus J (1993) Cell division inhibitors SulA and MinCD prevent formation of the FtsZ ring. J Bacteriol 175: 1118–1125.
56. Hengge-Aronis R (2002) Recent insights into the general stress response regulatory network in Escherichia coli. J Mol Microbiol Biotechnol 4: 341–346.
57. Hecker M, Volker U (2001) General stress response of Bacillus subtilis and other bacteria. Adv Microb Physiol 44: 35–91.
58. Mittenhuber G (2002) A phylogenomic study of the general stress response sigma factor sigmaB of Bacillus subtilis and its regulatory proteins. J Mol Microbiol Biotechnol 4: 427–452.
59. Alvarez-Martinez CE, Baldini RL, Gomes SL (2006) A caulobacter crescentus extracytoplasmic function sigma factor mediating the response to oxidative stress in stationary phase. J Bacteriol 188: 1835–1846.
60. Lourenco RF, Gomes SL (2009) The transcriptional response to cadmium, organic hydroperoxide, singlet oxygen and UV-A mediated by the sigmaE-ChrR system in Caulobacter crescentus. Mol Microbiol 72: 1159–1170.
61. Sauviac L, Philippe H, Phok K, Bruand C (2007) An extracytoplasmic function sigma factor acts as a general stress response regulator in Sinorhizobium meliloti. J Bacteriol 189: 4204–4216.
62. Staron A, Sofia HJ, Dietrich S, Ulrich LE, Liesegang H, et al. (2009) The third pillar of bacterial signal transduction: classification of the extracytoplasmic function (ECF) sigma factor protein family. Mol Microbiol 74: 557–581.
63. Evinger M, Agabian N (1977) Envelope-associated nucleoid from Caulobacter crescentus stalked and swarmer cells. J Bacteriol 132: 294–301.
64. Stephens C, Reisenauer A, Wright R, Shapiro L (1996) A cell cycle-regulated bacterial DNA methyltransferase is essential for viability. Proc Natl Acad Sci U S A 93: 1210–1214.
65. Tsai JW, Alley MR (2001) Proteolysis of the Caulobacter McpA chemoreceptor is cell cycle regulated by a ClpX-dependent pathway. J Bacteriol 183: 5001–5007.

66. Shi L, Adkins JN, Coleman JR, Schepmoes AA, Dohnkova A, et al. (2006) Proteomic analysis of Salmonella enterica serovar typhimurium isolated from RAW 264.7 macrophages: identification of a novel protein that contributes to the replication of serovar typhimurium inside macrophages. J Biol Chem 281: 29131–29140.
67. Schutzer SE, Liu T, Natelson BH, Angel TE, Schepmoes AA, et al. (2010) Establishing the proteome of normal human cerebrospinal fluid. PLoS One 5: e10980.
68. Monroe ME, Tolic N, Jaitly N, Shaw JL, Adkins JN, et al. (2007) VIPER: an advanced software package to support high-throughput LC-MS peptide identification. Bioinformatics 23: 2021–2023.
69. Du X, Callister SJ, Manes NP, Adkins JN, Alexandridis RA, et al. (2008) A computational strategy to analyze label-free temporal bottom-up proteomics data. J Proteome Res 7: 2595–2604.
70. Polpitiya AD, Qian WJ, Jaitly N, Petyuk VA, Adkins JN, et al. (2008) DAnTE: a statistical tool for quantitative analysis of -omics data. Bioinformatics 24: 1556–1558.
71. Thanbichler M, Shapiro L (2006) MipZ, a spatial regulator coordinating chromosome segregation with cell division in Caulobacter. Cell 126: 147–162.
72. Christen B, Fero MJ, Hillson NJ, Bowman G, Hong SH, et al. (2010) High-throughput identification of protein localization dependency networks. Proc Natl Acad Sci U S A 107: 4681–4686.
73. Domian IJ, Reisenauer A, Shapiro L (1999) Feedback control of a master bacterial cell-cycle regulator. Proc Natl Acad Sci U S A 96: 6648–6653.
74. Miller JH (1972) Experiments in molecular genetics. Cold Spring Harbor, N.Y.: Cold Spring Harbor Laboratory. xvi, 466 p.
75. Marks ME, Castro-Rojas CM, Teiling C, Du L, Kapatral V, et al. (2010) The genetic basis of laboratory adaptation in Caulobacter crescentus. J Bacteriol 192: 3678–3688.
76. Karp PD, Paley SM, Krummenacker M, Latendresse M, Dale JM, et al. (2010) Pathway Tools version 13.0: integrated software for pathway/genome informatics and systems biology. Brief Bioinform 11: 40–79.
77. Irizarry RA, Hobbs B, Collin F, Beazer-Barclay YD, Antonellis KJ, et al. (2003) Exploration, normalization, and summaries of high density oligonucleotide array probe level data. Biostatistics 4: 249–264.
78. Tusher VG, Tibshirani R, Chu G (2001) Significance analysis of microarrays applied to the ionizing radiation response. Proc Natl Acad Sci U S A 98: 5116–5121.
79. Bailey TL, Williams N, Misleh C, Li WW (2006) MEME: discovering and analyzing DNA and protein sequence motifs. Nucleic Acids Res 34: W369–373.

4

Dynamics of Soil Organic Carbon and Microbial Biomass Carbon in Relation to Water Erosion and Tillage Erosion

Nie Xiaojun[1,2], **Zhang Jianhui**[1]*, **Su Zhengan**[1]

1 Key Laboratory of Mountain Surface Processes and Ecological Regulation, Chinese Academy of Sciences, Institute of Mountain Hazards and Environment, Chinese Academy of Sciences and Ministry of Water Conservancy, Chengdu, China, **2** School of Surveying and Land Information Engineering, Henan Polytechnic University, Jiaozuo, China

Abstract

Dynamics of soil organic carbon (SOC) are associated with soil erosion, yet there is a shortage of research concerning the relationship between soil erosion, SOC, and especially microbial biomass carbon (MBC). In this paper, we selected two typical slope landscapes including gentle and steep slopes from the Sichuan Basin, China, and used the ^{137}Cs technique to determine the effects of water erosion and tillage erosion on the dynamics of SOC and MBC. Soil samples for the determination of ^{137}Cs, SOC, MBC and soil particle-size fractions were collected on two types of contrasting hillslopes. ^{137}Cs data revealed that soil loss occurred at upper slope positions of the two landscapes and soil accumulation at the lower slope positions. Soil erosion rates as well as distribution patterns of the <0.002-mm clay shows that water erosion is the major process of soil redistribution in the gentle slope landscape, while tillage erosion acts as the dominant process of soil redistribution in the steep slope landscape. In gentle slope landscapes, both SOC and MBC contents increased downslope and these distribution patterns were closely linked to soil redistribution rates. In steep slope landscapes, only SOC contents increased downslope, dependent on soil redistribution. It is noticeable that MBC/SOC ratios were significantly lower in gentle slope landscapes than in steep slope landscapes, implying that water erosion has a negative effect on the microbial biomass compared with tillage erosion. It is suggested that MBC dynamics are closely associated with soil redistribution by water erosion but independent of that by tillage erosion, while SOC dynamics are influenced by soil redistribution by both water erosion and tillage erosion.

Editor: David L. Kirchman, University of Delaware, United States of America

Funding: The funding for our study: the National Natural Science Foundation of China (41001157, 41271242), the 135 Strategic Program of the Institute of Mountain Hazards and Environment, CAS (SDS-135-1206), and the Doctoral Program Foundation of Henan Polytechnic University (B2010-47). The funders had no role in study design, data collection and analysis, decision to publish, or preparation of the manuscript.

Competing Interests: The authors have declared that no competing interests exist.

* E-mail: zjh@imde.ac.cn

Introduction

Soil erosion has a severe impact on soil organic carbon (SOC) pools, and as a consequence affects the soil carbon cycle. Historically, water erosion was assumed to be a major soil disturbance in agricultural slope landscapes. Numerous studies have emphasized the influences of water erosion on SOC removal and carbon sequestration in agricultural slope landscapes [1–6]. Since the widespread application of ^{137}Cs (i.e., Caesium-137) technique to assess soil erosion in the 1990s, field evidence from different regions of the world has documented that tillage erosion (i.e. the net movement of soil downslope in response to the action of farm implements) is another major soil disturbance in agricultural slope landscapes with intensive tillage [5,7–11]. The impacts of tillage erosion on within-field variability of SOC have recently been addressed [12–14]. For example, on steep hillslopes of the Sichuan Basin, China, there is a complete depletion of SOC at summit positions and remarkable accumulation at bottom positions after 15-yr tillage erosion [14].

Although the physical process of SOC transfer on the slope under water and tillage erosion has been well known, the SOC biochemical process (e.g. carbon mineralization), mostly linked with the carbon cycle, remains unclear in eroding landscapes. Polyakov and Lal (2004) reported that no significant differences in SOC mineralization exist between the eroding and control soils, but there is a 26% higher CO_2-efflux in the deposited soils compared with those without deposition [15]. In contrast, a recent study conducted by Hemelryck et al. (2011) shows that SOC mineralization in soils with deposition does not apparently differ from soils without deposition [16]. Van Oost et al. (2007) concluded that SOC mineralization associated with transport are relatively minor and that most deposited carbon is effectively preserved [17].

Indeed, it is difficult to make clear the relationship between erosion and SOC mineralization without considering the type of erosion event and soil microbial conditions. In fact, soil microbes participate in all of the biochemical process of SOC, thus microbial biomass affects the emission of CO_2 under soil erosion. Soil microbial biomass carbon (MBC) generally making up 1–5% of total SOC [18], acts as the most active component in the biochemical process of SOC turnover [19] and responds more rapidly to soil disturbance than does SOC [20]. Relatively, few studies have dealt with relationships between MBC dynamics and soil erosion [21]. In the process of water erosion, selective entrainment of fine soil particles is controlled by the transport capacity of the overland flow [22], and the labile SOC fraction adhered to fine soil particles is easily mineralized [23]. In contrast, tillage erosion unselectively moves the soil downslope [14,24], with no transport-related mineralization of SOC in the process of tillage erosion [25]. Given different mechanisms of soil transport, it would be expected that the two different processes of soil erosion may exert different impacts on MBC and the resultant SOC mineralization.

Caesium-137 (^{137}Cs) is a well-established tracer of soil erosion [26]. By comparing ^{137}Cs inventories of the study sites with those of reference sites, one can determine whether erosion (less ^{137}Cs in study sites than in reference sites) or deposition (more ^{137}Cs in study sites than in reference sites) has occurred. The relationships among patterns of soil erosion, SOC and some physio-chemical properties have been established for a number of fields and small agricultural watersheds in different climate zones [5,27–29], showing that this method is a valuable tool to investigate the erosion-carbon relationship. In this study, we use the same methodology to evaluate the relationship between erosion and carbon dynamics on cultivated slopes of the Sichuan Basin, China, where both water erosion and tillage erosion have occurred.

This study aims to (1) compare the difference in soil erosion between the two different slope landscapes (i.e., gentle and steep slopes), (2) examine spatial patterns of SOC and MBC distribution over eroded slopes, and (3) elucidate effects of water erosion and tillage erosion on the dynamics of SOC and MBC.

Materials and Methods

Study Area

The study area was located in Jianyang County of the Sichuan Basin, southwestern China (30°04′28″–30°39′00″N, 104°11′34″–104°53′36″E). This region is typical of the hilly areas of Sichuan (400–587 m above sea level) and has a humid subtropical climate (mean annual temperature 17°C; mean annual rainfall of 872 mm with 90% of precipitation between May and October). Most of the hillslopes have been dissected into slope segments with different slope gradients during previous long-term agricultural practices to minimize soil erosion and optimize farming operations. With the hoe, a predominant tillage implement in this area, one major tillage operation per year always starts at the bottom of the slope and moves upslope step by step, but at every step the tillage direction is always downslope (i.e., always pulling down). Due to long-term intensive hoeing, those in the upper parts of the slope are characterized by a thin soil layer, underlain by bedrock, whereas deeper soils are found in the lower parts. The soils in the study area, derived from purple mudstone and sandstone of Jurassic Age, are classified as Orthic Regosols in the FAO soil taxonomy. Soil texture is clay loam (27% clay, 29% silt, 44% sand), containing 1.0–2.5% organic matter and with a pH of 7.9–8.2. Crop rotation involves [wheat (*Triticum aestivum* L.), maize (*Zea mays* L.)/sweet potato (*Ipomoea batatas*)] and fertilizers (N: 330 kg ha^{-1} year^{-1} as urea and ammonium bicarbonate, P_2O_5:166 kg ha^{-1} year^{-1} as superphosphate) are uniformly applied on the cultivated slopes. Residue management is also uniformly on the slopes in the study area. For wheat and maize, the residues are cut by hand as low as possible and removed, retained residues with about 10-cm height above the ground are mixed into the tillage layer by manual hoeing. For sweet potato, all the residues are removed from the field.

Soil Sampling

Soil sampling was conducted on two types of contrasting hillslopes. Two neighboring cultivated slopes with a length of 44 and 53 m and a corresponding gradient of 7% and 4%, respectively, were considered as two replicates for gentle slope landscapes (Table 1). Another two neighboring slopes with a length of 17 and 22 m and a corresponding slope gradient of 26% and 19%, respectively, were considered as two replicates for steep slope landscapes (Table 1). Samples were collected using a soil corer (6.8-cm diameter) along the downslope transect at 10-m intervals on the two gentle slopes and at 5-m intervals on the two steep slopes. For steep slope landscapes, there were 4, 1 and 4 samples, respectively, in the upper, middle and lower positions. For gentle slope landscapes, 4, 3 and 4 samples were collected in the upper, middle and lower positions, respectively. The coordinates of each sampling point as well as elevation were measured using a survey-grade Differential Global Positioning System (DGPS).

For each sampling point, five soil cores were collected. Two cores for the determination of ^{137}Cs inventory were taken to the bedrock (depths range from 20 to 40 cm depending on thickness of total soil layer at different slope positions) and were then bulked to make a composite sample. In the context of this study, soil depth reached a maximum of 40 cm, and therefore sampling to this depth has all ^{137}Cs in soil profiles contained in samples. Another three core samples for the determination of SOC, MBC, and soil particle-size fractions were collected from the till layer (0–20 cm). One of major considerations for this is that soil redistribution by water and tillage happened mainly in the till layer. Another is that soil microbes are the most active in the till layer where a large portion of organic matter such as crop residues are concentrated, thus the effects of soil erosion on the biochemical process of SOC can be expected to be dominant. After the three soil cores were taken, they were mixed into a composite sample and then divided into two subsamples. One was sieved (2-mm) and immediately stored at 4°C in plastic bags loosely tied to ensure sufficient aeration and to prevent moisture loss until MBC analysis. The other was air-dried and sieved (2-mm) for analyses of SOC and particle-size fractions.

Laboratory Analysis

Soil samples for ^{137}Cs determination were air-dried, crushed and passed through a 2-mm mesh sieve to remove gravels. Samples of the <2 mm particle-size fraction were packed into plastic (PVC) beakers with a 320-cm^3 volume, and ^{137}Cs activity was measured using a hyperpure lithium-drifted germanium detector (HpC –40% efficiency) coupled with a Nuclear Data 6700 multichanel γ-ray spectrophotometer. Caesium-137 was detected at 662 KeV, and count time for each sample ranged from 40,000 to 60,000 s, providing a measurement precision of ±5%. The contents of ^{137}Cs were originally expressed as a unit mass basis (Bq kg^{-1}) and were then converted into an area basis (Bq m^{-2}) using the total weight of the bulked core sample and the cross-sectional area of the sampling device. Soil bulk densities (kg m^{-3}) were determined using oven-dried weight and sample volume.

SOC was determined using wet oxidation with $K_2Cr_2O_7$ and soil particle-size fractions were determined by pipette method following H_2O_2 treatment to destroy organic matter and dispersion of soil suspensions in Na-hexametaphosphate [30]. MBC was determined by fumigation with ethanol-free $CHCl_3$ and extraction with K_2SO_4 [31]. Measurement precision for SOC, MBC, and soil particle-size fractions is ±2%, ±4%, and ±2%, respectively.

Calculation of Soil Erosion Rates

The application of the ^{137}Cs technique and tillage erosion models provides a new perspective to assess the contribution of

Table 1. Total soil erosion rates, tillage erosion rates and water erosion rates.

	Gentle slope landscape		Steep slope landscape	
	4%-slope	7%-slope	19%-slope	26%-slope
Total length (m)	53	44	22	17
Sample number	6	5	5	4
Total soil erosion rate (t ha^{-1} yr^{-1})	24.41	28.01	39.77	44.83
Tillage erosion rate (t ha^{-1} yr^{-1})	6.96	9.29	26.16	39.67
Percentage of total erosion (%)	29	33	66	88
Water erosion rate (t ha^{-1} yr^{-1})	17.45	18.72	13.61	5.16
Percentage of total erosion (%)	71	67	34	12

water and tillage to total soil erosion [5,9–11]. For an eroding site where a total ^{137}Cs inventory A (Bq m^{-2}) is less than the reference inventory A_0 (Bq m^{-2}) at year t (date), soil erosion rates can be expressed as [32]:

$$Y = \frac{10dB}{P}\left[1-\left(1-\frac{A_0-A}{A_0}\right)^{1/(t-1963)}\right] \quad (1)$$

where Y is the soil erosion rate (t ha^{-1} yr^{-1}), d the sampling depth (m), B the bulk density of soil (kg m^{-3}) and P the particle size correction factor (assumed to be $P = 1.0$). According to the previous study conducted in the study area [5], a local ^{137}Cs reference inventory was calculated as 1318 Bq m^{-2} with radioactivity decay corrected to the year 2008.

Based on a large number of experimental observations conducted previously in the study area, an empirical model of tillage erosion has been well established by Zhang et al. (2004) [33]. We adopted the model in this study because of similar terrain attributes such as the shape (linear), ranges of gradient and slope length, and even similar soil properties. In the empirical model, tillage erosion rates increases with increasing slope gradient and is inversely proportional to the slope length, expressed as:

$$R = (k_3 + k_4 S)10/L_d \quad (2)$$

where R is the tillage erosion rate (t ha^{-1} yr^{-1}), k_3 and k_4 are the soil transport coefficient (kg m^{-1} yr^{-1}), S is the slope gradient (m m^{-1}) and L_d is the downslope parcel length (m). Tillage transport coefficients in the study area reach 30.72 and 141.28 kg m^{-1} yr^{-1}, respectively, for k_3 and k_4 [33]. Water erosion rates were obtained by the differences between total soil erosion rates derived from ^{137}Cs data and tillage erosion rates [5,11].

Data Analysis

Statistical analyses were carried out using SPSS 11.0 for Windows version (SPSS Inc., US, 2002). Linear regression analysis was used to test the correlations between ^{137}Cs, SOC and MBC. One-way analysis of variance was used to test the significance of differences in MBC/SOC ratios between gentle and steep slope landscapes, and between upper and lower positions of the two slope landscapes.

Ethics Statement

In this study, soil sampling and sample determinations conducted were permitted by the local government (i.e. Jianyang County People's Government). We also obtained a permission from the local government for reporting research results to the public.

Results

Soil Redistribution

For steep slope landscapes, tillage erosion rates were 26.16 and 39.67 t ha^{-1} yr^{-1} and accounted for 66% and 88% of total erosion rates for the 19%- and 26%-slopes, respectively (Table 1). For gentle slope landscapes, tillage erosion rates were 9.29 and 6.96 t ha^{-1} yr^{-1} and contributed only 33% and 29% to total erosion rates for the 7%- and 4%-slopes, respectively (Table 1). The results indicate that tillage erosion dominates the process of soil redistribution in steep slope landscapes, whereas water erosion is a major process of soil redistribution in gentle slope landscapes. Differences in dominant processes of soil erosion between the two slope landscapes were also supported by the distribution of the <0.002-mm particle-size fraction on the eroded slopes. The fine particle fraction gradually increased downslope on each gentle slope, but such a trend of the fine particle was not observed on the two steep slopes (Figure 1). This result suggests that soil fine particles are preferentially transported downslope mainly by water in gentle slope landscapes, while the soils are unselectively or

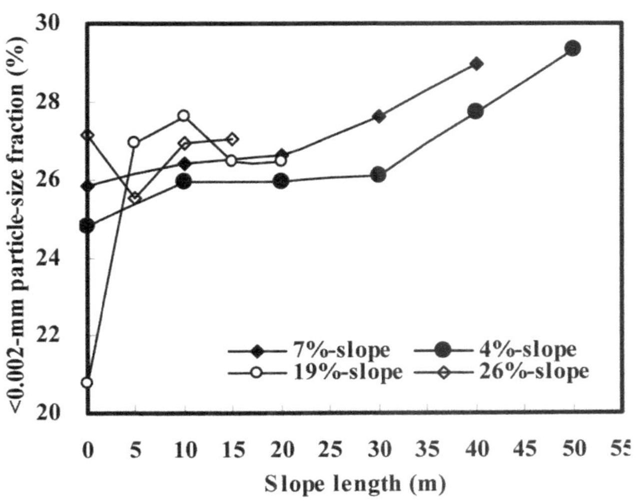

Figure 1. Distribution of the <0.002-mm particle-size fraction on eroded slopes.

entirely transported downslope due to the dominant process of tillage erosion in steep slope landscapes.

Figure 2 shows a similar ^{137}Cs distribution pattern in gentle and steep slope landscapes, i.e., ^{137}Cs inventory increased along the downslope transect. For gentle slope landscapes, ^{137}Cs inventory increased from 861 to 1336 Bq m^{-2} and 786 to 1289 Bq m^{-2} from the upper to lower positions of the 7%- and 4%-slopes, respectively (Table 2). For steep slope landscapes, ^{137}Cs inventory ranged from lows of 402 and 668 Bq m^{-2} at upper slope positions to highs of 1179 and 1002 Bq m^{-2} at lower slope positions of the 19%- and 26%-slopes, respectively (Table 2). The results suggest that soil loss occurs at upper slope positions and soil accumulation at lower slope positions.

SOC and MBC Dynamics

In gentle slope landscapes, both SOC and MBC contents increased downslope in a roughly consecutive increment (Figure 3a). SOC contents averaged 12.99 and 12.42 g kg^{-1} at lower slope positions of the 7%- and 4%-slopes with an increase of 44% and 31%, respectively, compared with those at respective upper slope positions (Table 2). From the upper to lower slope positions, MBC contents changed from 182.13 to 217.80 mg kg^{-1} with an increase of 20% on the 7%-slope, and from 168.78 to 221.13 mg kg^{-1} with an increase of 31% on the 4%-slope (Table 2). These patterns are in agreement with the distribution pattern in ^{137}Cs inventory that mirrors soil redistribution rates on the two gentle slopes. Correlation analysis also showed that inventories of SOC (kg m^{-2}) and MBC (g m^{-2}) were significantly correlated with ^{137}Cs inventory (Bq m^{-2}) in the landscape ($r = 0.82$, $P < 0.01$ and $r = 0.88$, $P < 0.01$, respectively, for SOC and MBC; see Figures 4, 5). As a consequence, it is clear that the dynamics of SOC and MBC are closely associated with soil redistribution in gentle slope landscapes.

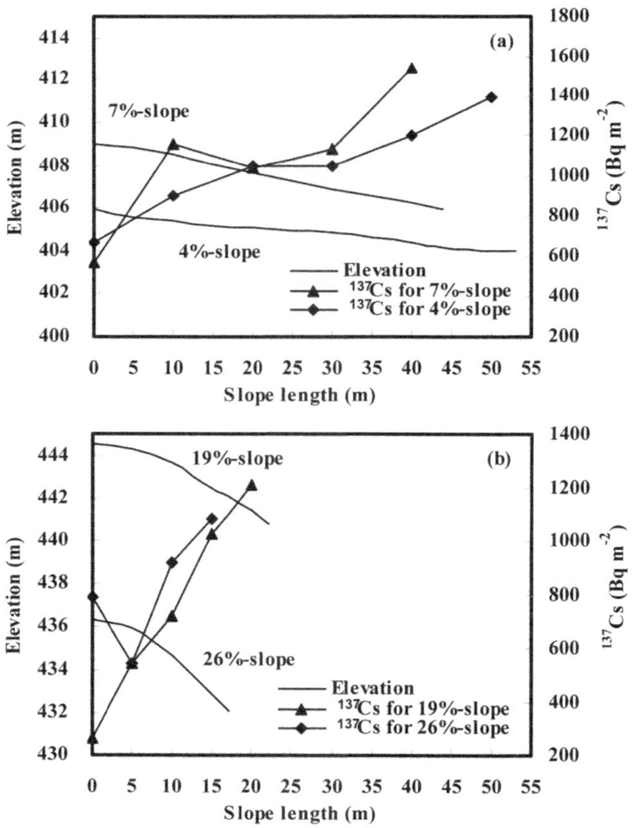

Figure 2. Distribution of ^{137}Cs inventories over eroded slopes.
(a) gentle slope landscape; (b) steep slope landscape.

Different distribution patterns between SOC and MBC contents were observed in steep slope landscapes, showing a rough increase in SOC content but a discrete pattern in MBC content along the downslope transect (Figure 3b). From the upper to lower slope positions, SOC contents changed from 7.33 to 10.15 g kg^{-1} with an increase of 38% on the 19%-slope, and 9.33 to 11.82 g kg^{-1} with an increase of 27% on the 26%-slope (Table 2). An increase in MBC content was observed only at middle positions of the 19%-slope, but little change in MBC between the upper and lower positions of each steep slope (Table 2). The distribution pattern in SOC content is consistent with ^{137}Cs inventory on each steep slope. A significant correlation ($r = 0.84$, $P < 0.05$) between SOC (kg m^{-2}) and ^{137}Cs (Bq m^{-2}) was also found in steep slope landscapes (Figure 4). However, MBC distribution could not be explained by ^{137}Cs data on each steep slope, as a result of no correlation ($r = 0.25$, $P > 0.05$) between MBC (g m^{-2}) and ^{137}Cs (Bq m^{-2}) in steep slope landscapes (Figure 5). Thus it is suggested that SOC dynamics are remarkably impacted by soil redistribution in steep slope landscapes, whereas MBC dynamics are independent of soil redistribution in the landscape. The close relationship between the <0.002-mm fine particle and MBC contents in gentle slope landscapes dominated by water erosion ($r = 0.86$–0.93, $P = 0.02$–0.03; data not shown) also showed that the microbial biomass decreased with decreasing soil fine particles under water erosion. However, no significant correlations between the two parameters ($r = -0.22$–0.79, $P = 0.11$–0.78; data not shown) were found in steep slope landscapes, implying that the microbial biomass is scarcely impacted by tillage erosion which produces unselective soil particle movement on the slope.

Relationship between MBC and SOC

A close correlation ($r = 0.91$, $P < 0.01$) between MBC and SOC contents (mass basis) was found in gentle slope landscapes, whereas the two were not correlated in steep slope landscapes ($r = 0.18$, $P > 0.05$) (Figure 6a). When the contents of SOC and MBC were expressed on a unit area basis (kg m^{-2} and g m^{-2}, respectively), a similar correlation ($r = 0.95$, $P < 0.01$) between the two variables was also found in gentle slope landscapes, but there was still no correlation ($r = 0.18$, $P > 0.05$) in steep slope landscapes (Figure 6b). In this study, additional information on the relationship between MBC and SOC was also obtained from changes in the ratio of MBC (g kg^{-1}) to SOC (g kg^{-1}). The MBC/SOC ratio was slightly greater at upper positions than at lower positions of each slope (Table 2). However, there were no statistically significant differences in MBC/SOC ratios between upper and lower positions of the two slope landscapes ($P > 0.05$). The results indicate the difference in the MBC fraction per unit of SOC between the upper and lower positions was insignificant in the landscape. In terms of the whole slope, the MBC/SOC ratio in steep slope landscapes averaged 2.77×10^{-2}, and was significantly greater than that (mean 1.87×10^{-2}) in gentle slope landscapes ($P < 0.05$). This showed a smaller MBC fraction per unit of SOC in gentle slope landscapes than in steep slope landscapes, suggesting that MBC is influenced by different processes of soil redistribution in the landscape.

Table 2. Descriptive statistics of ^{137}Cs inventories, SOC and MBC contents in different positions for gentle and steep slope landscapes.

Gentle slope landscape	**4%-slope**			**7%-slope**		
Slope position	Upper	Middle	Lower	Upper	Middle♀	Lower
Slope length	0–10	20–30	30–53	0–10	20	30–44
Sample number	2	2	2	2	1	2
Mean ^{137}Cs (Bq m^{-2})	786	1051	1296	861	1040	1336
Mean SOC (g kg^{-1})	9.33	9.84	12.42	9.20	9.39	12.99
Mean MBC (mg kg^{-1})	168.78	202.67	221.13	182.13	180.95	217.80
Mean MBC/SOC ratio	1.81×10^{-2}	2.06×10^{-2}	1.78×10^{-2}	1.99×10^{-2}	1.93×10^{-2}	1.69×10^{-2}
Steep slope landscape	**19%-slope**			**26%-slope**		
Slope position	Upper	Middle♀	Lower	Upper	Lower	
Slope length	0–5	10	10–22	0–10	10–17	
Sample number	2	1	2	2	2	
Mean ^{137}Cs (Bq m^{-2})	402	723	1117	668	1002	
Mean SOC (g kg^{-1})	7.33	8.49	10.15	9.33	11.82	
Mean MBC (mg kg^{-1})	242.45	308.44	234.41	257.09	273.62	
Mean MBC/SOC ratio	3.28×10^{-2}	3.63×10^{-2}	2.32×10^{-2}	2.83×10^{-2}	2.32×10^{-2}	

♀single value in the position

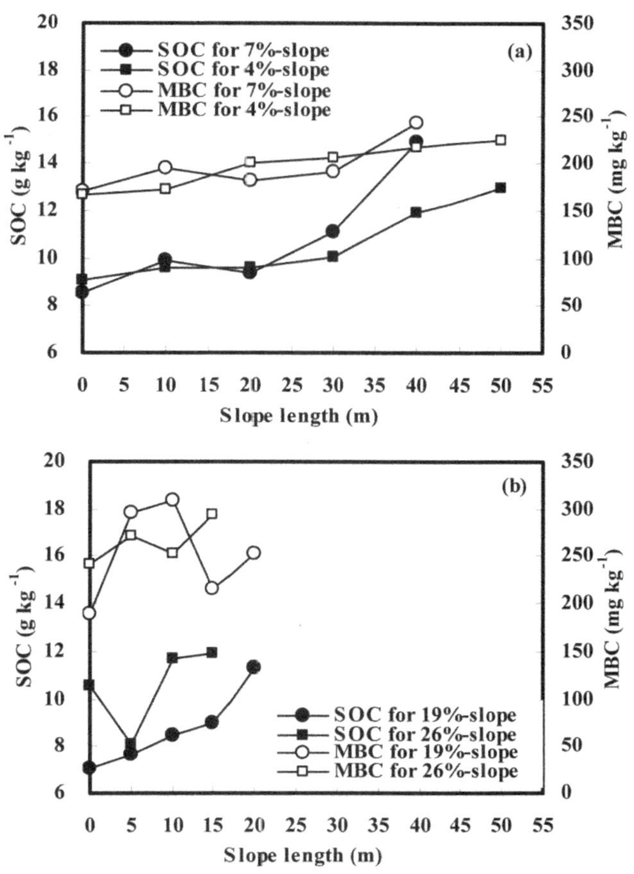

Figure 3. Distribution of SOC and MBC contents over eroded slopes. (a) gentle slope landscape; (b) steep slope landscape.

Discussion

The distribution pattern of SOC content was consistent with redistribution patterns of ^{137}Cs inventory in the two slope landscapes. The result agrees with previous studies [5,11,28], further confirming a close relationship between SOC translocation and soil redistribution by water and tillage. The MBC distribution pattern was in agreement with soil redistribution in gentle slope landscapes but independent of soil redistribution in steep slope landscapes. This is attributed to impacts of water-induced soil redistribution on SOC and MBC in gentle slope landscapes, and impacts of tillage-induced soil redistribution in steep slope landscapes. The difference in the relationship between MBC and SOC under the disturbances of water and tillage erosion

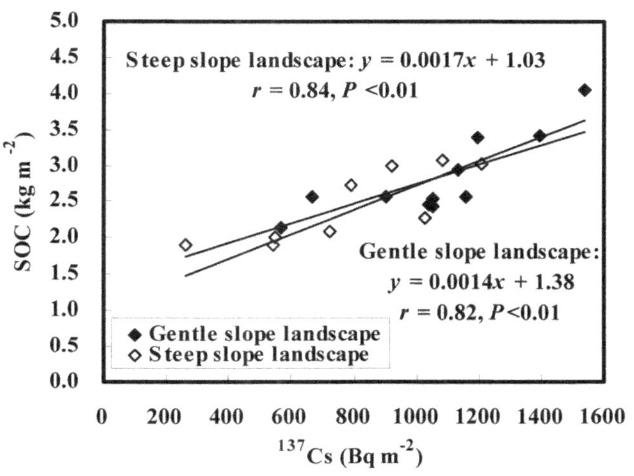

Figure 4. SOC contents vs. ^{137}Cs inventories in gentle and steep slope landscapes.

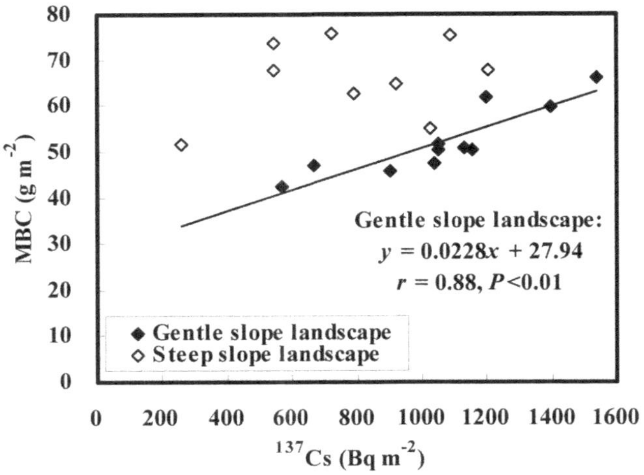

Figure 5. MBC contents vs. ^{137}Cs inventories in gentle and steep slope landscapes.

differed from the results of previous studies. A close correlation between MBC and SOC has been reported under various anthropic disturbances such as plant cover, fertilization, organic amendments, and tillage [34–37]. It is, therefore, that the dominant erosion process should be first considered when one assesses the dynamics of MBC and SOC in eroded slope landscapes.

In the two slope landscapes, lower SOC contents were found at upper slope positions with soil loss than at lower slope positions with soil accumulation. The result indicates the consistent physical process of SOC depletion in upper slope positions and SOC accumulation in lower slope positions regardless of water- or tillage-induced soil transport. However, changes in MBC contents were inconsistent between upper and lower positions in the two slope landscapes. On the two gentle slopes, MBC contents were apparently 20% and 31% higher in lower slope positions than upper slope positions, thus indicating that the physical transfer of MBC is related to soil redistribution. On the two steep slopes, however, there was little difference in MBC content between upper and lower positions, suggesting a negligible impact of soil redistribution on the physical dynamics of MBC. This similarity of MBC content between the two positions in steep slope landscapes could be attributed to unselective removal and the dilution effect due to the incorporation of the soil from deeper layers (e.g. subsoils) into the till layer by tillage (i.e. MBC contents in the newly formed surface soil decrease compared to the original surface soil, due to poor MBC in the subsoils) [14,17]. In the study area, the farmers have to till into subsoils, parent material, or even bedrocks to offset soil loss in the till layer. The soil translocated from the upslope moves entirely downslope and subsequently deposits at lower slope positions under consecutive tillage operations. Consequently, little impact on soil microbial biomass was present in the till layer in the line of slope, as a result of the redistribution of low fertility soils. No statistically significant difference ($P>0.05$) in MBC/SOC ratios between upper and lower positions in the two slope landscapes showed that the erosion-induced SOC depletion did not apparently decrease the ratio of MBC to SOC, and the deposition-induced SOC accumulation did not obviously increase this ratio either.

The MBC/SOC ratio represents the contribution of microbial biomass to organic carbon in soil [38], and is a more useful assessment index for soil health than either MBC or SOC [39].

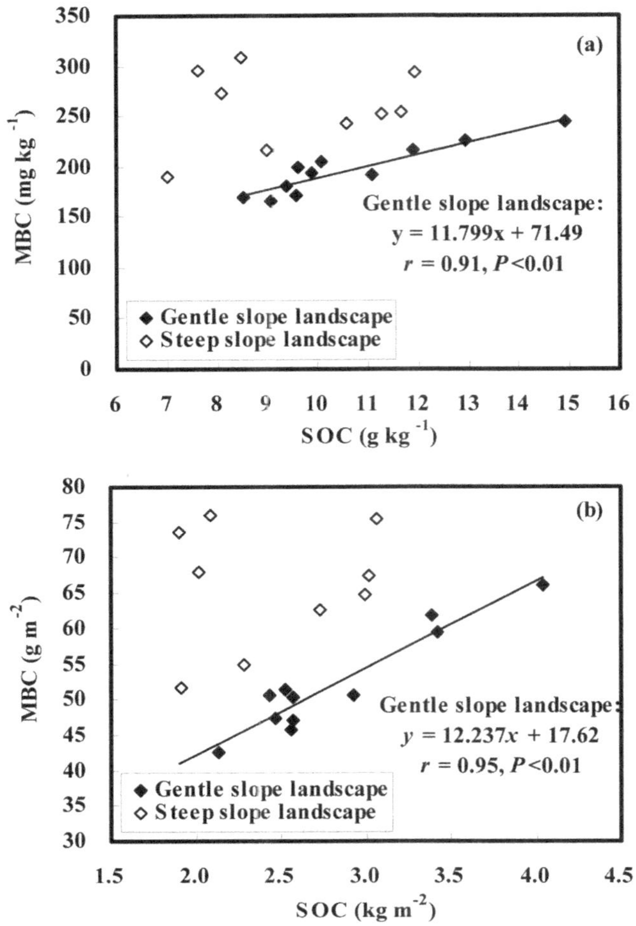

Figure 6. Relationship between MBC and SOC in gentle and steep slope landscapes. (a) mass basis; (b) area basis.

The lower mean of MBC/SOC ratios in gentle slope landscapes than in steep slope landscapes ($P<0.05$) shows a smaller MBC contribution to SOC pools in gentle slope landscapes. This also implies that water erosion with selective removal of soil fine particles has a more severe disturbance to soil microbial biomass than tillage erosion with unselective soil movement. The difference in MBC/SOC between the two slope landscapes is probably associated with different responses of microbial biomass to distinctive changes in soil fine particles. Insam et al. (1989) reported that the clay content of soil has an important influence on the variance in MBC/SOC [40]. It is well known that soil clay (i.e. the <0.002-mm fine particle) has a strong adhesion to organic matter due to its large surface area, and thus is regarded as the mineral binding agents to form soil aggregates. In the process of water erosion, soil clay is preferentially removed by runoff, thus disrupting soil aggregates [5,41]. Accordingly, microbes incorporated into initial soil aggregates die due to their exposure to an unprotected circumstance [23], which results in a decline in soil microbial biomass. It should be noted that the same tillage operation including tillage intensity (one time per year) and tillage tool (hoe) is performed in the two slope landscapes, although tillage reduces aggregate stability by exposing the encapsulated SOC binding agents to mineralization [42]. As a result, impacts of aggregate breakdown by tillage would be similar between the two slope landscapes. In this sense, the difference in soil microbial biomass between the two landscapes resulted from different processes of soil redistribution (dominated by water erosion or by tillage erosion), excluding direct impacts of aggregate breakdown due to tillage operations.

Conclusion

In water-eroded slope landscapes, both SOC and MBC contents showed a significant correlation with soil redistribution rates. In tillage-eroded slope landscapes, SOC contents also exhibited a similar relationship, whereas MBC contents did not reveal any clear pattern. The MBC/SOC ratio was lower on the water-eroded slope than on the tillage-eroded slope, while the contribution of MBC to SOC pools was similar between the erosion-induced SOC depletion and SOC accumulation sites irrespective of landscape types. It is clear that water erosion has a severe disturbance to soil microbial biomass compared with tillage erosion. We conclude that SOC dynamics are associated closely with soil redistribution by water erosion and by tillage erosion as well, while MBC dynamics are significantly influenced by water-induced soil redistribution, but independent of tillage-induced soil redistribution.

Author Contributions

Conceived and designed the experiments: ZJH. Performed the experiments: NXJ SZA. Analyzed the data: NXJ ZJH. Contributed reagents/materials/analysis tools: NXJ. Wrote the paper: NXJ ZJH.

References

1. Verity GE, Anderson DW (1990) Soil erosion effects on soil quality and yield. Can J Soil Sci 70: 471–484.
2. Mabit L, Bernard C (1998) Relationship between soil ^{137}Cs inventories and chemical properties in a small intensively cropped watershed. Earth Planet Sci 327: 527–532.
3. Ritchie JC, McCarty GW (2003) ^{137}Caesium and soil carbon in a small agricultural watershed. Soil Till Res 69: 45–51.
4. Shukla MK, Lal R (2005) Erosional effects on soil organic carbon stock in an on-farm study on Alfisols in west central, Ohio. Soil Till Res 81: 173–181.
5. Zhang JH, Quine TA, Ni SJ, Ge FL (2006) Stocks and dynamics of SOC in relation to soil redistribution by water and tillage erosion. Global Change Biol 12: 1834–1841.
6. Nie XJ, Wang XD, Liu SZ, Liu HJ, Gu SX (2010) ^{137}Cs tracing dynamics of soil erosion, organic carbon and nitrogen in sloping farmland converted from original grassland in Tibetan plateau. Appl Radiat Isot 68: 1650–1655.
7. Lindstrom MJ, Nelson WW, Schumacher TE (1992) Quantifying tillage erosion rates due to moldboard plowing. Soil Till Res 24: 243–255.
8. Lobb DA, Kachanoski RG, Miller MH (1995) Tillage translocation and tillage erosion on shoulder slope landscape positions measured using ^{137}Cs as a tracer. Can J Soil Sci 75: 211–218.
9. Govers G, Quine TA, Desmet PJJ, Walling DE (1996) The relative contribution of soil tillage and overland flow erosion to soil redistribution on agricultural land. Earth Surf Proc Land 21: 929–946.
10. Quine TA, Walling DE, Chakela OK, Mandiringana OT, Zhang X (1999) Rates and patterns of tillage and water erosion on terraces and contour strips: Evidence from caesium-137 measurements. Catena 36: 115–142.
11. Li Y, Lindstrom MJ (2001) Evaluating soil quality–soil redistribution relationship on terraces and steep hillslope. Soil Sci Soc Am J 65: 1500–1508.
12. Heckrath G, Djurhuus J, Quine TA, Van Oost K, Govers G, et al. (2005) Tillage erosion and its effect on soil properties and crop yield in Denmark. J Environ Qual 34: 312–324.
13. Papiernik SK, Lindstrom MJ, Schumacher JA, Farenhorst A, Stephans KD, et al. (2005) Variation in soil properties and crop yield across an eroded prairie landscape. J Soil Water Conserv 60: 388–395.
14. Zhang JH, Nie XJ, Su ZA (2008) Soil profile properties in relation to soil redistribution by intense tillage on a steep hillslope. Soil Sci Soc Am J 72: 1767–1773.
15. Polyakov VO, Lal R (2004) Soil erosion and carbon dynamics under simulated rainfall. Soil Sci 169: 590–599.
16. Hemelryck HV, Govers G, Van Oost K, Merckx R (2011) Evaluating the impact of soil redistribution on the in situ mineralization of soil organic carbon. Earth Surf. Process. Landforms 36: 427–438.
17. Van Oost K, Quine TA, Govers G, Gryze SD, Six J, et al. (2007) The impact of agricultural soil erosion on the global carbon cycle. Science 318: 626–629.
18. Jenkinson DS, Ladd JN (1981) Microbial biomass in soil, measurement and turn over. In: Paul EA, Ladd JN, editor. Soil Biochemistry. New York: Marcel Dekker press, 415–471.
19. Marinari S, Mancinelli R, Campiglia E, Grego S (2006) Chemical and biological indicators of soil quality in organic and conventional farming systems in Central Italy. Ecol Indic 6: 701–711.
20. Bergstrom DW, Monreal CM, King DJ (1998) Sensitivity of soil enzyme activities to conservation practices. Soil Sci Soc Am J 62: 1286–1295.
21. Mabuhay JA, Nakagoshi N, Isagi Y (2004) Influence of erosion on soil microbial biomass, abundance and community diversity. Land Degrad Dev 15: 183–195.
22. Schiettecatte W, Gabriels D, Cornelis WM, Hofinan G (2008) Impact of deposition on the enrichment of organic carbon in eroded sediment. Catena 72: 340–347.
23. Schlesinger WH (1995) Soil respiration and changes in soil carbon stocks. In: Biotic Woodwell GM, Mackenzie FT, editor. Feedback in the Global Climatic System: Will the Warming Feed the Warming? New York: Oxford University Press, 159–168.
24. De Alba S, Lindstrom M, Schumacher TE, Malo DD (2007) Soil landscape evolution due to soil redistribution by tillage: a new conceptual model of soil catena evolution in agricultural landscapes. Catena 58: 77–100.
25. Van Oost K, Govers G, Quine TA, Heckrath G (2004) Comment on 'Managing soil carbon'. Science 305: 1567b.
26. Walling DE, Quine TA (1991) The use of caesium-137 measurements to investigate soil erosion on arable fields in the UK: potential applications and limitations. J Soil Sci 42: 147–165.
27. Quine TA, Zhang Y (2002) An investigation of spatial variation in soil erosion, soil properties, and crop production within an agricultural field in Devon, United Kingdom. J Soil Water Conserv 57: 55–65.
28. Ritchie JC, McCarty GW (2003) ^{137}Caesium and soil carbon in a small agricultural watershed. Soil Till Res 69: 45–51.
29. Mabit L, Bernard C, Makhlouf M, Laverdière MR (2008) Spatial variability of erosion and soil organic matter content estimated from ^{137}Cs measurements and geostatistics. Geoderma 145: 245–251.
30. Liu GS (1996) Soil Physical and Chemical Analysis and Description of Soil Profiles. Beijing: Chinese Standard Press, 123–166 (in Chinese).
31. Vance ED, Brookes PC, Jenkinson DS (1987) An extraction method for measuring soil microbial biomass-C. Soil Biol Biochem 19: 703–707.
32. Zhang XB, Higgitt DL, Walling DE (1990) A preliminary assessment of the potential for using caesium-137 to estimate rates of soil erosion in the Loess Plateau of China. Hydrolog Sci J 35: 267–276.
33. Zhang JH, Lobb DA, Li Y, Liu GC (2004) Assessment for tillage translocation and tillage erosion by hoeing on the steep land in hilly areas of Sichuan, China. Soil Till Res 75: 99–107.
34. García C, Hernández T, Roldán A, Martin A (2002) Effect of plant cover decline on chemical and microbiological parameters under Mediterranean climate. Soil Biol Biochem 34: 635–642.
35. Melero S, Porras JCR, Herencia JF, Madejon E (2006) Chemical and biochemical properties in a silty loam soil under conventional and organic management. Soil Till Res 90: 162–170.
36. Madejón E, Moreno F, Murillo JM, Pelegrín F (2007) Soil biochemical response to long-term conservation tillage under semi-arid Mediterranean conditions. Soil Till Res 94: 346–352.
37. Vineela C, Wani SP, Srinivasarao CH, Padmaja B, Vittal KPR (2008) Microbial properties of soils as affected by cropping and nutrient management practices in several long-term manurial experiments in the semi-arid tropics of India. Appl Soil Ecol 40: 165–173.
38. Sparling GP (1992) Ratio of microbial biomass to soil organic carbon as a sensitive indicator of changes in soil organic matter. Aust J Soil Res 30: 195–207.
39. Anderson TH, Domsch KH (1989) Ratios of microbial biomass carbon to total organic carbon in arable soils. Soil Bio Biochem 21: 471–479.
40. Insam H, Parkinson D, Domsch K (1989) Influence of macroclimate on the soil microbial biomass. Soil Bio Biochem 21: 211–221.
41. Lal R (2003) Soil erosion and the global carbon budget. Environ Int 29: 437–450.
42. Hajabbasi MA, Hemmat A (2000) Tillage impacts on aggregate stability and crop productivity in a clay-loam soil in central Iran. Soil Till Res 56: 205–212.

5

Phosphoglycerate Mutases Function as Reverse Regulated Isoenzymes in *Synechococcus elongatus* PCC 7942

Jiri Jablonsky[1]*, Martin Hagemann[2], Doreen Schwarz[2], Olaf Wolkenhauer[1]

1 Department of Systems Biology and Bioinformatics, University of Rostock, Rostock, Germany, **2** Department of Plant Physiology, University of Rostock, Rostock, Germany

Abstract

Phosphoglycerate-mutase (PGM) is an ubiquitous glycolytic enzyme, which in eukaryotic cells can be found in different compartments. In prokaryotic cells, several PGMs are annotated/localized in one compartment. The identification and functional characterization of PGMs in prokaryotes is therefore important for better understanding of metabolic regulation. Here we introduce a method, based on a multi-level kinetic model of the primary carbon metabolism in cyanobacterium *Synechococcus elongatus* PCC 7942, that allows the identification of a specific function for a particular PGM. The strategy employs multiple parameter estimation runs in high CO_2, combined with simulations testing a broad range of kinetic parameters against the changes in transcript levels of annotated PGMs. Simulations are evaluated for a match in metabolic level in low CO_2, to reveal trends that can be linked to the function of a particular PGM. A one-isoenzyme scenario shows that PGM2 is a major regulator of glycolysis, while PGM1 and PGM4 make the system robust against environmental changes. Strikingly, combining two PGMs with reverse transcriptional regulation allows both features. A conclusion arising from our analysis is that a two-enzyme PGM system is required to regulate the flux between glycolysis and the Calvin-Benson cycle, while an additional PGM increases the robustness of the system.

Editor: Nestor V. Torres, Universidad de La Laguna, Spain

Funding: The work presented here has been supported by the German Research Foundation (DFG) as part of PROMICS research group 1186. The funders had no role in study design, data collection and analysis, decision to publish, or preparation of the manuscript.

Competing Interests: The authors have declared that no competing interests exist.

* E-mail: jiri.jablonsky@gmail.com

Introduction

Phosphoglycerate-mutases (PGMs) are a group of non-homologous glycolytic enzymes [1], having independent evolutionary origins, catalyzing the reversible conversion of 3-phosphoglycerate (3PGA) to 2-phosphoglycerate (2PGA). PGMs can be found in different compartments in the eukaryotic cells fulfilling different tasks [2]. PGMs are also present in prokaryotes [3] such as cyanobacteria, in which the different PGM isoenzymes are localized in one compartment. PGMs can be divided into two analogous subgroups, cofactor-independent and cofactor (2,3-bisphosphoglycerate)-dependent enzymes [3]. The occurrence of these enzyme types seems to be scattered among prokaryotes. Based on our own analysis, the presence of PGM proteins from these two groups is unpredictable among cyanobacteria, as has been reported before for other bacteria [1]. This scattered occurrence seems to imply that the type of PGM has no significant role in the regulation of the system. However, the occurrence of more than one PGM in one compartment certainly needs some regulation or functional specialization, unknown so far.

The preferred substrate of PGM is 3-phosphoglycerate (3PGA). In cyanobacteria, 3PGA is made by photosynthetic CO_2 fixation, in which 3PGA represents the first stable carbon-fixation product of the Calvin-Benson cycle. The majority of 3PGA is used for the regeneration of the Calvin-Benson cycle, CO_2 acceptor molecule ribulose-bisphophate. Excess carbon is taken out of the cycle and stored as glycogen or shuttled into the primary carbon metabolism via glycolysis (Fig. 1). The flux of 3PGA in the compartmented cell of land plants is regulated by phosphate translocator (chloroplast ↔ cytosol), which is not present in cyanobacteria. In the non-compartmented cyanobacterial cell, PGMs could be the key regulator of the flux out of the Calvin-Benson cycle. Marked changes in the relative flux of carbon through the Calvin-Benson cycle and its export into the glycolytic path have been observed in cyanobacteria exposed to excess or limited amounts of inorganic carbon [4,5].

Steady state metabolomic and transcriptomic analysis with the model cyanobacterium *Synechococcus elonagtus* PCC 7942 (hereafter *Synechococcus*) indicated that cells shifted to low CO_2 availability reduce Calvin-Benson cycle activity, stop glycogen storage, and increase the export of organic carbon through the glycolysis [6]. In the *Synechococcus* genome (http://genome.kazusa.or.jp/cyanobase/SYNPCC7942) four PGM isoenzymes are annotated (Fig. 2). These annotations are mainly based on sequence similarities determined with the BLAST algorithm [7]. However, automated BLAST-based annotation can provide false functional assignments [8]. In the case of *Synechococcus*, the occurrence of isoenzymes in one compartment suggests a regulatory/specialized role of these isoenzymes in central metabolism. Moreover, in the cyanobacterial carbon metabolism, PGMs have a cardinal position at the crossroads of Calvin-Benson cycle and associated carbon metabolism via glycolysis (Fig. 1). Therefore, it is important to

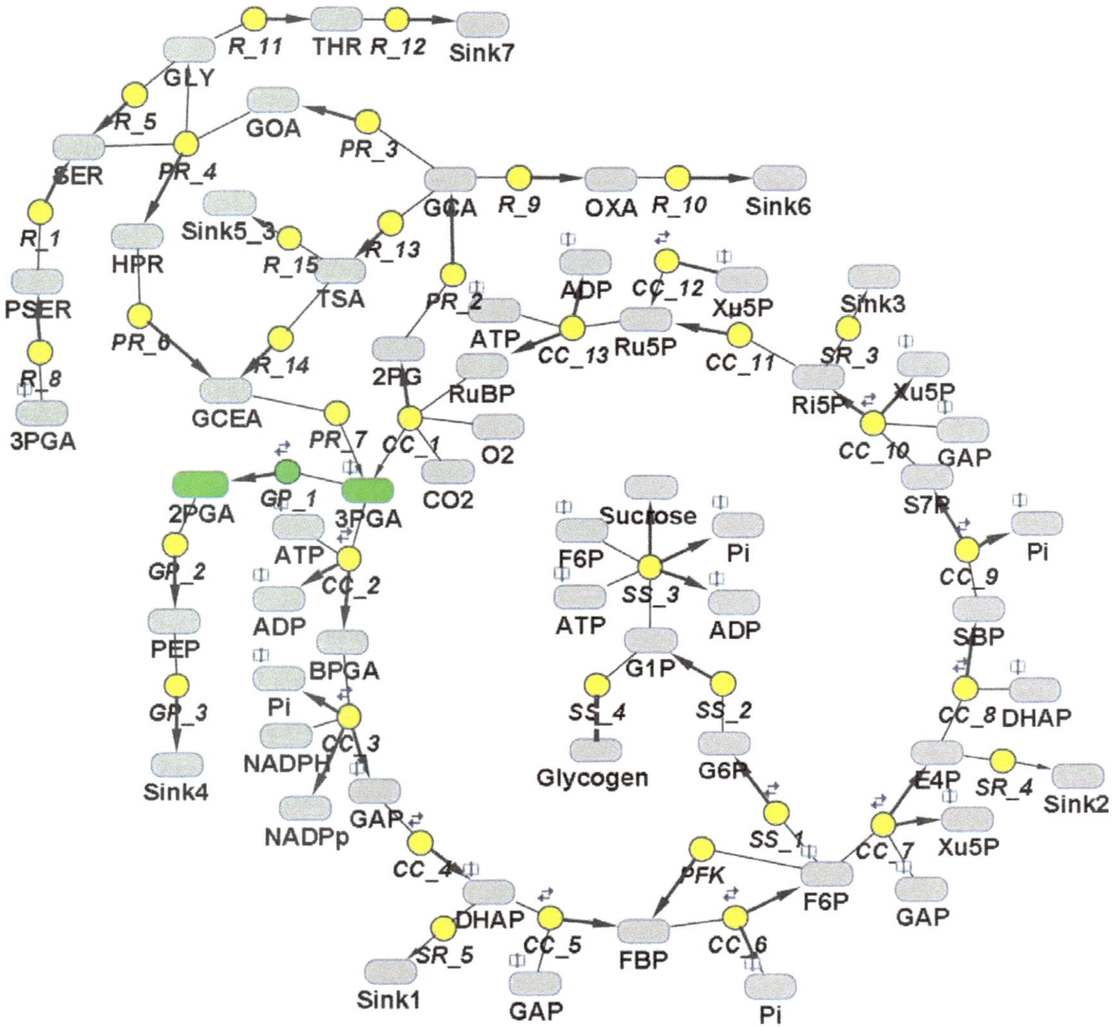

Figure 1. Scheme of the primary carbon metabolism, encoded as a kinetic model of *Synechococcus elongatus* PCC 7942. The model includes the Calvin-Benson cycle, sucrose and glycogen synthesis, photorespiratory pathways, glycolysis and sink reactions, representing the adjacent pathways. Green color represents the reaction catalyzed by phosphoglycerate mutase (PGM) and indicates its cardinal position in the crossroads of metabolic pathways (need for complex model). Note: The reactions are described in the model by reversible and irreversible Michaelis-Menten kinetics; reversibility of particular reaction is indicated by two little arrows.

understand the regulation of PGMs and thereby to validate their annotation.

We here propose the use of kinetic modeling for such analysis. Kinetic modeling is a standard method for predicting the behavior of biological systems [9]. However, reactions catalyzed by isoenzymes are commonly described by single enzymatic kinetics, which would not explain hidden regulatory mechanisms. In order to address this challenge, we have designed a multi-level kinetic model of carbon core metabolism of *Synechococcus*. The model not only helps in the validation of the annotation but also in the

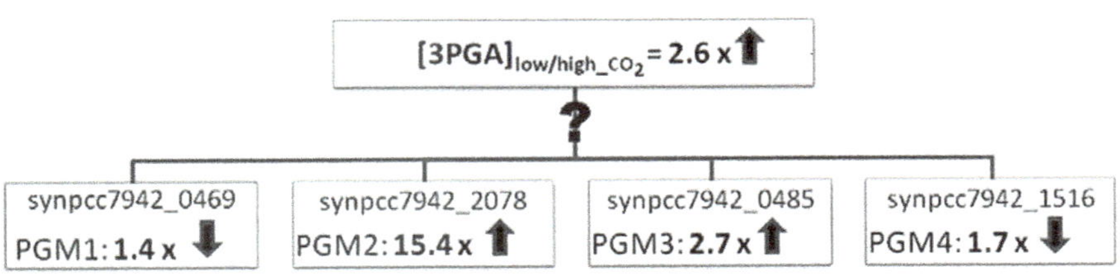

Figure 2. Comparison of fold changes in concentration for 3PGA and in expression levels of the four annotated PGM isoenzymes in cells of *Synechococcus elongatus* PCC 7942 after shifting from high to low CO_2 level. Note: synpcc7942_0485 probably represents a gene encoding a phosphoserine phosphatase but we cannot exclude if it functions as PGM.

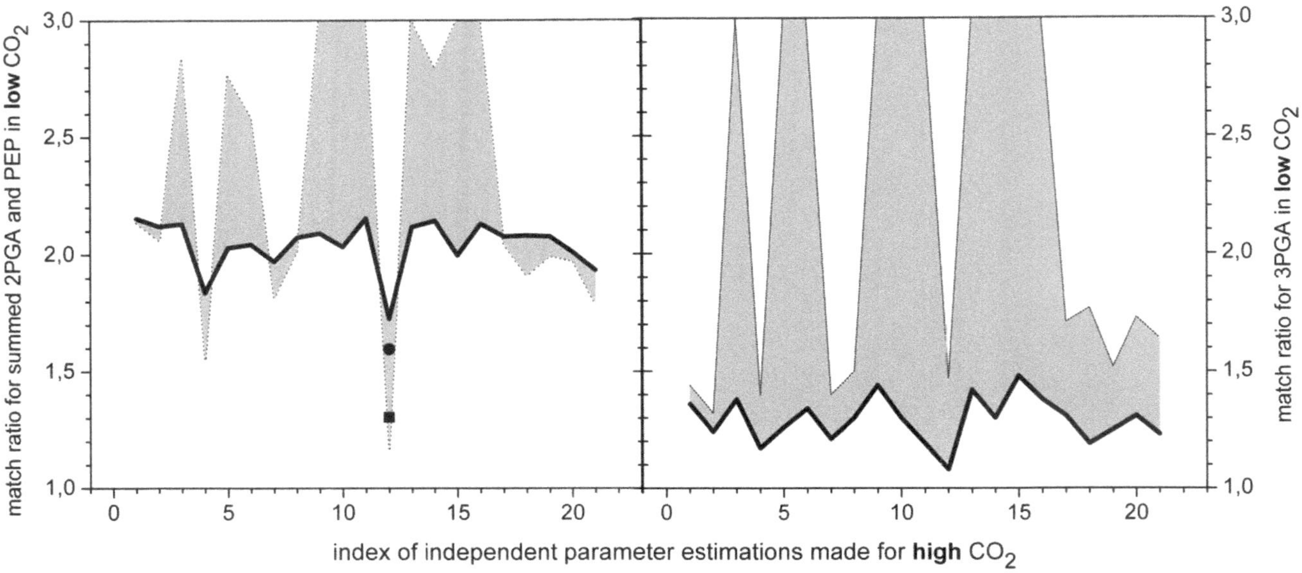

Figure 3. Quality for the match of simulated and measured data in low CO_2 for single PGM scenario in dependence on (i) varying PGM activities fitted in high CO_2 and (ii) regulation by transcript amounts. Figures represent the match between simulated and measured data in low CO_2 in dependence of estimated kinetic parameters (V_{max} and k_M values for preferred substrate and product) in high CO_2 for a single PGM scenario.The V_{max}, fitted to steady state in high CO_2, was modified by the amount of PGM isoforms taken from the changes in mRNA values (one by one) after shift from high to low CO_2. Results are shown for randomly chosen set of twenty parameters estimation runs for PGM and the best fit (Nr. 12). The **left figure** shows the match for preferred product of PGM, 2-phosphoglycerate (2PGA). The black solid line shows the impact of 1.7-fold down-regulated enzyme, corresponding to PGM4; gray contours indicate the difference in matching the data if PGM is 15.4-fold up-regulated (corresponding to PGM2). In order to illustrate the impact of transcriptomic changes for the other two annotated PGMs, the results for 1.4-fold down-regulated (circle) and 2.7-fold up-regulated (square) isoenzymes are presented in the case of the best fit. The **right figure** shows the match for preferred substrate of PGM, 3-phosphoglycerate (3PGA); colors/lines have the same meaning as for the figure on the left. Notes: 1) the top boundary of axis y shows results equal or worse than match ratio equals to 3, 2) The match ratio is calculated as X/Y where X(Y) is a higher(lower) number from a pair of simulated and experimental values for a particular data point, 3) Fit from high CO_2 was included (saved as a result) if the difference between simulated and measured data was smaller than 15%, 4) each point represents an independent simulation run compared to experimental data - the lines improve the perception for the differences in match ratios and have no other meaning.

explanation of regulatory mechanism for PGM isoforms in one compartment.

Materials and Methods

The model was developed and simulations were executed using the SimBiology toolbox of MATLAB (Mathworks Inc.). The routine employed for parameter estimation was a hybrid genetic algorithm (ga_hybrid, Mathworks Inc.). The model is available in the supplement (Model S1).

Systems Biology Workflow

The kinetic model, a successor of the corrected Zhu model [10,11] of the Calvin-Benson cycle, was redesigned for cyanobacteria, extended (photorespiratory pathways, glycolysis) and validated on available metabolic data from *Synechococcus el.* PCC 7942 [6]. The constraints of the model are: 1) The ATP · (ADP +ATP)$^{-1}$ ratio was maintained in the physiological range 0.74–0.76 [12], both in high and low CO_2 steady states and 2) the biomass production in high CO_2 was calculated from growth rate data. On average, *Synechococcus* shows a 3.4-fold higher growth at 5% CO_2 (defined as high CO_2, [6]) compared to ambient air CO_2 (defined as low CO_2, [6]).

The parameter estimation for up to four annotated isoenzymes of PGM was run in high CO_2 steady state and the result (kinetic parameters) stored if the difference between the simulated and experimental metabolic data was lower than 15%. We also assumed that, in the case of isoenzymes in non-compartmented prokaryotic cell, a change in gene expression in the steady state is equal to changes in protein activity, i.e., minimal or no post-translation modifications. For the unidentified part of carbon metabolism - reaction catalyzed by PGM - four transcriptomic profiles (four PGM isoforms) were tested for combinations of one to four isoforms and the quality of match with metabolic data in low CO_2 was stored for each profile.

Experimental Data

Relative transcriptomic and metabolomics data of high CO_2 as well as low CO_2 *Synechococcus* cells were taken from [6]. A considering of two environmental conditions was necessary to known and implement the changes in transcriptomic level of PGMs in the model and helpful for constraining the model by doubling the amount of metabolic data. For 3PGA, cellular concentrations were calculated using the data from [13]. These authors reported that cyanobacteria contain 1300 nmol 3PGA/g fresh weight. According to our calibration of cyanobacterial fresh weight and total cellular volume, this amount corresponds to 5.39 mM 3PGA in the total cell volume in low CO_2 grown cells. The concentration of 2-phosphoglycerate was recalculated on the basis of known ratio to 3PGA [6].

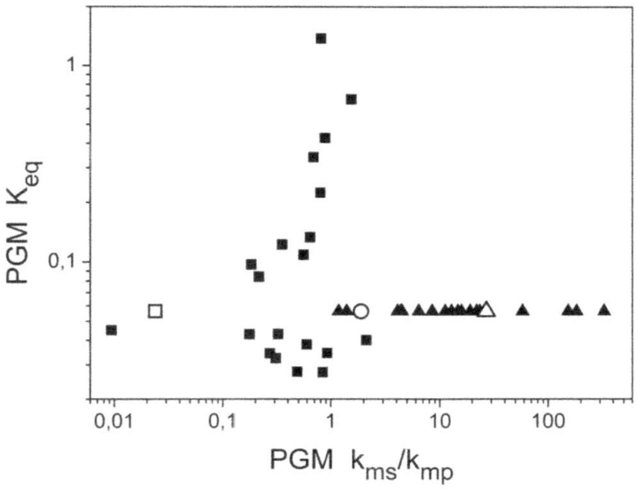

Figure 4. Parameter space of the equilibrium constant and ratio of k_M constants for single PGM and dual PGMs scenarios. The values k_{ms} and k_{mp} indicate the k_M values for the preferred substrate and product (reversibility), respectively. Keq indicates the equilibrium constant. **Solid squares** denote independent runs of parameter estimation, presented in Fig. 3, for single PGM scenario. **The open square** denotes the best fit for single PGM scenario. **Solid triangles** denote independent runs of parameter estimation, presented in Fig. 5, for dual reverse regulated PGMs scenario for PGM beta. **The open triangle** denotes the best fit for dual reverse regulated PGMs scenario for PGM beta. **The open circle** denotes the best fit for triple PGMs scenario for PGM gamma. This analysis shows how multiple sets of kinetic parameters match experimental data in one steady state for unconstrainted parameter estimation.

Results and Discussion

When One Enzyme is not Enough

The implementation of isoenzymes within a kinetic model, when localized in one compartment, deals with two extremes. First, for only one enzyme and the model not being constrained by a lack of data, there are still non-identifiability problems due to multiple sets of parameter value combinations that are equally able to fit the data related to a particular state. Second, considering two or more enzymes, the combinational explosion in the parameter space leads to computational requirements that make it virtually impossible to fit the model to sets of experimental data for different states of the system.

Various studies have focused on the response of cyanobacteria to changes in the environment, by considering a range of omics data [6,14,15], which allow constraining the model. On the basis of these results, we were aiming to develop a method that would permit the identification and analysis of isoenzymes in a one-compartment system and within a reasonable amount of time (days to weeks). We here particularly focus on PGMs because of their key position in regulating the flux of carbon through the Calvin-Benson cycle and its export into glycolysis in cyanobacteria. This position of PGMs in the metabolic network makes it necessary to establish a complex model (Fig. 1).

We started with the simplest scenario, encoding only one PGM, which is a common approach to un-constrained kinetic modeling. Although this approach cannot reveal anything about the regulation of catalyzed reactions in the un-constrained model, it is yet unknown if the outcome based on the constrained model can reveal some additional information about the system regulation or not. We therefore run a set of independent parameter estimations in high CO_2 for a single PGM and used these values in numerical simulations. The quality of the estimates is evaluated by comparing the quality of match of simulations with metabolic data, separately for preferred substrate, 3-phosphoglycerate (3PGA) and product, 2-phosphoglycerate (2PGA), in low CO_2 (for details see Materials and Methods).

The results are striking. The analysis reveals two trends, showing that one PGM provides a match either for preferred substrate or preferred product (Fig. 3) but never for both. Moreover, these trends indicate an impact of transcriptomic regulation, which affects V_{max} in the enzymatic kinetics that describes the PGM. The first trend, exhibiting robustness in the system behavior (Fig. 3, black solid line), is associated with down-regulated PGM4. The other one (Fig. 3, gray contours), connected with up-regulated PGM2, shows a high sensitivity towards variations in the kinetic parameters (K_{eq} and k_M). This pattern was tested for a broad range of K_{eq} and k_M values (Fig. 4, solid squares). Furthermore, we can observe both, a very good and bad match, with experimental data (Fig. 3, gray contours), implying a major role in the kinetic regulation of the system. Taken together, single PGM cannot regulate the interconversion between 3PGA and 2PGA, however, the single isoenzyme scenario has provided valuable information about the requirements for homeostasis of this metabolic crossroad.

Reverse Regulated Versus Co-regulated Isoenzymes

We have shown that the one-isoenzyme scenario cannot keep the expected balance between 3PGA and 2PGA (Fig. 3). We can either get a robust system, which is however out of the physiological range, or a system that is very sensitive to changes in the parameter values. The natural next step in the analysis is to test for two PGMs in one compartment. This approach requires metabolic data from two steady states and transcriptomic data describing the shift between these two states, in our case, from high to low CO_2. However, parameter estimation for two isoenzymes, running for two steady states of metabolism, has enormous computational demands. At this point, data collected from the single PGM scenario, based on the constrained model, proved to be very useful.

The single PGM scenario suggests that it is the up-regulated PGM2 which is likely to provide the kinetic regulation for the reaction due to high sensitivity with respect to kinetic parameter values. We therefore tested the dual PGM scenario with two PGMs, denoted as alpha and beta. We employed the parameter values from the best fit provided by the single PGM scenario for alpha PGM (Fig. 4, open square) and varied the parameters for beta PGM, see Fig. 4 (solid triangles); the equilibrium constant is the same for both PGMs. Moreover, for the sake of comparison, we have added the robust case from the single PGM scenario (1.7-fold down-regulated PGM4) and another dual PGM scenario in which both PGMs were co-down-regulated (PGM1 and PGM4).

The comparison with other scenarios clearly shows a large improvement, both in robustness against varying parameter values for the beta PGM and in the ability to describe the experimental data if we proceed from single PGM scenario (Fig. 5, dashed line), over dual co-regulated PGMs (Fig. 5, dotted line) to dual reverse regulated PGMs (Fig. 5, solid line). It is therefore the combination of two PGMs with reverse transcriptomic regulation, PGM2 and PGM4, giving the closest match with experimental data. Moreover, if we have a look at the estimated parameters for the best fit for this scenario (Fig. 4, open square and open triangle), the Michaelis constants for preferred substrate and product for alpha and beta PGMs in Table 1 have nearly interchanged values. This is an expected outcome from a evolutionary point of view for two

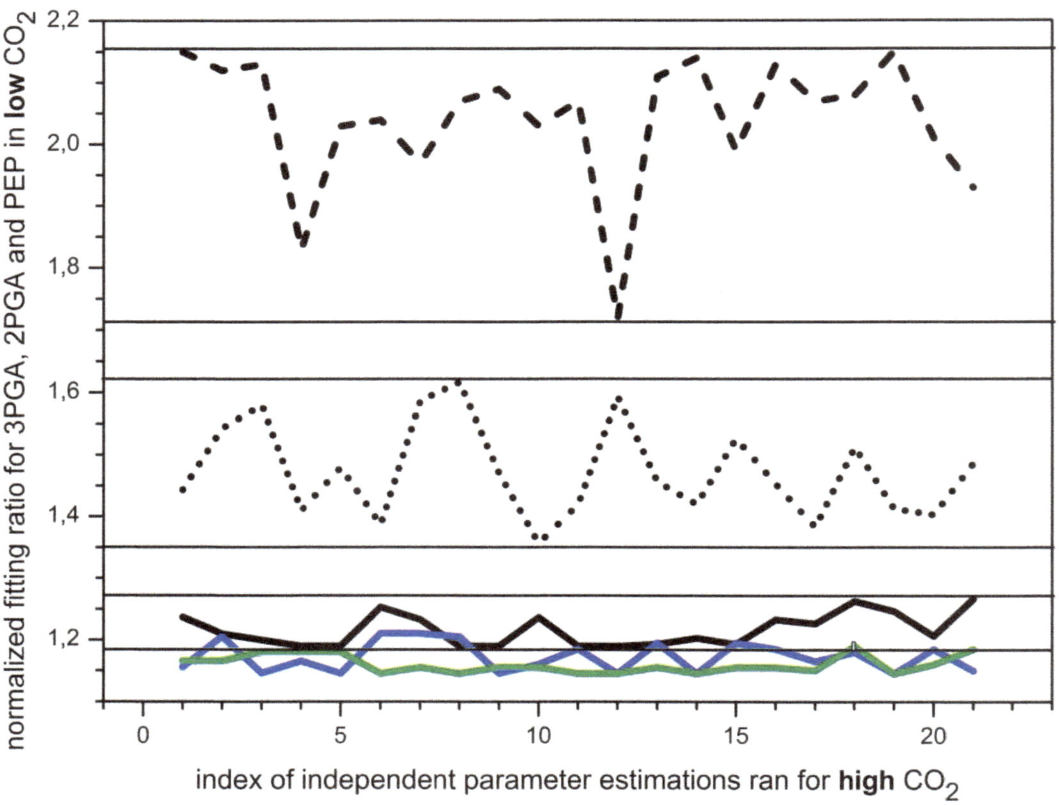

Figure 5. Impact of reverse and similar transcriptional regulation of PGM isoenzyme genes in one compartment. For the purpose of comparison, the dashed line represents the single enzyme scenario from Fig. 3, 1.7-fold down-regulated PGM4 (merged solid lines from Figure 3). Further, two dual PGMs scenarios are presented. K_{eq} and k_M values for alpha PGM are from the best fit for single PGM scenario (Fig. 4, open square) and V_{max} for alpha PGM and all kinetic parameters for PGM beta were estimated to fit the experimental data in high CO_2 steady state. Alpha and beta stand for two PGMs in dual PGMs scenario. The dotted line shows the cooperation of co-regulated PGM4 and PGM1 (1.7- and 1.4-fold down-regulation, respectively). The solid line indicates the case in which reverse regulation of PGM2 and PGM4 takes a place (15.4-fold up-regulation and 1.7-fold down-regulation). The blue line indicates the scenario of three PGMs in which PGM1 (1.4-fold down-regulation) is considered as gamma PGM. The k_M values for beta PGM are taken from the best fit of dual reverse PGMs scenario (Fig. 4, open triangle). The green line indicates another triple PGMs scenario where k_M values, both for beta and gamma PGMs, were varied. Note: each point represents an independent simulation run compared to experimental data - the lines improve the perception for the differences in match ratios and have no other meaning.

enzymes catalyzing the same reaction in one compartment, thus supporting the applied approach. In summary, our in silico experiment led us to the conclusion that reverse gene regulation for two isoenzymes in one compartment, together with the opposite affinity for the substrate and product, provides very good explanation of the measured metabolic data.

Table 1. Estimated kinetic parameters for the best fit describing the reverse regulated dual PGMs scenario.

kinetic parameter	value	unit
Keq	0.056	dimensionless
Vmax_alpha	0.050	$mM \cdot s^{-1}$
kms_alpha	0.042	mM
kmp_alpha	1.772	mM
Vmax_beta	0.027	$mM \cdot s^{-1}$
kms_beta	1.257	mM
kmp_beta	0.047	mM

Activity of PGMs is in the model described by reversible Michaelis-Menten kinetics. Note: V_{max} values are normalized to the activity of RuBisCO. The routine employed for parameter estimation was a hybrid genetic algorithm.

The results clearly show high robustness of the system (Fig. 5, solid line) against varying the kinetic parameters (Fig. 4, solid triangles), which is equivalent to noise from fluctuating metabolite concentrations in a single cell. Hence the results support our initial assumption that system of isoenzymes does not require an additional regulatory mechanism, for instance the post-translational modifications. This was implicit in the assumed 1:1 ratio between the transcript level and protein activity for isoenzymes in the non-compartmented cell.

The Case of Two other Annotated PGMs

We have shown that two reverse regulated PGMs are able to explain the experimental data (Fig. 5), i.e., to regulate the interconversion between 3PGA and 2PGA and thus the flux between the Calvin-Benson cycle and glycolysis. This raises the questions whether there is any benefit of having more than two PGMs in one compartment. In order to test the triple PGMs scenario, we have taken the best fits for alpha and beta PGMs (Fig. 4, open square and open triangle) and varied the k_M values for gamma PGM. Our analysis of the triple PGMs scenario shows that there is a negligible improvement in robustness and only

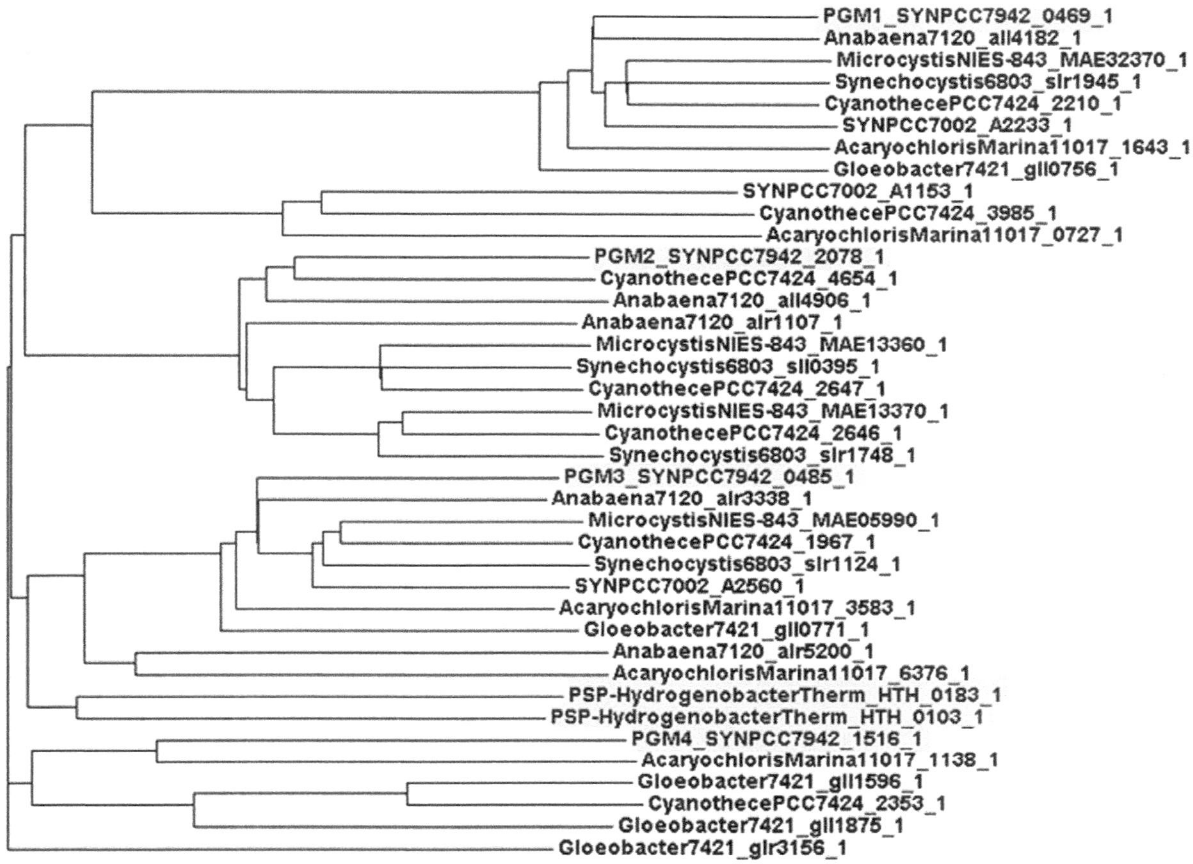

Figure 6. Grouping of PGMs and PSPs by using cluster analysis. ClustalW2 2.1 (http://www.ebi.ac.uk/Tools/) was employed as a tool for protein alignment analysis, codon table for bacteria was selected. PGMs 1–4 (*Synechococcus*) and two PSPs (*Hydrogenobacter thermophilus*) are highlighted. PGM3 is clustered with two PSPs.

a small improvement in matching experimental data (Fig. 5, blue line). The k_M values for gamma PGM, in the case of the best fit, is in between the k_M values of PGMs alpha and beta (Fig. 4, open circle). There was however no improvement in the biomass production (our estimation of growth speed), which implies that only two PGMs are necessary to control the flux between the Calvin-Benson cycle and glycolysis in the controlled environment. Since two-component response regulators have been proposed for the glycolytic pathway [16], this further supports our conclusion that only two PGMs are essential for the regulation of the system.

Recently, two assumed members of PGM family, cofactor dependent PGMs annotated in *Hydrogenobacter thermophilus*, were identified in a wet lab experiment as a phosphoserine phosphatases [17]. It is known that phosphoserine phosphatase (PSP) is essential for serine and glycine metabolism [18,19] but no PSP is annotated for *Synechococcus*. The alternative pathways for serine and glycine synthesis in cyanobacteria were proposed [20,21], however, we tested if one or more of the PGMs annotated in *Synechococcus* can be identified as PSP. Cluster analysis of PGM-like proteins shows that each PGM belongs to different protein family (Fig. 6) but, interestingly, PSPs in *Hydrogenobacter thermophilus* are in the same cluster as PGM3 (Fig. 6). Moreover, we have detected conservative regions between that PSPs and PGM3. It is therefore likely that we can identify PGM3 as PSP, however, we cannot entirely exclude that this enzymes functions as PGM as well. Finally, PGM3 is up-regulated in low CO_2 (Fig. 3) which lead to speculation on contribution of photorespiration for serine biosynthesis as suggested [20].

As for the PGM1 and PGM4, due to almost the same expression in low CO_2 (Fig. 2), we do not have sufficient data to clearly distinguish between these two down-regulated enzymes. We however showed that stronger down-regulation gives higher robustness to the system, as indicated in Fig. 3. There might also be other reasons why there are three, or even four, PGMs in *Synechococcus*. For instance, it is known that the majority of proteins have more than one function [22].

The simultaneous parameter estimation for two PGMs for more steady states is very difficult and nearly impossible for three PGMs. One can therefore assume a rather small likelihood for the right combination of substrate-product affinities for every PGM to occur during evolution. Moreover, our analysis shows that the system in not very sensitive to varying separately the affinity of beta or gamma PGM although the best fit fulfills the expectation regarding the affinities for the product/substrate for two isoenzymes (Tab. 1). Therefore, an occurrence of several PGMs could be explained, from the evolutionary point of view, as the simplest means to achieve the robust response of the regulated system. As we have shown, the number of isoenzymes has a big impact on the system robustness (Fig. 5). Indirect support for such speculation is our analysis for triple PGMs scenario where k_M values, both for beta and gamma PGMs, were varied. The results demonstrate even higher robustness of the system (Fig. 5, green line) than the case of fixed beta PGM and varied gamma PGM (Fig. 5, blue line). Furthermore, the mean quality of match with experimental data improved by 5.6% in comparison to dual reverse PGMs scenario. Taken together, these results might imply that in the case of essential metabolic crossroads, the occurrence of more than two isoenzymes catalyzing the same reaction, especially in non-controlled environment, work as a buffer keeping the homeostasis of the system.

Despite the fact that our analysis cannot provide a clear cut answer why there are more PGMs, the presented approach contributes to an understanding of enzymatic regulation and provides a ration approach to identify the roles of particular, even non-homologous, isoenzymes.

Supporting Information

Model S1 The model of primary carbon metabolism in cyanobacterium *Synechococcus elongatus* PCC 7942 is provided and encoded in SBML L2V4 and set up for triple PGMs scenario, based on the best fits of alpha, beta and gamma PGM (Fig. 4).

Author Contributions

Jointly discussed ideas and concepts: JJ MH DS OW. The modeling was done by: JJ. Contributed to the writing of the final manuscript: OW MH. Provided the experimental data: DS MH. Read and approved the final manuscript: JJ MH DS OW. Conceived and designed the experiments: MH. Performed the experiments: MH DS. Analyzed the data: JJ MH. Contributed reagents/materials/analysis tools: MH OW. Wrote the paper: JJ.

References

1. Foster JM, Davis PJ, Raverdy S, Sibley MH, Raleigh EA, et al. (2010) Evolution of Bacterial Phosphoglycerate Mutases: Non-Homologous Isofunctional Enzymes Undergoing Gene Losses, Gains and Lateral Transfers. PLoS ONE 5.
2. Carreras J, Mezquita J, Bosch J, Bartrons R, Pons G (1982) Phylogeny and ontogeny of the phosphoglycerate mutases-IV. Distribution of glycerate-2,3-P2 dependent and independent phosphoglycerate mutases in algae, fungi, plants and animals. Comp Biochem Physiol B 71: 591–597.
3. Fraser HI, Kvaratskhelia M, White MF (1999) The two analogous phosphoglycerate mutases of *Escherichia coli*. FEBS Lett 455: 344–348.
4. Huege J, Goetze J, Schwarz D, Bauwe H, Hagemann M, et al. (2011) Modulation of the Major Paths of Carbon in Photorespiratory Mutants of *Synechocystis*. PLoS ONE 6.
5. Young JD, Shastri AA, Gregory S, Morgan JA (2011) Mapping photoautotrophic metabolism with isotopically nonstationary C-13 flux analysis. Metabolic Engineering 13: 656–665.
6. Schwarz D, Nodop A, Hüge J, Purfürst S, Forchhammer K, et al. (2011) Metabolic and transcriptomic phenotyping of inorganic carbon acclimation in the cyanobacterium *Synechococcus elongatus* PCC 7942. Plant Physiology 155: 1640–1655.
7. Zhang Z, Schaffer AA, Miller W, Madden TL, Lipman DJ, et al. (1998) Protein sequence similarity searches using patterns as seeds. Nucleic Acids Research 26: 3986–3990.
8. Ashkenazi S,Snir R and Ofran Y (2012) Assessing the Relationship between Conservation of Function and Conservation of Sequence Using Photosynthetic Proteins. Bioinformatics 28: 3203–3210.
9. Heinrich R, Schuster (1996) The Regulation of Cellular Systems, Springer, Berlin, Germany.
10. Zhu XG, de Sturler E, Long SP (2007) Optimizing the distribution of resources between enzymes of carbon metabolism can dramatically increase photosynthetic rate: a numerical simulation using an evolutionary algorithm. Plant Physiol 145: 513–526.
11. Jablonsky J, Bauwe H, Wolkenhauer H (2011) Modelling the Calvin-Benson cycle. BMC Systems Biology 5: 185.
12. Kallas T, Castenholz RW (1982) Internal pH and ATP-ADP pools in the cyanobacterium *Synechococcus* sp During exposure to growth inhibiting low pH. Journal of Bacterology 149: 229–236.
13. Takahashi H, Uchimiya H, Hihara Y (2008) Difference in metabolite levels between photoautotrophic and photomixotrophic cultures of *Synechocystis* sp. PCC 6803 examined by capillary electrophoresis electrospray ionization mass spectrometry. J Exp Bot 59: 3009–3018.
14. Eisenhut M, von Wobeser EA, Jonas L, Schubert H, Ibelings BW, et al. (2007) Long- term response toward inorganic carbon limitation in wild type and glycolate turnover mutants of the cyanobacterium *Synechocystis sp.* strain PCC 6803. Plant Physiol 144: 1946–1959.
15. Hackenberg C, Hüge J, Engelhardt A, Wittink F, Laue M, et al. (2012) Low-carbon acclimation in carboxysome-less and photorespiratory mutants of the cyanobacterium *Synechocystis sp.* strain PCC 6803. Microbiology 158: 398–413.
16. Tabei Y, Okada K, Tsuzuki M (2007) Sll1330 controls the expression of glycolytic genes in *Synechocystis sp.* PCC 6803. BBRC 355: 1045–1050.
17. Chiba Y, Oshima K, Arai H, Ishii M, Igarashi Y (2012) Discovery and Analysis of Cofactor-dependent Phosphoglycerate Mutase Homologs as Novel Phosphoserine Phosphatases in *Hydrogenobacter thermophilus*. Journal of Biological Chemistry 287: 11934–11941.
18. Borkenhagen LF, Kennedy EP (1959) The enzymatic exchange of L-serine with O- phospho-L-serine catalyzed by a specific phosphatase. J Biol Chem 234: 849–53.
19. Ponce-de-Leon MM, Pizer LI (1972) Serine biosynthesis and its regulation in Bacillus subtilis. J Bacteriol 110: 895–904.
20. Knoop H, Zilliges Y, Lockau W, Steuer R (2010) The Metabolic Network of *Synechocystis* sp. PCC 6803: Systemic Properties of Autotrophic Growth. Plant Physiology 154: 410–422.
21. Steuer R, Knoop H, Machne R (2012) Modelling cyanobacteria: from metabolism to integrative models of phototrophic growth. J Exp Bot 63: 2259–2274.
22. Zaretsky JZ, Wreschner DH (2008) Protein multifunctionality: principles and mechanisms. Translational oncogenomics 3: 99–136.

6

Impact of Environmental Parameters on Marathon Running Performance

Nour El Helou[1,2,3]*, Muriel Tafflet[1,4], Geoffroy Berthelot[1,2], Julien Tolaini[1], Andy Marc[1,2], Marion Guillaume[1], Christophe Hausswirth[5], Jean-François Toussaint[1,2,6]

1 IRMES (bioMedical Research Institute of Sports Epidemiology), INSEP, Paris, France, **2** Université Paris Descartes, Sorbonne Paris Cité, Paris, France, **3** Faculté de Pharmacie, Département de Nutrition, Université Saint Joseph, Beirut, Lebanon, **4** INSERM, U970, Paris Cardiovascular Research Center – PARCC, Paris, France, **5** Research Department, INSEP, Paris, France, **6** Hôtel-Dieu Hospital, CIMS, AP-HP, Paris, France

Abstract

Purpose: The objectives of this study were to describe the distribution of all runners' performances in the largest marathons worldwide and to determine which environmental parameters have the maximal impact.

Methods: We analysed the results of six European (Paris, London, Berlin) and American (Boston, Chicago, New York) marathon races from 2001 to 2010 through 1,791,972 participants' performances (all finishers per year and race). Four environmental factors were gathered for each of the 60 races: temperature (°C), humidity (%), dew point (°C), and the atmospheric pressure at sea level (hPA); as well as the concentrations of four atmospheric pollutants: NO_2 – SO_2 – O_3 and PM_{10} ($\mu g.m^{-3}$).

Results: All performances per year and race are normally distributed with distribution parameters (mean and standard deviation) that differ according to environmental factors. Air temperature and performance are significantly correlated through a quadratic model. The optimal temperatures for maximal mean speed of all runners vary depending on the performance level. When temperature increases above these optima, running speed decreases and withdrawal rates increase. Ozone also impacts performance but its effect might be linked to temperature. The other environmental parameters do not have any significant impact.

Conclusions: The large amount of data analyzed and the model developed in this study highlight the major influence of air temperature above all other climatic parameter on human running capacity and adaptation to race conditions.

Editor: Alejandro Lucia, Universidad Europea de Madrid, Spain

Funding: The authors have no support or funding to report.

Competing Interests: The authors have declared that no competing interests exist.

* E-mail: nour.elhelou@insep.fr

Introduction

Like most phenotypic traits, athletic performance is multifactorial and influenced by genetic and environmental factors: exogenous factors contribute to the expression of the predisposing characteristics among best athletes [1,2]. The marathon is one of the most challenging endurance competitions; it is a mass participation race held under variable environmental conditions and temperatures sometimes vary widely from start to finish [3–5]. Warm weather during a marathon is detrimental for runners and is commonly referenced as limiting for thermoregulatory control [3,6]. More medical complaints of hyperthermia (internal temperature ≥39°C) occur in warm weather events, while hypothermia (internal temperature ≤35°C) sometimes occurs during cool weather events [3].

In addition, participating in an outdoor urban event exposes athletes to air pollution which raises concerns for both performance and health [7]. Runners could be at risk during competitions as they are subject to elevated ventilation rate and increased airflow velocity amplifying the dose of inhaled pollutants and carrying them deeper into the lungs [7–9]. They switch from nasal to mouth breathing, bypassing nasal filtration mechanisms for large particles. Both might increase the deleterious effects of pollutants on health and athletic performance [8,10]. Exposure to air pollution during exercise might be expected to impair an athlete's performance in endurance events lasting one hour or more [7,10].

The relationship between marathon performance decline and warmer air temperature has been well established. Vihma [6] and Ely et al. [11,12] found a progressive and quantifiable slowing of marathon performance as WBGT (Wet Bulb Globe Temperature) increases, for men and women of wide ranging abilities. Ely et al. [13] as well as Montain et al. [14] also found that cooler weather (5–10°C) was associated with better ability to maintain running velocity through a marathon race compared to warmer conditions especially by fastest runners; weather impacted pacing and the impact was dependent on finishing position. Marr and Ely [9] found significant correlations between the increase of WBGT and PM_{10}, and slower marathon performance of both men and women; but they did not find significant correlations with any other pollutant.

Previous studies have mostly analysed the performances of the top 3 males and females finishers as well as the 25th-, 100th-, and 300th- place finishers [11,13–16]. Here we targeted exhaustiveness and analysed the total number of finishers in order to quantify the effect of climate on the full range of runners.

The objectives of this study were 1) to analyse all levels of running performance by describing the distribution of all marathons finishers by race, year and gender; 2) to determine the impact of environmental parameters: on the distribution of all marathon runners' performance in men and women (first and last finishers, quantiles of distribution); and on the percentage of runners withdrawals. We then modelled the relation between running speed and air temperature to determine the optimal environmental conditions for achieving the best running performances, and to help, based on known environmental parameters, to predict the distribution and inform runners on possible outcomes of running at different ambient temperatures. We tested the hypothesis that all runners' performances distributions may be similar in all races, and may be similarly affected by temperature.

Methods

Data Collection

Marathon race results were obtained from six marathons included in the « IAAF Gold Labeled Road Races » and « World Marathon Majors »: Berlin, Boston, Chicago, London, New York and Paris. From 2001 to 2010 (available data are limited before 2001) the arrival times in hours: minutes: seconds, of all finishers were gathered for each race. These data are available in the public domain on the official internet website of each city marathon, and on marathon archives websites [17] and complementary data when needed from official sites of each race. Written and informed consent was therefore not required from individual athletes. The total number of collected performances was 1,791,972 for the 60 races (10 years × 6 marathons), including 1,791,071 performances for which the gender was known. We also gathered the total number of starters in order to calculate the number and the percentage of non-finishers (runner withdrawal) per race.

Hourly weather data corresponding to the race day, time span and location of the marathon were obtained from "weather underground website" [5]. Four climatic data were gathered for each of the 60 races: air temperature (°C), air humidity (%), dew-point temperatures (°C), and atmospheric pressure at sea level (hPA). Each of these parameters was averaged for the first 4 hours after the start of each race. Hourly air pollution data for the day, time span and location of each race were also obtained through the concentrations of three atmospheric pollutants: NO_2 – SO_2 – O_3 ($\mu g.m^{-3}$) from the Environmental Agency in each state (the Illinois Environmental Protection Agency for Chicago maratho'n, the Massachusetts Department of environmental Protection for Boston marathon and the New York State Department of Environmental Conservation for New York marathon), and the Environmental agency websites of the three European cities [18–20]. All pollutants values were averaged for the first 4 hours after the start of each race.

Concurrent measurements of air pollution for all ten race years (2001–2010) were only available for 3 pollutants, because air pollution monitoring sites typically measure only a subset of pollutants and may not have been operational in all years. In addition, particulate matters PM_{10} were collected in Paris and Berlin, but there were not enough measurements in the other four cities races days.

Data Analysis and selection

Men and women performances were analysed separately. For each race and each gender every year, we fitted the Normal and log-Normal distributions to the performances and tested the normality and log normality using the Kolmogorov-Smirnov D statistic. We rejected the null hypothesis that the sample is normally or log–normally distributed when p values <0.01.

The following statistics (performance levels) were determined for all runners' performances distribution of each race, every year and for each gender:

- the first percentile of the distribution (P1), representing the elite of each race.
- the winner.
- the last finisher.
- the first quartile of the distribution (Q1), representing the 25th percentile of best performers of the studied race.
- the median.
- the inter quartile range (IQR), representing the statistical dispersion, being equal to the difference between the third and first quartiles.

A Spearman correlation test was performed between each performance level and climate and air pollution parameters, in order to quantify the impact of weather and pollution on marathon performances. Spearman correlation tests were also performed between each environmental parameter. The year factor was not included because we previously demonstrated that for the past ten years, marathon performances were now progressing at a slower rate [21].

Temperature and running speed

We modelled the relation between running speed of each performance level for each gender and air temperature, using a second degree polynomial quadratic model, which seems appropriate to depict such physiological relations [22–24].

The second degree polynomial equation was applied to determine the optimal temperature at which maximal running speed is achieved for each level of performance for each gender, and then used to calculate the speed decrease associated with temperature increase and decrease above the optimum.

We similarly modelled the relation between air temperature and the percentage of runners' withdrawal.

All analyses were performed using the MATLAB and SAS software.

Results

The total numbers of starters and finishers of the 6 marathons increased over the 10 studied years (Figure 1). Marathons characteristics are described in supplementary data (Table S1). The race with the least number of finishers was Boston 2001 with 13381 finishers and the highest number was seen in New York 2010 with 44763 finishers.

Three marathons were held in April, the other three during fall. Air temperatures ranged from 1.7°C (Chicago 2009) to 25.2°C (Boston 2004) (Table 1).

Performance distribution

For all 60 studied races, the women and men's performance distributions were a good approximation of the "log normal" and "normal" distributions (p-values of Kolmogorov-Smirnov statistics ≥ 0.01).

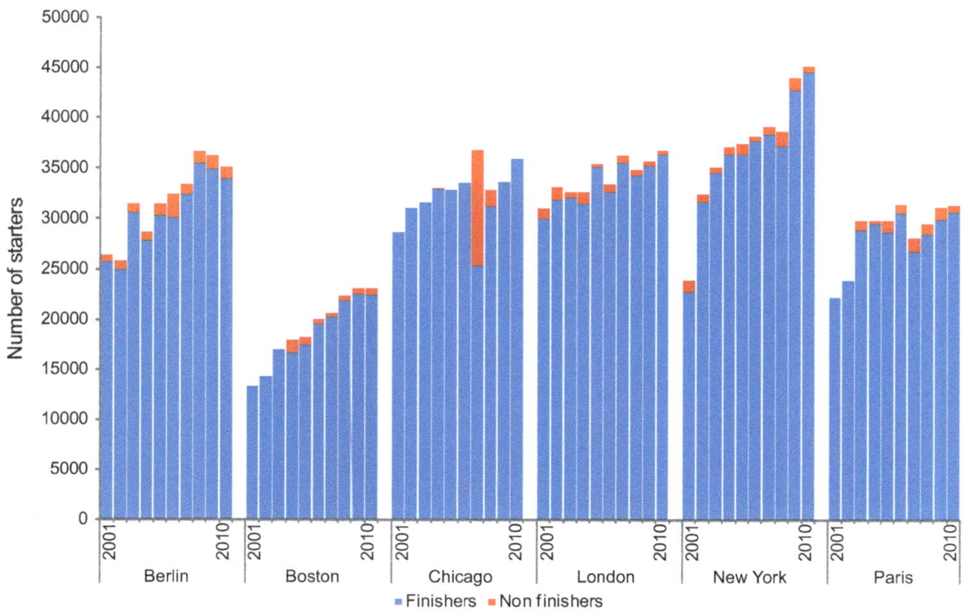

Figure 1. Number of starters and finishers by marathon and year (missing data points for Boston, Chicago and Paris marathons).

Figure 2 illustrates examples of 4 races' performances distribution fit: men's performances distribution of two races in Paris (2002: T° = 7.6°C; and 2007: T° = 17.4°C) and Chicago (2002: T° = 5.4°C; and 2007: T° = 25°C).

We notice a stable gap between male and female performances at all levels in all marathons, women being on average 10.3% ± 1.6% (mean ± standard deviation) slower than men (Table S1); mean female winners are 9.9% ± 1.5% slower than male winners, mean female median is 9.9% ± 1.6% than male median, and mean female Q1 are 11.1% ± 1.5% slower that male Q1.

Correlations

Spearman correlations results are displayed in Table 2, detailed correlations by marathon are available in supplementary data (Table S2).

The environmental parameter that had the most significant correlations with marathons performances was air temperature: it was significantly correlated with all performance levels in both male and female runners.

Humidity was the second parameter with a high impact on performance; it was significantly correlated with women's P1 and men's all performance levels.

The dew point and atmospheric pressure only had a slight influence ($p<0.1$) in men's P1 and women's P1 respectively, and did not affect the other performance levels.

Concerning the atmospheric pollutants, NO_2 had the most significant correlation with performance: it was significantly correlated with Q1, IQR and the median for both genders. Sulfur dioxide (SO_2) was correlated with men's P1 ($p<0.01$) and had a slight influence ($p<0.1$) on men's Q1. Finally ozone (O_3) only had a slight influence ($p<0.1$) on men's Q1. In the marathon by marathon analysis, ozone (O_3) had the most significant correlation with performance (Table S2): it was significantly correlated with all performance levels (P1, Q1, IQR and the median) of the Berlin and Boston (except men's IQR) marathon for both genders. It also affected Chicago (men's P1, Q1, and men's median), and New York (women's Q1) marathons.

Temperature and running speed

When temperature increased above an optimum, performance decreased. Figure 3 describes the relationship between marathons running speeds and air temperature, fit through a quadratic second degree polynomial curve for women's P1 and men's Q1 of all 60 races.

For each performance level the speed decrease associated with temperature increase and decrease is presented in supplementary data (Table S3).

For example the optimal temperature at which women's P1 maximal running speed was attained was 9.9°C, and an increase of 1°C from this optimal temperature will result in a speed loss of 0.03%. The optimal temperatures to run at maximal speed for men and women, varied from 3.8°C to 9.9°C according to each level of performance (Table S3).

Warmer air temperatures were associated with higher percentages of runners' withdrawal during a race (Figure 4). After testing linear, quadratic, exponential and logarithmic fits, the quadratic equation was the best fit ($r^2 = 0.36$; $p<0.0001$) for modelling the percentage of runners withdrawals associated with air temperature (Figure 4):

$$\%withdrawals = -0.59 \times t^{\circ}C + 0.02 \times t^{\circ}C^2 + 5.75$$

Discussion

Our study is the first to our knowledge to analyse the exhaustiveness of all marathon finishers' performances in the three major European (Berlin, Paris and London, which were not previously analysed) and three American marathons. Previous studies have mostly analysed American marathons including Chicago, Boston and New York that are analysed in the present paper [9,11–15], but they have only included the performances of the top 3 males and females finishers as well as the 25th-, 100th-, and 300th- place finishers [11,13–15]. In the present study we

Table 1. Average and range values of all weather and pollution parameters for the six marathons.

Marathon	Parameter	N	Mean	Std Dev	Minimum	Maximum
Berlin Run in September; Starts 9am	Temperature (°C)	10	14.9	3.2	11.3	21.3
	Dew Point (°C)	10	10.6	1.8	5.8	12.3
	Humidity (%)	10	78.0	14.5	55.0	98.5
	Atmospheric pressure (hPA)	10	1017.0	6.3	1003.0	1029.0
	NO_2 (μg.m^{-3})	10	26.5	4.0	20.8	33.2
	O_3 (μg.m^{-3})	10	41.0	17.3	21.2	81.8
	PM_{10} (μg.m^{-3})	8	25.1	11.4	7.6	46.5
	SO_2 (μg.m^{-3})	10	5.0	3.1	1.1	10.7
Boston Run in April; Starts 10am	Temperature (°C)	10	11.8	5.1	8.0	25.2
	Dew Point (°C)	10	3.9	3.8	−2.1	10.2
	Humidity (%)	10	62.6	19.9	28.3	91.0
	Atmospheric pressure (hPA)	10	1013.0	12.4	981.6	1029.0
	NO_2 (μg.m^{-3})	10	29.3	10.3	14.6	50.5
	O_3 (μg.m^{-3})	10	73.5	25.7	18.5	122.7
	PM_{10} (μg.m^{-3})	0				
	SO_2 (μg.m^{-3})	10	7.0	2.9	1.6	12.1
Chicago Run in October; Starts 7:30am	Temperature (°C)	10	12.1	7.5	1.7	25.0
	Dew Point (°C)	10	4.9	7.6	−5.9	19.0
	Humidity (%)	10	62.8	8.1	52.3	79.2
	Atmospheric pressure (hPA)	10	1022.0	6.4	1012.0	1031.0
	NO_2 (μg.m^{-3})	10	27.9	13.0	9.7	52.0
	O_3 (μg.m^{-3})	10	57.1	15.1	35.9	84.0
	PM_{10} (μg.m^{-3})	2	26.7	11.6	15.3	38.0
	SO_2 (μg.m^{-3})	9	6.5	3.1	2.1	12.4
London Run in April; Starts 9:30am	Temperature (°C)	10	12.4	3.2	9.5	19.1
	Dew Point (°C)	10	6.0	2.9	0.8	10.7
	Humidity (%)	10	66.9	16.7	42.9	86.1
	Atmospheric pressure (hPA)	10	1010.0	12.5	976.4	1020.0
	NO_2 (μg.m^{-3})	10	44.8	14.5	22.8	72.2
	O_3 (μg.m^{-3})	9	51.4	17.1	35.0	92.3
	PM_{10} (μg.m^{-3})	2	27.8	14.5	13.7	41.9
	SO_2 (μg.m^{-3})	10	4.5	2.8	0.0	8.8
New York Run in November; Starts 10am	Temperature (°C)	10	12.5	4.1	7.1	18.4
	Dew Point (°C)	10	2.3	6.4	−5.6	12.8
	Humidity (%)	10	51.1	12.1	36.5	79.8
	Atmospheric pressure (hPA)	10	1020.0	7.8	1009.0	1034.0
	NO_2 (μg.m^{-3})	9	55.1	17.2	21.9	77.3
	O_3 (μg.m^{-3})	10	32.6	12.3	11.1	53.8
	PM_{10} (μg.m^{-3})	10	5.0	0.0	5.0	5.0
	SO_2 (μg.m^{-3})	9	19.7	12.2	4.8	42.4
Paris Run in April; Starts 8:45am	Temperature (°C)	10	9.2	3.2	4.8	17.4
	Dew Point (°C)	10	4.2	4.1	−3.6	13.4
	Humidity (%)	10	72.4	10.1	45.9	85.4
	Atmospheric pressure (hPA)	10	1019.0	6.2	1005.0	1026.0
	NO_2 (μg.m^{-3})	10	43.0	13.7	23.4	73.1

Table 1. Cont.

Marathon	Parameter	N	Mean	Std Dev	Minimum	Maximum
	O_3 ($\mu g.m^{-3}$)	10	66.9	9.8	55.2	82.1
	PM_{10} ($\mu g.m^{-3}$)	10	37.9	32.6	16.6	132.7
	SO_2 ($\mu g.m^{-3}$)	10	6.4	3.7	1.5	12.2

analysed the total number of finishers in order to exhaustively quantify the effect of climate on runners from all performance levels. Updating and extending earlier results, this study still concludes that the main environmental factor influencing marathon performance remains temperature. The pattern of performance reduction with increasing temperature is analogous in men and women, suggesting no apparent gender differences. In addition the mean gap between male and female performances is the same across all marathons and all performance levels (Table 1). This is consistent with our previous work that showed that the gender gap in athletic performance has been stable for more than 25 years, whatever the environmental conditions [25].

The more the temperature increases, the larger the decreases in running speeds (Table S3). This is supported by the increased percentage of runners' withdrawals when races were contested in very hot weather (Figure 4), and by the significant shift of the race's results through the whole range of performance distribution (Figure 2). The significant effect of air temperature on the median values (Table 2) also suggests that all runners' performances are similarly affected by an increase in air temperature, as seen in Figure 2 showing performances distribution of races in Paris and Chicago with different air temperatures: the significant shift of performance towards the right concerns all runners categories, from the elite to the less trained competitors. In addition the percentage of runner's withdrawals in Chicago 2007 was the highest (30.74%) among all 60 studied races (Figure 1 and Figure 4). Roberts [26] reported that organisers tried to interrupt the race 3.5 h after the start. This was not successful as most of the finishers crossed the finish line much later (up to 7 h after the start); 66 runners were admitted to the hospital (12 intensive care cases with hydration disorders, heat shock syndromes and 1 death). During the 2004 Boston Marathon (T° = 22.5°C) more than 300 emergency medical calls were observed, consequently the race's start time changed from noon to 10 am in order to decrease heat stress and related casualties [26]. The 2007 London Marathon was hot by London standards (air T° = 19.1°C vs. an average of 11.6°C for the nine other years analysed in our study), 73 hospitalisations were recorded with 6 cases of severe electrolyte imbalance and one death, the total average time (all participants' average) was 17 min slower than usual. In contrast, the number of people treated in London 2008 in cool and rainy conditions (T° = 9.9°C), was 20% lower [26]. Our results showed that the percentage of runners' withdrawals from races significantly increases with increasing temperature (Figure 4). The acceptable upper limit for competition judged by the American College of Sports Medicine (ACSM) is a WBGT of 28°C, but it may not reflect the safety profile of unacclimatized, non-elite marathon runners [3,26–28]. Roberts [26] stated that marathons should not be allowed to start for non-elite racers at a WBGT of 20.5°C. Our results suggest that there is no threshold but a continuous process

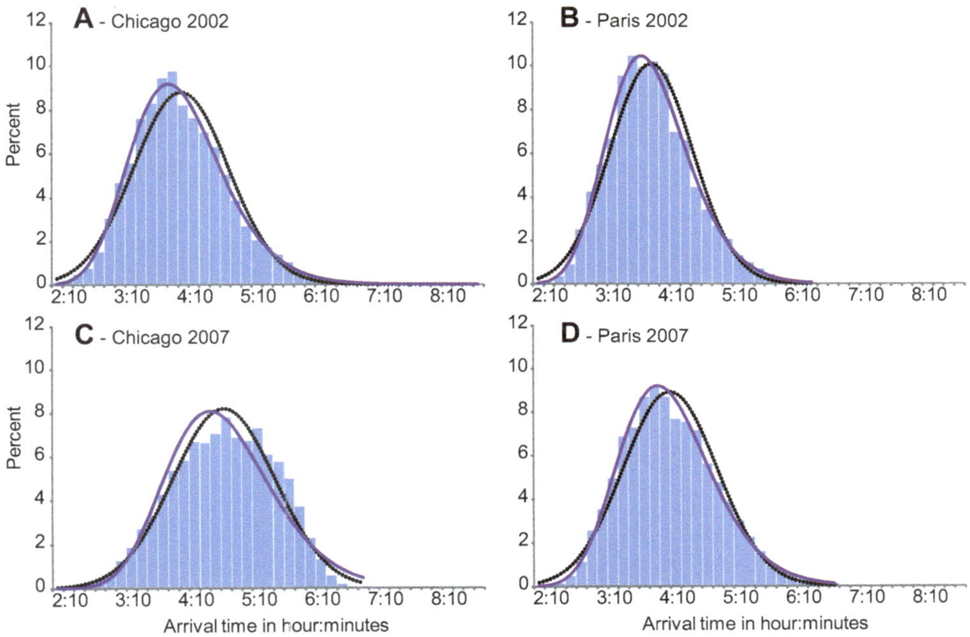

Figure 2. Distribution of performances: example of men's performances distribution for Chicago (in 2002: T°C = 5.4°C; and in 2007: T°C = 25°C); and Paris (in 2002: T°C = 7.6°C; and in 2007: T°C = 17.4°C).

Table 2. Spearman correlations results between all marathons performance levels and environmental parameters: $ = p<0.1; * = p<0.05; ** = p<0.01; *** = p<0.001.

Parameter	Gender	P1	Median	Q1	IQR
Temperature	Women	0.31*	0.30*	0.35**	0.15
	Men	0.48***	0.40***	0.44***	0.25$
Dew Point	Women	0.14	0.18	0.21	0.01
	Men	0.25$	0.19	0.20	0.10
Humidity	Women	−0.3*	−0.16	−0.19	−0.21
	Men	−0.34**	−0.28*	−0.32*	−0.19
Atm. Pressure	Women	0.22$	0.06	0.07	0.06
	Men	0.13	0.04	0.06	0.06
NO2	Women	0.11	0.40**	0.43***	0.33*
	Men	0.25$	0.38**	0.35**	0.27*
O3	Women	0.01	−0.15	−0.11	−0.20
	Men	−0.05	−0.21	−0.24$	−0.11
PM10	Women	0.08	0.15	0.25	0.03
	Men	0.10	0.10	0.09	0.16
SO2	Women	0.21	0.13	0.21	0.02
	Men	0.37**	0.20	0.25$	0.04

P1: first percentile, Q1: first quartile, IQR: Inter Quartile Range.

on both side of an optimum: the larger the gap from the optimal temperature, the lower the tolerance and the higher the risk. In fact, in environments with high heat and humidity, not only is performance potentially compromised, but health is also at risk [29]; both are similarly affected. As soon as WBGT is higher than 13°C the rate of finish line medical encounters and on-course marathon dropouts begin to rise [26] as similarly seen in our study in Figure 4.

Warm weather enhances the risk of exercise induced hyperthermia; its first measurable impact is the reduction of physical performance [4,14,29–31] as it is detrimental for the cardiovascular, muscular and central nervous systems [32,33]. More recent work suggested that central fatigue develops before any elevation in body temperature occurs: evidence supported that subjects would subconsciously reduce their velocity earlier after the start of an exercise in hot environment, when internal temperatures are still lower than levels associated with bodily harm. Exercise is thus homeostatically regulated by the decrease of exercise intensity (decrease of running performance and heat production) in order to prevent hyperthermia and related catastrophic failures [34,35]. On the other hand, cool weather is associated with an improved ability to maintain running velocity and power output as compared to warmer conditions, but very cold conditions also tend to reduce performance [29,36,37].

Among the studied races' winners, men's marathon world record was beaten in Berlin in 2007 and 2008 (Haile Gebrselassie in 02:03:59), as well as women's marathon world record, beaten in London 2003 (Paula Radcliffe in 02:15:25). The winners' speeds couldn't be affected in the same way than the other runners by air temperature and the other environmental parameters, because top performances can fluctuate from year to year due to numerous factors, such as prize money, race strategies, or overall competition [11]. Another explanation is that, in all of our 60 studied races, 89.5% of male winners were of African origin (57.9% from Kenya; 21.1% from Ethiopia; and 10.5% from Eritrea, Morocco and South Africa); as well as 54.5% of female winners (27.3% from Kenya and 27.3% from Ethiopia- data not shown). African runners might have an advantage over Caucasian athletes, possibly due to a unique combination of the main endurance factors such as maximal oxygen uptake, fractional utilization of VO_{2max} and running economy [38]. They might also perform better in warm environments as they are usually thinner than Caucasian runners (smaller size and body mass index) producing less heat with lower rates of heat storage [38–40]. Psychological factors may also play a role; some hypothesis suggested that regardless of the possible existence of physiological advantages in East African runners, belief that such differences exist may create a background that can have significant positive consequences on performance [41,42].

Genetics and training influence the tolerance for hyperthermia [4,38,43]. Acclimatisation involving repeated exposures to exercise in the heat also results in large improvements in the time to fatigue. Optimal thermoregulatory responses are observed in runners who have been acclimatized to heat and who avoid thirst before and during the race. Their best performances might be less influenced by temperature as winners had been more acclimatized to it [4,29,30,44]. The avoidance of thirst sensation rather than optimum

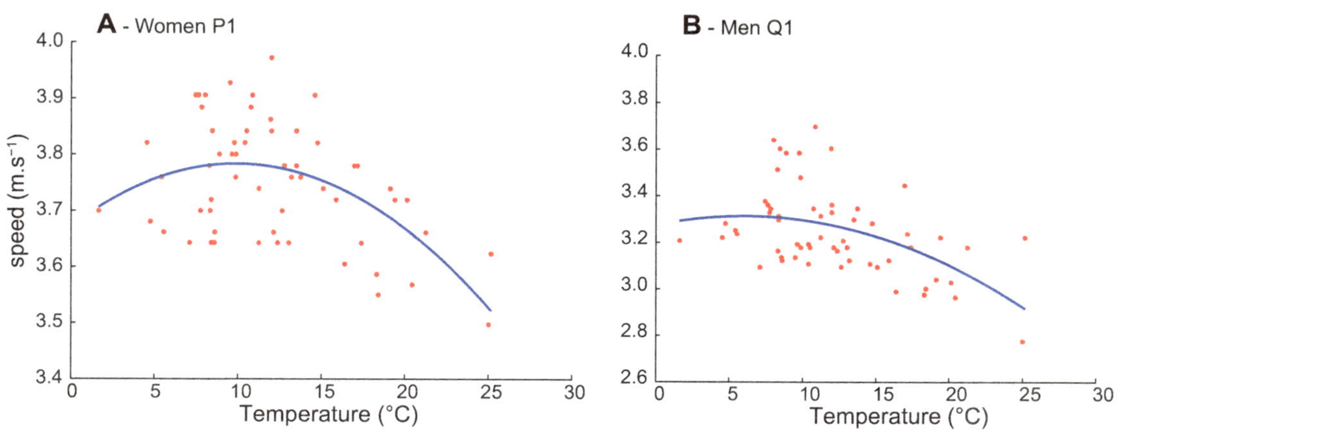

Figure 3. Quadratic second degree polynomial fit for Women's P1 running speeds vs. air temperature, $r^2=0.27$; $p<0.001$; max = 9.9°C. B) Men's Q1 running speeds vs. air temperature, $r^2=0.24$; $p<0.001$; max = 6°C.

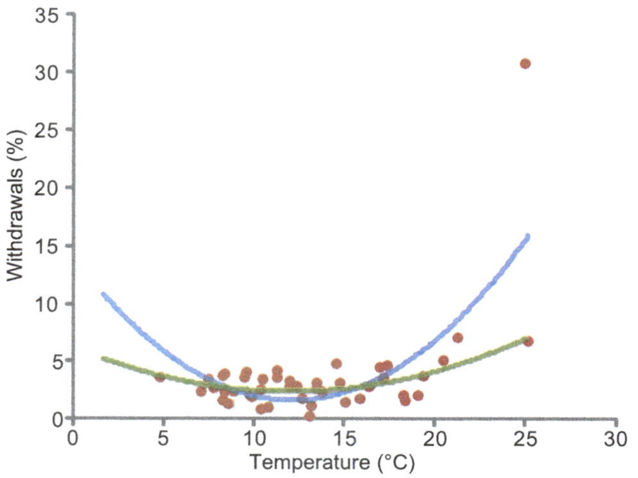

Figure 4. Relationship between air temperature and the percentage of runners' withdrawals, modeled with a quadratic fit (blue curve, $r^2 = 0.36$; $p<0.0001$). The green curve represents the quadratic fit without the maxima (Chicago 2007: 30.74% withdrawals at a race temperature of 25°C).

hydration prevents the decline in running performance [45]; contradicting the idea that dehydration associated with a body weight loss of 2% during an exercise will impair performance, recent studies reported that Haile Gebrselassie lost 10% of his body weight when he established his world record [45–47].

Previous studies suggested that the impact of weather on speed might depend on running ability, with faster runners being less limited than slower ones [6,13,14,29]. This could be attributable to a longer time of exposition to the environmental conditions of slower runners during the race [11]. Also, slower runners tend to run in closer proximity to other runners with clustering formation [48,49], which may cause more heat stress as compared with running solo [50]. These elements, however, are not supported after analyzing the full range of finisher's data; at a population level, temperature causes its full effect whatever the initial capacity. Differences in fitness relative to physiological potential may also contribute to differences in performance times and ability to cope with increasing heat stress [11,48,49].

There was a strong correlation of running speed with air temperature (Figure 3). The maximal average speeds were performed at an optimal temperature comprised between 3.8°C and 9.9°C depending on the performance level (Table S3); small increases in air temperatures caused marathon performances to decline in a predictable and quantifiable manner. On the other hand, large decreases in air temperatures under the optimum also reduce performances. These optimal temperatures found in the present study are comprised in the optimal temperature range of 5–10°C WBGT found in previous studies [14]; other studies stated that a weather of 10–12°C WBGT is the norm for fast field performance and reported a decrease of performance with increasing WBGT [12,27,51,52]. Best marathon times and most marathon world records were achieved in cool environmental temperatures (10–15°C) and have been run in the early morning during spring and fall [12]. Analysing Gebrselassie's performances in Berlin reveals that they follow the same trend, with both World Records obtained at the lowest temperatures (14°C in 2007 and 13°C in 2008, *vs.* 18°C in 2009 and 22°C in 2006 when he also won these two races without beating the world record).

The relationship between running speed and air temperature defined in our study (Figure 3) is similar to the relationship found between mortality and air temperature (asymmetrical U-like pattern) in France defined by Laaidi et al [53], where mortality rates increase with the lowest and the highest temperatures. A "thermal optimum" occurs in between, where mortality rates are minimal [53]. The great influence that temperature has on performance is comparable to the influence it has on mortality, suggesting that both sports performance and mortality are thermodynamically regulated. This also emphasizes the utility of prevention programs, the assessment of public health impacts and acclimatization before participating in hot marathons [53]. Similar correlations were also found between temperature and swimming performance in juvenile southern catfish [22], and between increases in summer water temperature and elevated mortality rates of adult sockeye salmon [23]; suggesting that physiological adaptations to temperature, similarly occur in various taxons, but vary within specific limits that depend on species and will modify performances.

Air pollution and performance

The measured levels of pollution had no impact on performance, except for ozone (Table S2) and NO_2 (Table 2). Assessing the effect of any single air pollutant separately is not simple; it is not isolated in the inhaled air, but rather combined with other parameters. Therefore any possible influence might probably be due to a combination of components. In addition most marathons are held on Sunday mornings, when urban transport activity and its associated emissions are low, and photochemical reactions driven by solar radiation have not yet produced secondary pollutants such as ozone [9]. This is the most probable explanation to our results, confirming previous studies. Among the air pollutants analysed in the present study, ozone and NO_2 had the greatest effect on decreasing marathon performances (Table S2). Ozone concentrations on the ground increase linearly with air temperature [7,8,10]; thus the effect of ozone in our study may be mainly associated with the temperature effect, as seen in Berlin and Chicago. However ozone and other pollutants effects are known to be detrimental to exercise performance only when exposure is sufficiently high. Many studies showed no effect of air pollutants on sports performance [9]. Some of them showed that $PM_{2.5}$ and aerosol acidity were associated with acute decrements in pulmonary function, but these changes in pulmonary function were unlikely to result in clinical symptoms [54]. Others showed that chronic exposure to mixed pollutants during exercise may result in decreased lung function, or vascular dysfunction, and may compromise performance [55]. During the marathons studied here, concentrations of air pollutants never exceeded the limits set forth by national environmental agencies (US Environmental Protection Agency- EPA; AirParif; European Environmental Agency- EEA) or the levels known to alter lung function in laboratory situations [9].

Conclusions

Air temperature is the most important factor influencing marathon running performance for runners of all levels. It greatly influences the entire distribution of runners' performances as well as the percentage of withdrawals. Running speed at all levels is linked to temperature through a quadratic model. Any increase or decrease from the optimal temperature range will result in running speed decrease. Ozone also has an influence on performance but its effect might be linked to the temperature impact. The model developed in this study could be used for further predictions, in order to evaluate expected performance variations with changing weather conditions.

Supporting Information

Table S1 Time values of different descriptive statistics and their variability by marathon and gender. [1] Value of the described statistic for all performances of all year together, hour:min:sec [2] Standard deviation of the described statistic for all performances of each year, hour:min:sec [3] IQR: Inter Quartile Range.

Acknowledgments

We thank the Centre National de Développement du Sport and the Ministry of Health, Youth and Sport. We thank INSEP teams for their full support. We thank Mrs Karine Schaal for carefully reviewing the manuscript.

Author Contributions

Conceived and designed the experiments: JFT NEH GB AM. Analyzed the data: JT GB AM NEH MG MT. Wrote the paper: NEH GB JFT. Reviewed the paper: CH JFT.

References

1. Lippi G, Favaloro EJ, Guidi GC (2008) The genetic basis of human athletic performance. Why are psychological components so often overlooked? J Physiol 586(Pt 12):3017; author reply 3019–3020.
2. Macarthur DG, North KN (2005) Genes and human elite athletic performance. Hum Genet 116(5): 331–339.
3. Cheuvront SN, Haymes EM (2001) Thermoregulation and marathon running, biological and environmental influences. Sports Med 31(10): 743–762.
4. Kenefick RW, Cheuvront SN, Sawka MN (2007) Thermoregulatory function during the marathon. Sports Med 37(4–5): 312–315.
5. Weather Underground website. Internet weather service. Weather data from each marathon race. Available: http://www.wunderground.com/history/. Accessed 2011 Mar, 30.
6. Vihma T (2010) Effects of weather on the performance of marathon runners. Int J Biometeorol 54(3): 297–306.
7. Shephard RJ (1984) Athletic performance and urban air pollution. Can Med Assoc J 131(2): 105–109.
8. Chimenti L, Morici G, Paterno A, Bonanno A, Vultaggio M, et al. (2009) Environmental conditions, air pollutants, and airway cells in runners: A longitudinal field study. J Sports Sci 27(9): 925–935.
9. Marr LC, Ely MR (2010) Effect of air pollution on marathon running performance. Med Sci Sports Exerc 42(3): 585–591.
10. Lippi G, Guidi GC, Maffulli N (2008) Air pollution and sports performance in Beijing. Int J Sports Med 29: 696–698.
11. Ely MR, Cheuvront SN, Roberts WO, Montain SJ (2007) Impact of weather on marathon-running performance. Med Sci Sports Exerc 39(3): 487–493.
12. Ely MR, Cheuvront SN, Montain SJ (2007) Neither cloud cover nor low solar loads are associated with fast marathon performance. Med Sci Sports Exerc 39(11): 2029–2035.
13. Ely MR, Martin DE, Cheuvront SN, Montain SJ (2008) Effect of ambient temperature on marathon pacing is dependent on runner ability. Med Sci Sports Exerc 40(9): 1675–1680.
14. Montain SJ, Ely MR, Cheuvront SN (2007) Marathon performance in thermally stressing conditions. Sports Med 37(4–5): 320–323.
15. Martin DE, Buoncristiani JF (1999) The effects of temperature on marathon runners' performance. Chance 12(4): 20–24.
16. Trapasso LM, Cooper JD (1989) Record performances at the Boston Marathon: biometeorological factors. Int J Biometeorol 33(4): 233–237.
17. Online worldwide athletic results database website. Marathons races results. Available: http://www.athlinks.com. Accessed 2011 Apr, 30.
18. AirParif website. Air pollution data for Paris retrieved (March – May 2009). Available: http://www.airparif.com.
19. Station Database of the Environmental Agency. Air pollution data for Berlin retrieved (June 24, 2009). Available: http://www.env-it.de/stationen/public/language.do;jsessionid=FB278996EE26B0351076A5D974C8BD04?language=en.
20. LondonAir website. Air pollution data for London retrieved (May 26, 2009). Available: http://www.londonair.org.uk/london/asp/default.asp.
21. Berthelot G, Tafflet M, El Helou N, Len S, Escolano S, et al. (2010) Athlete atypicity on the edge of human achievement: Performances stagnate after the last peak, in 1988. PLoS ONE, 5 (1), e8800: DOI: 10.1371/journal.pone.0008800.
22. Zeng LQ, Cao ZD, Fu SJ, Peng JL, Wang YX (2009) Effect of temperature on swimming performance in juvenile southern catfish (Silurus meridionalis). Comp Biochem Physiol A Mol Integr Physiol 153(2): 125–130.
23. Eliason EJ, Clark TD, Hague MJ, Hanson LM, Gallagher ZS, et al. (2011) Differences in thermal tolerance among sockeye salmon populations. Science 332(6025): 109–112.
24. Kirschbaum MUF, Watt MS (2011) Use of a process-based model to describe spatial variation in Pinus radiate productivity in New Zealand. Forest Ecology and Management 262: 1008–1019.
25. Thibault V, Guillaume M, Berthelot G, El Helou N, Schaal K, et al. (2010) Women and men in sport performance: the gender gap has not evolved since 1983. J Sports Sci Med 9: 214–223.
26. Roberts WO (2010) Determining a "do not start" temperature for a marathon on the basis of adverse outcomes. Med Sci Sports Exerc 42(2): 226–232.
27. Zhang S, Meng G, Wang Y, Li J (1992) Study of the relationships between weather conditions and the marathon race, and of meteorotropic effects on distance runners. Int J Biometeorol 36: 63–68.
28. Armstrong LE, Epstein Y, Greenleaf JE, Haymes EM, Hubbard RW, et al. (1996) American College of Sports Medicine position stand. Heat and cold illnesses during distance running. Med Sci Sports Exerc 28(12): i– x.
29. Maughan RJ, Watson P, Shirreffs SM (2007) Heat and cold, what does the environment do to the marathon runner? Sports Med 37(4–5): 396–399.
30. Hargreaves M (2008) Physiological limits to exercise performance in the heat. J Sci Med Sport 11(1): 66–71.
31. Walters TJ, Ryan KL, Tate LM, Mason PA (2000) Exercise in the heat is limited by a critical internal temperature. J Appl Physiol 89: 799–806.
32. Coyle EF (2007) Physiological regulation of marathon performance. Sports Med 37(4-5): 306–311.
33. González-Alonso J (2007) Hyperthermia Impairs Brain, Heart and Muscle Function in Exercising Humans. Sports Med 37(4–5): 371–373.
34. Tucker R, Rauch L, Harley YX, Noakes TD (2004) Impaired exercise performance in the heat is associated with an anticipatory reduction in skeletal muscle recruitment. Pflugers Arch 448(4): 422–430.
35. Tucker R, Marle T, Lambert EV, Noakes TD (2006) The rate of heat storage mediates an anticipatory reduction in exercise intensity during cycling at a fixed rating of perceived exertion. J Physiol 574(Pt 3): 905–915.
36. Nimmo M (2004) Exercise in the cold. J Sports Sci 22: 898–915.
37. Weller AS, Millard CE, Stroud MA, Greenhaff PL, Macdonald IA (1997) Physiological responses to a cold, wet, and windy environment during prolonged intermittent walking. Am J Physiol 272(1 Pt 2): R226–R233.
38. Larsen HB (2003) Kenyan dominance in distance running. Comp Biochem Physiol A Mol Integr Physiol 136(1): 161–170.
39. Marino FE, Lambert MI, Noakes TD (2004) Superior performance of African runners in warm humid but not in cool environmental conditions. J Appl Physiol 96: 124–130.
40. Marino FE, Mbambo Z, Kortekaas E, Wilson G, Lambert MI, et al. (2000) Advantages of smaller body mass during distance running in warm, humid environments. Pflügers Arch 441(2–3): 359–367.
41. Hamilton B (2000) East African running dominance: what is behind it? Br J Sports Med 34(5): 391–394.
42. Baker J, Horton S (2003) East African running dominance revisited: a role for stereotype threat? Br J Sports Med 37(6): 553–555.
43. Sawka MN, Young A (2006) Physiological systems and their responses to conditions of heat and cold. In: Tipton CM, ed. American College of Sports Medicine's Advanced exercise physiology. Philadelphia (PA): Lippincott Williams and Wilkins. 535–563 p.
44. Zouhal H, Groussard C, Vincent S, Jacob C, Abderrahman AB, et al. (2009) Athletic performance and weight changes during the "Marathon of Sands" in athletes well-trained in endurance. Int J Sports Med 30: 516–521.
45. Goulet ED (2011) Effect of exercise-induced dehydration on time-trial exercise performance: a meta-analysis. Br J Sports Med 45(14): 1149–1156.
46. Zouhal H, Groussard C, Minter G, Vincent S, Cretual A, et al. (2011) Inverse relationship between percentage body weight change and finishing time in 643 forty-two-kilometre marathon runners. Br J Sports Med 45(14): 1101–1105.
47. Beis LY, Wright-Whyte M, Fudge B, Noakes T, Pitsiladis YP (2012) Drinking Behaviors of Elite Male Runners During Marathon Competition. Clin J Sport Med. March [Epub ahead of print] doi: 10.1097/JSM.0b013e31824a55d7.
48. Alvarez-Ramirez J, Rodriguez E (2006) Scaling properties of marathon races. Physica A: Stat Mech Appl 365(2): 509–520.

49. Alvarez-Ramirez J, Rodriguez E, Dagduga L (2007) Time-correlations in marathon arrival sequences. Physica A: Stat Mech Appl 380: 447–454.
50. Dawson NJ, De Freitas CR, Mackey WJ, Young AA (1987) The stressful microclimate created by massed fun runners. Transactions of the Menzies Foundation 14: 41–44.
51. Galloway SDR, Maughan RJ (1997) Effects of ambient temperature on the capacity to perform prolonged cycle exercise in man. Med Sci Sports Exerc 29(9): 1240–2149.
52. Buoncristiani JF, Martin DE (1983) Factors affecting runners' marathon performance. Chance 6(4): 24–30.
53. Laaidi M, Laaidi K, Besancenot JP (2006) Temperature-related mortality in France, a comparison between regions with different climates from the perspective of global warming. Int J Biometeorol 51(2): 145–153.
54. Korrick SA, Neas LM, Dockery DW, Gold DR, Allen GA, et al. (1998) Effects of ozone and other pollutants on the pulmonary function of adult hickers. Environ Health Perspect 106: 93–99.
55. Rundell KW (2012) Effect of air pollution on athlete health and performance. Br J Sports Med. Epub ahead of print.

7

Exposure to Ozone Modulates Human Airway Protease/ Antiprotease Balance Contributing to Increased Influenza A Infection

Matthew J. Kesic[1]*, Megan Meyer[1,4], Rebecca Bauer[1,2], Ilona Jaspers[1,2,3,4]

1 Center for Environmental Medicine, Asthma, and Lung Biology, University of North Carolina Chapel Hill, North Carolina, United States of America, 2 Curriculum in Toxicology, University of North Carolina Chapel Hill, North Carolina, United States of America, 3 Department of Pediatrics University of North Carolina Chapel Hill, North Carolina, United States of America, 4 Department of Microbiology and Immunology University of North Carolina Chapel Hill, North Carolina, United States of America

Abstract

Exposure to oxidant air pollution is associated with increased respiratory morbidities and susceptibility to infections. Ozone is a commonly encountered oxidant air pollutant, yet its effects on influenza infections in humans are not known. The greater Mexico City area was the primary site for the spring 2009 influenza A H1N1 pandemic, which also coincided with high levels of environmental ozone. Proteolytic cleavage of the viral membrane protein hemagglutinin (HA) is essential for influenza virus infectivity. Recent studies suggest that HA cleavage might be cell-associated and facilitated by the type II transmembrane serine proteases (TTSPs) human airway trypsin-like protease (HAT) and transmembrane protease, serine 2 (TMPRSS2), whose activities are regulated by antiproteases, such as secretory leukocyte protease inhibitor (SLPI). Based on these observations, we sought to determine how acute exposure to ozone may modulate cellular protease/antiprotease expression and function, and to define their roles in a viral infection. We utilized our *in vitro* model of differentiated human nasal epithelial cells (NECs) to determine the effects of ozone on influenza cleavage, entry, and replication. We show that ozone exposure disrupts the protease/antiprotease balance within the airway liquid. We also determined that functional forms of HAT, TMPRSS2, and SLPI are secreted from human airway epithelium, and acute exposure to ozone inversely alters their expression levels. We also show that addition of antioxidants significantly reduces virus replication through the induction of SLPI. In addition, we determined that ozone-induced cleavage of the viral HA protein is not cell-associated and that secreted endogenous proteases are sufficient to activate HA leading to a significant increase in viral replication. Our data indicate that pre-exposure to ozone disrupts the protease/antiprotease balance found in the human airway, leading to increased influenza susceptibility.

Editor: Andrew Pekosz, Johns Hopkins University - Bloomberg School of Public Health, United States of America

Funding: The project described was in part supported by grant number HL095163 from the National Heart, Lung, and Blood Institute (NHLBI), National Institutes of Health, a grant from the Flight Attendant Medical Research Institute (FAMRI), as well as a grant from the Environmental Protection Agency (CR829522) (all to IJ). This research was also supported by grant number T32 ES007126-26 National Institute of Environmental Health Sciences Curriculum of Toxicology Training Grant. The funders had no role in study design, data collection and analysis, decision to publish, or preparation of the manuscript.

Competing Interests: The authors have declared that no competing interests exist.

* E-mail: kesic@email.unc.edu

Introduction

Influenza A virus is responsible for the seasonal epidemics and reoccurring pandemics, which represents a worldwide threat to global public health [1]. It is a major cause of morbidity and mortality worldwide, as during the recent H1N1 pandemic. In the U.S. alone, over 100,000 individuals are hospitalized and over 20,000 people die every year due to influenza virus infection and related diseases [2,3]. Despite large-scale efforts in vaccination and antiviral therapies, the morbidity and mortality rates associated with influenza infections have not significantly changed in recent years [4,5]. In the context of potentially pandemic respiratory viral infections, it is important to identify molecular targets/pathways for therapeutic intervention to protect susceptible sub-populations. Epidemiologic studies show that exposure to inhaled oxidants such as cigarette smoke, diesel exhaust, and ozone can modulate immune function and increase susceptibility to respiratory viral infections [6,7,8,9,10,11]. Despite the extensive study into different factors influencing susceptibility to infection, the mechanism(s) by which inhaled oxidants modify viral pathogenesis are very complex and have yet to be fully elucidated.

The ability of oxidants to cause lung injury/dysfunction is dependent, in part, on the delicate equilibrium that exists between oxidants and antioxidants. Oxidative stress is caused by an imbalance between the production of reactive oxygen species (ROS) and the body's ability to readily detoxify reactive intermediates. Ozone is one of the most abundant components of air pollution in urban areas, and has been shown to be a potent inducer of oxidative stress causing airway inflammation and increased respiratory morbidities [8,9,12]. The Mexico City Metropolitan Area (MCMA), one of the most densely populated areas in the world, experiences high levels of air pollutants such as environmental ozone and particulate matter (PM) [13,14,15]. The MCMA is located 2240 m above sea level and is surrounded by mountains. Due to this geographical location, there is less available oxygen, making combustion less efficient, which produces more polycyclic aromatic hydrocarbon (PAH) pollutants. For these reasons, Mexico City receives higher levels of environmental ozone and various other types of photochemical smog [13]. It was here that the first influenza pandemic of the 21st century emerged in March 2009. It was responsible for an estimated ~258,698 laboratory confirmed cases and roughly ~1,370 deaths by December 2009 [1,16,17]. The normal "Flu Season" occurs during the colder half of the year in each hemisphere [18]. Interestingly, this outbreak emerged during the dry and warmer months when ozone levels were significantly higher [19,20].

Possible mechanisms by which oxidative stress alters the airway environment leading to broadened cellular tropism and/or susceptibility to viral infections have been proposed. [21,22,23] The relationship between oxidative stress and the cellular protease/antiprotease balance has been considered to be a major contributor in the development of numerous airway pathologies [24]. Early studies demonstrated that the cleavage of influenza HA is essential for viral penetration into host cells and that this mechanism was mediated by cellular trypsin-like serine proteases. These proteases, in turn, are regulated by mucus antiproteases, such as SLPI and α_1-antitrypsin (A1AT) [25,26,27,28,29]. Recent studies have identified specific cellular proteases and antiproteases that may be involved in influenza infection, which include transmembrane protease serine 2 (TMPRSS2), human airway trypsin-like protease (HAT), and secretory leukocyte proteinase inhibitor (SLPI) [30,31,32,33,34]. The expression of these proteases in the lung are necessary for cleavage of the viral HA surface protein, thus allowing viral fusion and entry into the host cell. Studies have shown a correlation between inflammation and oxidative stress which alters expression of these proteases and antiproteases [29,35]. Specifically, HAT has been shown to be released into the airway fluids under inflammatory conditions, particularly in asthmatics [31,36]. In addition, a murine study showed SLPI expression was suppressed and protease expression was increased in Nrf2-deficient mice which led to increased inflammation, further demonstrating a relationship between oxidative stress and protease expression [37]. Oxidative stress derived from cigarette smoke or addition of reactive oxygen intermediates has been shown to decrease antiprotease activity [35,38] and a murine study demonstrated that decreased antiprotease expression increased influenza infectivity [39]. The imbalance of protease/antiprotease expression caused by oxidative stress has been attributed to both an increase in inflammatory cells, which release proteolytic enzymes capable of destroying lung tissue, and a functional deficit of antiproteases due to oxidation of their active site [35,40]. In addition to influenza, studies have shown that regulated proteolysis is required for the spread/ propagation of many human viruses, including Human immunodeficiency virus (HIV), Nipah, Ebola, severe acute respiratory syndrome coronavirus (SARS-CoV), and metapneumoviruses [41,42,43,44,45].

Although it has been shown that oxidative stress increases severity of viral infections, the exact mechanism as to how and why this happens and the role of ozone exposure in these responses are not fully understood. While the deleterious health effects of both ozone and influenza have been well documented, few studies have looked at the effects of ozone in the context of an influenza infection. Herein, we used our established cell culture model of differentiated human nasal epithelial cells (NECs) [46], exposed them to 0.4ppm of ozone, and determined the effects on susceptibility to an influenza A infection. We show that there is a delicate balance of proteases and antiproteases present in the human airway surface liquid. Additionally, we found exposure to ozone disrupts this equilibrium by enhancing protease secretion while inhibiting antiprotease expression leading to increased susceptibility to viral infection. We directly show that ozone increases soluble protease expression and that they are functional for cleaving the viral HA protein and activation of the intact influenza virion.

Materials and Methods

Ethics Statement

Primary human nasal epithelial cells (NECs) were obtained from healthy adult volunteers. This protocol was approved by the University of North Carolina School of Medicine Institutional Review Board for Biomedical Research. In addition, written informed consent was provided by each study participants and/or their legal guardians.

Nasal epithelial cell cultures, cell lines, and growth conditions

Primary human nasal epithelial cells (NECs) were obtained as previously described [6]. Briefly, NECs were obtained from healthy adult volunteers by gently stroking the inferior surface of the turbinate several times with a Rhino-Probe curette (Arlington Scientific, Arlington, TX), which was inserted through a nasoscope. This protocol was approved by the University of North Carolina School of Medicine Institutional Review Board for Biomedical Research. NEC were expanded to passage 2 in bronchial epithelial growth medium (BEGM, Cambrex Bioscience Walkersville, Inc., Walkersville, MD) and then plated on collagen-coated filter supports with a 0.4 µM pore size (Trans-CLR; Costar, Cambridge, MA) and cultured in a 1:1 mixture of bronchial epithelial cell basic medium (BEBM) and DMEM-H with SingleQuot supplements (Cambrex), bovine pituitary extracts (13 mg/ml), bovine serum albumin (BSA, 1.5 µg/ml), and nystatin (20 units). Upon confluency, all-trans retinoic acid was added to the basolateral medium to establish air liquid interface (ALI) culture conditions (removal of the apical medium) to promote differentiation. Mucociliary differentiation was achieved after 18–21 days post-ALI. We obtained the Madin Darby canine kidney (MDCK) cell line from the American Type Culture Collection (ATCC, Manassas, VA). The BEAS-2B cell line was derived by transforming human bronchial cells with an adenovirus 12-simian virus 40 construct [47]. We obtained our BEAS-2B cells from the American Type Culture Collection (ATCC, Manassas, VA). BEAS-2B cells were grown in keratinocyte basal medium (KBM) supplemented with 30 µg/ml bovine pituitary extract, 5 ng/ml human epidermal growth factor, 500 ng/ml hydrocortisone, 0.1 mM ethanolamine, 0.1 mM phosphoethanolamine, and 5 ng/ml insulin.

The primary human NECs were exposed under ALI conditions to filtered air or 0.4 ppm ozone for 4 h in the exposure chambers operated by the U.S. Environmental Protection Agency, Environmental Public Health Division. In noted experiments, 10 mM of a cell-permeable form of reduced glutathione, glutathione-ethylester (GSH-ET) (Sigma-Aldrich St. Louis. Mo) or 1 µM of EGCG (Sigma-Aldrich St. Louis. Mo) was added to the basolateral medium 30 min prior to ozone exposure, similar to our previous studies [6,46].

Virus-like particle (VLP) entry assay

Construction of VLP expression plasmids. To generate the β-lactamase-M1 fusion expression plasmid (pCAGGS-β-lacM1 PR8) the influenza A/Puerto Rico/8/34/Mount Sinai (H1N1) M1 sequence was PCR amplified from the M1 expression vector pDZ-M (which was kindly provided by Dr. Adolfo Garcia-

Sastre, Mount Sinai School of Medicine) and inserted into the pCAGGS vector [48,49,50]. The β-lactamase gene was PCR amplified from pcDNA3.1 and fused N-terminally to M1 within pCAGGS to create a modified β-lactamase-M1 fusion. In the modified β-lactamase, the first 24 amino acids were excluded to remove a secretion signal and His 24 was substituted with Asp to create an optimal Kozak consensus sequence. The pcDNA3.1-β-lactamase construct has been described previously [51]. The influenza A/Puerto Rico/8/34/Mount Sinai (H1N1) HA (pCAGGS-HA) and NA (pDZ-NA) over-expression vectors were generously provided by Dr. Aldolfo Garcia-Sastre and have been described previously [48,49,50,52].

Production of VLPs. To generate influenza A/PR/8/H1N1 β-lactamaseM1VLPs (PR8 β-lacM1 VLPs), 293T cells were co-transfected with 3 μg each of pCAGGS-HA, pDZ-NA, and pCAGGS-β-lacM1 PR8 using FuGENE® HD (Roche Applied Science, Indianapolis, IN) according to manufacturer's instructions. The supernatant containing the VLPs were collect 48 h post-transfection and clarified of floating cell debris by centrifugation at 3,000 rpm for 10 min. The VLPs were concentrated once by low-speed centrifugation through an Amicon Ultra 100 kD centrifuge filter unit (Millipore; Billerica, MA), and the retentates were aliquoted and stored at −80°C.

VLP entry assay. The VLP entry assay was performed as previously described [6] with modifications. Briefly, target cells were exposed to filtered air or 0.4 ppm ozone for 4 h. 24 h post-exposure, VLPs were added to the apical surface in a total volume of 100 μl and incubated at 37°C for 3 h. The cells were washed twice with HBSS to remove unbound virus and infected cells were detected by using GeneBLAzer™ FRET *in vivo* cell-based assay system substrate CCF2-AM according to manufacturer's recommendations (Invitrogen). Samples were lysed in HBSS by freeze-thaw method repeatedly treated for 3 cycles (frozen in liquid nitrogen for 3 min and thawed in a 65°C water bath for 3 min). Viral entry was quantified by using the POLARstar OPTIMA plate reader (BMG LABTECH, Inc.). All experiments were performed in triplicate in three independent experiments.

Infection with Influenza A

We used influenza A/Bangkok/1/79 H3N2 serotype (which was kindly provided by Dr. Melinda Beck, University of North Carolina). We also obtained influenza A/Malaya/302/1954 H1N1 serotype from the American Type Culture Collection (ATCC, Manassas, VA) for Western blot analysis of the cleavage products of the viral HA protein. Virus was propagated in 10-day-old embryonated hen's eggs. The virus was collected in the allantoic fluid and titered by 50% tissue culture infectious dose in Madin-Darby canine kidney cells and hemagglutination as described before [53]. Stock virus was aliquoted and stored at −80°C until use. Unless otherwise indicated, for infection about 5×10^5 cells were infected with approximately 128 hemagglutination units (HAU) of influenza A, which resulted in approximately 10% of the cells being infected with influenza 24 h post-infection. Total RNA, total protein, basolateral supernatants, and apical washes were collected 24 h post-infection.

Immunoblotting

Apical surface liquid was collected at 24 h post-exposure by washing the apical surface of the cells with HBSS and total protein concentrations were determined by Bradford protein assay (Bio-Rad). Cell lysates were prepared at 24 h post-infection in Passive Lysis Buffer (Promega, Madison, WI) with a protease inhibitor mixture (Cocktail Set III; Calbiochem, San Diego, CA) as well as phosphatase inhibitors (0.5 mM NaVO4, 1 mM β-glycerophosphate) on ice for 30 min. After centrifugation, total protein concentrations were determined by Bradford protein assay (Bio-Rad). Both the apical supernatants and cellular lysates were subjected to 12% sodium dodecyl sulfate-polyacrylamide gel electrophoresis (SDS-PAGE) and transferred to nitrocellulose (Schleicher & Schuell Biosciences, Keene, NH). Proteins were detected using specific antibodies to HAT, TMPRSS2, and SPLI (1:1,000; Santa Cruz Biotechnology, Santa Cruz, CA) or Influenza A virus hemagglutinin H1 antibody (1:1,000; Abcam, Cambridge, MA). β-actin (1:2,000; US Biological, Swampscott, MA), which was used as a loading control. Antigen-antibody complexes were stained with anti-rabbit or anti-mouse, horseradish peroxidase (HRP)-conjugated antibody (1:2000, Santa Cruz Biotechnology) and detected with SuperSignal West Pico Chemiluminescent Substrate (Pierce, Rockford, IL). Densitometry was performed using a Fujifilm LAS-3000 imager (Fuji Film Global Tokyo, Japan).

Influenza HA cleavage

Apical surface liquid was collected by washing the apical surface of the cells with HBSS and total protein concentrations were determined by Bradford protein assay (Bio-Rad). 50 μg of total protein were immunoprecipated overnight at 4°C with 200 ng of anti-TMPRSS2, anti-HAT, or IgG isotype as a control, followed by 1 h incubation with 50 μl of protein G-agrose beads (Pierce, Rockford, IL). Cleared supernatants were incubated with 50 ul of Influenza A/Malaya/302/1954 H1N1 for 30 min at 32°C. Virus was subjected to 12% sodium dodecyl sulfate-polyacrylamide gel electrophoresis (SDS-PAGE) and transferred to nitrocellulose (Schleicher & Schuell Biosciences, Keene, NH). Proteins were detected using specific antibodies to HAT, TMPRSS2, and SPLI (1:1,000; Santa Cruz Biotechnology, Santa Cruz, CA) or Influenza A virus hemagglutinin H1 antibody (1:1,000; Abcam, Cambridge, MA).

Influenza virus titer

Influenza virus titers in apical washes were determined by 50% tissue-culture infectious dose (TCID50) in Madin Darby canine kidney (MDCK) cells and evaluated using agglutination of red blood cells as an indicator according to a modified protocol described before [54]. Briefly, MDCK cells grown in round-bottom 96-well plates were inoculated with virus-containing samples diluted in serum-free DMEM containing 20 μg/ml trypsin using log10 dilutions. After 3 days incubation, a suspension of human erythrocytes (0.5%) was added to each well. Wells exhibiting hemagglutination were considered positive and virus titers were expressed as log $TCID_{50}$.

Protease activity assay

A modification to the above protocol was performed to test secreted protease activity. In these experiments, NECs were exposed to either 0.4ppm of ozone or air for 4 h. 24 h post-exposure, apical supernatants were collected and protein concentrations were determined by Bradford protein assay. 100 μg of protein within the apical supernatant, which contained the endogenous secreted proteases, was incubated with 80 μl of virus for 30 min at room temperature. MDCK cells grown in round-bottom 96-well plates were inoculated with the above samples diluted in serum-free DMEM or serum-free DMEM containing 20 μg/ml trypsin, as a positive control, using 10-fold dilutions. After 3 days incubation, a suspension of human erythrocytes (0.5%) was added to each well. Wells exhibiting hemagglutination were considered positive and virus titers were expressed as log $TCID_{50}$. 10 mM Phenylmethyl sulfonyl floride (PMSF) (Sigma-

Aldrich St. Louis Mo) was used to inhibit trypsin protease activity in control samples.

Protease inhibition assay

Recombinant human SLPI (rhSLPI) (R&D Systems Minneapolis MN) was added to serum-free DMEM containing 20 μg/ml trypsin and incubated for 60 min at room temperature prior to addition of 80 μl of virus. MDCK cells grown in round-bottom 96-well plates were inoculated with the above samples diluted in serum-free DMEM using 10-fold dilutions. After 3 days incubation, a suspension of human erythrocytes (0.5%) was added to each well. Wells exhibiting hemagglutination were considered positive and virus titers were expressed as log $TCID_{50}$. All experiments were performed in triplicate in three independent experiments.

RT-PCR

Total RNA was extracted using TRizol (Invitrogen) according to the supplier's instruction. First-strand cDNA synthesis and real-time RT-PCR was performed as described previously [46,53] using commercially available primers and probes for HA (inventoried Taqman Gene Expression Assays) purchased from Applied Biosystems (Foster City, CA).

Measurement of Lactate Dehydrogenase (LDH) and Interleukin-6 (IL-6) levels

NECs were exposed to filtered air or 0.4 ppm ozone for 4 h. 24 h post exposure, basolateral supernatants were collected for the measurement of LDH concentration according to the manufacturer's recommendations (Takara Bio Inc. Madison, WI). Released LDH concentration was expressed as optical density. Measurements of IL-6 concentration were determined according to the manufacturer's recommendations (BD Biosciences San Diego, CA). Released IL-6 concentration was expressed as pg/ml.

SLPI trans-activation reporter assay

To measure trans-activation of the SLPI promoter, 1.2×10^5 BEAS-2B were cotransfected with either 0.1 μg of SLPI-luciferase reporter plasmid (which was kindly provided by Dr. Rosalia Simmen, Arkansas Children's Hospital Research Institute) along with 0.02 μg of Thymidine kinase-*Renilla* luciferase by using Fugene 6 (Roche) according to the manufacturer's recommendations. After 24 h, cells were treated with 1 μM EGCG or DMSO. 8 h post-treatment cells were pelleted and lysed in passive lysis buffer (Promega, Madison, WI). The trans-activation activity was measured as luciferase light units as described previously [55]. All transfection experiments were performed in triplicate and normalized for transfection efficiency by using Renilla Luciferase.

Statistical Analysis

Unpaired Student's t-test and One-way ANOVA were used for determination of statistically significant differences. The use of the term significant in text refers to a comparison of values for which $p<0.05$.

Results

Ozone exposure enhances viral replication

Our group has recently demonstrated that oxidative stress increases susceptibility to influenza virus and that these responses may be mediated via increased viral entry [6,46]. Therefore, we wanted to evaluate how exposure to ozone affects viral entry and replication. We chose to examine influenza infection in human NECs exposed to 0.4ppm ozone for 4 h, based on our group's recent *in vivo* [12,56] and *in vitro* [9] studies. To assess the effects of ozone on NECs, we determined LDH and IL-6 levels in basolateral supernatants as markers of cytotoxicity and inflammation, respectively (Figure 1A and 1B). As expected, exposure to 0.4ppm of ozone for 4 h causes a significant increase in both inflammation and cytotoxicity [57]. Subsequently, 24 h post-exposure, cells were infected with influenza A/Bangkok/1/79 or a mock control. 24 h post-infection total RNA was subjected to real time RT-PCR to quantitate the influenza viral hemagglutinin transcript (HA) (Figure 1C). As depicted, we saw a significant increase in the amount of viral *HA* mRNA produced in the cells previously exposed to ozone as compared to the filtered air control. Similarly, by analyzing the apical washes for influenza viral titers 24 h post-infection, we saw a significant increase in viral titers in ozone exposed cells as compared to the control (Figure 1D). These results demonstrate that pre-exposure to ozone significantly increased viral replication in NECs.

Ozone exposure does not affect cellular antiviral responses

To determine the potential mechanism(s) mediating the enhancement of viral replication following acute ozone exposure, we assessed whether innate antiviral immune response mediators were modulated. Specifically, we analyzed the expression of interferon-β (IFN-β), interferon-α (IFN-α), retinoic acid inducible gene I (RIG-I), and Toll-Like recptor-3 (TLR-3). Human NECs were exposed to ozone and 24 h post-exposure were infected with influenza A. 24 h post-infection total RNA was isolated and RT-PCR was performed to quantitate the amount of each cellular transcript. Ozone alone had no significant effect on mRNA expression in any of the four antiviral genes as compared to the air control (Figure 2A–D). As expected, we found substantial inductions of these four antiviral mediators 24 h post-infection with influenza A, but without significant differences in ozone exposed compared to control cells. These data demonstrate that exposure to ozone does not alter baseline expression of host antiviral immune response genes, and does not interfere with their induction in response to influenza infection.

Antioxidant supplementation suppresses viral replication

Since we have shown that ozone exposure did not abrogate antiviral responses, we wanted to determine if the increase in influenza virus infection is due to the oxidative stress pathway/ mechanism induced by ozone exposure. To test this hypothesis, we treated our NECs with either a potent phase-II antioxidant inducer or a direct antioxidant. We have previously shown that oxidative stress increases susceptibility to influenza infection [6]. Previous reports have shown that *Nrf2* gene expression and protein expression can be induced via antioxidant supplementation [58,59,60], more specifically by the addition of the polyphenolic catechin, epigallocatechin-3-gallate (EGCG). This compound has the ability to induce *Nrf2* activation which in turn up-regulates the expression of multiple phase-II antioxidants [6,61,62]. Based on our previous published data [6,46], we determined that 1 μM of EGCG or 10 mM of GSH-ET is sufficient to detect an increase in Nrf2 expression or counteract oxidative stress induced by diesel exhaust exposure in NECs. These concentrations also correlate with levels of EGCG that are achievable *in vivo* [63,64,65]. NECs were treated with either 1 μM EGCG, 10 mM GSH-ET, or DMSO as a vehicle control for 30 min prior to a 4 h 0.4ppm ozone exposure. 24 h post-ozone exposure cells were infected with influenza A virus and total RNA was isolated 24 h post-infection to determine the level of influenza HA transcripts by RT-PCR. As shown in Figure 3A, exposure to ozone significantly increases viral

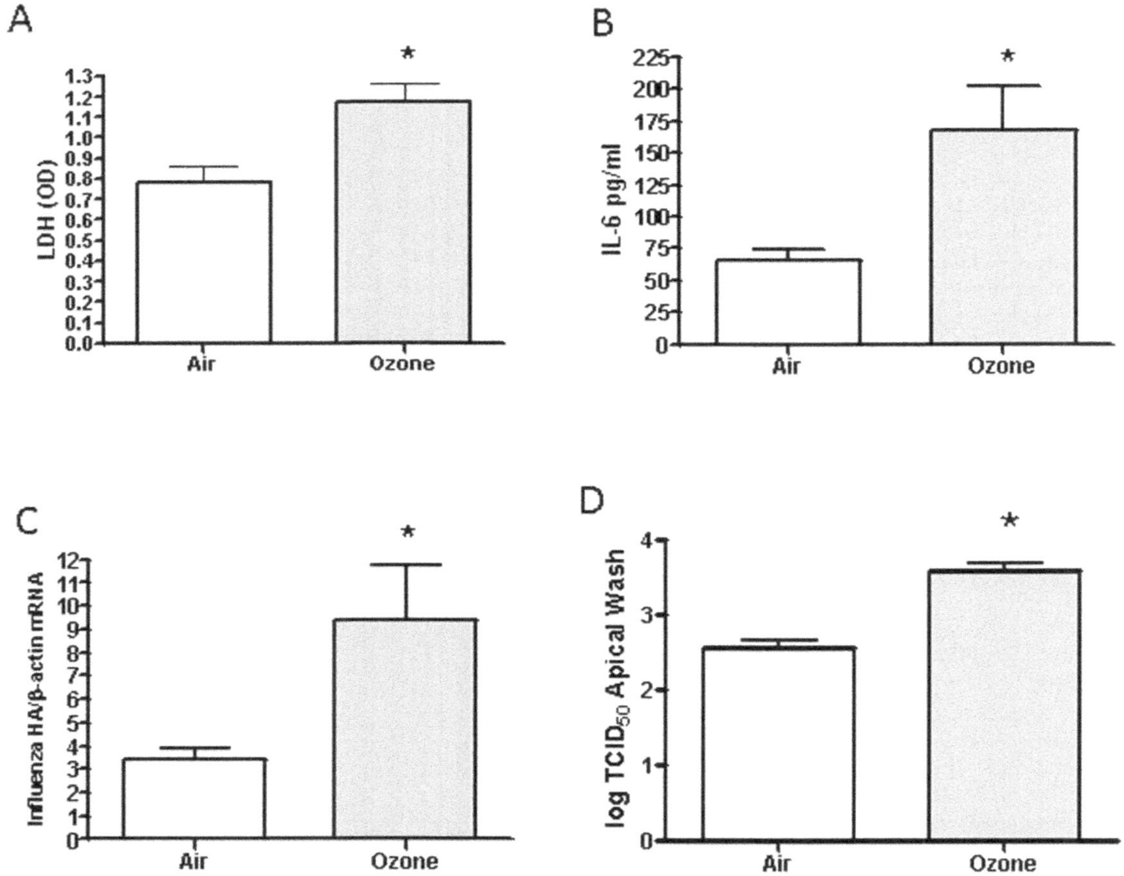

Figure 1. Exposure to ozone enhances influenza infection. Cultures of differentiated human epithelial cells were exposed to either 0.4ppm of ozone or air for 4 h. **A**) Basolateral supernatants collected 24 h post-exposure were analyzed for LDH levels. **B**) Basolateral supernatants collected 24 h post-exposure were analyzed for IL-6 levels. **C**) 24 h post-exposure, cells were infected with influenza A/Bangkok/1/79. Total RNA was isolated and subjected to RT-PCR to quantify Influenza *HA* transcripts and normalized to the expression of *β-actin*. **D**) 24 h post-exposure, cells were infected with influenza A/Bangkok/1/79. Apical supernatants collected 24 h post-infection were analyzed for determination of the viral titer indicated in log $TCID_{50}$. NECs were obtained from six healthy volunteers (n = 6), and each experiment was performed in triplicate. Asterisk indicates statistical significance between test sample and control, * $p<0.05$.

HA expression. Interestingly, the addition of either EGCG or GSH-ET significantly inhibited viral HA transcription as compared to the vehicle control. We again collected apical washes to perform viral titer assays. Our results, shown on a log scale, indicate that both EGCG and GSH-ET significantly reduced viral replication (Figure 3B). Taken together, these results demonstrate that antioxidants, either through Nrf2 activation (EGCG) or by direct addition of an antioxidant (GSH-ET), counteract the oxidative stress induced by ozone exposure and can suppress influenza virus replication in NECs.

Ozone exposure increases viral entry

Ozone not only triggers intracellular oxidative stress, but has also been shown to directly affect and modify the cellular lipid membrane [66,67] and transmembrane molecules [68]. Since the influenza virion must first bind and enter the target cell prior to replication, we hypothesize that the increased susceptibility to influenza after ozone exposure most likely happens early in the virus lifecycle, upstream of viral replication. To determine which step(s) in the virus life cycle are affected by ozone exposure, we employed an enzymatic virus-like particle (VLP) assay. This assay quantitatively measures the amount of virus that enters the cells. Similar to our previous work [6], cells were infected with a VLP that only express the hemagglutinin (HA), neuraminidase (NA), and matrix (M) proteins along with a functional β-lactamase reporter fusion (PR8 β-lacM1 VLP) prior to loading with the fluorogenic substrate CCF2-AM . In cells in which the PR8 β-lacM1 VLP has entered, the CCF2-AM substrate will be cleaved, disrupting FRET of the substrate, and resulting in increased CCF2 emission at 447 nm. To determine the effects of ozone exposure on virus entry, NECs and the cell line MDCK as a control were exposed to air or ozone and 24 h post-exposure, cells were infected with PR8 β-lacM1 VLP. In this experiment, MDCK cells are used as a control cell line which does not produce proteases required to activate the viral HA protein. Figure 4 shows that exposure to ozone results in a significant increase in viral entry in the NECs alone. In summary, these data demonstrate that acute exposure to ozone enhances viral entry, consistent with the increased viral replication shown in Figure 1C.

Ozone exposure disrupts host protease/antiproteases balance

As stated earlier, recent reports have identified two endogenous human trypsin-like serine proteases (TMPRSS2 and HAT) that

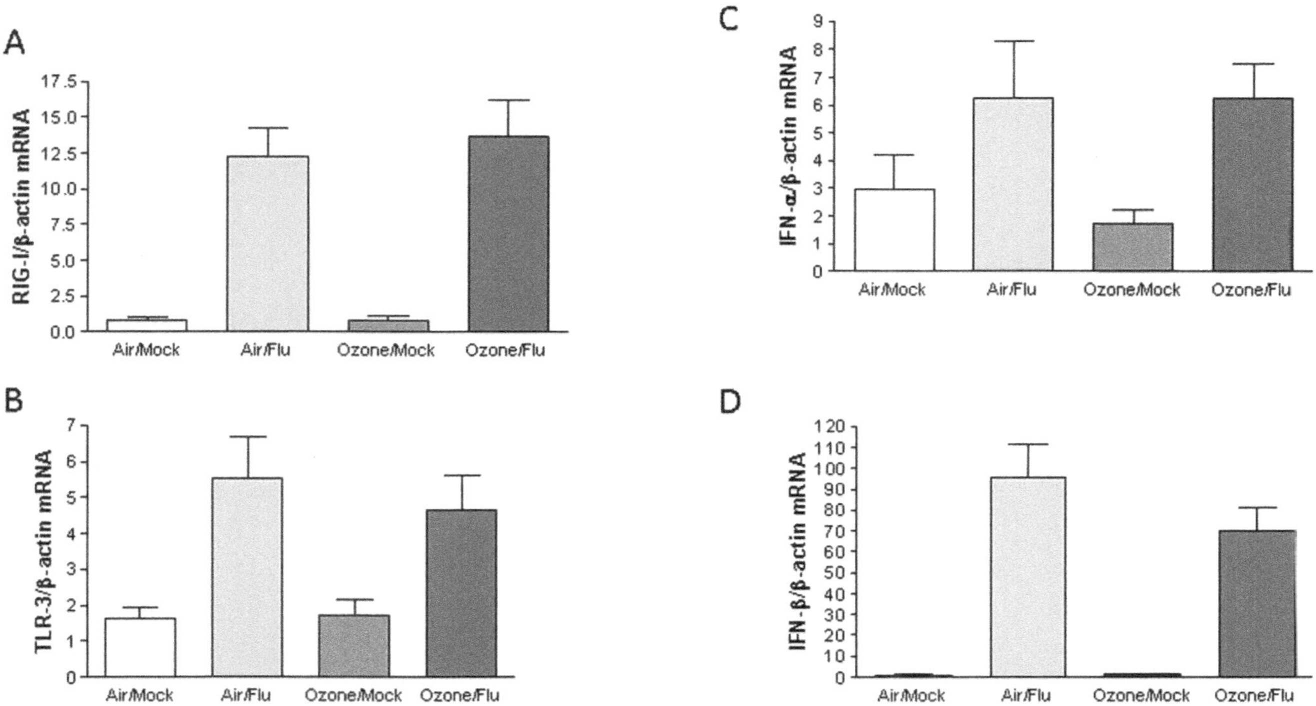

Figure 2. Exposure to ozone does not affect cellular antiviral responses. Cultures of differentiated human epithelial cells were exposed to either 0.4ppm of ozone or air for 4 h. 24 h post-exposure, cells were infected with influenza A/Bangkok/1/79. 24 h post-infection **A)** Total RNA was isolated and subjected to RT-PCR to quantify RIG-I transcripts and normalized to the expression of *β-actin*. **B)** TLR-3 transcripts and normalized to the expression of *β-actin*. **C)** IFN-α transcripts and normalized to the expression of *β-actin*. **D)** IFN-β transcripts and normalized to the expression of *β-actin*. NECs were obtained from six healthy volunteers (n = 6), and each experiment was performed in triplicate. Asterisk indicates statistical significance between test sample and control, * $p<0.05$.

possess the ability to cleave influenza virus *in vitro* [30,32,33,69]. Viral HA proteolytic cleavage is required for viral fusion and entry into the host cell. Proteases are regulated by antiproteases, such as SLPI and to a lesser extent A1AT. Studies have shown that a disruption of the protease/antiprotease balance is a hallmark of numerous lung diseases and pathologies including COPD, emphysema, asthma, and cancer [40,70,71,72]. Hennet et al, demonstrated a link between oxidative stress and protease expression which led to increased influenza infection in mice [39]. Linking this work with our current hypothesis, we investigated the protease/antiprotease balance on the epithelial surface in the context of an acute ozone exposure. For these experiments we exposed our NECs to either air or ozone for 4 h. 24 h post-exposure, both apical surface liquid and cell lysates were collected to characterize secreted/soluble and membrane-bound intracellular levels of HAT, TMPRSS2, and SLPI. Western blots revealed that intracellular levels of SLPI decreased slightly with ozone exposure, while HAT and TMPRSS2 expression remained relatively unchanged (Figure 5A). Although it is well known that SLPI is secreted and is present in normal human airway surface liquid, only recently was the transmembrane protease, HAT, found to be secreted in an *in vitro* overexpression system [33]. We found the levels of the secreted forms of HAT, TMPRSS2, and SLPI to be significantly changed by ozone exposure (Figure 5B). Under normal conditions, SLPI is in greater abundance than the proteases, yet 24 h post ozone exposure, we found a decrease in SLPI but a significant increase in both HAT and TMPRSS2. It is worth noting that the time at which these samples were taken corresponds to the exact time post ozone exposure when the cells were infected with influenza in the data shown in Figures 1, 2 and 3, thus representing the levels of protease and antiprotease present at the time of infection. To determine if disruption in the protease/antiprotease equilibrium is due to oxidative stress imposed by the ozone exposure, we treated our NECs with the potent Nrf2 inducer EGCG prior to ozone challenge. Figure 5C shows that EGCG did increase levels of SLPI which corresponds with a decrease in protease expression. Finally, Figure 5D demonstrate that the addition of 1 μM EGCG significantly increases the trans-activation of the SLPI promoter which correlates with the increase in SLPI protein expression displayed in Figure 5C. Taken together, these results indicate that ozone-induced oxidative stress disrupts the protease/antiprotease balance, in favor of protease activity. Furthermore, we demonstrate that the equilibrium can be restored with the induction of Nrf2 via antioxidant supplementation resulting in an increase in the antiprotease SLPI.

Secreted proteases are functional for hemagglutinin cleavage

We characterized the cleavage products of the viral HA protein to determine if the proteases produced by acute ozone exposure can cleave an intact influenza virion. We first investigated where HA cleavage takes place. NECs were exposed to air or ozone and 24 h post-exposure, cells were infected with Influenza A/Malaya/302/1954 H1N1. The cell lysates were analyzed by Western blotting for detection of the HA cleavage products. Figure 6A shows that the majority of the viral HA protein within the cell is cleaved and there is not a significant change with ozone exposure. In addition, Western blot analysis of apical washings incubated with virus showed that HA0 was cleaved and that there was an

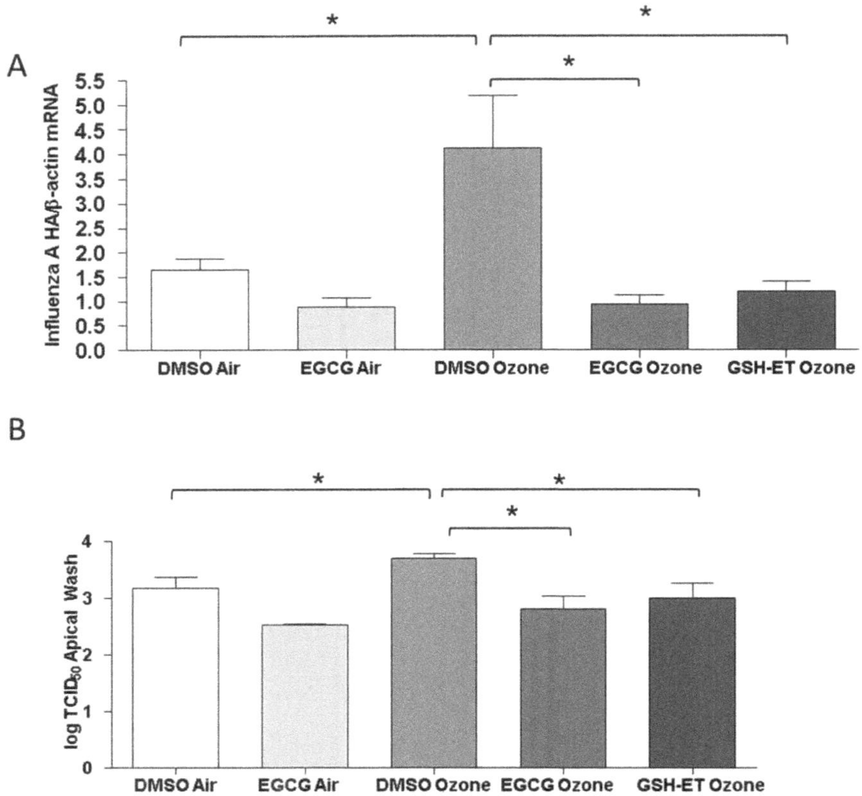

Figure 3. Antioxidant supplementation suppresses viral replication. Cultures of differentiated human epithelial cells were treated with EGCG, GSH-ET, or DMSO as a vehicle control for 30 min before exposure to either 0.4ppm of ozone or air for 4 h. **A**) 24 h post-exposure, cells were infected with influenza A/Bangkok/1/79. Total RNA was isolated and subjected to RT-PCR to quantify Influenza *HA* transcripts and normalized to the expression of *β-actin*. **B**) 24 h post-exposure, cells were infected with influenza A/Bangkok/1/79. Apical supernatants collected 24 h post-infection were analyzed for determination of the viral titer indicated in log $TCID_{50}$. NECs were obtained from 6 healthy volunteers (n = 6), and each experiment was performed in triplicate. Asterisk indicates statistical significance between test sample and control, * $p<0.05$.

increase in the cleavage of HA from the ozone exposed cells as measured by densitometry. Quantification of the H2 cleavage products demonstrated that ozone exposure leads to enhanced

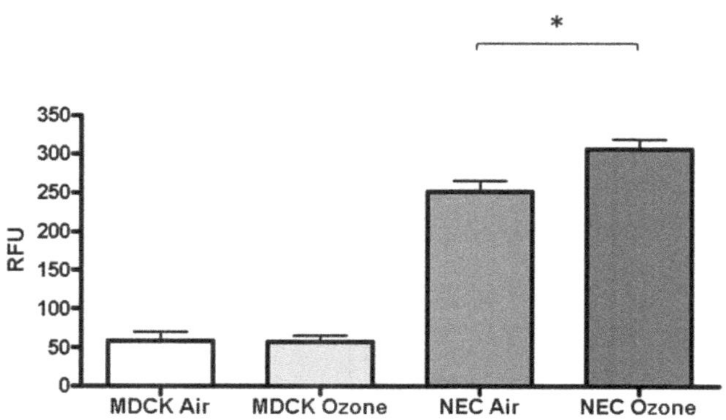

Figure 4. Ozone exposure increases viral entry. Cultures of differentiated human epithelial cells and MDCK cells used as controls, were exposed to either 0.4ppm of ozone or air for 4 h. 24 h post ozone exposure cells were infected with PR8 β-lacM1 VLP. At 4 h post-infection, cells were loaded with CCF2-AM substrate and assayed for viral entry. Data from a representative experiment performed in triplicate. Asterisk indicates statistical significance between test sample and control, * $p<0.05$.

proteolytic activation of the virion. This data demonstrate that HA cleavage can take place in both the intra- and extracellular environment and that ozone exposure increases the proteolytic activation of the virus in the airway (Figure 6B). To characterize the function of TMPRSS2 and HAT in HA cleavage, we depleted the proteases via immunoprecipitation prior to addition of the virus. Figure 6C shows the apical washes from NECs that were immunoprecitipated with anti-HAT, anti-TMPRSS2, and IgG as a control. The cleared supernatants were incubated with Influenza A/Malaya/302/1954 H1N1. As shown in Figure 6D, there is little difference in the cleavage of the HA protein when the protease are depleted separately, but we demonstrate a significant decrease in the cleavage of HA when both proteases were removed. This data show that the protease(s) found in the apical surface liquid are able to cleave an intact influenza A virion. In addition we show that there are additional proteases secreted into the airway liquid space that are able to activate the virus.

Secreted proteases are functional and can activate Influenza A virions to propagate a viral infection

Although we show that SLPI, HAT, and TMPRSS2 are secreted by NECs, the question remains as to whether they are functional. To test functionality and the role they play in a viral infection, we employed a modified infectivity viral titer assay. These experiments are very similar to the viral titer assays mentioned earlier, but with a few modifications. Similar to the

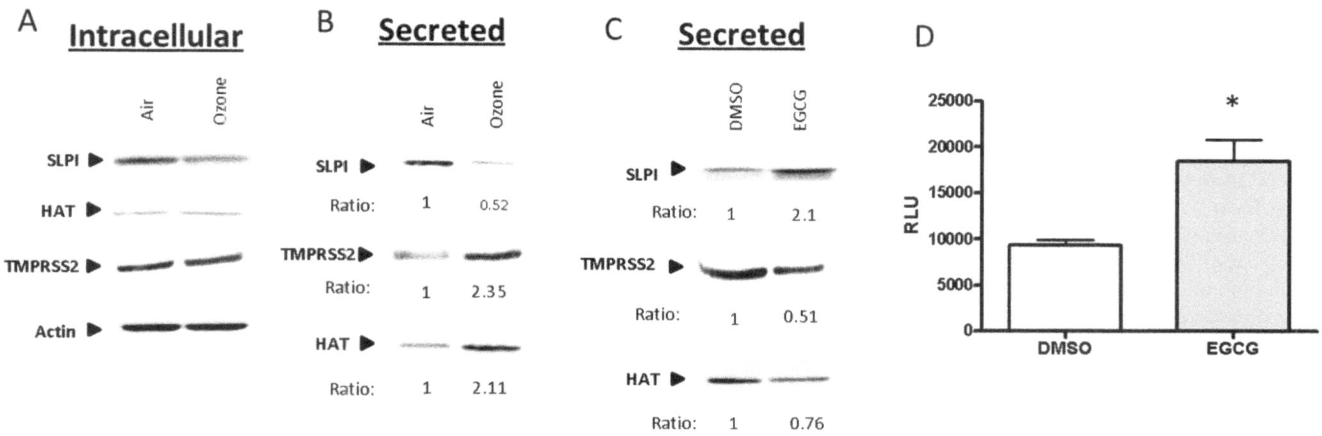

Figure 5. Ozone exposure modulates cellular protease and anti-protease levels. Cultures of differentiated human epithelial cells were exposed to either 0.4ppm of ozone or air for 4 h. 24 h post-exposure, samples were collected and subjected to Western blot analysis to detect SLPI, HAT, TMPRSS2, and β-actin. **A**) Total cellular lysates were analyzed for intracellular SLPI, TMPRSS2, and HAT protein expression by Western blot. Membrane was stripped and analyzed for β-actin as a loading control. **B**) 24 h post-exposure, apical supernatants were analyzed for secreted SLPI, TMPRSS2, and HAT protein levels. **C**) Cultures of differentiated human epithelial cells were treated with EGCG or DMSO as a vehicle control for 30 min before exposure to either 0.4ppm of ozone or air for 4 h. 24 h post-exposure, apical supernatants were analyzed for secreted SLPI, TMPRSS2, and HAT protein levels. **D**) Function activity of the SLPI promoter was determined using a dual-luciferase reporter assay. BEAS-2B cells were co-transfected with SLPI-Luc and TK-Renilla. Cells were treated with EGCG or DMSO as a vehicle control. Cells were harvested 8 h post-treatment and analyzed for luciferase expression. The values represent luciferase production from a representative experiment performed in triplicate. NECs were obtained from four healthy volunteers (n = 4). Densitometry was used to quantitate the amounts of protein, and the numbers below the gel indicate the ozone/air or EGCG/DMSO sample ratio.

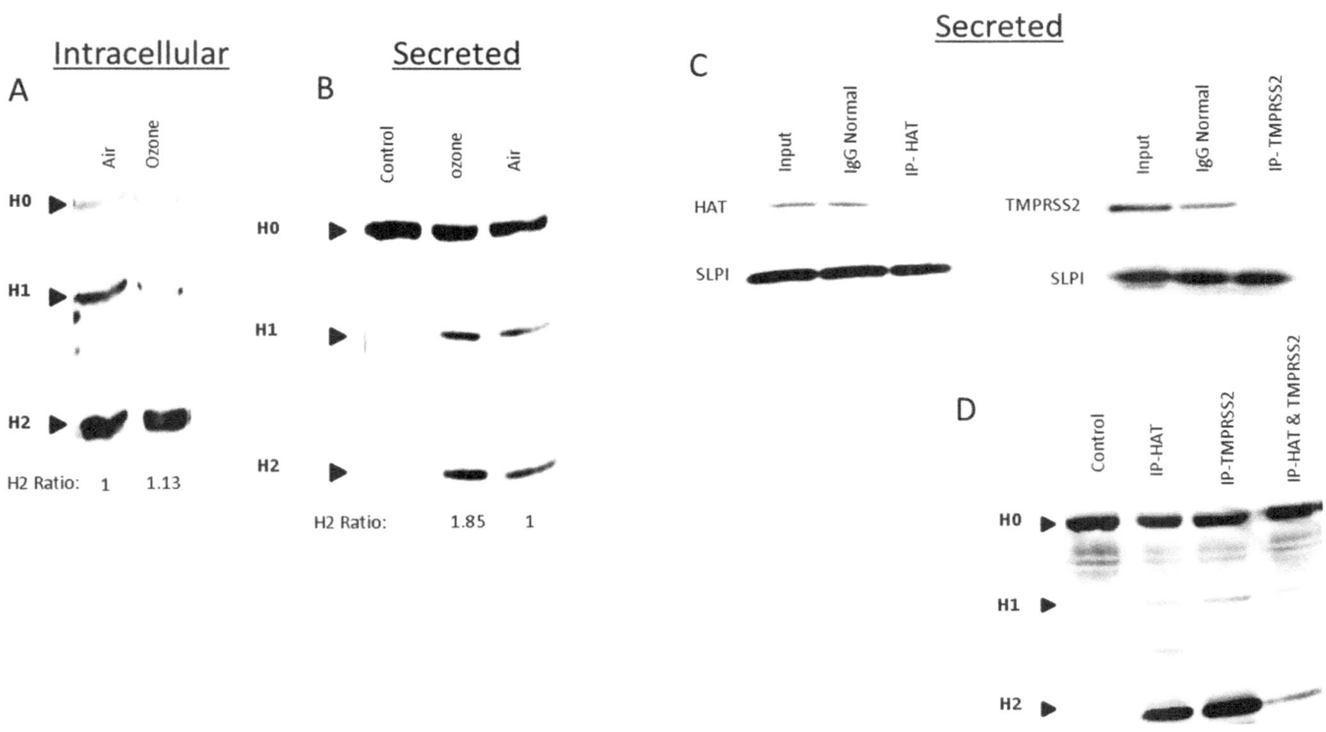

Figure 6. Secreted proteases are functional for Hemagglutinin cleavage. Cultures of differentiated human epithelial cells were exposed to either 0.4ppm of ozone or air for 4 h. A) 24 h post-exposure, cells were infected with influenza A/Malaya/302/1954. 24 h post-infection, total cellular lysates were analyzed for intracellular cleavage of the viral HA protein by Western blot. B) 24 h post-exposure, apical supernatants were collected and incubated with influenza A/Malaya/302/1954. Cleavage of the viral HA protein was analyzed by Western blot. C) Cultures of differentiated human epithelial cells were exposed to either 0.4ppm of ozone or air for 4 h. 24 h post-exposure, apical supernatants were immunoprecipitated (IP) with anti- HAT, anti- TMPRSS2, or an isotype control (IgG). Protein levels were analyzed by Western blot. D) Apical supernatants of differentiated human epithelial cells, were immunoprecipitated (IP) with anti- HAT, anti- TMPRSS2, or an isotype control (IgG) followed by incubation with influenza A/ Malaya/302/1954. Cleavage of the viral HA protein was analyzed by Western blot. NECs were obtained from four healthy volunteers (n = 4). Densitometry was used to quantitate the amounts of cleaved H2 protein, and the numbers below the gel indicate the ozone/air ratio.

assays before, we utilized the well-characterized MDCK cell line which does not produce the proper proteases to activate the influenza virion, which can be overcome by the addition of exogenous trypsin. In contrast to the previous titer experiments above, we did not add exogenous trypsin to enable multicycle replication of influenza virus in these cells. These experiments allowed us to test if secreted proteases present in the apical surface liquid from NECs exposed to air or ozone were able to facilitate multiple rounds of viral replication in MDCKs. We show that the secreted proteases present in the apical supernatants from NECs are functional and can propagate viral entry and replication. We also demonstrate that the ozone-exposed NEC supernatants displayed significantly greater infection kinetics than the air control (Figure 7A). This correlates with the increase in protease expression post ozone exposure (Figure 5B), viral entry (Figure 4), and viral replication (Figure 1C and 1D). In addition we show that this is a protease-mediated event, since the addition of a known serine protease inhibitor, PMSF, significantly inhibits viral replication.

To demonstrate that SLPI has the ability to protect airway epithelium from influenza infection by inhibiting proteolytic cleavage, we incubated various amounts of recombinant human SLPI (rhSLPI) to our traditional viral titer assay. Briefly, we incubated our viral titer media, serum-free DMEM containing 20 μg/ml of exogenous trypsin, with varying amounts of rhSLPI prior to addition of influenza. As shown in Figure 7B, addition of rhSLPI significantly inhibits multicycle viral replication. We conclude that the addition of rhSLPI directly inhibits trypsin-mediated cleavage of viral HA leading to decreased infection and replication. Taken together, these data demonstrate that exposure to ozone disrupts the protease/antiprotease balance resulting in greater secretion of endogenous proteases leading to increased viral cleavage/activation culminating in enhanced viral entry and replication.

Discussion

Numerous epidemiological studies link increased ambient ozone levels with respiratory viral infections [1,73,74]. However, the role of ozone exposure in viral replication and pathogenesis remains to be fully defined. Because ambient ozone is a major source of oxidative stress in the airway epithelium and this is the primary site for influenza infections, it is important to determine the role ozone exposure plays in viral infections. Although it has been shown that ozone exposure can modulate many aspects of the immune response [8], how these alterations influence a viral infection have yet to be elucidated. Recent studies have demonstrated the importance of the protease/antiprotease balance in the context of normal lung homeostasis, and have shown that oxidative stress can disrupt this delicate equilibrium [40,75]. It has been well documented that many lung pathologies are dependent on the regulatory interplay between oxidative stress and protease expression [40,70,71]. Although there is speculation that Mexico City's increased influenza-related morbidity and mortality rates were due to exposures to higher levels of ozone, little research has been done looking at the effects of pre-exposure to ozone in the context of an influenza infection in humans. Herein, we demonstrate that acute ozone exposure disrupts the protease/antiprotease balance resulting in increased secreted protease expression in primary human epithelial cells, resulting in increased

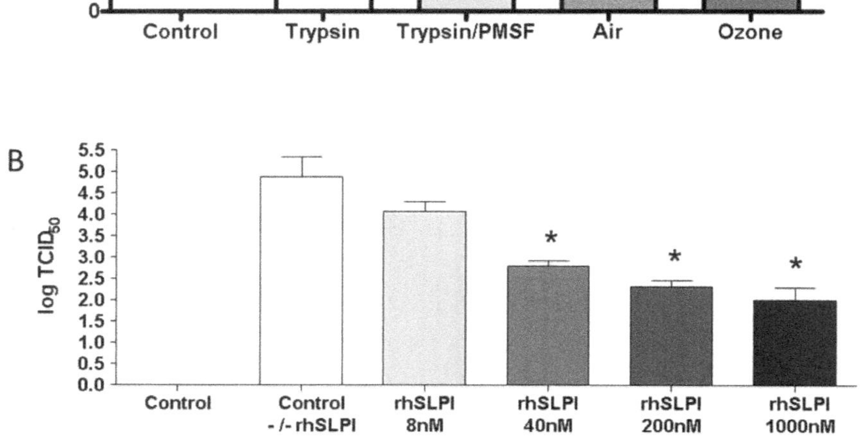

Figure 7. Secreted proteases are functional and can activate virions to propagate a viral infection. **A**) Cultures of differentiated human epithelial cells were exposed to either 0.4ppm of ozone or air for 4 h. 24 h post-exposure apical supernatants were incubated with influenza A/Bangkok/1/79 for 30 min, and then analyzed for determination of the viral titer indicated in log $TCID_{50}$. **B**) Increasing amounts of rhSLPI (8 nM, 40 nM, 200 nM, 1000 nM) were incubated with media containing 20 μg/ml trypsin for 60 min, then analyzed for determination of the viral titer indicated in log $TCID_{50}$. NECs were obtained from four healthy volunteers (n = 4), and each experiment was performed in triplicate. Data from a representative experiment performed in triplicate. Asterisk indicates statistical significance between test sample and control, * $p<0.05$.

HA cleavage and activation of the influenza virion, which correlated with enhanced viral entry and replication.

Numerous host cell-dependent factors can affect and control influenza virus attachment and uptake by: (i) proteolytic cleavage of viral HA by host cell-derived serine proteases [30,33], (ii) host cell derived innate immune defense molecules aimed at inhibiting the infectious virions [25,76], and (iii) antiviral mediators limiting viral replication and shedding of virus particles [6,52]. We first focused on potential effects of ozone on host antiviral defense responses. Our data showed that exposure to ozone elicited the typical pro-inflammatory and cellular damage responses as seen with increased release of IL-6 and LDH. We found that the same exposure regimen also resulted in significant increase in viral replication. We initially hypothesized that this could be due to a disruption in one of the cellular antiviral response mechanism/ pathways, and assessed the transcription levels of 4 classical antiviral mediators; RIG-I, IFN-α, IFN-β, and TLR-3. Similar to our previous studies using exposure to diesel exhaust [46], we found that ozone did not decrease expression of these mediators or their response when challenged with influenza. In our previous study [46], we found that exposure to an oxidant pollutant increased influenza virus attachment and/or entry, but did not distinguish between these two steps in the viral infection cycle. Therefore, we focused our attention on steps upstream of viral replication to identify mechanisms by which oxidants, such as ozone, affect influenza infectivity.

To dissect specific points in the virus life cycle upstream of viral replication that could determine the role ozone exposure plays in viral susceptibility, which ultimately dictates viral pathogenesis and outcome, we utilized our previously described enzymatic virus-like particle (VLP) assay [6]. Our data demonstrated that ozone exposure significantly increased influenza virus entry. This is consistent with our previous work showing that exposure to other oxidant pollutants and suppression of Nrf2 increases influenza virus entry [6,46]. Similar to these previous studies, data shown here demonstrated that the mechanism(s) through which ozone exposure alters influenza virus entry is mediated by oxidative stress. Since ozone directly interacts with the apical surface of the respiratory epithelium we hypothesize that ozone either directly, or via the induction of biologically active ozone reaction products, modifies either the expression or function of surface proteins, as previously been demonstrated for human surfactant protein A (SP-A) [77]. Among the most prominent proteins on the apical surface potentially regulating infection are the proteases. To date, at least five different proteases have been identified in the airways of animals and humans [73]. Despite many years of effort, the exact protease(s) that activates influenza has yet to be identified. Numerous groups have reported that the family of type II transmembrane serine proteases and trypsin-like proteases are responsible for viral cleavage [26,30,33,79]. Previous studies have examined these proteases in the context of influenza infections, specifically TMPRSS2 and HAT. Although these studies demonstrated that these proteases are capable of cleavage, the experiments were performed in cell lines such as MDCKs [33] or the colorectal adenocarcinoma cell line Caco-2 [32], neither of which are natural target cells for influenza. Since the cellular protease determines the tropism as well as efficiency of viral replication, we utilized fully differentiated human primary nasal epithelial cells (NECs) in this study, a natural host cell for influenza virus. We show that HAT and TMPRSS2 are not only present, but secreted from NECs (Figure 5B), and that ozone exposure enhances the release of the proteases into the apical compartment. This is consistent with reports showing that oxidative stress increases cellular protease activity [38]. Since proteases in the upper airway are inhibited by SLPI [78], we focused on this antiprotease and its expression in relation to HAT and TMPRSS2. Similar to previous reports, we report that SLPI is detected in the secreted airway and oxidative stress reduces the expression and function of the protein (Figure 5B) [35,80]. These data are consistent with our hypothesis that influenza virions can become activated in the airway surface liquid prior to binding to the epithelial surface. Virion activation in the airway prior to binding would eliminate the need for a membrane-bound protease thus broadening viral tropism from airway epithelium, that contain the necessary proteases, to any mammalian cell, since *2,6*-linked sialic acid is expressed by most mammalian cells ranging from lymphoid to glial cells [81,82].This could explain the increased morbidity and mortality associated with the Mexico City 2009 H1N1 pandemic which resulted in increased infection of multiple cell types including distal airway epithelium and immune cells including alveolar macrophages [83,84] as compared to previous seasonal outbreaks.

EGCG has been shown to have potent antioxidant capabilities; in part by inducing the expression of a number of antioxidant enzymes [85]. *In vitro* and *in vivo* studies have shown that this supplement induces the expression of phase II antioxidant genes such as GSTM1, which was associated with Nrf2-ARE signaling [58,86]. Whether and how activation of antioxidants expression is involved in the potential antiviral effects of EGCG is not known. Previous studies have indicated that it directly binds influenza virus and therefore prevents attachment and entry into host cells [63,87]. However, these studies were conducted in MDCK cells, which are not a natural host cell for influenza and require addition of exogenous proteases to achieve viral entry [63]. Iizuka et al, demonstrated that Nrf2-knockout mice display protease/antiprotease imbalance leading to increased susceptibility to cigarette smoke-induced emphysema [37]. In addition, the same group showed that Nrf2 expression exerts a protective effect through the transcriptional activation of antiproteases [88]. Our data demonstrate that supplementation of differentiated nasal epithelial cells with EGCG from the basolateral side (to eliminate direct interaction with the virus during infection) significantly increases SLPI production (Figure 5C). It is worthy to note that we show a direct inverse relationship between the levels of secreted SLPI and secreted protease expression, suggesting a potential novel antiviral mechanism by which EGCG could be protective against influenza infections.

We next determined that the secreted proteases are functional. We demonstrate that proteases present in the apical surface liquid were able to activate the virus (Figure 6B) and produce multiple rounds of viral replication (Figure 7A). In addition, apical supernatants from ozone-exposed epithelial cells had significantly higher viral titers, which correlate with the increased protease expression displayed in Figure 5B. In the final experiment we utilized rhSLPI and show a dose dependent protective response (Figure 7B). This demonstrates that that the antiprotease SLPI has a protective affect against viral activation and replication.

In conclusion, this is the first study to demonstrate that secreted proteases from primary human respiratory nasal epithelium proteolytically activate influenza virions and that exposure to a common ambient air oxidant pollutant increases these effects. We speculate that individuals with increased airway oxidative stress, whether due to an underlying medical condition or exposure to high levels of air pollutants, would develop disruption of the epithelial protease/antiprotease balance and become more susceptible to infections with viruses like influenza, SARS-CoV, and metapneumovirus. Disruption in the protease/antiprotease balance has been described in several respiratory diseases including COPD, emphysema, and asthma [70,71,89]. This mechanism could be exploited as a novel anti-viral therapy in combination with conventional antiviral treatments. Antiproteases, such as SLPI appear to protect the respiratory epithelium from influenza virus infection and nutritional supplements may increase the antiprotease:protease ratio. Taken together, the data shown here provide further support to the concept that targeting HA-activating proteases and decreasing oxidative stress to arrest viral infection and spread in conjunction with antiviral agents is a promising potential therapeutic intervention approach against influenza.

Acknowledgments

We thank Ms. Luisa E. Brighton for her expert technical assistance and Dr. Philip Bromberg, M.D., for his critical review of the manuscript.

Author Contributions

Conceived and designed the experiments: MK MM IJ. Performed the experiments: MK MM RB. Analyzed the data: MK MM RB IJ. Wrote the paper: MK.

References

1. Chowell G, Echevarria-Zuno S, Viboud C, Simonsen L, Tamerius J, et al. (2011) Characterizing the epidemiology of the 2009 influenza A/H1N1 pandemic in Mexico. PLoS medicine 8: e1000436.
2. Monto AS (2004) Occurrence of respiratory virus: time, place and person. The Pediatric infectious disease journal 23: S58–64.
3. Thompson WW, Shay DK, Weintraub E, Brammer L, Cox N, et al. (2003) Mortality associated with influenza and respiratory syncytial virus in the United States. JAMA : the journal of the American Medical Association 289: 179–186.
4. Lambert LC, Fauci AS (2010) Influenza vaccines for the future. The New England journal of medicine 363: 2036–2044.
5. Thompson WW, Shay DK, Weintraub E, Brammer L, Bridges CB, et al. (2004) Influenza-associated hospitalizations in the United States. JAMA : the journal of the American Medical Association 292: 1333–1340.
6. Kesic MJ, Simmons SO, Bauer R, Jaspers I (2011) Nrf2 expression modifies influenza A entry and replication in nasal epithelial cells. Free radical biology & medicine 51: 444–453.
7. Cho HY, Imani F, Miller-DeGraff L, Walters D, Melendi GA, et al. (2009) Antiviral activity of Nrf2 in a murine model of respiratory syncytial virus disease. American journal of respiratory and critical care medicine 179: 138–150.
8. Jakab GJ, Spannhake EW, Canning BJ, Kleeberger SR, Gilmour MI (1995) The effects of ozone on immune function. Environmental health perspectives 103 Suppl 2: 77–89.
9. Wu W, Doreswamy V, Diaz-Sanchez D, Samet JM, Kesic M, et al. (2011) GSTM1 modulation of IL-8 expression in human bronchial epithelial cells exposed to ozone. Free radical biology & medicine 51: 522–529.
10. Jakab GJ, Bassett DJ (1990) Influenza virus infection, ozone exposure, and fibrogenesis. The American review of respiratory disease 141: 1307–1315.
11. Razani-Boroujerdi S, Singh SP, Knall C, Hahn FF, Pena-Philippides JC, et al. (2004) Chronic nicotine inhibits inflammation and promotes influenza infection. Cellular immunology 230: 1–9.
12. Hernandez ML, Lay JC, Harris B, Esther CR, Jr., Brickey WJ, et al. (2010) Atopic asthmatic subjects but not atopic subjects without asthma have enhanced inflammatory response to ozone. The Journal of allergy and clinical immunology 126: 537–544 e531.
13. Garcia-Suastegui WA, Huerta-Chagoya A, Carrasco-Colin KL, Pratt MM, John K, et al. (2011) Seasonal variations in the levels of PAH-DNA adducts in young adults living in Mexico City. Mutagenesis 26: 385–391.
14. O'Neill MS, Loomis D, Borja Aburto VH, Gold D, Hertz-Picciotto I, et al. (2004) Do associations between airborne particles and daily mortality in Mexico City differ by measurement method, region, or modeling strategy? Journal of exposure analysis and environmental epidemiology 14: 429–439.
15. Borja-Aburto VH, Castillejos M, Gold DR, Bierzwinski S, Loomis D (1998) Mortality and ambient fine particles in southwest Mexico City, 1993–1995. Environmental health perspectives 106: 849–855.
16. Dawood FS, Jain S, Finelli L, Shaw MW, Lindstrom S, et al. (2009) Emergence of a novel swine-origin influenza A (H1N1) virus in humans. The New England journal of medicine 360: 2605–2615.
17. Fraser C, Donnelly CA, Cauchemez S, Hanage WP, Van Kerkhove MD, et al. (2009) Pandemic potential of a strain of influenza A (H1N1): early findings. Science 324: 1557–1561.
18. Cannell JJ, Zasloff M, Garland CF, Scragg R, Giovannucci E (2008) On the epidemiology of influenza. Virology journal 5: 29.
19. Gold DR, Damokosh AI, Pope CA, III, Dockery DW, McDonnell WF, et al. (1999) Particulate and ozone pollutant effects on the respiratory function of children in southwest Mexico City. Epidemiology 10: 8–16.
20. Torres-Jardon R, Garcia-Reynoso JA, Jazcilevich A, Ruiz-Suarez LG, Keener TC (2009) Assessment of the ozone-nitrogen oxide-volatile organic compound sensitivity of Mexico City through an indicator-based approach: measurements and numerical simulations comparison. Journal of the Air & Waste Management Association 59: 1155–1172.
21. Ciriolo MR, Palamara AT, Incerpi S, Lafavia E, Bue MC, et al. (1997) Loss of GSH, oxidative stress, and decrease of intracellular pH as sequential steps in viral infection. The Journal of biological chemistry 272: 2700–2708.
22. Schwarz KB (1996) Oxidative stress during viral infection: a review. Free radical biology & medicine 21: 641–649.
23. Cho HY, Kleeberger SR (2010) Nrf2 protects against airway disorders. Toxicology and applied pharmacology 244: 43–56.
24. Keller JN (2006) The many nuances of oxidative stress and proteolysis. Antioxidants & redox signaling 8: 119–120.
25. Kido H, Beppu Y, Imamura Y, Chen Y, Murakami M, et al. (1999) The human mucus protease inhibitor and its mutants are novel defensive compounds against infection with influenza A and Sendai viruses. Biopolymers 51: 79–86.
26. Kido H, Yokogoshi Y, Sakai K, Tashiro M, Kishino Y, et al. (1992) Isolation and characterization of a novel trypsin-like protease found in rat bronchiolar epithelial Clara cells. A possible activator of the viral fusion glycoprotein. The Journal of biological chemistry 267: 13573–13579.
27. Sakai K, Kawaguchi Y, Kishino Y, Kido H (1993) Electron immunohistochemical localization in rat bronchiolar epithelial cells of tryptase Clara, which determines the pneumotropism and pathogenicity of Sendai virus and influenza virus. The journal of histochemistry and cytochemistry : official journal of the Histochemistry Society 41: 89–93.
28. Ying QL, Simon SR (2000) DNA from bronchial secretions modulates elastase inhibition by alpha(1)-proteinase inhibitor and oxidized secretory leukoprotease inhibitor. American journal of respiratory cell and molecular biology 23: 506–513.
29. Cavarra E, Bartalesi B, Lucattelli M, Fineschi S, Lunghi B, et al. (2001) Effects of cigarette smoke in mice with different levels of alpha(1)-proteinase inhibitor and sensitivity to oxidants. American journal of respiratory and critical care medicine 164: 886–890.
30. Bottcher E, Matrosovich T, Beyerle M, Klenk HD, Garten W, et al. (2006) Proteolytic activation of influenza viruses by serine proteases TMPRSS2 and HAT from human airway epithelium. Journal of virology 80: 9896–9898.
31. Yasuoka S, Ohnishi T, Kawano S, Tsuchihashi S, Ogawara M, et al. (1997) Purification, characterization, and localization of a novel trypsin-like protease found in the human airway. American journal of respiratory cell and molecular biology 16: 300–308.
32. Bertram S, Glowacka I, Blazejewska P, Soilleux E, Allen P, et al. (2010) TMPRSS2 and TMPRSS4 facilitate trypsin-independent spread of influenza virus in Caco-2 cells. Journal of virology 84: 10016–10025.
33. Bottcher-Friebertshauser E, Freuer C, Sielaff F, Schmidt S, Eickmann M, et al. (2010) Cleavage of influenza virus hemagglutinin by airway proteases TMPRSS2 and HAT differs in subcellular localization and susceptibility to protease inhibitors. Journal of virology 84: 5605–5614.
34. Bottcher-Friebertshauser E, Stein DA, Klenk HD, Garten W (2011) Inhibition of influenza virus infection in human airway cell cultures by an antisense peptide-conjugated morpholino oligomer targeting the hemagglutinin-activating protease TMPRSS2. Journal of virology 85: 1554–1562.
35. Cavarra E, Lucattelli M, Gambelli F, Bartalesi B, Fineschi S, et al. (2001) Human SLPI inactivation after cigarette smoke exposure in a new in vivo model of pulmonary oxidative stress. American journal of physiology Lung cellular and molecular physiology 281: L412–417.
36. Szabo R, Wu Q, Dickson RB, Netzel-Arnett S, Antalis TM, et al. (2003) Type II transmembrane serine proteases. Thrombosis and haemostasis 90: 185–193.
37. Iizuka T, Ishii Y, Itoh K, Kiwamoto T, Kimura T, et al. (2005) Nrf2-deficient mice are highly susceptible to cigarette smoke-induced emphysema. Genes to cells : devoted to molecular & cellular mechanisms 10: 1113–1125.
38. Vogelmeier C, Biedermann T, Maier K, Mazur G, Behr J, et al. (1997) Comparative loss of activity of recombinant secretory leukoprotease inhibitor and alpha 1-protease inhibitor caused by different forms of oxidative stress. The European respiratory journal : official journal of the European Society for Clinical Respiratory Physiology 10: 2114–2119.
39. Hennet T, Peterhans E, Stocker R (1992) Alterations in antioxidant defences in lung and liver of mice infected with influenza A virus. The Journal of general virology 73(Pt 1): 39–46.
40. Skrzydlewska E, Farbiszewski R, Gacko M (1997) [The effect of protein oxidation modification on protease-antiprotease balance and intracellular proteolysis]. Postepy higieny i medycyny doswiadczalnej 51: 443–456.
41. Shulla A, Heald-Sargent T, Subramanya G, Zhao J, Perlman S, et al. (2011) A transmembrane serine protease is linked to the severe acute respiratory syndrome coronavirus receptor and activates virus entry. Journal of virology 85: 873–882.

42. Matsuyama S, Nagata N, Shirato K, Kawase M, Takeda M, et al. (2010) Efficient activation of the severe acute respiratory syndrome coronavirus spike protein by the transmembrane protease TMPRSS2. Journal of virology 84: 12658–12664.
43. Chandran K, Sullivan NJ, Felbor U, Whelan SP, Cunningham JM (2005) Endosomal proteolysis of the Ebola virus glycoprotein is necessary for infection. Science 308: 1643–1645.
44. Matsuyama S, Taguchi F (2009) Two-step conformational changes in a coronavirus envelope glycoprotein mediated by receptor binding and proteolysis. Journal of virology 83: 11133–11141.
45. Simmons G, Reeves JD, Rennekamp AJ, Amberg SM, Piefer AJ, et al. (2004) Characterization of severe acute respiratory syndrome-associated coronavirus (SARS-CoV) spike glycoprotein-mediated viral entry. Proceedings of the National Academy of Sciences of the United States of America 101: 4240–4245.
46. Jaspers I, Ciencewicki JM, Zhang W, Brighton LE, Carson JL, et al. (2005) Diesel exhaust enhances influenza virus infections in respiratory epithelial cells. Toxicological sciences : an official journal of the Society of Toxicology 85: 990–1002.
47. Reddel RR, Ke Y, Gerwin BI, McMenamin MG, Lechner JF, et al. (1988) Transformation of human bronchial epithelial cells by infection with SV40 or adenovirus-12 SV40 hybrid virus, or transfection via strontium phosphate coprecipitation with a plasmid containing SV40 early region genes. Cancer research 48: 1904–1909.
48. Niwa H, Yamamura K, Miyazaki J (1991) Efficient selection for high-expression transfectants with a novel eukaryotic vector. Gene 108: 193–199.
49. Quinlivan M, Zamarin D, Garcia-Sastre A, Cullinane A, Chambers T, et al. (2005) Attenuation of equine influenza viruses through truncations of the NS1 protein. Journal of virology 79: 8431–8439.
50. Neumann G, Watanabe T, Ito H, Watanabe S, Goto H, et al. (1999) Generation of influenza A viruses entirely from cloned cDNAs. Proceedings of the National Academy of Sciences of the United States of America 96: 9345–9350.
51. Manicassamy B, Rong L (2009) Expression of Ebolavirus glycoprotein on the target cells enhances viral entry. Virology journal 6: 75.
52. Marsh GA, Hatami R, Palese P (2007) Specific residues of the influenza A virus hemagglutinin viral RNA are important for efficient packaging into budding virions. Journal of virology 81: 9727–9736.
53. Jaspers I, Zhang W, Brighton LE, Carson JL, Styblo M, et al. (2007) Selenium deficiency alters epithelial cell morphology and responses to influenza. Free radical biology & medicine 42: 1826–1837.
54. Farag-Mahmod FI, Wyde PR, Rosborough JP, Six HR (1988) Immunogenicity and efficacy of orally administered inactivated influenza virus vaccine in mice. Vaccine 6: 262–268.
55. Reed KL, Badinga L, Davis DL, Chung TE, Simmen RC (1996) Porcine endometrial glandular epithelial cells in vitro: transcriptional activities of the pregnancy-associated genes encoding antileukoproteinase and uteroferrin. Biology of reproduction 55: 469–477.
56. Hernandez ML, Harris B, Lay JC, Bromberg PA, Diaz-Sanchez D, et al. (2010) Comparative airway inflammatory response of normal volunteers to ozone and lipopolysaccharide challenge. Inhalation toxicology 22: 648–656.
57. Gabrielson EW, Yu XY, Spannhake EW (1994) Comparison of the toxic effects of hydrogen peroxide and ozone on cultured human bronchial epithelial cells. Environmental health perspectives 102: 972–974.
58. Nair S, Barve A, Khor TO, Shen GX, Lin W, et al. (2010) Regulation of Nrf2- and AP-1-mediated gene expression by epigallocatechin-3-gallate and sulforaphane in prostate of Nrf2-knockout or C57BL/6J mice and PC-3 AP-1 human prostate cancer cells. Acta pharmacologica Sinica 31: 1223–1240.
59. Shinkai Y, Sumi D, Fukami I, Ishii T, Kumagai Y (2006) Sulforaphane, an activator of Nrf2, suppresses cellular accumulation of arsenic and its cytotoxicity in primary mouse hepatocytes. FEBS letters 580: 1771–1774.
60. Reddy NM, Kleeberger SR, Cho HY, Yamamoto M, Kensler TW, et al. (2007) Deficiency in Nrf2-GSH signaling impairs type II cell growth and enhances sensitivity to oxidants. American journal of respiratory cell and molecular biology 37: 3–8.
61. Na HK, Surh YJ (2008) Modulation of Nrf2-mediated antioxidant and detoxifying enzyme induction by the green tea polyphenol EGCG. Food and chemical toxicology : an international journal published for the British Industrial Biological Research Association 46: 1271–1278.
62. Zhu H, Jia Z, Zhang L, Yamamoto M, Misra HP, et al. (2008) Antioxidants and phase 2 enzymes in macrophages: regulation by Nrf2 signaling and protection against oxidative and electrophilic stress. Experimental biology and medicine 233: 463–474.
63. Wu CC, Hsu MC, Hsieh CW, Lin JB, Lai PH, et al. (2006) Upregulation of heme oxygenase-1 by Epigallocatechin-3-gallate via the phosphatidylinositol 3-kinase/Akt and ERK pathways. Life sciences 78: 2889–2897.
64. Van Amelsvoort JM, Van Hof KH, Mathot JN, Mulder TP, Wiersma A, et al. (2001) Plasma concentrations of individual tea catechins after a single oral dose in humans. Xenobiotica; the fate of foreign compounds in biological systems 31: 891–901.
65. Ye L, Dinkova-Kostova AT, Wade KL, Zhang Y, Shapiro TA, et al. (2002) Quantitative determination of dithiocarbamates in human plasma, serum, erythrocytes and urine: pharmacokinetics of broccoli sprout isothiocyanates in humans. Clinica chimica acta; international journal of clinical chemistry 316: 43–53.
66. Friedman M, Madden MC, Samet JM, Koren HS (1992) Effects of ozone exposure on lipid metabolism in human alveolar macrophages. Environmental health perspectives 97: 95–101.
67. Pryor WA, Squadrito GL, Friedman M (1995) The cascade mechanism to explain ozone toxicity: the role of lipid ozonation products. Free radical biology & medicine 19: 935–941.
68. Qu F, Qin XQ, Cui YR, Xiang Y, Tan YR, et al. (2009) Ozone stress down-regulates the expression of cystic fibrosis transmembrane conductance regulator in human bronchial epithelial cells. Chemico-biological interactions 179: 219–226.
69. Bertram S, Glowacka I, Muller MA, Lavender H, Gnirss K, et al. (2011) Cleavage and activation of the severe acute respiratory syndrome coronavirus spike protein by human airway trypsin-like protease. Journal of virology 85: 13363–13372.
70. Abboud RT, Vimalanathan S (2008) Pathogenesis of COPD. Part I. The role of protease-antiprotease imbalance in emphysema. The international journal of tuberculosis and lung disease : the official journal of the International Union against Tuberculosis and Lung Disease 12: 361–367.
71. Cox LA, Jr. (2009) A mathematical model of protease-antiprotease homeostasis failure in chronic obstructive pulmonary disease (COPD). Risk analysis an official publication of the Society for Risk Analysis 29: 576–586.
72. Dillard GH, Chanutin A (1949) The protease and antiprotease of plasmas of patients with cancer and other diseases. Cancer research 9: 665–668.
73. Becker S, Soukup JM, Reed W, Carson J, Devlin RB, et al. (1998) Effect of ozone on susceptibility to respiratory viral infection and virus-induced cytokine secretion. Environmental toxicology and pharmacology 6: 257–265.
74. Purvis MR, Miller S, Ehrlich R (1961) Effect of atmospheric pollutants on susceptibility to respiratory infection. I. Effect of ozone. The Journal of infectious diseases 109: 238–242.
75. Bogadel'nikov IV, Kubyshkin AV, Gavalova NG (1990) Changes in protease inhibitor balance of the bronchoalveolar secretion in children with inflammatory lung diseases. Pediatriia. pp 17–20.
76. Garcia-Sastre A (2011) Induction and evasion of type I interferon responses by influenza viruses. Virus research 162: 12–18.
77. Mikerov AN, Umstead TM, Gan X, Huang W, Guo X, et al. (2008) Impact of ozone exposure on the phagocytic activity of human surfactant protein A (SP-A) and SP-A variants. American journal of physiology Lung cellular and molecular physiology 294: L121–130.
78. Kido H, Okumura Y, Yamada H, Le TQ, Yano M (2007) Proteases essential for human influenza virus entry into cells and their inhibitors as potential therapeutic agents. Current pharmaceutical design 13: 405–414.
79. Hooper JD, Clements JA, Quigley JP, Antalis TM (2001) Type II transmembrane serine proteases. Insights into an emerging class of cell surface proteolytic enzymes. The Journal of biological chemistry 276: 857–860.
80. Fischer BM, Pavlisko E, Voynow JA (2011) Pathogenic triad in COPD: oxidative stress, protease-antiprotease imbalance, and inflammation. International journal of chronic obstructive pulmonary disease 6: 413–421.
81. Liu CK, Wei G, Atwood WJ (1998) Infection of glial cells by the human polyomavirus JC is mediated by an N-linked glycoprotein containing terminal alpha(2–6)-linked sialic acids. Journal of virology 72: 4643–4649.
82. Gagneux P, Cheriyan M, Hurtado-Ziola N, van der Linden EC, Anderson D, et al. (2003) Human-specific regulation of alpha 2–6-linked sialic acids. The Journal of biological chemistry 278: 48245–48250.
83. Mukhopadhyay S, Philip AT, Stoppacher R (2010) Pathologic findings in novel influenza A (H1N1) virus ("Swine Flu") infection: contrasting clinical manifestations and lung pathology in two fatal cases. American journal of clinical pathology 133: 380–387.
84. Mauad T, Hajjar LA, Callegari GD, da Silva LF, Schout D, et al. (2010) Lung pathology in fatal novel human influenza A (H1N1) infection. American journal of respiratory and critical care medicine 181: 72–79.
85. Cho HY, Reddy SP, Debiase A, Yamamoto M, Kleeberger SR (2005) Gene expression profiling of NRF2-mediated protection against oxidative injury. Free radical biology & medicine 38: 325–343.
86. Zhu M, Chen Y, Li RC (2000) Oral absorption and bioavailability of tea catechins. Planta medica 66: 444–447.
87. Nakayama M, Suzuki K, Toda M, Okubo S, Hara Y, et al. (1993) Inhibition of the infectivity of influenza virus by tea polyphenols. Antiviral research 21: 289–299.
88. Ishii Y, Itoh K, Morishima Y, Kimura T, Kiwamoto T, et al. (2005) Transcription factor Nrf2 plays a pivotal role in protection against elastase-induced pulmonary inflammation and emphysema. Journal of immunology 175: 6968–6975.
89. Venkatasamy R, Spina D (2007) Protease inhibitors in respiratory disease: focus on asthma and chronic obstructive pulmonary disease. Expert review of clinical immunology 3: 365–381.

8

Spatial Differences in East Scotia Ridge Hydrothermal Vent Food Webs: Influences of Chemistry, Microbiology and Predation on Trophodynamics

William D. K. Reid[1]*, Christopher J. Sweeting[1], Ben D. Wigham[2], Katrin Zwirglmaier[3¤], Jeffrey A. Hawkes[4], Rona A. R. McGill[5], Katrin Linse[3], Nicholas V. C. Polunin[1]

1 School of Marine Science and Technology, Newcastle University, Newcastle upon Tyne, United Kingdom, **2** Dove Marine Laboratory, School of Marine Science and Technology, Newcastle University, Cullercoats, United Kingdom, **3** British Antarctic Survey, Natural Environment Research Council, High Cross, Madingley Road, Cambridge, United Kingdom, **4** Ocean and Earth Science, University of Southampton, National Oceanography Centre Southampton, Southampton, United Kingdom, **5** Natural Environment Research Council Life Sciences Mass Spectrometry Facility, Scottish Universities Environmental Research Centre, East Kilbride, United Kingdom

Abstract

The hydrothermal vents on the East Scotia Ridge are the first to be explored in the Antarctic and are dominated by large peltospiroid gastropods, stalked barnacles (*Vulcanolepas* sp.) and anomuran crabs (*Kiwa* sp.) but their food webs are unknown. Vent fluid and macroconsumer samples were collected at three vent sites (E2, E9N and E9S) at distances of tens of metres to hundreds of kilometres apart with contrasting vent fluid chemistries to describe trophic interactions and identify potential carbon fixation pathways using stable isotopes. $\delta^{13}C$ of dissolved inorganic carbon from vent fluids ranged from −4.6‰ to 0.8‰ at E2 and from −4.4‰ to 1.5‰ at E9. The lowest macroconsumer $\delta^{13}C$ was observed in peltospiroid gastropods (−30.0‰ to −31.1‰) and indicated carbon fixation via the Calvin-Benson-Bassham (CBB) cycle by endosymbiotic gamma-Proteobacteria. Highest $\delta^{13}C$ occurred in *Kiwa* sp. (−19.0‰ to −10.5‰), similar to that of the epibionts sampled from their ventral setae. *Kiwa* sp. $\delta^{13}C$ differed among sites, which were attributed to spatial differences in the epibiont community and the relative contribution of carbon fixed via the reductive tricarboxylic acid (rTCA) and CBB cycles assimilated by *Kiwa* sp. Site differences in carbon fixation pathways were traced into higher trophic levels e.g. a stichasterid asteroid that predates on *Kiwa* sp. Sponges and anemones at the periphery of E2 assimilated a proportion of epipelagic photosynthetic primary production but this was not observed at E9N. Differences in the $\delta^{13}C$ and $\delta^{34}S$ values of vent macroconsumers between E2 and E9 sites suggest the relative contributions of photosynthetic and chemoautotrophic carbon fixation (rTCA v CBB) entering the hydrothermal vent food webs vary between the sites.

Editor: Simon Thrush, National Institute of Water & Atmospheric Research, New Zealand

Funding: The research was funded by the Natural Environment Research Council (NERC) through ChEsSO consortium grant NE/DO1249X/1 and studentships NE/F010664/1 (WDKR) and NE/H524922 (JAH). Sample analysis was funded via NERC Life Sciences Mass Spectrometry Facilities grant LSMSFBRIS043_04/10_R_09/10. The funders had no role in the study design, data collection and anlaysis, decision to publish, or preparation of the manuscript.

Competing Interests: The authors have declared that no competing interests exist.

* E-mail: wdkreid@gmail.com

¤ Current address: Technische Universitaet Muenchen, Wissenschaftszentrum Weihenstephan, Limnologische Station ffeldorf, Iffeldorf, Germany

Introduction

Deep-sea hydrothermal vents are chemically reducing habitats occurring on mid-ocean and back-arc spreading centres, seamounts, volcanic hotspots and off-axis ridge settings [1,2,3]. They are distinct from the surrounding deep sea with respect to environmental conditions, the energy sources sustaining life and their biological communities [4,5]. High densities of organisms are found to thrive at the interface where hot, mineral-rich fluids discharge from the seafloor and mix with colder, oxygenated seawater. The hot fluids emitted from the seafloor may differ in pH and are enriched in reduced gases (e.g. H_2S, CH_4, H_2) and metals (e.g. Fe^{2+}, Cu, Mn) relative to seawater [6]. Microorganisms oxidise the reduced species in vent fluids and utilise the energy released to fix CO_2 or other single carbon compounds (e.g. CO, CH_4) into cellular material [7]. This results in microbial chemosynthesis replacing photosynthetic primary production at the base of the food chain [7].

Sulfide oxidation appears to be the principal energy acquisition pathway, which microorganisms use to drive carbon fixation [3,7,8]. The most important carbon fixation pathways at the base of the metazoan hydrothermal vent food webs are the Calvin-Benson-Bassham (CBB) and reductive tricarboxylic acid (rTCA) cycles [9,10,11]. Methane oxidation (methanotrophy) is a further carbon fixation process at hydrothermal vents with CH_4 of thermogenic, biogenic or magmatic origin available depending on the host substrate [3,12]. Epipelagic photosynthetic primary production may also provide some nutrition to vent macroconsumers, although the relative contribution to vent fauna is thought to be negligible [12,13]. Macroconsumers utilise the vent organic carbon through endo- and episymbiotic relationships, consumption of free-living microorganisms either from various surfaces or the water column and indirectly through predation and scavenging [14,15,16].

The relative contributions of different carbon sources and complexity of hydrothermal vent food webs vary globally depending on the species present, the geological host substrate and the vent fluid chemistry [17,18,19]. The first Antarctic hydrothermal vent communities were discovered recently on the East Scotia Ridge (ESR), a back-arc spreading centre in the Atlantic sector of the Southern Ocean [20,21]. The two basalt-hosted vent fields occur on the ridge segments E2 and E9, which lack the characteristic alvinocarid shrimps, bathymodiolid mussels and siboglinid worms found at Atlantic, Indian and Pacific hydrothermal vents, respectively [21]. Instead, biomass at the ESR vents is dominated by anomuran crabs (*Kiwa* sp.), stalked barnacles (*Vulcanolepas* sp.) and large peltospiroid gastropods [22], indicating a new biogeographic province [21]. Furthermore, there are differences in the end-member vent fluid chemistry between the E2 and E9 vent fields as well as within field between northern (E9N) and southern (E9S) areas of E9 [21].

Stable isotopes of carbon ($^{13}C/^{12}C$ expressed as $\delta^{13}C$), nitrogen ($^{15}N/^{14}N$ expressed as $\delta^{15}N$) and sulfur ($^{34}S/^{32}S$ expressed as $\delta^{34}S$) have been used to examine hydrothermal vent community trophodynamics [23,24]. $\delta^{13}C$ can be used to characterise the various carbon sources utilised by vent macroconsumers [25]. This is done by comparing the expected carbon fractionation between dissolved inorganic carbon (DIC) and the macroconsumer's tissue. Enzymatic reactions catalysed by the ribulose-1,5-biphosphate carboxylase/oxygenase form I (RuBisCO form I) of the CBB cycle (22‰ to 30‰: [26,27,28]) exhibit greater fractionation than those of the rTCA cycle (2‰ to 14‰: [29,30,31]). Once organic material is incorporated into the macroconsumer food web, carbon trophic discrimination ($\Delta^{13}C$) is small, ranging from 0 to 1.5‰ between the food source and consumer [32]. $\delta^{34}S$ also identifies energy sources (sulfur trophic discrimination, −1‰ to 2‰: [32]). The large difference in $\delta^{34}S$ between seawater sulphate and sulfides at hydrothermal vents [33] results in organic matter of photosynthetic (~16‰ to 19‰) and chemosynthetic (−9‰ to 10‰) origin having distinctive $\delta^{34}S$ values [34,35]. The greater trophic discrimination (2‰ to 5‰) in $\delta^{15}N$ between consumer and food source provides information on the trophic position of an organism relative to a primary consumer [32]. Therefore, the isotopic value of a vent macroconsumer is the product of the following factors: (1) the inorganic substrate and its isotopic value used by the chemoautotroph; (2) the isotopic discrimination processes occurring during metabolic reactions involving inorganic substrates to create organic compounds (e.g. CBB or rTCA cycles) by the chemoautotroph; (3) food source-macroconsumer trophic interactions (e.g. endosymbiont-host, predator-prey) that occur as a function of (1) and (2); and (4) the physiology associated with the macroconsumer's isotopic trophic discrimination.

The goal of the present research was to investigate intra- and inter-site patterns in the trophic assemblages of macroconsumers occurring at hydrothermal vents on the ESR using $\delta^{13}C$, $\delta^{34}S$ and $\delta^{15}N$. Specifically, the aims were to: (1) compare $\delta^{13}C_{DIC}$ among vent sites and thus establish difference in the isotopic inorganic substrates used by chemoautotrophs; (2) compare $\delta^{13}C$, $\delta^{15}N$ and $\delta^{34}S$ between vent and benthic non-vent fauna to assess any photosynthetic inputs into the hydrothermal vent food web; (3) investigate differences in trophic structures among the three sites; and (4) assess which species are driving any differences in trophic structure. The investigation provides a unique opportunity to examine differences in trophic structure at the scale of tens of metres to 100s of kilometres in a newly discovered hydrothermal vent biogeographical province.

Materials and Methods

Ethics Statement

Permits for the fieldwork were granted by the United Kingdom Foreign and Commonwealth Office. This study met the ethical requirements of the affiliated research institutions for research utilising animal tissues. No animal husbandry or laboratory controlled experiments were part of the research that required permits from the UK Home Office. The fish were collected at a water depth of 2500 m, which meant that they were dead when they arrived on deck as a result of changes in pressure. This was the case with the majority of the animals dissected within this study. The research also adhered to the Inter Ridge code of conduct for sampling hydrothermal vents (http://www.interridge.org/IRStatement).

Study Sites

The E2 and E9 vent fields are situated approximately 440 km apart at 56° 05.35′S, 30° 19.20′W and 60° 02.50′ S, 29° 58.93′ W, respectively (Fig. 1). E2 is at a depth of ~2600 m and seafloor topography is complex with a series of terraced features and lobed pillow basalts filling a major north-south steep-sided fissure [21]. The main high-temperature and diffusive venting occurred at an intersection between this fissure and an east-west running fault or scarp [21]. E9 was located at ~2400 m depth and its topography was relatively flat with sheet lava, a series of lava drain back features and collapsed pillow basalts. A series of north-south fissures were found with venting mainly occurring on the most western [21,22]. The end-member fluid chemistry exiting chimneys differed between the northern and southern sections of E9 [21], therefore E9N and E9S are here considered to be separate sites. Ambient seabed water temperatures were 0.0°C at E2 and between −0.1°C and −1.3°C at E9 [21].

Sample Collection and Ship-board Processing

Samples were collected onboard the R.R.S *James Cook* during the 2010 austral summer (7 January to 21 February) using the remotely operated vehicle (ROV) *Isis*. High temperature and diffuse flow fluids were collected for DIC using titanium samplers, equipped with an inductively coupled link high temperature sensor. The nozzle of the titanium sampler was inserted into the chimney orifice for high temperature fluid samples and once the temperature reading became stable the fluid was collected. For diffuse flow samples, a circular titanium housing was placed over the area of diffuse venting to minimise the entrainment of seawater. Once the diffuse flow was visible exiting the top of the housing, the titanium sampler was inserted into the opening and the diffuse flow sample was collected once the temperature reading was stable. On board, an aliquot for stable isotope analysis of DIC was sampled to exclude air and poisoned with mercury chloride.

Vent macroconsumers were collected by suction sampler or scoop with species separated into a series of acrylic chambers or perspex boxes to avoid predation or contamination. Six species were collected at all three sites. No female *Kiwa* sp. were collected from E9N or E9S. Fish and pycnogonids were collected using large collapsible and small metal baited traps deployed from the ROV. Non-vent macroconsumers were collected from metres to tens of metres away from active venting where there were no obvious signs of hydrothermal influence, i.e. no bacterial mat, and where temperature was consistent with local Antarctic bottom water. Non-vent samples were collected on separate dives from those for vent fauna to avoid contamination. Only one non-vent species was collected from E2 and sampling was limited to the areas adjacent to E9N because of ROV operational time

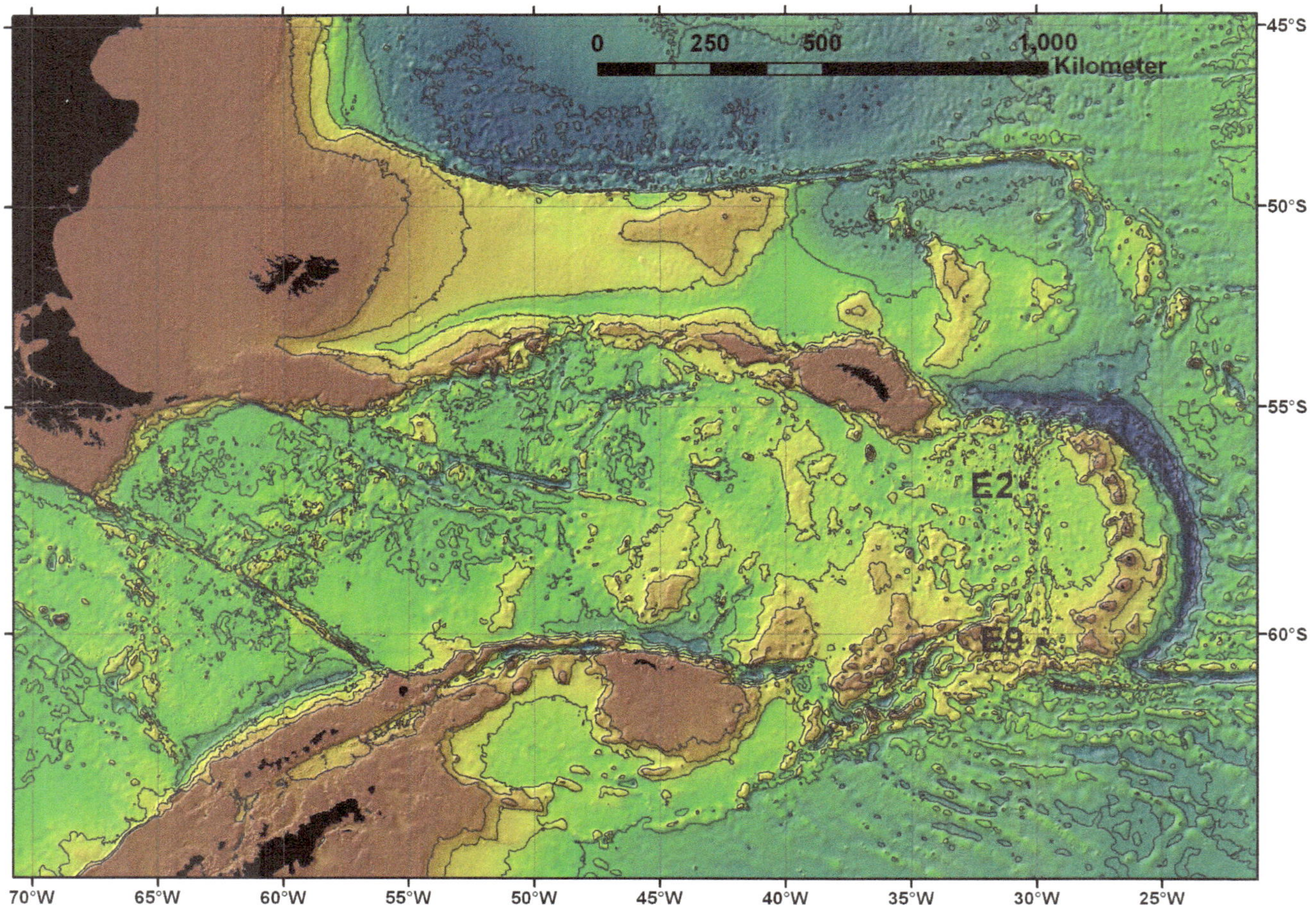

Figure 1. Bathymetric map illustrating positions of the E2 and E9 vent sites (black circles). The vent sites are located at the northern and southern ends of the East Scotia Ridge (ESR), located in the Atlantic sector of the Southern Ocean. The map shows the position of the ESR in relation to South America and the Antarctic Peninsula.

constraints. Potential food sources were collected by scraping material from rocks collected by ROV manipulators and epibionts from the ventral setae of the decapod *Kiwa* sp. Particulate suspended material was collected from the acrylic chambers, which was sampled incidentally during faunal collection. Samples were sorted on board to the lowest possible taxonomic resolution. The majority of the vent species are undescribed to date.

Faunal samples were frozen at -80°C whole or after dissection, depending on their size, for stable isotope analysis. Muscle was removed from the chelipeds of *Kiwa* sp., foot dissected from Peltospiroidea sp., tube feet removed from the asteroids Stichasteridae sp. and *Freyella* cf *fragilissima* and tentacles removed from the anemones. Legs were removed from the pycnogonids *Colossendeis* cf. *concedis* and *C.* cf. *elephantis*, while *Sericosura* spp. was sampled whole. The gastropods Provannidae sp. 1 and 2, *Lepetodrilus* sp., and juvenile Peltospiroidea sp. (<7 mm shell length), and the stalked barnacle *Vulcanolepas* sp. were removed from their shells and sampled whole. White muscle tissue was dissected from the anterior dorso-lateral region of the zoarcid fish.

Sample Processing Onshore

Each end-member and diffuse flow DIC sample was prepared for isotopic analysis by removing a 1 mL water sample and transferring it into a separate vial. The headspace was flushed with helium, phosphoric acid was injected into the vial and then the contents were vortex mixed. The samples were then left to react for 24 hours to ensure complete conversion of all DIC to CO_2 for isotopic analysis. The CO_2 was then analysed by continuous-flow isotope ratio mass spectrometry (IRMS) using a Europa Scientific 20–20 IRMS by Iso-Analytical (Crewe, United Kingdom). Samples were run in duplicate and the mean is reported. An internal reference gas (IA-R060, $\delta^{13}C = -36.08‰ \pm$ SD 0.13) was used to determine the $\delta^{13}C_{DIC}$ values and is traceable to the International Atomic Energy Agency standard, NBS-19. Concentrations of CH_4 in the water samples were insufficient for isotope analysis.

Faunal tissue samples were freeze dried and ground to a homogenous powder using a pestle and mortar. Aliquots of fauna, particulate suspended material and material scraped from rocks were tested for carbonates prior to analysis with 0.1 N HCl. If the sample effervesced, this indicated carbonates were present and it was subsequently acidified by further addition of HCl until the effervescence ceased. Samples were re-dried at 50°C for 48 hours. If the sample did not effervesce, no acidification was carried out. Aliquots for $\delta^{13}C$ analysis were not lipid extracted. Any confounding lipid effects due to metabolic processes would not affect the interpretation of the ultimate carbon sources of the vent fauna described by $\delta^{13}C$ because of the large differences in the $\delta^{13}C$ values of trophic end-members.

Approximately 0.7 mg of powder was weighed into a tin capsule for carbon and nitrogen IRMS. For sulfur, 2 mg of sample and 4 mg of the catalyst vanadium pentoxide were weighed into each tin capsule. Dual stable carbon and nitrogen isotope ratios were measured by continuous-flow IRMS using a Costech Elemental Analyser interfaced with Thermo Finnigan Delta Plus XP (Natural Environment Research Council, Life Sciences Mass Spectrometry Facility, SUERC, East Kilbride, United Kingdom). Two laboratory standards were analysed for every ten samples in each analytical sequence. These alternated between paired alanine standards, differing in $\delta^{13}C$ and $\delta^{15}N$, and an internal laboratory gelatin standard. Sulfur was analysed by Iso-Analytical using a SERCON Elemental Analyser coupled to a Europa Scientific 20–20 IRMS. Laboratory standards of barium sulphate (two sets of differing $\delta^{34}S$) and silver sulfide were used for calibration and drift correction. An internal standard of whale baleen was used for quality control (n = 28, 16.34‰ ± SD 0.21). Stable isotope ratios were expressed in delta (δ) notation as parts per thousand/permil (‰). All internal standards are traceable to the following international standards: v-PDB (Pee Dee Belemnite), AIR (atmospheric nitrogen) and NBS-127 (barium sulphate), IAEA-S-1 (silver sulfide) and IAEA-SO-5 (barium sulphate). An external reference material of freeze dried and ground deep-sea fish white muscle (*Antimora rostrata*) was also analysed ($\delta^{13}C$, n = 24, −18.94‰ ± SD 0.09; $\delta^{15}N$, n = 24, 13.11‰ ± SD 0.38; $\delta^{34}S$, n = 30, 18.20‰, ± SD 0.59).

Data Analysis

Data were assessed for normality using a Shapiro-Wilk test before statistical tests examining spatial patterns in trophic structure and species stable isotope values. Homogeneity, or otherwise, of variances is ecologically informative, for example in identifying distinct energy sources at the base of the food web [36]. Inter-site differences in trophic structure were examined using a Fligner-Killeen test for homogeneity of variance to assess differences in the spread of the mean stable isotope values of each species. Inter-site differences in species were analysed using a one-way ANOVA followed by Tukey's honest significant difference (HSD) when variance was homogeneous among sites. Welch's ANOVA followed by t-tests were used when there was heterogeneity of variance among sites because it uses adjusted degrees of freedom to protect against Type I errors when variances are unequal [37]. A Bonferroni correction ($p = 0.05/n$) was used for multiple comparisons. When data were not normally distributed, a two sample Wilcoxon test was used. All statistics were preformed in R version 12.13.1 [38].

Results

Dissolved Inorganic Carbon Stable Isotope Values

Mean (± SD) $\delta^{13}C_{DIC}$ of high temperature and diffuse flow fluids are summarised in Table 1. $\delta^{13}C_{DIC}$ of high temperature samples collected at two E2 locations were −4.7‰ (±0.0) (max temperature 351.0°C) and −2.5‰ (±0.1) (max temperature 323.0°C). At E9N, $\delta^{13}C_{DIC}$ from separate orifices of the same chimney structure were −4.6‰ (±0.0) (max temperature 380.2°C) and −4.5‰ (±0.0) (max temperature 357.0°C). No high temperature fluids were collected from E9S for $\delta^{13}C_{DIC}$ analysis because the pressure was too high within the titanium samples to safely and accurately collect a representative sample. Diffuse flow samples from amongst *Kiwa* sp. and anemones, at E2, had $\delta^{13}C_{DIC}$ values of 0.8‰ (±0.1) (max temperature 19.9°C) and 0.2‰ (±0.2) (max temperature 3.5°C), respectively. A single diffuse flow sample collected from amongst an aggregation of *Kiwa* sp. at E9N had a $\delta^{13}C_{DIC}$ value of 1.5‰ (±0.1) (max temperature 12.6°C). At E9S, diffuse flow samples from amongst *Kiwa* sp. had a $\delta^{13}C_{DIC}$ value of 0.9‰ (±0.1) (max temperature 19.9°C), while a sample taken from a mixed aggregation of *Kiwa* sp. and peltospiroid gastropods had $\delta^{13}C_{DIC}$ value of 0.1‰ (±0.1) (max temperature 5.0°C).

Comparison between Vent and Benthic Non-vent Macro-consumers at E9N

At E9N, mean $\delta^{13}C$ and $\delta^{15}N$ values of vent fauna overlapped with non-vent benthic fauna (Welch's t-test, $\delta^{13}C$ DF = 10.59, t = 0.66, p = 0.52; Welch's t-test, $\delta^{15}N$ DF = 10.42, t = −0.30, p = 0.76; Fig. 2, Tables 2 & 3) while mean $\delta^{34}S$ values differed between non-vent benthic fauna and vent fauna (Welch's t-test, DF = 12.56, t = −9.08, p<0.01) (Fig. 3, Tables 2 & 3).

Intra- and Inter-site Differences in Community Trophodynamics

Eleven, ten and seven species were collected at E2, E9N and E9S respectively for stable isotope analysis (Table 3). The ranges of mean $\delta^{13}C$ values of the vent fauna differed amongst the three sites (Fligner-Killeen test, DF = 2, χ^2 = 6.46, p<0.05). E2 had the narrowest $\delta^{13}C$ range (−29.9‰ to −19.0‰), whereas at E9N and E9S $\delta^{13}C$ ranged from −31.4‰ to −9.9‰ and −30.0‰ to −10.5‰, respectively (Fig. 2). Across the three sites Peltospiroidea sp. had the lowest values while *Kiwa* sp. had the highest $\delta^{13}C$ values (Fig. 2, Table 3), and *Lepetodrilus* sp., *Vulcanolepas* sp., *Pacmanactis* sp. and *Colossendeis* spp. all had intermediate $\delta^{13}C$ values (Fig. 2, Table 3). However, there was no overall difference in mean $\delta^{13}C$ values among sites for the combined data across species (Welch's ANOVA, DF = 2.00, F = 0.59, p = 0.56). The range and mean $\delta^{34}S$ values (Fig. 3, Table 3) did not differ among sites (Fligner-Killeen test, DF = 2, χ^2 = 0.84, p = 0.65; ANOVA, DF = 2, 26, F = 1.94, p = 0.16), however *Kiwa* sp. had the lowest $\delta^{34}S$ at E2 and E9S while *Lepetodrilus* sp. had the lowest $\delta^{34}S$ values at E9N (Fig. 3, Table 3). The highest vent fauna $\delta^{34}S$ values were in *Pacmanactis* sp. (E2), *Vulcanolepas* sp. (E9N) and *Sericosura* spp. (E9S) (Fig. 3, Table 3). Neither the range nor the mean $\delta^{15}N$ values differed among sites (Fligner-Killeen test, DF = 2, χ^2 = 0.40, p = 0.83; ANOVA, DF = 2, 26, F = 1.19, p = 0.31). The provannid gastropods at E2 and E9S had the lowest $\delta^{15}N$ values while Peltospiroidea sp. had the lowest values at E9N (Fig. 2, Table 3).

Table 1. $\delta^{13}C$ values of dissolved inorganic carbon (DIC) sampled from high temperature and diffuse flow venting from the E2 and E9 ridge segments of the East Scotia Ridge, Southern Ocean.

Site	Temperature (°C)	$\delta^{13}C$ DIC
E2	351.0	−4.7 (0.0)
	323.0	−2.5 (0.1)
	19.9	0.8 (0.1)
	3.5	0.2 (0.2)
E9N	380.2	−4.7 (0.0)
	357.0	−4.7 (0.0)
	12.6	1.5 (0.1)
E9S	19.9	0.9 (0.1)
	5.0	0.1 (0.1)

Standard deviations are in parentheses.

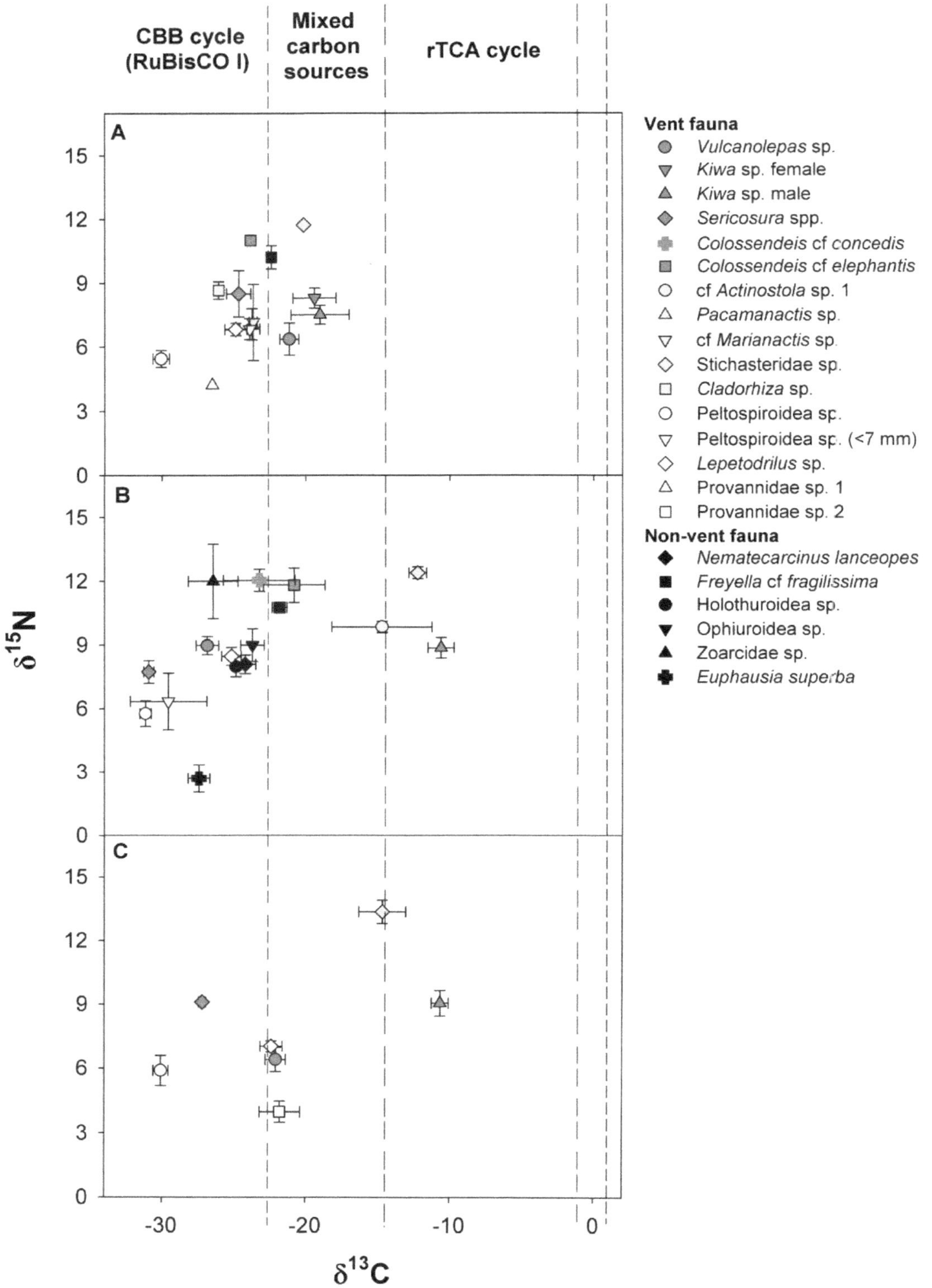

Figure 2. $\delta^{13}C$ and $\delta^{15}N$ values of macroconsumers collected from the East Scotia Ridge, Southern Ocean. The values represent means (± standard deviations) for hydrothermal vent and non-vent macroconsumers from the three sample sites: (a) E2, (b) E9N and (c) E9S. Dashed vertical lines represent potential ranges of $\delta^{13}C$ values indicative of carbon sources sustaining macroconsumers at the ESR: triple dashed line represents the Calvin-Benson-Bassham (CBB) cycle utilising form I RuBisCO, double dashed line represents the reductive tricarboxylic acid (rTCA) cycle, mixed carbon sources occur between the triple and double dashed line and the continuous dashed line represents the approximate $\delta^{13}C$ values of the dissolved inorganic carbon from the diffuse flow areas.

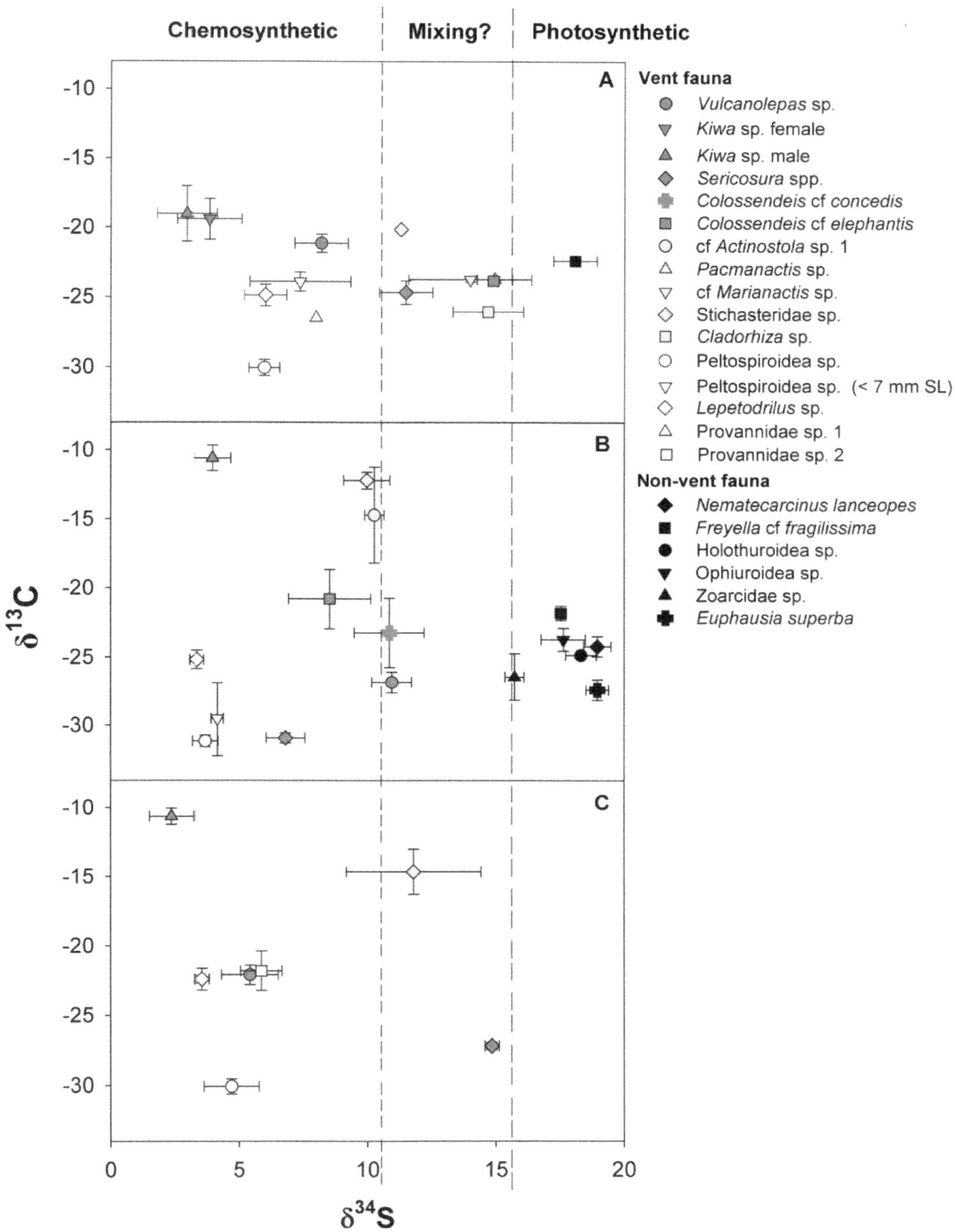

Figure 3. $\delta^{13}C$ and $\delta^{34}S$ values of macroconsumers collected from the East Scotia Ridge, Southern Ocean. The values represent means (± standard deviations) for hydrothermal vent and non-vent macroconsumers from the three sample sites: (a) E2, (b) E9N and (c) E9S. $\delta^{34}S$ values between the triple and double dashed lines represent potential areas of isotopic mixing between chemosynthetic and photosynthetic food sources.

The stichasterid sp. consistently had the highest $\delta^{15}N$ values relative to the other vent fauna at each site (Fig. 2, Table 3).

Spatial Differences in Macroconsumer Trophodynamics

Vulcanolepas sp. exhibited spatial differences in $\delta^{13}C$, $\delta^{15}N$ and $\delta^{34}S$ but there was no consistent pattern in isotopic differences among sites (Table 4) Male and female *Kiwa* sp. at E2 did not differ in $\delta^{13}C$ but males were lower in $\delta^{15}N$ and $\delta^{34}S$ than females (Table 5). Male *Kiwa* sp. showed spatial differences in each stable isotope (Table 4). $\delta^{13}C$ of the males showed a greater range (Fligner-Killeen test, $DF = 2$, $\chi^2 = 10.91$, $p<0.01$) and lower values at E2 than E9N and E9S (Table 3 & 4). The epibionts attached to the ventral surface of male *Kiwa* sp. also exhibited a greater spread of $\delta^{13}C$ values at E2 than E9S (F-test, $DF = 4, 3$, $F = 244.46$,

Table 2. Mean δ^{13}C, δ^{15}N and δ^{34}S values (‰) of non-vent deep-sea fauna collected from the E2 and E9 ridge segments of the East Scotia Ridge, Southern Ocean. Standard deviations are in parentheses.

Taxonomic group	Species	Site	N	δ^{13}C	δ ^{34}S	δ ^{15}N
Crustacea						
Decapoda	*Nematocarcinus lanceopes*	E9	5	−24.2 (0.7)	18.9 (0.5)	8.1 (0.4)
	Euphausia superba	E9	3	−27.4 (0.8)	19.0 (0.1)	2.7 (0.7)
Echinodermata						
Asteroidea	*Freyella* cf *fragilissima*	E2	3	−22.4 (0.3)	18.0 (0.8)	10.2 (0.6)
	Freyella cf *fragilissima*	E9	2	−21.8 (0.5)	17.5 (0.2)	10.8 (0.2)
Holothuroidea	Holothuroidea sp.	E9	3	−24.9 (0.0)	18.3 (0.6)	8.0 (0.4)
Ophiuroidea	Ophiuroidea sp.	E9	3	−23.7 (0.8)	17.6 (0.9)	9.0 (0.8)
Vertebrata						
Osteichthys	Zoarcidae sp.	E9	4	−26.5 (1.7)	15.7 (0.4)	12.0 (1.8)

Table 3. Mean δ^{13}C, δ^{15}N and δ^{34}S values (‰)of hydrothermal vent fauna collected from the E2 and E9 ridge segments of the East Scotia Ridge, Southern Ocean.

Taxonomic group	E2				E9N				E9S			
	N	δ^{13}C	δ^{34}S	δ^{15}N	N	δ^{13}C	δ^{34}S	δ^{15}N	N	δ^{13}C	δ^{34}S	δ^{15}N
Cirripedia												
Vulcanolepas sp.	22	−21.1 (0.6)	8.2 (1.0)	6.3 (0.7)	23	−26.9 (0.8)	11.0 (0.8)	9.0 (0.4)	23	−22.1 (0.8)	5.4 (1.1)	6.4 (0.6)
Decapoda												
Kiwa sp. female	20	−19.4 (1.5)	3.9 (1.3)	8.2 (0.5)	0	–	–	–	0	–	–	–
Kiwa sp. Male	18	−19.0 (2.0)	3.0 (1.2)	7.5 (0.5)	22	−10.6 (0.9)	4.0 (0.7)	8.9 (0.5)	30	−10.7 (0.6)	2.4 (0.9)	9.1 (0.6)
Pycnogonida												
Sericosura spp.	6	−24.7 (0.9)	11.9 (0.4)	8.5 (1.3)	9	−30.9 (0.5)	6.8 (0.8)	7.7 (0.5)	2	−27.2 (0.3)	14.9 (0.3)	9.1 (0.1)
Colossendeis cf *concedis*	0	–	–	–	6	−23.3 (2.5)	10.9 (1.4)	12.1 (0.5)	0	–	–	–
Colossendeis cf *elephantis*	1	−23.8	14.9	11.0	3	−20.8 (2.2)	8.5 (1.6)	11.8 (0.8)	0	–	–	–
Anthozoa												
cf *Actinostola* sp. 1	0	–	–	–	4	−14.7 (3.5)	10.3 (0.4)	9.9 (0.3)	0	–	–	–
Pacmanactis sp.	5	−23.8 (0.2)	14.9 (0.7)	7.1 (0.7)	0	–	–	–	0	–	–	–
cf *Marianactis* sp.	5	−23.7 (0.3)	14.0 (2.4)	7.2 (1.8)	0	–	–	–	0	–	–	–
Asteroidea												
Stichasteridae sp.	1	−20.2	11.3	12.3	5	−12.2 (0.6)	10.0 (0.9)	12.4 (0.4)	5	−14.7 (1.6)	11.8 (2.6)	13.4 (0.6)
Gastropoda												
Peltospiroidea sp.	19	−30.1 (0.6)	6.0 (0.6)	5.4 (0.4)	22	−31.2 (0.4)	3.7 (0.5)	5.8 (0.6)	15	−30.1 (0.5)	4.7 (1.1)	5.9 (0.7)
Peltospiroidea sp (<7 mm)	4	−23.9 (0.7)	7.4 (2.0)	6.8 (0.5)	5	−29.6 (2.7)	4.2 (0.2)	6.4 (1.3)	0	–	–	–
Provannidae sp. 1	1	−26.5	8.0	4.2	0	–	–	–	0	–	–	–
P-rovannidae sp. 2	0	–	–	–	0	–	–	–	4	−21.8 (1.4)	5.9 (0.8)	4.0 (0.5)
Lepetodrilus sp.	5	−24.9 (0.8)	6.4 (0.5)	6.8 (0.3)	4	−25.2 (0.7)	3.4 (0.3)	8.5 (0.4)	4	−22.4 (0.8)	3.6 (0.3)	7.0 (0.3)
Cladorhizidae												
Cladorhiza sp.	5	−26.1 (0.4)	14.7 (1.4)	8.7 (0.4)	0	–	–	–	0	–	–	–
Potential food sources												
Particulate suspended material	3	−23.2 (5.4)	10.0 (1.1)	−0.1 (4.9)	0	–	–	–	0	–	–	–
Rock scrapings	0	–	–	–	1	−23.2	0.8	2.4	1	−31.1	–	1.9
K-iwa n. sp episymbiont	5	−18.9 (5.3)	7.5 (0.3)	3.3 (1.5)	0	–	–	–	5	−9.9 (0.3)	6.6 (0.2)	5.2 (0.8)

Standard deviations are in parentheses and - indicates no data.

$p<0.01$) as well as lower $\delta^{13}C$ but higher $\delta^{15}N$ values at E2 than E9S (Table 3 & 5). *Sericosura* spp. $\delta^{13}C$ and $\delta^{34}S$ values varied amongst sites but $\delta^{15}N$ values scarcely did (Table 4). $\delta^{13}C$ and $\delta^{34}S$ values were lowest at E9N but highest at E2 for $\delta^{13}C$ and E9S for $\delta^{34}S$ (Table 3). Peltospiroidea sp. showed spatial differences in $\delta^{13}C$ and $\delta^{34}S$ but not in $\delta^{15}N$ (Table 4). $\delta^{34}S$ values differed among all sites but E9N $\delta^{13}C$ values were lower than those at E9S and E2 (Table 4). Stichasteridae sp. revealed differences between all sites for $\delta^{13}C$ and $\delta^{15}N$ but for $\delta^{34}S$ only between E2 and E9N (Table 5).

Discussion

This study described the trophic structure of a new vent biogeographical province recently discovered on the ESR in the Southern Ocean [21]. In addressing this aim, the study shared the challenges of preceding work in characterising energy sources, separating isotopic overlap and mixing of energy sources, and following energy sources into subsequent predator-prey relationships. However, the tri-isotope approach and integration of both vent chemistry and microbiology, here, provided a more holistic understanding of vent trophic ecology at within- and among-vent field scales.

Intra-site Trophic Interactions and Energy Sources

Scarcity of $\Delta^{13}C$ estimates between inorganic carbon and cellular biomass for primary producers at hydrothermal vents [27,30] makes interpretation of the origin of organic carbon fixed within the hydrothermal vent system and assimilated by macroconsumers tentative for species not within a symbiotic or known predator-prey relationship. Diffuse flow $\delta^{13}C_{DIC}$ of approximately 1‰ at the ESR vent fields suggests ESR vent macroconsumers with $\delta^{13}C$ values <-22‰ are assimilating carbon fixed via the CBB cycle because the net fractionation associated with fixing inorganic into organic carbon for RuBisCO form I ranges from -22‰ to -30‰ [26,27]. Peltospiroidea sp. housed an endosymbiotic gamma-Proteobacteria (K. Zwirglmaier unpublished data) and is within the $\delta^{13}C$ range expected for carbon fixed via RuBisCO form I at all three locations. Molluscs containing a single strain of endosymbiotic gamma-Proteobacteria living in other biogeographical vent provinces include some species of bathymodiolid mussels, vesicomyid clams and *Ifremeria* gastropods, all of which have $\delta^{13}C$ values between -37‰ and -27‰ [39,40]. Other species of ESR vent macroconsumers, which had $\delta^{13}C$ values <-22‰ included *Vulcanolepas* sp. (E2 and E9S), *Sericusora* spp., E2 anemones and *Lepetodrilus* sp. These species consume free-living bacteria [14,23] so organic carbon fixed via other carbon fixation pathways cannot be ruled out as part of their assimilated diet.

Vent macroconsumers inhabiting the hottest areas of the hydrothermal vent tolerable to metazoan life, including rimicarid shrimps, polychaetes *Alvinella* spp. and *Riftia pachyptila* and some alvinoconchid gastropods, tend to assimilate rTCA-fixed carbon from their diet [11,41,42] and have $\delta^{13}C$ values >-16‰ [14,25,43,44]. As $\delta^{13}C_{DIC}$ is approximately 1‰ at the ESR sites, vent macroconsumers utilising carbon fixed via the rTCA cycle would have had $\delta^{13}C$ values >-13‰; assuming a -2‰ to -14‰ net fractionation between the inorganic substrate and organic product catalysed by the enzymes involved in the rTCA cycle [29,30]. *Kiwa* sp. living at E9N and E9S, along with its epibionts, had $\delta^{13}C$ values that were >-12‰ and are found in areas close to discharging vent fluids [22]. This potentially indicates the epibionts living on *Kiwa* sp. ventral setae were fixing carbon via the rTCA cycle. *Kiwa* sp. was also ^{15}N-enriched by between 3.8‰ and 4.2‰ relative to its epibionts, suggesting the epibionts were an important food source. A similar episymbiotic relationship between the ESR kiwid is therefore hypothesised to that of *Kiwa puravida*, for which lipid, stable isotope and behavioural analyses indicate the harvesting of epibiont bacteria [45]. Stichasteridae sp. (~-13‰) and cf *Actinostola* sp. (~-14‰) also appeared to be assimilating carbon indicative of the rTCA cycle at E9N and E9S.

Several vent macroconsumers fell within the range of $\delta^{13}C$ values indicative of mixed carbon sources. Those within the $\delta^{13}C$ -22‰ to -15‰ range may consume free-living bacteria or are predators or scavengers that utilise a number of trophic pathways. At the ESR hydrothermal vents, *Lepetodrilus* sp., Provannidae sp. 2, *Vulcanolepas* sp., *Kiwa* sp., Stichasteridae sp. and *Colossendeis* cf. *elephantis* fell into this range at one or more sites. Related species of *Lepetodrilus* sp., Provannidae sp. 2 and *Vulcanolepas* sp. are all thought to consume free-living bacteria at other vents sites [14,23]. Such feeding can result in consuming heterogeneous bacterial communities, which have multiple pathways for carbon fixation and elemental cycling [9,46,47]. The biological cycling of carbon is very complex at hydrothermal vents because of the multiple single carbon substrates for carbon fixation (e.g. CO_2, CH_4, CO), spatial variability in the $\delta^{13}C$ value of the substrate and various microbial primary producers associated with different carbon fixation pathways [7,11,48]. Furthermore, the incorporation of photosynthetic derived carbon as particulate or dissolved organic matter is possible and may provide some nutrition to vent macroconsumers [12,13]. Therefore, complex isotopic mixes of food sources are available to these species.

Table 4. Results of ANOVA and *post-hoc* Tukey honest significant differences tests for the differences in stable isotope values of vent fauna among the three sites on the East Scotia Ridge.

Species	$\delta^{13}C$				$\delta^{34}S$				$\delta^{15}N$			
	DF	F	p	*Post-hoc*	DF	F	p	*Post-hoc*	F	DF	p	*Post-hoc*
Vulcanolepas sp.	2, 63	403.18	<0.01	E9N<E9S<E2	2, 63	176.16	<0.01	E9S<E2=E9S	2, 63	138.26	<0.01	E2=E9S<E9N
Kiwa sp. male	2, 31.36	147.29	<0.01	E2< E9S=E9N*	2, 66	19.52	<0.01	E2=E9S<E9N	2, 66	52.64	<0.01	E2< E9N=E9S
Sericosura spp.	2, 15	215.00	<0.01	E9N<E9S<E2	2, 15	100.61	<0.01	E9N<E2< E9S	2, 15	3.39	0.06	NA
Peltospiroidea sp.	2, 52	29.50	<0.01	E9N<E9S=E2	2, 52	49.26	<0.01	E9N<E9S<E2	2, 52	2.90	0.06	NA
Lepetodrilus sp.	2, 10	17.41	<0.01	E2=E9N<E9S	2, 10	31.99	<0.01	E9N=E9S<E2	2, 10	32.10	<0.01	E2=E9S<E9N

*Welch's ANOVA with *post hoc* analysis by t-test with Bonferroni correction ($p=0.05/3=0.017$).

Table 5. Results of t-tests for between-sites differences in stable isotope values of vent fauna at the East Scotia Ridge.

Species	Comparison	$\delta^{13}C$			$\delta^{34}S$			$\delta^{15}N$		
		DF	t	p	DF	t	p	DF	t	p
Kiwa sp.	E2 female v male	36	−0.50	0.62	36	2.23	<0.05	36	5.13	<0.01
Kiwa sp. Epibionts	E2 v E9S	4.05	−3.81	<0.05*	6.93	4.57	<0.01*	2	na	<0.05$^{\Psi}$
Stichasteridae sp.	E2 v E9N	4	28.92	<0.01	4	−3.29	<0.05	4	5.14	<0.01
	E2 v E9S	4	7.28	<0.01	4	0.385	0.59	4	5.94	<0.01
	E9N v E9S	8	3.64	<0.01	8	−1.76	0.11	8	−3.49	<0.05
Colossendeis cf *elphantis*	E2 v E9N	2	2.45	0.13	2	−6.93	<0.05	2	1.71	0.23

*Welch's t-test,
$^{\Psi}$Wilcoxon test.

The majority of ESR vent macroconsumers had $\delta^{34}S$ values less than or equal to the 10‰ threshold, indicating chemosynthetic food sources [49]. Species exceeding the 10‰ value occurred mainly at E2 in the anemones *Pacamanactis* sp. and cf *Marianactis* sp, the sponge *Cladorhiza* sp., the pycnogonids *C. elephantis* and the stichasterid seastar along with *Sericosura* spp. and stichasterid seastar at E9S. All had $\delta^{34}S$ values between 10‰ and 16‰. Mixing of epipelagic photosynthetic and hydrothermal vent chemosynthetic production sources at these sites cannot be ruled out.

Determining intra-site differences in food sources and trophic interactions using $\delta^{34}S$ is challenging for macroconsumers with $\delta^{34}S$ values <10‰ because the $\delta^{34}S$ values of inorganic substrates and the net fractionation effect between inorganic substrates and products for primary producers and consumers are uncertain. At E9, $\delta^{34}S$ appeared to increase from macroconsumers living closest to vent openings and within diffuse flow areas (i.e. *Kiwa* sp., Peltospiroidea sp. and *Lepetodrilus* sp.) to those in the periphery (i.e. anemones, stichasterid seastars and *Colossendeis* spp.). It is unclear why an increase in $\delta^{34}S$ occurred from the centre of the vent to the periphery: it may be the result of changes in sulfide speciation [50] or other sulfur sources with increasing distance from the vent opening [33], differences in levels of sulfide exposure [50], incorporation of epipelagic photosynthetic primary production or a combination of the above.

Stichasterid seastars, cf *Actinostola* sp. and *Colossendeis* spp. consistently had the highest $\delta^{15}N$ values of all the ESR vent macroconsumers, which suggested they occupied the highest trophic positions of those predators sampled. Behavioural observations [22] and $\delta^{13}C$ values indicated that *Kiwa* sp. is consumed by stichasterid seastar and cf *Actinostola* sp. 1 but only the stichasterid seastar had $\delta^{15}N$ values indicative of a higher trophic position than *Kiwa* sp. In the case of *Colossendeis* spp., feeding on anemones occurs at the ESR vent sites [22] and at E2 all three stable isotopes indicated a strong predator-prey link. At E9N there was a large difference in $\delta^{13}C$ and $\delta^{34}S$ between cf *Actinostola* sp. 1 and the two species of *Colossendeis* as well as lower $\delta^{15}N$ in these pycnogonids compared to cf *Actinostola* sp. 1. This suggests that at E9N the feeding incidents between cf *Actinostola* sp. 1 and *Colossendeis* spp. are either rare or stable isotope values of *Colossendeis* spp. are strongly affected by isotopic mixing of different energy sources ($\delta^{13}C$ and $\delta^{34}S$) and feeding over multiple trophic positions ($\delta^{15}N$).

It is evident from the ESR hydrothermal vent food webs that predators may have similar or lower $\delta^{15}N$ values than their prey. Calculating trophic position assuming taxon specific nitrogen trophic discrimination factors [23] or applying the more universal value of 3.4‰ [12] was not undertaken within this study because they may have provided erroneous results. Establishing a suitable $\delta^{15}N$ baseline is problematic because: the macroconsumer with the lowest $\delta^{15}N$ differed among locations, is confounded by the use of different tissues (e.g. whole animals, muscle) to construct the food webs [32] and the observed high $\delta^{15}N$ variability in potential food sources. Compound-specific amino acid stable isotope analysis may provide higher resolution information on the organic nitrogen compounds assimilated by vent macroconsumers because the isotopic values of different amino acids record trophic and basal source information [51,52]. Thus it may circumvent some of the limitations of bulk $\delta^{15}N$ analysis and provide a better understanding of nitrogen cycling at hydrothermal vents.

Spatial Patterns in Macroconsumer Trophodynamics

Large spatial differences in $\delta^{13}C$ values for *Kiwa* sp., Stichasteridae sp. and *Sericosura* spp. were attributed primarily to differences in carbon fixation pathways at the base of the food web, which is in turn transferred to higher trophic positions. $\delta^{13}C$ values of *Kiwa* sp. differed by ~9‰ between E2 and E9S as did that of associated *Kiwa* sp. epibionts. Also, epsilon-Proteobacteria dominated the epibiont community at E9 with gamma-Proteobacteria largely absent, compared to a mix of gamma- and epsilon-Proteobacteria at E2 (K. Zwirglmaier unpublished data). All epsilon-Proteobacteria to date use the rTCA cycle to fix carbon while gamma-Proteobacteria predominantly use the CBB cycle [11]. *Riftia pachyptila* has similar differences in $\delta^{13}C$ among vent sites, but this is attributed to its endosymbionts shifting between rTCA and CBB cycles [53] rather than changes in the microbial community it consumes. Alvinoconchid gastropods have $\delta^{13}C$ values that differ by >20‰ among vent fields, which relates to whether epsilon- or gamma-Proteobacteria are the endosymbionts [54]. It is unclear why *Kiwa* sp. epibiont diversity is different between E2 and E9. At other hydrothermal vent locations differences in vent fluid chemical composition influences microbial communities [46,55] and it may be similar at the ESR vent fields. The difference in carbon fixation appeared to be transferred through *Kiwa* sp. to the predatory stichasterid seastar. Such a predator-prey interaction may also explain the large difference in $\delta^{13}C$ values between E2 and E9N in *Sericosura* spp. At E2 *Sericosura* spp. were collected from amongst anemones that had $\delta^{13}C$ values indicative of a mixed carbon source but at E9 they were collected from amongst peltospiroid gastropods dependent on CBB fixed carbon, although *Sericosura* spp. were not observed directly feeding on either anemones or Peltospiroidea sp.

Relatively small differences in stable isotope values were observed among sites in Peltospiroidea sp., *Lepetodrilus* sp. and *Vulcanolepas* sp. To date, Peltospiroidea sp. contains a single strain of gamma-Proteobacteria endosymbiont (K. Zwirglmaier unpublished data), which means spatial differences in $\delta^{13}C$ and $\delta^{34}S$ are unlikely to be the result of differences in the type of endosymbiont [55]. The differences were potentially a result of site-specific variations in the $\delta^{13}C_{DIC}$ and inorganic $\delta^{34}S$ values used by the endosymbionts during chemoautotrophy or physiological temperature-related effects on isotopic discrimination. Small differences among sites for the grazer *Lepetodrilus* sp. and suspension feeder *Vulcanolepas* sp. are harder to explain because of the various factors that are likely to influence their food source. $\delta^{13}C$ values indicated these organisms consume a mixed diet of free-living microbes and particulate material. However, differences in $\delta^{13}C$ values within sites may be related to the organism's distribution within the vent field [56] and in turn the composition of the microbial community [46], the stable isotope values of the inorganic substrate used during chemoautotrophy [57] and temperature effects on trophic discrimination. *Lepetodrilus* sp. and *Vulcanolepas* sp. were collected from single points within each vent site and, therefore, it is not clear whether the difference in stable isotope values among sites is greater or less than that within sites.

Because of the snap-shot nature of this study, it is difficult to identify factors that caused the spatial differences in the *Kiwa* sp. epibiont communities that resulted in a greater range of $\delta^{13}C$ values at the E9 sites compared to E2. Higher concentrations of dissolved sulfides in vent fluids may favour the rTCA pathway resulting in increasing numbers of organisms with $\delta^{13}C$ values greater than −16‰ [12]. On the ESR, E9 has higher hydrogen sulfide and lower chloride concentrations than E2 meaning that there are greater concentrations of available gases for microbial primary production due to phase separation [6,21]. Higher concentrations of reduced compounds and gases may be one of the drivers of the differences in trophic structure at the ESR vents. However, hydrothermal vent communities also undergo changes in community composition with age [58] and fluctuating hydrothermal activity [59], which will have an effect on trophic structure. As data presented here were obtained concurrently with the discovery of the new biogeographical province it is not possible to determine whether the communities at E2 and E9 represent different successional stages, are a product of varying chemistry or a mix of such processes.

Conclusion

Trophic structure differed substantially between the E2 and E9 vents fields, and only slightly between E9N and E9S. $\delta^{13}C_{DIC}$ of the end-member fluid and diffuse flow samples were similar among the sites but large differences in the $\delta^{13}C$ values of some vent macroconsumers indicated spatial variations in the way microbes were fixing carbon at the base of the food chain. $\delta^{13}C$ values >−13‰ at the E9N and E9S suggest that the relative contribution to the macroconsumer food web of carbon fixed via the rTCA cycle is likely to be greater than at E2. The greater range of $\delta^{34}S$ values at E2 and E9S indicated a potentially greater influence of epipelagic photosynthetic primary production than at E9N. The greater contribution of rTCA fixed carbon at the E9 vent field may ultimately be related to differences in vent fluid, but more work is required to link vent fluid chemistry with microbial primary production and the related trophic structure at hydrothermal vents.

Acknowledgments

We thank the officers, crew, ROV team & scientists onboard JC42 for assistance in sample collection and sorting. We also thank Veerle Huvenne for producing the bathymetric map.

Author Contributions

Conceived and designed the experiments: WDKR CJS BDW KZ JAH KL NVCP. Performed the experiments: WDKR CJS KZ JAH KL. Analyzed the data: WDKR CJS BDW KZ JAH RARM NVCP. Contributed reagents/materials/analysis tools: WDKR CJS KZ JAH RARM. Wrote the paper: WDKR CJS BDW KZ JAH RARM KL NCVP. Participants on JC042: WDKR CJS JAH KZ KL.

References

1. Kelley DS, Karson JA, Blackman DK, Fruh-Green GL, Butterfield DA, et al. (2001) An off-axis hydrothermal vent field near the Mid-Atlantic Ridge at 30°N Nature 412: 145–149.
2. Staudigel H, Hart SR, Pile A, Bailey SE, Baker ET, et al. (2006) Vailulu'u seamount, Samoa: Life and death on an active submarine volcano. Proc Natl Acad Sci USA 103: 6448–6453.
3. Tunnicliffe V, Juniper SK, Sibuet M (2003) Reducing environments of the deep-sea floor. In: Tyler PA, editor. Ecosystems of the Deep Ocean. Amsterdam: Elsevier Science. 81–110.
4. Cavanaugh CM, Gardiner SL, Jones ML, Jannasch HW, Waterbury JB (1981) Prokaryotic cells in the hydrothermal vent tube worm *Riftia pachyptila* Jones - possible chemoautotrophic symbionts. Science 213: 340–342.
5. German CR, Ramirez-Llodra E, Baker MC, Tyler PA (2011) Deep-water chemosynthetic ecosystem research during the Census of Marine Life decade and beyond: A proposed deep-ocean road map. PLoS ONE 6: 16.
6. German CR, Von Damm KL (2003) Hydrothermal Processes. In: Elderfield H, editor. The ocean and marine geochemistry. Oxford: Elsevier. 181–222.
7. Karl DM (1995) Ecology of free-living, hydrothermal vent microbial communities. In: Karl DM, editor. The microbiology of deep-sea hydrothermal vents. Boca Raton: CRC Press Inc. 35–124.
8. McCollom TM, Shock EL (1997) Geochemical constraints on chemolithoautotrophic metabolism by microorganisms in seafloor hydrothermal systems. Geochim Cosmochim Acta 61: 4375–4391.
9. Campbell BJ, Cary SC (2004) Abundance of reverse tricarboxylic acid cycle genes in free-living microorganisms at deep-sea hydrothermal vents. Appl Environ Microbiol 70: 6282–6289.
10. Desbruyeres D, Alaysedanet AM, Ohta S, Antoine E, Barbier G, et al. (1994) Deep-sea hydrothermal communities in the southwestern Pacific back-arc basins (the North Fiji and Lau Basins)- composition, microdistribution and food web. Mar Geol 116: 227–242.
11. Hugler M, Sievert SM (2011) Beyond the Calvin Cycle: Autotrophic carbon fixation in the ocean. Ann Rev Mar Sci. Palo Alto: Annual Reviews. 261–289.
12. De Busserolles F, Sarrazin J, Gauthier O, Gelinas Y, Fabri MC, et al. (2009) Are spatial variations in the diets of hydrothermal fauna linked to local environmental conditions? Deep Sea Res Part II Top Stud Oceanogr 56: 1649–1664.
13. Riou V, Colaco A, Bouillon S, Khripounoff A, Dando PR, et al. (2010) Mixotrophy in the deep sea: a dual endosymbiotic hydrothermal mytilid assimilates dissolved and particulate organic matter Mar Ecol Prog Ser 405: 187–201.
14. Colaco A, Dehairs F, Desbruyeres D (2002) Nutritional relations of deep-sea hydrothermal fields at the Mid-Atlantic Ridge: a stable isotope approach. Deep Sea Res Part I Oceanogr Res Pap 49: 395–412.
15. Rau GH, Hedges JI (1979) Carbon-13 depletion in a hydrothermal vent mussel - suggestion of a chemosynthetic food source. Science 203: 648–649.
16. Van Dover CL, Fry B (1994) Microorganisms as food resources at deep-sea hydrothermal vents. Limnol Oceanogr 39: 51–57.
17. Levin LA, Mendoza GF, Konotchick T, Lee R (2009) Macrobenthos community structure and trophic relationships within active and inactive Pacific hydrothermal sediments. Deep Sea Res Part II Top Stud Oceanogr 56: 1632–1648.
18. Limen H, Juniper SK (2006) Habitat controls on vent food webs at Eifuku Volcano, Mariana Arc. Cah Biol Mar 47: 449–455.
19. Van Dover CL (2002) Trophic relationships among invertebrates at the Kairei hydrothermal vent field (Central Indian Ridge). Mar Biol 141: 761–772.
20. German CR, Livermore RA, Baker ET, Bruguier NI, Connelly DP, et al. (2000) Hydrothermal plumes above the East Scotia Ridge: an isolated high-latitude back-arc spreading centre. Earth Planet Sci Lett 184: 241–250.
21. Rogers AD, Tyler PA, Connelly DP, Copley JT, James R, et al. (2012) The discovery of new deep-sea hydrothermal vent communities in the Southern Ocean and implications for biogeography. PLoS Biol 10: e1001234.

22. Marsh L, Copley JT, Huvenne VAI, Linse K, Reid WDK, et al. (2012) Microdistribution of faunal assemblages at deep-sea hydrothermal vents in the Southern Ocean. PLoS ONE 7: e48348.
23. Bergquist DC, Eckner JT, Urcuyo IA, Cordes EE, Hourdez S, et al. (2007) Using stable isotopes and quantitative community characteristics to determine a local hydrothermal vent food web. Mar Ecol Prog Ser 330: 49–65.
24. Van Dover CL, Fry B (1989) Stable isotopic compositions of hydrothermal vent organisms. Mar Biol 102: 257–263.
25. Fisher CR, Childress JJ, Macko SA, Brooks JM (1994) Nutritional interactions in Galapagos Rift hydrothermal vent communities - inferences from stable carbon and nitrogen isotope analyses. Mar Ecol Prog Ser 103: 45–55.
26. Guy RD, Fogel ML, Berry JA (1993) Photosynthetic fractionation of the stable isotopes of oxygen and carbon. Plant Physiology 101: 37–47.
27. Robinson JJ, Scott KM, Swanson ST, O'Leary MH, Horken K, et al. (2003) Kinetic isotope effect and characterization of form II RuBisCO from the chemoautotrophic endosymbionts of the hydrothermal vent tubeworm *Riftia pachyptila*. Limnol Oceanogr 48: 48–54.
28. Roeske CA, O'Leary MH (1984) Carbon isotope effects on the enzyme-catalyzed carboxylation of ribulose bisphosphate. Biochemistry 23: 6275–6284.
29. House CH, Schopf JW, Stetter KO (2003) Carbon isotopic fractionation by Archaeans and other thermophilic prokaryotes. Org Geochem 34: 345–356.
30. Suzuki Y, Sasaki T, Suzuki M, Nogi Y, Miwa T, et al. (2005) Novel chemoautotrophic endosymbiosis between a member of the Epsilonproteobacteria and the hydrothermal-vent gastropod *Alviniconcha* aff. *hessleri* (Gastropoda: Provannidae) from the Indian Ocean. Appl Environ Microbiol 71: 5440–5450.
31. Wirsen CO, Sievert SM, Cavanaugh CM, Molyneaux SJ, Ahmad A, et al. (2002) Characterization of an autotrophic sulfide-oxidizing marine *Arcobacter* sp that produces filamentous sulfur. Appl Environ Microbiol 68: 316–325.
32. Michener RH, Kaufman L (2007) Stable isotope ratios as tracers in marine food webs: an update. In: Michener RH, Lajtha K, editors. Stable isotopes in ecology and environmental science. 2nd ed. Singapore: Blackwell Publishing. 238–282.
33. Herzig PM, Hannington MD, Arribas A Jr (1998) Sulfur isotopic composition of hydrothermal precipitates from the Lau back-arc: implications for magmatic contributions to seafloor hydrothermal systems. Mineralium Deposita 33: 226–237.
34. Erickson KL, Macko SA, Van Dover CL (2009) Evidence for a chemoautotrophically based food web at inactive hydrothermal vents (Manus Basin). Deep Sea Res Part II Top Stud Oceanogr 56: 1577–1585.
35. Reid WDK, Wigham BD, McGill RAR, Polunin NVC (2012) Elucidating trophic pathways in benthic deep-sea assemblages of the Mid-Atlantic Ridge north and south of the Charlie-Gibbs Fracture Zone. Mar Ecol Prog Ser 463: 89–103.
36. Layman CA, Arrington DA, Montana CG, Post DM (2007) Can stable isotope ratios provide for community-wide measures of trophic structure? Ecology 88: 42–48.
37. Quinn GP, Keough MJ (2002) Experimental Design and Data Analysis for Biologists. Cambridge: Cambridge University Press. 537 p.
38. R Core Team (2011) R: A language and environment for statistical computing. R Foundation for Statistical Computing, Vienna, Austria. ISBN 3-900051-07-0, URL http://www.R-project.org/.
39. Brooks JM, Kennicutt MC, Fisher CR, Macko SA, Cole K, et al. (1987) Deep-sea hydrocarbon seep communities - evidence for energy and nutritional carbon sources. Science 238: 1138–1142.
40. Childress JJ, Fisher CR (1992) The biology of hydrothermal vent animals-physiology, biochemistry and autotrophic symbioses. Oceanogr Mar Biol 30: 337–441.
41. Campbell BJ, Engel AS, Porter ML, Takai K (2006) The versatile epsilon-Proteobacteria: key players in sulphidic habitats. Nat Rev Microbiol 4: 458–468.
42. Campbell BJ, Stein JL, Cary SC (2003) Evidence of chemolithoautotrophy in the bacterial community associated with *Alvinella pompejana*, a hydrothermal vent polychaete. Appl Environ Microbiol 69: 5070–5078.
43. Levesque C, Juniper SK, Marcus J (2003) Food resource partitioning and competition among alvinellid polychaetes of Juan de Fuca Ridge hydrothermal vents. Mar Ecol Prog Ser 246: 173–182.
44. Suzuki Y, Kojima S, Sasaki T, Suzuki M, Utsumi T, et al. (2006) Host-symbiont relationships in hydrothermal vent gastropods of the genus Alviniconcha from the Southwest Pacific. Appl Environ Microbiol 72: 1388–1393.
45. Thurber AR, Jones WJ, Schnabel K (2011) Dancing for food in the deep sea: bacterial farming by a new species of yeti crab. PLoS ONE 6: e26243.
46. Flores GE, Campbell JH, Kirshtein JD, Meneghin J, Podar M, et al. (2011) Microbial community structure of hydrothermal deposits from geochemically different vent fields along the Mid-Atlantic Ridge. Environ Microbiol 13: 2158–2171.
47. Takai K, Nunoura T, Horikoshi K, Shibuya T, Nakamura K, et al. (2009) Variability in microbial communities in black smoker chimneys at the NW Caldera vent field, Brothers Volcano, Kermadec Arc. Geomicrobiol J 26: 552–569.
48. Nakagawa S, Takai K (2008) Deep-sea vent chemoautotrophs: diversity, biochemistry and ecological significance. FEMS Microbiol Ecol 65: 1–14.
49. Vetter RD, Fry B (1998) Sulfur contents and sulfur-isotope compositions of thiotrophic symbioses in bivalve molluscs and vestimentiferan worms. Mar Biol 132: 453–460.
50. Luther GW, Rozan TF, Taillefert M, Nuzzio DB, Di Meo C, et al. (2001) Chemical speciation drives hydrothermal vent ecology. Nature 410: 813–816.
51. Chikaraishi Y, Ogawa NO, Kashiyama Y, Takano Y, Suga H, et al. (2009) Determination of aquatic food-web structure based on compound-specific nitrogen isotopic composition of amino acids. Limnol Oceanogr Methods 7: 740–750.
52. McClelland JW, Montoya JP (2002) Trophic relationships and the nitrogen isotopic composition of amino acids in plankton. Ecology 83: 2173–2180.
53. Markert S, Arndt C, Felbeck H, Becher D, Sievert SM, et al. (2007) Physiological proteomics of the uncultured endosymbiont of *Riftia pachyptila*. Science 315: 247–250.
54. Suzuki Y, Sasaki T, Suzuki M, Tsuchida S, Nealson KH, et al. (2005) Molecular phylogenetic and isotopic evidence of two lineages of chemoautotrophic endosymbionts distinct at the subdivision level harbored in one host-animal type: The genus *Alviniconcha* (Gastropoda : Provannidae). FEMS Microbiol Lett 249: 105–112.
55. Trask JL, Van Dover CL (1999) Site-specific and ontogenetic variations in nutrition of mussels (*Bathymodiolus* sp.) from the Lucky Strike hydrothermal vent field, Mid-Atlantic Ridge. Limnol Oceanogr 44: 334–343.
56. Levesque C, Juniper SK, Limen H (2006) Spatial organization of food webs along habitat gradients at deep-sea hydrothermal vents on Axial Volcano, Northeast Pacific. Deep Sea Res Part I Oceanogr Res Pap 53: 726–739.
57. Levesque C, Limen H, Juniper SK (2005) Origin, composition and nutritional quality of particulate matter at deep-sea hydrothermal vents on Axial Volcano, NE Pacific. Mar Ecol Prog Ser 289: 43–52.
58. Shank TM, Fornari DJ, Von Damm KL, Lilley MD, Haymon RM, et al. (1998) Temporal and spatial patterns of biological community development at nascent deep-sea hydrothermal vents (9 degrees 50′ N, East Pacific Rise). Deep Sea Res Part II Top Stud Oceanogr 45: 465–515.
59. Cuvelier D, Sarrazin J, Colaco A, Copley JT, Glover AG, et al. (2011) Community dynamics over 14 years at the Eiffel Tower hydrothermal edifice on the Mid-Atlantic Ridge. Limnol Oceanogr 56: 1624–1640.

9

IL-17A Modulates Oxidant Stress-Induced Airway Hyperresponsiveness but Not Emphysema

Mariona Pinart[1], Min Zhang[1], Feng Li[1], Farhana Hussain[1], Jie Zhu[1], Coen Wiegman[1], Bernard Ryffel[2], Kian Fan Chung[1]*

1 Airways Disease, National Heart & Lung Institute, Imperial College, London, United Kingdom, **2** Université d'Orléans and CNRS, UMR 6218 Molecular Immunology and Embryology, Orléans, France

Abstract

IL-17A induces the release of pro-inflammatory cytokines and of reactive oxygen species which could lead to neutrophilic inflammation. We determined the role of IL-17 receptor (IL-17R) signalling in oxidant-induced lung emphysema and airway hyperresponsiveness. IL-17R$^{-/-}$ and wild-type C57/BL6 mice were exposed to ozone (3 ppm; 3 hours) for 12 times over 6 weeks. Bronchial responsiveness to acetylcholine was measured, and lungs were retrieved. Mean linear intercept (L_m) and isometric contractile responses of intrapulmonary airways to acetylcholine were determined. In wild-type mice but not in IL-17R$^{-/-}$, chronic ozone exposure caused airway hyperresponsiveness. The increase in L_m after chronic ozone exposure of wild-type mice was also observed in IL-17R$^{-/-}$ mice. The increased maximal contractile response to acetylcholine seen in airways of wild-type mice exposed to ozone was abolished in IL-17R$^{-/-}$ mice. p38-mitogen-activated protein kinase (MAPK) and dexamethasone-dependent increase in contractile response was reduced in airways from IL-17R$^{-/-}$ ozone-exposed mice. Lung inflammation scores were not altered in IL-17R$^{-/-}$ mice exposed to ozone compared to wild-type mice. The increased release of IL-17 and IL-1β, and the activation of p38 MAPK in the lungs of ozone-exposed mice was reduced in IL-17R$^{-/-}$ mice. IL-17R signalling underlies the increase in airway hyperresponsiveness seen after ozone exposure, mediated by the increased contractility of airway smooth muscle. The emphysema and lung inflammation induced by ozone is not dependent on IL-17.

Editor: Chuen-Mao Yang, Chang Gung University, Taiwan

Funding: This work was supported by Project Grant 083905 from the Wellcome Trust. MP and FL were supported by Long-term Fellowships from the European Respiratory Society. Dr Mariona Pinart is the recipient of a European Respiratory Society/Marie Curie Joint Postdoctoral Research Fellowship - Number MC [0935]-2009. The research leading to these results has received funding from the European Respiratory Society and the European Community's Seventh Framework Programme FP7/2007-2013 - Marie Curie Actions under grant agreement RESPIRE, PCOFUND-GA-2008-229571. The funders had no role in study design, data collection and analysis, decision to publish, or preparation of the manuscript.

Competing Interests: The authors have declared that no competing interests exist.

* E-mail: f.chung@imperial.ac.uk

Introduction

Cigarette smoking is the most commonly encountered risk factor for chronic obstructive pulmonary disease (COPD) and is a potent inducer of oxidative stress, which plays an important role in the pathogenesis of COPD by activating pathways that lead to chronic inflammation and emphysema as demonstrated in mouse models of cigarette exposure [1,2]. Further evidence for a role for oxidative stress comes from the observation that, following cigarette smoke exposure, Nrf-2 knockout mice that express lower levels of antioxidant gene expression, were more susceptible to developing emphysema and lung inflammation [3].The crucial role of oxidant stress in the induction of COPD and emphysema is also supported by the observation that chronic exposure of mice to ozone, a ubiquitous oxidizing and toxic air pollutant generated photochemically from nitric oxides and hydrocarbons, led to the development of emphysema-like lung injury with alveolar enlargement and chronic lung inflammation [4]. In addition, exposure to ozone also increases the contractility of the airways and causes bronchial hyperrresponsiveness to constrictor agents such as acetylcholine [5].

Interleukin-17 (IL-17, also known as IL-17A), is produced by CD4+ Th17 cell, cytotoxic T-cells, invariant natural killer T-cells, lymphoid tissue-induced cells and γδT cells [6]. The IL-17 receptor (IL-17R) family comprises five receptor subunits, of which IL-17RA is the largest member and is necessary for IL-17A-mediated signal transduction [7]. IL-17A induces the release of the pro-inflammatory cytokines, CXCL-8, CXCL1 (GRO-α), KC, G-CSF and GM-CSF from airway epithelial cells, smooth muscle cells and macrophages, and thereby orchestrates neutrophilic inflammation and release of reactive oxygen species [6,8].

The role of IL-17 in COPD has been reinforced by the report that over-expression of IL-17 in murine lung epithelium induced a COPD-like phenotype [9]. Administration of IL-17A into the airways increased neutrophil and chemokine expression [10]. IL-17A$^+$ cells in the submucosa and IL-17 levels in the sputum were reported to be increased in COPD patients [11]. These observations indicate that IL-17 may play a role in COPD. Indeed, in a recent study of cigarette exposure in mice, the induction of emphysema was found to be partly dependent on IL-17 [12]. The induction of airway hyperresponsiveness by ozone exposure has also been shown to be dependent on IL-17 [13]. However, the role of IL-17 on the inflammatory response and emphysema induced by ozone are not known. We hypothesised that

IL-17A may play an important role in airway hyperresponsiveness, pulmonary inflammation and emphysema induced by chronic exposure to ozone. We also studied the potential role of IL-17 on the direct contractile response of intrapulmonary airways to acetylcholine.

Results

In vivo Airway Responsiveness

There were no significant differences in the baseline lung resistance (R_L) values following PBS challenge in the four groups of mice. Air-exposed IL-17R$^{-/-}$ mice showed a non-significant higher responsiveness to ACh compared with air-exposed C57/BL6 mice (Fig. 1). Airway hyperresponsiveness (AHR) to ACh was induced in chronic ozone-exposed C57/BL6 mice compared with air-exposed mice ($-\log PC_{100}$ ozone: -1.510 ± 0.088 vs. air: -2.055 ± 0.126; $p<0.01$; Fig. 1). However, IL-17R$^{-/-}$ mice exposed to ozone did not exhibit AHR to ACh compared with IL-17R$^{-/-}$ air-exposed mice ($-\log PC_{100}$ ozone: -1.713 ± 0.086 vs air: -1.722 ± 0.160; Fig. 1).

Isometric Contractile Response of Intrapulmonary Airways

Isometric contractile responses to ACh (i.e E_{max}) were similar in C57/BL6 mice and IL-17R$^{-/-}$ mice exposed to air. E_{max} was increased in ozone-exposed C57/BL6 mice compared with air-exposed mice (9.44±0.28 vs 6.86±0.23 mN, respectively; $p<0.05$), but pEC_{50} values remained unchanged (5.30±0.06 vs 5.34±0.08) (Fig. 2A). However, E_{max} was not significantly changed in ozone-exposed IL-17R$^{-/-}$ mice compared with air-exposed mice (7.39±0.30 vs 7.25±0.37 mN, respectively) and pEC_{50} values were not different (pEC_{50}:5.24±0.08 vs 5.33±0.11; Fig. 2B).

Effects of p38-MAPK Inhibitor and Dexamethasone ex-vivo on Contractile Responses

In order to understand how IL-17 can modulate the contractile process, we studied the contractile responses of the intrapulmonary airways in the presence of a selective p38 MAPK inhibitor, SB239063 (10^{-6} M), and of a corticosteroid, dexamethasone (10^{-6} M), as previously described [14]. In the presence of SB239063, there was a reduction in the maximal contractility response in both mouse strains exposed to air or to ozone. Comparing the airways from ozone–exposed mice from the C57/BL6 and IL-17R$^{-/-}$ mice, we found a p38-dependent portion of the contractility attributable to IL-17 (Fig. 3C, D).

We also studied the effect of dexamethasone on the contractile responses *ex-vivo*. In C57/Bl6 mice, dexamethasone inhibited the maximal contractile responses to acetylcholine non-significantly, although in IL-17R$^{-/-}$ mice, this inhibition was significant in air-exposed mice (Fig. 4). The effect of dexamethasone on ozone-exposed airways was different in the C57/Bl6 and IL-17R$^{-/-}$ mice. Hence, dexamethasone inhibited ozone-induced enhancement of the contractile response in C57/Bl6 mice but there was no such inhibition in IL-17R$^{-/-}$ mice exposed to ozone.

Mean Linear Intercept (L_m) and Emphysema Scores

There was no significant difference in L_m in air-exposed C57/BL6 mice and IL-17R$^{-/-}$ mice (50.31±1.67 vs 53.61±2.12). L_m was increased in ozone-exposed C57/BL6 mice compared with air-exposed mice (Fig. 5A), indicating that ozone-exposed mice showed an increase in alveolar size and therefore developed an emphysematous-like pattern (ozone: 66.79±2.78 vs air: 50.31±1.67, $P<0.001$). Similarly, L_m was increased in ozone-exposed IL-17R$^{-/-}$ mice compared to air-exposed mice (ozone: 65.16±2.04 vs air: 53.61±2.12) (Fig. 5A).

The baseline emphysema score was similar in air-exposed C57/BL6 mice and IL-17R$^{-/-}$ mice. The emphysema score was increased in both ozone-exposed C57/BL6 and IL-17R$^{-/-}$mice. However, IL-17R$^{-/-}$ mice exposed to ozone had a higher emphysema score compared with C57/BL6 mice (IL-17R$^{-/-}$:2.15±0.26 vs C57/BL6:0.98±0.21, $p<0.01$) (Fig. 4B). Representative photomicrographs in Fig. 5C show alveolar enlargement occurring in the ozone-exposed mice.

Inflammation Scores

There was no significant difference in the lung inflammation score at baseline in the air-exposed C57/BL6 and IL-17R$^{-/-}$ mice. In the IL-17R$^{-/-}$ mice exposed to ozone, the inflammatory score was not significantly different from the air-exposed mice; however, this may be accounted for by the slightly higher baseline inflammation score in the IL-17R$^{-/-}$ mice (Fig. 6). The inflammation score significantly increased in ozone-exposed C57/BL6 mice but not in IL-17R$^{-/-}$ mice (Fig. 5).

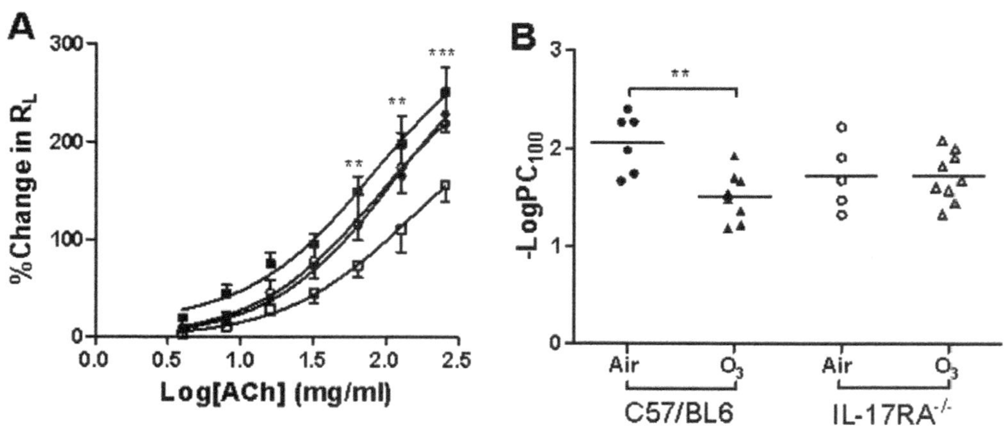

Figure 1. Concentration-response curves to acetylcholine (ACh) and –log provocative concentration of ACh required to increase lung resistance (R_L) by 100% from baseline (PC_{100}). C57/BL6 and IL-17R$^{-/-}$ mice were exposed to air or to ozone. Data is expressed as mean ± S.E.M. **$p<0.01$; ***$p<0.001$, compared with to air-exposed mice.

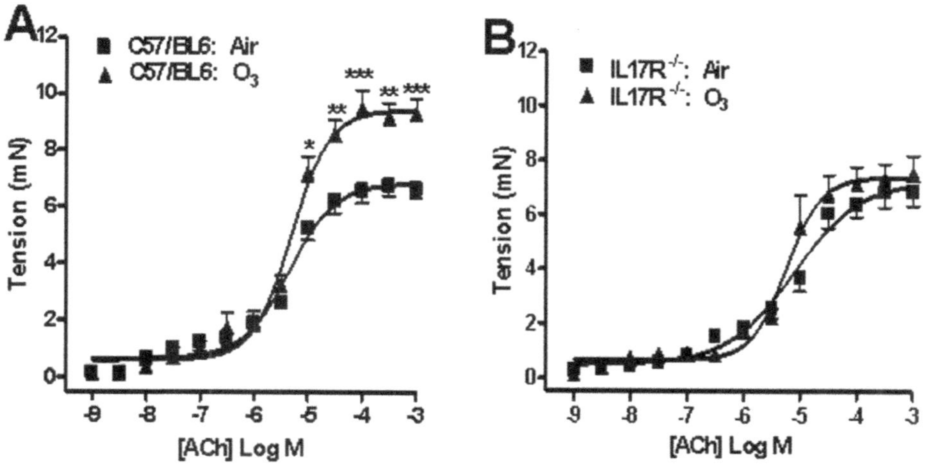

Figure 2. Acetylcholine (ACh)-induced isometric bronchial contractile tension. Air- and ozone-exposed C57/BL6 mice (6 in air- and 9 in ozone-exposed) and $IL17RA^{-/-}$ mice (6 in air- and 5 in ozone-exposed) were studied. Data expressed as mean ±S.E.M. $^{*}p<0.05$, $^{**}p<0.01$, $^{***}p<0.001$, compared with air-exposed mice.

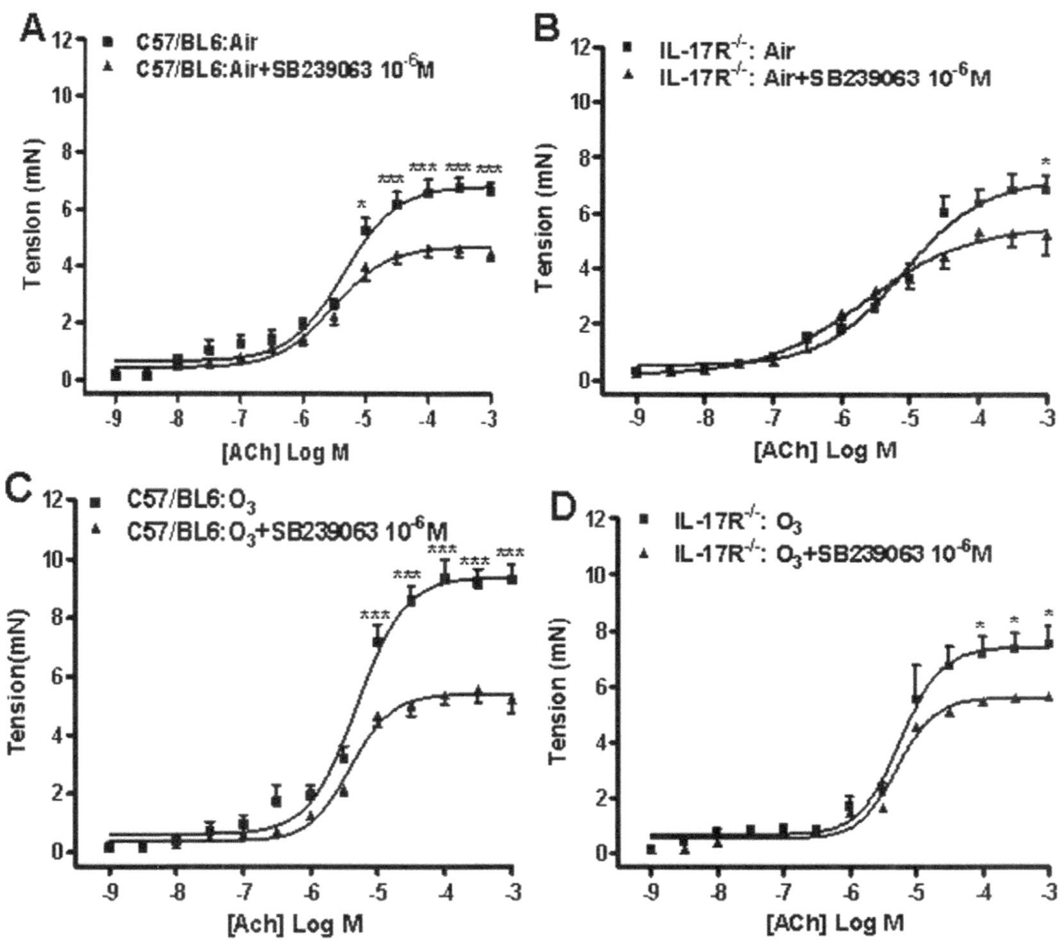

Figure 3. Effect of SB239063 (10^{-6} M) on acetylcholine (ACh)-induced bronchial contractile responses. Air-exposed C57/BL6 mice (n = 6; Panel A) and $IL-17R^{-/-}$ mice (n = 6; Panel B) and ozone-exposed C57/BL6 mice (n = 9; Panel C) and $IL-17R^{-/-}$ mice (n = 6; Panel D) were studied. Under each condition, the effect of SB239063 has been compared to responses in the absence of this inhibitor. Data presented as mean±SEM. $^{*}p<0.05$; $^{**}p<0.01$ compared with SB239063-treated tissues.

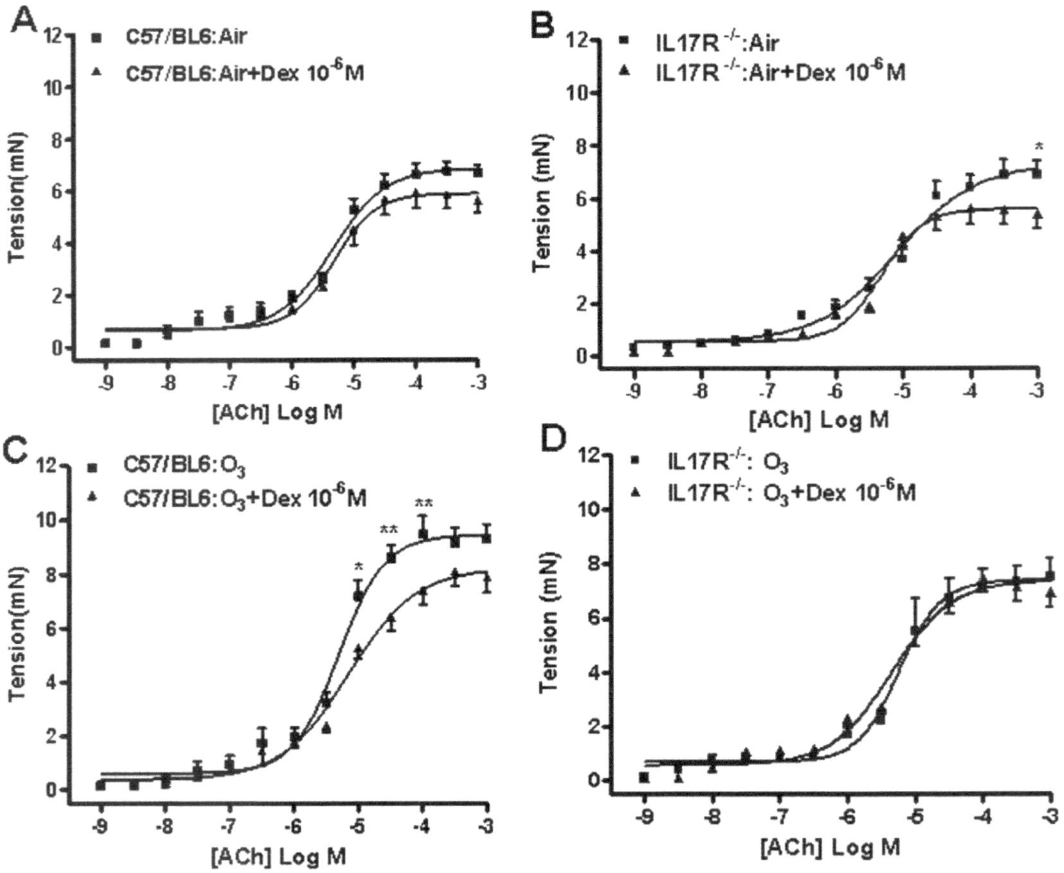

Figure 4. Effect of dexamethasone (10^{-6} M) on acetylcholine (ACh)-induced bronchial contractile responses. Air-exposed C57/BL6 mice (n = 6; Panel A) and IL-17R$^{-/-}$ mice (n = 6; Panel B) and ozone-exposed C57/BL6 mice (n = 9; Panel C) and IL-17R$^{-/-}$ mice (n = 6; Panel D) were studied. Under each condition, the effect of dexamethasone has been compared to responses in the absence of this inhibitor. Data presented as mean±SEM. *p<0.05 compared with dexamethasone-treated tissues.

Lung IL-17A, IL-1β and TNFα and Phosphorylated p38, JNK and ERK

In C57/Bl6 wild-type mice, lung levels of IL-17 were increased after ozone exposure (Fig. 7). Baseline levels of lung IL-17 in IL-17R$^{-/-}$ mice were similar to those in C57/Bl mice but the levels of lung IL-17 were less increased after ozone exposure (Fig. 7). Similarly, lung levels of IL-1β increased in C57/Bl6 mice exposed to ozone, but these levels were reduced in IL-17R$^{-/-}$ mice exposed to ozone. Lung levels of TNFα did not increase significantly after ozone exposure.

In C57/Bl6 mice, ozone exposure also led to an activation of p38 MAPK but not of ERK or JNK, since the levels of phosphorylated p38 were increased. On the other hand, in IL-17R$^{-/-}$ mice exposed to air, the amount of phosphorylated ERK and p38 measured was significantly reduced compared to C57/Bl6 mice. While ERK activity was increased by ozone exposure in IL-17R$^{-/-}$ mice, p38 activation was completely suppressed (Fig. 8).

Discussion

We have shown that repeated exposure to ozone induces lung emphysema and inflammation as previously reported, and in this study, we also demonstrated an increase in airway responsiveness to acetylcholine, accompanied by an *ex-vivo* increase in the maximal isometric contractile response to acetylcholine. The lung emphysema and inflammation, with the airway hyperresponsiveness are both irreversible when mice were studied up to 6 weeks after cessation of exposure to ozone (*unpublished data*). We now demonstrate that, in the IL-17 receptor knock-out mouse model, the airway hyperresponsiveness induced by multiple exposures to ozone is inhibited; we also found that this was accompanied by no enhancement of the maximal contractile response of bronchial airways. However, there was no inhibition of the emphysematous process and the chronic inflammatory response induced by chronic ozone exposure. Thus, we conclude that IL-17 is important for ozone-induced bronchial hyperresponsiveness but not for the induction of emphysema and inflammation. This dependent effect of airway hyperresponsiveness on IL-17 is likely to be a direct effect of IL-17 on airway smooth muscle. Our data is in agreement with previous studies using mice deficient in IL-17A or IL-17R. These studies reported that IL-17A was necessary for the development of airway hyperresponsiveness in an ovalbumin-induced asthma model [15,16]. Furthermore, a recent study using IL-17 knock-out mice reported that IL-17 was required for ozone-induced airway hyperresponsiveness through stimulation of the production of IL-17 by i-NKT cells and T cells in the lungs [13].

The mechanisms of ozone-induced airway hyperresponsiveness are complex. Our previous work indicated that a single ozone exposure could directly cause an increase in airway smooth muscle

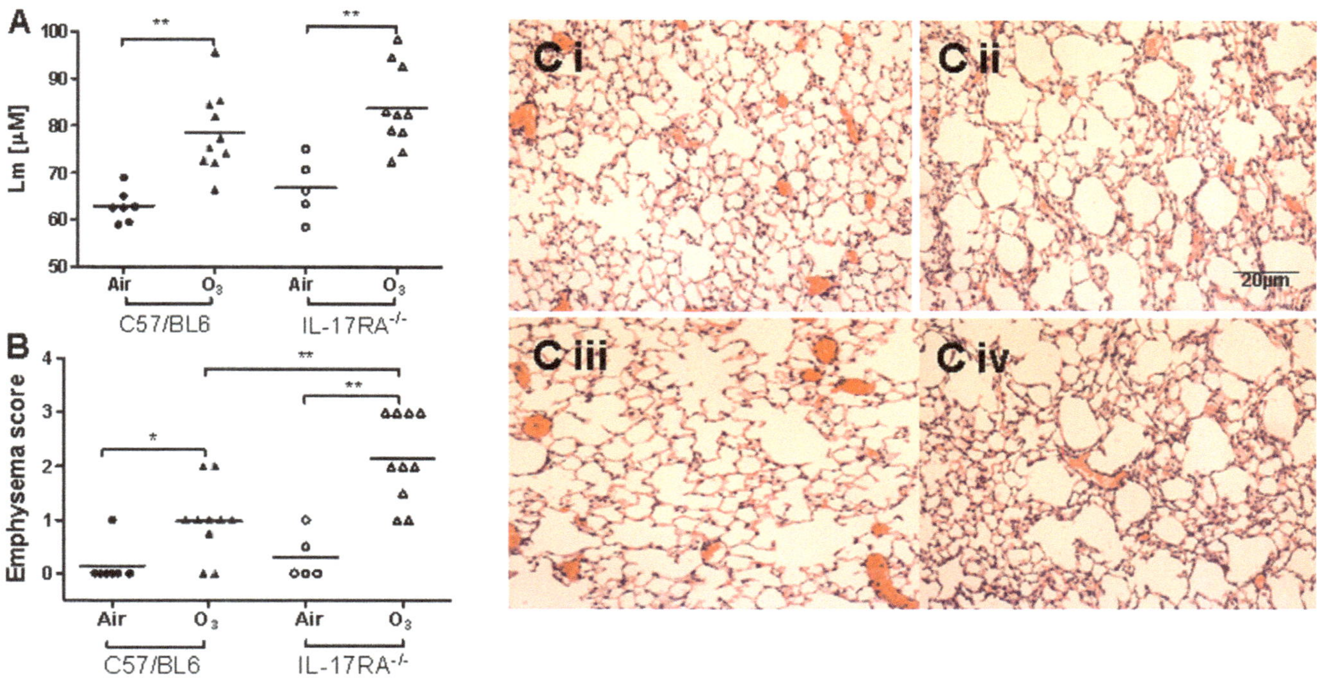

Figure 5. Mean linear intercept (L_m) in the lungs of air- and ozone exposed mice (Panel A). Lungs inflated at 25 cm of water were sectioned and stained with haematoxylin and eosin and microscopically assessed for L_m. Ozone exposed C57/BL6 mice and IL-17R$^{-/-}$ mice showed increased Lm (alveolar enlargement) compared with their appropriate air-exposed control mice. Emphysema score in the lungs of air- and ozone exposed mice (Panel B). Compared with air exposed mice, the emphysema score was increased in ozone-exposed C57/BL6 and IL-17R$^{-/-}$mice, while it was increased further in ozoneexposed IL-17R$^{-/-}$ mice. Data are expressed as means ± SEM. *$p<0.05$; **$p<0.01$; ***$p<0.001$. Representative histological sections of mouse lungs (Panel C i, ii, iii & iv). Lung sections were stained with haematoxylin and eosin after 6 weeks of exposure to ozone showing enlargement of alveolar spaces in C57/BL6 (Panel C ii) and IL-17A$^{-/-}$ mice (Panel C iv).

contractility to acetylcholine, which may contribute to the pathogenesis of airway hyperresponsiveness. Direct instillation of IL-17 into the airways of ovalbumin-sensitized and -challenged mice could trigger robust airway hyperresponsiveness [17]. IL-17 has also been shown to activate many common downstream signalling pathways, including NF-κB, and the MAPKs (mitogen-activated protein kinases) JNK (c-Jun N-terminal kinase), p38 and ERK (extracellular-signal-regulated kinase), together with other kinases including PI3K (phosphoinositide 3-kinase) and JAK (Janus kinase)/STATs [7]. IL-17 may increase AHR e through a direct action on the airway smooth muscle as IL-17 has been shown to activate p38 MAPK pathway in these cells [18]. We have previously shown the p38 MAPK/HSP27 is involved in the *ex-vivo* cholinergic agonist-induced increase in bronchial smooth muscle contractility following a single exposure to ozone [14]. We now show a similar response in the airways from mice chronically-exposed to ozone, with the development of concomitant emphysema. Our ex-vivo data of airway smooth muscle contractility indicates that IL-17 may contribute partly to the activation of p38 MAPK underlying the increase in contractility induced by ozone. Furthermore, in the current study, we demonstrate that IL-17A and IL-1β are produced in the lung after chronic exposure to ozone, and interestingly, this production was dependent partly on the presence of IL-17 receptor. This is particularly interesting for IL-17, which indicates an autoregulatory mechanism on the production of IL-17 through its own receptor. In addition, p38 MAPK activation is demonstrated after ozone exposure also, and this was abolished in IL-17R$^{-/-}$ mice. These observations in the lungs provide further support to the concept that IL-17 may contribute to the activation of p38 MAPK.

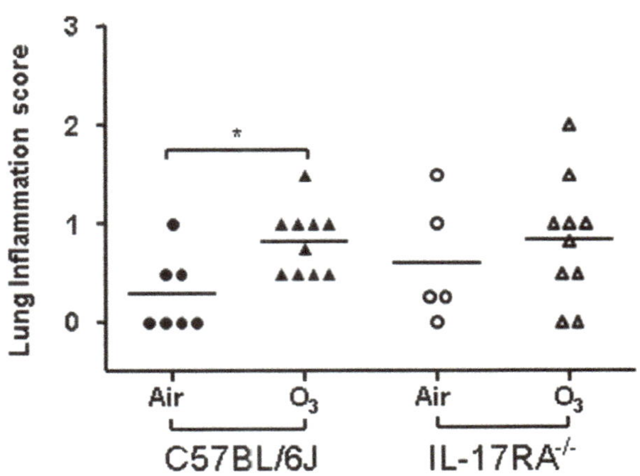

Figure 6. Inflammation score in the airways and lungs of air- and ozone-exposed mice. Compared with air exposed control mice, the inflammation score was increased significantly in ozone exposed C57/BL6 mice but not in IL-17R$^{-/-}$mice. Data are expressed as means ± SEM. *$p<0.01$.

Another aspect of interest in our study is the fact that dexamethasone inhibition of the maximal contractile response in IL-17R$^{-/-}$ mice is lost following exposure to chronic ozone exposure, an indication of corticosteroid insensitivity of the airway smooth muscle. This was not present in the wild-type mice, which indicates that IL-17A is important in determining the sensitivity of the airway smooth muscle to corticosteroids. This observation

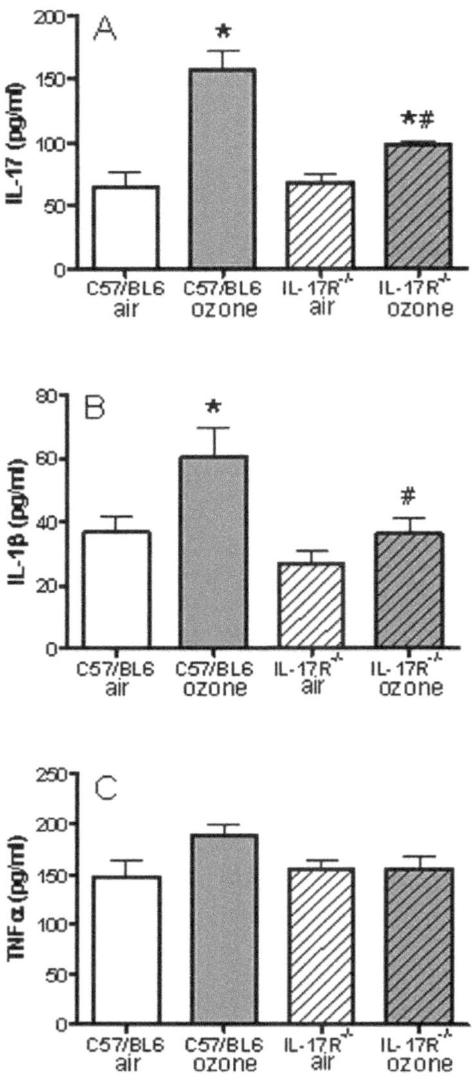

Figure 7. Levels of lung IL-17, IL-1β and TNFα. C57/Bl6 and IL-17R$^{-/-}$ mice were exposed to air or to ozone. Data shown as mean ± SEM for n = 5 in each group; *p<0.05 compared to air in the same species; #p<0.05 compared to C57/Bl6 mice with corresponding exposure.

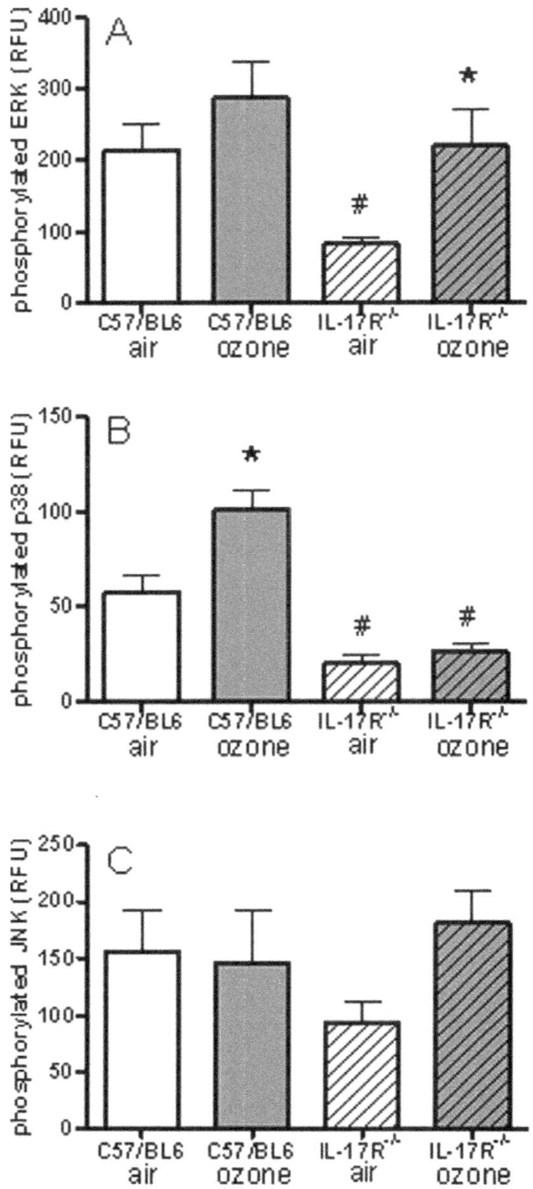

Figure 8. Lung levels of phosphorylated ERK1/2, p38 MAPK and JNK 1/2/3. C57/Bl6 and IL-17R$^{-/-}$ mice were exposed to air or to ozone. Data shown as mean ± SEM for n = 5 in each group; *p<0.05 compared to air in the same species; #p<0.05 compared to C57/Bl6 mice with corresponding exposure. RFU: Relative fluorescence unit.

contrasts with the recent reports that IL-17 may be involved in the induction of corticosteroid insensitivity in human epithelial cells [19] or could be the causative link between Th17 cells and glucocorticosteroid-insensitive allergic airway hyperrresponsiveness in the mouse [20]. It is likely that these differences relate to our unique challenge model of chronic ozone model.

IL-17 has been implicated in promoting pulmonary inflammation and tissue destruction, specifically through upregulation of MMP-9 and its effects as a neutrophil chemoattractant [21]. IL-17 stimulates mucin production by respiratory epithelial cells [22]. As both neutrophilic infiltration and mucus hypersecretion are characteristics of COPD, Th17 cells may play an important role in promoting these responses. There is evidence that IL-17R$^{-/-}$ mice failed to develop emphysema after 6 months of cigarette smoke exposure, indicating a role for IL-17 in the development of cigarette-smoke-induced emphysema [12]. Further evidence of the role for Th17 cells in cigarette smoke induced inflammation and airspace enlargement is the study showing that CCR6-positive Th17 cell population increased in bronchoalveolar lavage fluid of chronic smoke-exposed mice [11]. However, we were unable to observe any protective effect against ozone-induced emphysema in our IL-17R$^{-/-}$ mice, indicating that a specific oxidative stress stimulus such as ozone does not require IL-17 or Th-17 cells to induce emphysema. We hypothesise that components of the complex mixture that is cigarette smoke apart from direct oxidant gases are capable of acting as a Th17 adjuvant to activate metalloproteases such as MMP-12. The inflammation score tended to be higher in air-exposed IL-17R$^{-/-}$ mice compared with C57/BL6 mice, and was within the same range in both mouse species following chronic ozone exposure, leading us to conclude that the chronic lung inflammation is not IL-17 dependent.

In conclusion, in the chronic exposure model to an oxidant, ozone, the induced airway hyperresponsiveness and associated increase in maximal contractile responses to acetylcholine are dependent on IL-17, and this is likely to represent an IL-17-dependent activation of p38 MAPK. The induction of emphysema and lung inflammation in this ozone exposure model are not IL-17-dependent.

Materials and Methods

Mice and Ozone Exposure

Pathogen-free, 10–12 week old male C57/BL6 mice (Harlan, UK) and IL-17R-deficient (IL-17R$^{-/-}$) mice were housed within 'maximiser' filter-topped cages (Maximiser, Theseus caging system Inc., Hazelton, PA, USA), maintained in a temperature-controlled (23°C) facility with a strict 12-h light–dark cycle and were given free access to food and water. IL-17R$^{-/-}$ mice were on C57/BL6 genetic background, as previously described [23] and were obtained from Centre National de la Recherche Scientifique (CNRS), Orléans, France. The protocols were approved by the Imperial College Biosciences group and performed under a licence from the Home Office, UK, under the Animals (Scientific Procedures) Act 1986.

The two strains of mice (IL-17R$^{-/-}$ and C57/BL6) were exposed to ozone produced from an ozoniser (Model 500 Sander Ozoniser, Germany), mixed with air, for 3 hours at a concentration 2.5 parts per million (ppm) in a sealed Perspex container. Mice received ozone twice a week for 6 weeks. Control animals received medical air only over the equivalent period. Ozone concentration was continuously monitored with an ozone probe (ATi Technologies, Ashton-U-Lyne, UK).

Measurement of Airways Hyperresponsievness (AHR)

Twenty-four hours following the last ozone exposure, mice were anesthetized with an intraperitoneal injection of anesthetic solution containing midazolam (Roche Products Ltd., Welwyn Garden City, UK) and Hypnorm (0.315 mg/ml fentanyl citrate and 10 mg/ml fluanisone; Janssen Animal Health, Wantage, UK). Mice were tracheostomised and ventilated (MiniVent type 845, Hugo Sach Electronic, Germany; rate: 250 breaths/min and tidal volume: 250 µl), and were monitored in a whole body plethysmograph with a pneumotachograph connected to a transducer (EMMS, Hants, UK). Transpulmonary pressure was assessed via an esophageal catheter (EMMS, Hants, UK). Instantaneous calculation of pulmonary resistance (R_L) was obtained. Increasing concentrations of acetylcholine (ACh) (Sigma, Dorset, UK) (4–256 mg/ml) were administered with an Aeroneb® Lab Micropump Nebulizer (EMMS, Hants, UK), and R_L was recorded for a 3-min period following each concentration. R_L after each concentration was expressed as percentage change from baseline R_L measured following nebulized PBS (Sigma, Dorset, UK). The concentration of acetylcholine required to increase R_L by 100% from baseline was calculated (PC_{100}) and –log PC_{100} was taken as a measure of airway responsiveness.

Bronchial Ring Preparation and Myography

The lung was dissected out whole following terminal anaesthesia with pentobarbitone. The inferior lobe of the right lung was rapidly excised and immersed in physiological salt solution (PSS; NaCl, 119 mM; KCl, 4.7 mM; $CaCl_2$, 2.5 mM; $MgSO_4$, 1.17 mM; $NaHCO_3$, 25 mM; KH_2PO_4, 1.18 mM; EDTA, 0.027 mM; and glucose, 5.5 mM). Intrapulmonary bronchi were dissected under the microscope, parenchymal and connective tissues were carefully removed. Intact bronchi, 200–400 µm in diameter, 2 mm in length, were mounted on to metal prongs of a myograph (Myograph 610 M, Danish Myo Technology, Aarhus, Denmark), suspended in an organ bath, filled with 5 ml of PSS, bubbled with 95% oxygen and maintained at 37°C. Isometric tension was recorded and analysed using Chart software (AD Instruments Ltd., Hastings, U.K.). The optimal length for each bronchial ring was obtained through incremental radial stretch (passive tension) and repeated stimulation with 124 mM potassium PSS (active tension). Optimal length was taken at the point at which increased stretch ceased to increase active tension. Bronchi were then allowed to equilibrate for 30 min in PSS and 3 µM indomethacin was added into the organ bath to inhibit prostaglandin release. The bronchial contractile response in air- and ozone-exposed mice was generated with 10^{-9} M to 10^{-3} M of ACh. The concentration-response curves were fitted by nonlinear regression and with Hill equation (GraphPad Prism 4.03, San Diego, CA, USA) to provide an estimated maximal contraction (E_{max}) and the negative logarithm of the effective concentration to cause 50% of the maximal contractile response (pEC_{50}).

Histological and Morphometric Analysis

The left lung was inflated with fresh 4% paraformaldehyde and maintained with 25 cm of water pressure for at least 4 hours and then embedded in paraffin. Paraffin blocks were sectioned to expose the maximum surface area of lung tissue in the plane of the bronchial tree. Five µm sections were cut and stained with haematoxylin and eosin (H&E). All counts were performed by observers who were blinded as to the mice studied.

The mean linear intercept, a measure of interalveolar septal wall distance, was determined using a reticule with a Thurlbeck grid comprising of 5 lines (each 550 µM long), with 10 fields per section assessed at random. Two slides per mouse were coded and analyzed. Fields with airways or vessels were avoided by moving one field in any one direction. L_m was calculated by dividing the length of the line by the number of alveolar wall and grid line interceptions.

The severity of emphysematous change observed in the H&E stained lung sections was also scored on a 0–3 scale defined as: 0 = no emphysema; <1 = lung parenchyma is involved; 1 = lung parenchyma is involved with small enlargement of alveolar space; 2 = lung parenchyma is involved with medium enlargement of alveolar space; 3 = lung parenchyma is involved with pronounced enlargement of alveolar space.

The severity of inflammatory response observed in the H&E stained lung sections was scored on a 0–3 scale defined as: 0 = no inflammatory response, 1 = mild inflammation with foci of inflammatory cells in bronchial or vascular wall and in alveolar septa; 2 = moderate inflammation with patchy inflammation or localised inflammation in walls of bronchi or blood vessel and alveolar septa and less than 1/3 of lung cross-sectional area is involved; and 3 = severe inflammation with diffuse inflammatory cells in walls of bronchi or blood vessels, and alveoli septa; between one third to two thirds of the lung area is involved.

Lung protein isolation, MAP kinase phosphorylation status and cytokines assays. Lung tissue (30 mg) was homogenised using 1.4 mm Precellys Ceramic beads and Precellys 24 homogeniser (Peqlab, Erlangen, Germany) at 6800 rpm for 15 seconds. Cytosolic proteins were extracted with a hypotonic buffer (Active Motif, part #100505) and detergent (Active motif, part #100512) by centrifugation at 14000 rpm for 30 seconds at 4°C. Final protein concentration was determined using a protein assay reagent kit (Pierce Chemical, Rockford, IL, USA) and BSA standard (Sigma, Gillingham, UK).

MAP kinase protein phosphorylation was determined in cytoplasmic protein extracts using the PhosphoTracer ERK1/2 (pT202/Y204)+p38 MAPK (pT180/Y182)+JNK1/2/3 (pT183/Y185) Elisa Kit (Abcam, Cambridge, UK). The kit detects ERK1 and 2, p38 and JNK1, 2 and 3 only when phosphorylated on the indicated conserved threonine or tyrosine sites of each protein and was used according to manufacturer's instructions. Fluorescent data was normalised against total protein concentration from the same sample.

IL-17, IL-1β and TNFα concentrations were measured in lung homogenate supernatants with commercial available ELISA kits (R&D Systems Europe Ltd, Abingdon, UK) and were performed according to manufacturer's instructions.

Data Analysis

Data are presented as mean ± S.E.M. For multiple comparisons of different groups (% change in R_L), two-way analysis of variance was performed. Unpaired t-test was carried out for comparison between two individual groups. For data that did not comply with the normality and homoscedasticity assumptions, nonparametric t-test (U-Mann Whitney) analyses were carried out. All hypothesis testing was two-sided and a p value of less than 0.05 was accepted as significant.

Author Contributions

Performed the histological studies: JZ. Conceived and designed the experiments: KFC MP MZ. Performed the experiments: MP MZ FL FH CW. Analyzed the data: JZ MP. Contributed reagents/materials/analysis tools: BR. Wrote the paper: KFC MP.

References

1. Shapiro SD (2007) Transgenic and gene-targeted mice as models for chronic obstructive pulmonary disease. Eur Respir J 29: 375–378.
2. Tarasevicienе-Stewart L, Voelkel NF (2008) Molecular pathogenesis of emphysema. J Clin Invest 118: 394–402.
3. Rangasamy T, Cho CY, Thimmulappa RK, Zhen L, Srisuma SS, et al. (2004) Genetic ablation of Nrf2 enhances susceptibility to cigarette smoke-induced emphysema in mice. J Clin Invest 114: 1248–1259.
4. Triantaphyllopoulos K, Hussain F, Pinart M, Zhang M, Li F, et al. (2011) A model of chronic inflammation and pulmonary emphysema after multiple ozone exposures in mice. Am J Physiol Lung Cell Mol Physiol 300: L691–L700.
5. Williams AS, Leung SY, Nath P, Khorasani NM, Bhavsar P, et al. (2007) Role of TLR2, TLR4, and MyD88 in murine ozone-induced airway hyperresponsiveness and neutrophilia. J Appl Physiol 103: 1189–1195.
6. Vanaudenaerde BM, Verleden SE, Vos R, De Vleeschauwer SI, Willems-Widyastuti A, et al. (2011) Innate and adaptive interleukin-17-producing lymphocytes in chronic inflammatory lung disorders. Am J Respir Crit Care Med 183: 977–986. 201007-1196PP [pii];10.1164/rccm.201007-1196PP [doi].
7. Gaffen SL (2009) Structure and signalling in the IL-17 receptor family. Nat Rev Immunol 9: 556–567.
8. Linden A, Laan M, Anderson GP (2005) Neutrophils, interleukin-17A and lung disease. Eur Respir J 25: 159–172. 25/1/159 [pii];10.1183/09031936.04.00032904 [doi].
9. Park H, Li Z, Yang XO, Chang SH, Nurieva R, et al. (2005) A distinct lineage of CD4 T cells regulates tissue inflammation by producing interleukin 17. Nat Immunol 6: 1133–1141. ni1261 [pii];10.1038/ni1261 [doi].
10. Liang SC, Long AJ, Bennett F, Whitters MJ, Karim R, et al. (2007) An IL-17F/A heterodimer protein is produced by mouse Th17 cells and induces airway neutrophil recruitment. J Immunol 179: 7791–7799. 179/11/7791 [pii].
11. Doe C, Bafadhel M, Siddiqui S, Desai D, Mistry V, et al. (2010) Expression of the T helper 17-associated cytokines IL-17A and IL-17F in asthma and COPD. Chest 138: 1140–1147.
12. Chen K, Pociask DA, McAleer JP, Chan YR, Alcorn JF, et al. (2011) IL-17RA is required for CCL2 expression, macrophage recruitment, and emphysema in response to cigarette smoke. PLoS One 6: e20333.
13. Pichavant M, Goya S, Meyer EH, Johnston RA, Kim HY, et al. (2008) Ozone exposure in a mouse model induces airway hyperreactivity that requires the presence of natural killer T cells and IL-17. J Exp Med 205: 385–393. jem.20071507 [pii];10.1084/jem 20071507 [doi].
14. Li F, Zhang M, Hussain F, Triantaphyllopoulos K, Clark AR, et al. (2011) Inhibition of p38 MAPK-dependent bronchial contraction after ozone by corticosteroids. Eur Respir J 37: 933–942.
15. Nakae S, Komiyama Y, Nambu A, Sudo K, Iwase M, et al. (2002) Antigen-specific T cell sensitization is impaired in IL-17-deficient mice, causing suppression of allergic cellular and humoral responses. Immunity 17: 375–387.
16. Schnyder-Candrian S, Togbe D, Couillin I, Mercier I, Brombacher F, et al. (2006) Interleukin-17 is a negative regulator of established allergic asthma. J Exp Med 203: 2715–2725.
17. Wilson RH, Whitehead GS, Nakano H, Free ME, Kolls JK, et al. (2009) Allergic sensitization through the airway primes Th17-dependent neutrophilia and airway hyperresponsiveness. Am J Respir Crit Care Med 180: 720–730. 200904-0573OC [pii];10.1164/rccm.200904-0573OC [doi].
18. Rahman MS, Yamasaki A, Yang J, Shan L, Halayko AJ, et al. (2006) IL-17A induces eotaxin-1/CC chemokine ligand 11 expression in human airway smooth muscle cells: role of MAPK (Erk1/2, JNK, and p38) pathways. J Immunol 177: 4064–4071.
19. Zijlstra GJ, ten Hacken NH, Hoffmann RF, Van Oosterhout AJ, Heijink IH (2012) Interleukin-17A induces glucocorticoid insensitivity in human bronchial epithelial cells. Eur Respir J 39: 439–445.
20. McKinley L, Alcorn JF, Peterson A, Dupont RB, Kapadia S, et al. (2008) TH17 cells mediate steroid-resistant airway inflammation and airway hyperresponsiveness in mice. J Immunol 181: 4089–4097.
21. Prause O, Bozinovski S, Anderson GP, Linden A (2004) Increased matrix metalloproteinase-9 concentration and activity after stimulation with interleukin-17 in mouse airways. Thorax 59: 313–317.
22. Chen Y, Thai P, Zhao YH, Ho YS, DeSouza MM, et al. (2003) Stimulation of airway mucin gene expression by interleukin (IL)-17 through IL-6 paracrine/autocrine loop. J Biol Chem 278: 17036–17043. 10.1074/jbc.M210429200 [doi];M210429200 [pii].
23. Ye P, Rodriguez FH, Kanaly S, Stocking KL, Schurr J, et al. (2001) Requirement of interleukin 17 receptor signaling for lung CXC chemokine and granulocyte colony-stimulating factor expression, neutrophil recruitment, and host defense. J Exp Med 194: 519–527.

Latitudinal Gradients in Degradation of Marine Dissolved Organic Carbon

Carol Arnosti[1]*, Andrew D. Steen[1¤], Kai Ziervogel[1], Sherif Ghobrial[1], Wade H. Jeffrey[2]

1 Department of Marine Sciences, University of North Carolina, Chapel Hill, North Carolina, United States of America, 2 Center for Environmental Diagnostics and Bioremediation, University of West Florida, Pensacola, Florida, United States of America

Abstract

Heterotrophic microbial communities cycle nearly half of net primary productivity in the ocean, and play a particularly important role in transformations of dissolved organic carbon (DOC). The specific means by which these communities mediate the transformations of organic carbon are largely unknown, since the vast majority of marine bacteria have not been isolated in culture, and most measurements of DOC degradation rates have focused on uptake and metabolism of either bulk DOC or of simple model compounds (e.g. specific amino acids or sugars). Genomic investigations provide information about the potential capabilities of organisms and communities but not the extent to which such potential is expressed. We tested directly the capabilities of heterotrophic microbial communities in surface ocean waters at 32 stations spanning latitudes from 76°S to 79°N to hydrolyze a range of high molecular weight organic substrates and thereby initiate organic matter degradation. These data demonstrate the existence of a latitudinal gradient in the range of complex substrates available to heterotrophic microbial communities, paralleling the global gradient in bacterial species richness. As changing climate increasingly affects the marine environment, changes in the spectrum of substrates accessible by microbial communities may lead to shifts in the location and rate at which marine DOC is respired. Since the inventory of DOC in the ocean is comparable in magnitude to the atmospheric CO_2 reservoir, such a change could profoundly affect the global carbon cycle.

Editor: Meni Wanunu, Northeastern University, United States of America

Funding: This work was supported by the National Science Foundation (Dr. Arnosti, Dr. Jeffrey), EPA Graduate Fellowship (Dr. Steen), and the Deutsche Forschungsgemeinschaft Postdoctoral Fellowship (Dr. Ziervogel). The authors are also grateful to the Max-Planck Institute for Marine Microbiology (Bremen) for supporting the work on Svalbard. The funders had no role in study design, data collection and analysis, decision to publish, or preparation of the manuscript.

Competing Interests: The authors have declared that no competing interests exist.

* E-mail: arnosti@email.unc.edu

¤ Current address: Center for Geomicrobiology, Aarhus University, Herning, Denmark

Introduction

Marine DOC (dissolved organic carbon) is one of the largest actively cycling reservoirs of organic carbon on earth, comparable in magnitude to the atmospheric reservoir of CO_2 [1]; heterotrophic microbial communities play a key role in driving the DOC cycle [2–4]. DOC consists of many thousands of different compounds, and is operationally divided into labile, semi-labile, and recalcitrant fractions that are defined based on timescales of removal in bioassays or by direct measurement in the ocean [5]. Structural or mechanistic explanations for varying timescales of DOC degradation, however, are lacking [6]. Since heterotrophic bacteria are unable to transport directly into the cell most substrates with a molecular weight greater than 600 Da [7], hydrolysis via extracellular enzymes is required prior to substrate uptake. The requirement for enzymatic hydrolysis is therefore a promising starting point to search for mechanistic explanations of variations in the abilities of marine bacteria to utilize specific fractions of DOC as a substrate. The activities and structural specificities of polysaccharide-hydrolyzing enzymes are of particular importance in this respect, since carbohydrates constitute a large proportion of marine high molecular weight DOC: 54% of surface water DOC and 25% of DOC in the deep ocean [8]. Most of the rest of DOC is classified as 'uncharacterized' on a molecular basis, since lipids, amino acids, and amino sugars together constitute less than ca. 5% of the total [9].

Assessing the enzymatic capabilities of marine heterotrophic microbial communities can best be done directly in seawater, since cultured microbial isolates constitute only a small, unrepresentative fraction of extent marine microbes [10]. Most measurements of enzyme activity in seawater are based on small chromogenic or fluorogenic substrates (e.g. [11]), which provide very little information about enzymatic substrate specificities [12]. Genomic investigations can yield valuable insights into community and organism potential [13–15], but provide no information about the extent to which such potential might be expressed. Metatranscriptomic profiling is a promising route to investigate the extent to which genetic potential is realized, but assignment of sequences from the environment to functions such as specific enzyme activities is still problematic due to limitations in database coverage [16] and the vast structural and functional diversity of polysaccharide-hydrolyzing enzymes [17].

To gain insight into the capabilities of natural microbial communities to access polysaccharides, we measured extracellular enzyme activities in surface waters at 32 stations in the Atlantic, Pacific, Arctic, and Southern Oceans, as well as the Gulf of Mexico, spanning latitudes from 76°S to 79°N and a temperature range from −1.8°C to 29°C (Table 1; Fig. S1). We focused on

Table 1. Station locations and water temperatures.

Station	Latitude	Longitude	Water temp. (°C)
J	79.4N	11.1E	4
AB	77.4N	15.1E	4
P2	66.5N	168.1W	8.7
DO	38.4N	74.6W	13.5
CO	36.4N	74.8W	22
P10	35.5N	164.2W	24.2
GOM1	30.2N	87.4W	28
GOM11	29.6N	87W	28
GOM072	28.5N	89.4W	28.7
GOM073	28.3N	89.4W	28.9
P15	15.5N	161.4W	27.4
BOT12	15.1N	105.5W	29.1
BOT10	10.2N	99.4W	28.6
BOT8	5.4N	92.4W	27.4
T33	1.1N	83.5W	26
BOT7	0.005S	86W	24.1
P21	7.2S	168.4W	29.2
BOT5	8.2S	83.5W	19.8
BOT4	12.2S	81.4W	18
BOT3	17S	79.1W	17.3
T15	23.1S	79.2W	17.5
BOT1	26.3S	75W	15.2
P27	26.5S	173W	21.1
T3	39.2S	77.6W	11
G1	49.3S	174.4W	8.5
R1	56.3S	176.2E	6
R3	62.3S	178.4W	−1.5
M9	65S	176W	−1.7
G11B	74.3S	173.3W	0
R10C	76.1S	170.3E	−1.8
G9A	76.3S	179W	0
R13f	76.5S	177.5E	−1.8

direct detection of the hydrolysis of specific polysaccharides, rather than on investigations of genetic potential, in order to measure the abilities of microbial communities to access substrates irrespective of the multiplicity of enzyme(s) [18] —of known and unknown sequence—that might hydrolyze a specific substrate. Since determining the specific structures of marine dissolved carbohydrates is not possible using currently available analytical techniques [19], we used as substrates a suite of polysaccharides that are components of marine algae [20], and/or whose hydrolytic enzymes have been identified in the genomes of recently-sequenced marine bacteria [13,15,21]. These polysaccharides—laminarin, xylan, fucoidan, arabinogalactan, pullulan, and chondroitin sulfate—are structurally diverse, vary in monomer composition and linkage position, and because they are constituents of marine plankton, many are present in considerable quantities in the ocean. The production of laminarin by diatoms and *Phaeocystis* in the ocean, for example, has been estimated at 5 to15 billion metric tons annually [22]. A previous investigation at a few locations had shown evidence of spatial variations in microbial extracellular enzyme activities in surface ocean waters [23]. The present study, the culmination of a series of investigations carried out over the course of a decade, demonstrates that there are recognizable patterns in microbial potential to access DOC on large-scale gradients, mirroring emerging patterns of microbial biogeography in the ocean [24–28].

Results and Discussion

Differing abilities to hydrolyze a diverse range of substrates, as evident in our survey (Fig. 1), demonstrate functional differences among pelagic microbial communities. All substrates were hydrolyzed at only 4 of the 32 stations, all in the Gulf of Mexico. At the other 28 sites, the spectrum of hydrolysis ranged from one to five substrates. Only laminarin was hydrolyzed at every station; chondroitin and xylan were hydrolyzed at the majority (81% and 78%, respectively) of the stations, while pullulan, arabinogalactan, and fucoidan were hydrolyzed at 63%, 47%, and 34% of the stations, respectively. Summed hydrolysis rates were maximal at tropical/subtropical stations, and these summed rates as well as the spectrum of detected enzyme activities, decreased towards the poles (Figs. 1 and 2). The relative contribution of each enzyme activity to the sum at a given station (evenness of hydrolysis rates; Fig. 3) showed a similar pattern, with increasing evenness at higher temperatures (and lower latitudes).

The broad correlation between latitude and summed hydrolysis rates points to the relationship between summed hydrolysis rates and water temperature ($n = 35$, including 4 visits to a single station (Station J; Table 1), $r^2 = 0.64$, $p<0.005$), a relationship that could be due to the kinetic effect of temperature. This correlation, however, is driven primarily by the correlation with laminarin ($r^2 = 0.79$; $p<0.005$) and by the fact that a broader spectrum of enzyme activities is detected across the range of stations in lower-latitude waters (Figs. 1 and 2). Temperature was poorly correlated with hydrolysis of chondroitin ($r^2 = 0.15$; $p<0.05$), arabinogalactan and pullulan ($r^2 = 0.20$ and 0.21, respectively; $p<0.01$), and fucoidan ($r^2 = 0.23$, $p<0.005$). The correlation with xylan was stronger ($r^2 = 0.49$, $p<0.005$), but hydrolysis rates of the substrates at stations with similar temperatures varied greatly. At temperatures close to 28°C, for example, the variation in hydrolysis of xylan, fucoidan, arabinogalactan, and chondroitin was an order of magnitude or more (Table 1; Table S1; Fig. 1). Moreover, the differences in hydrolysis rate evenness (Fig. 3) are not explained solely by temperature, since a purely kinetic effect of temperature on hydrolysis rates would be expected to cause a generally proportionate change in the activities of all enzymes, and greatly extended incubation times do not markedly broaden the spectrum of substrates hydrolyzed at high latitude [29].

To the extent that these differences in enzymatic capabilities cannot be explained by temperature, they may derive from variations in microbial community composition that cascade into these functional differences. Arctic microbial communities differ in composition from their temperate counterparts [24,30–31]. Recent investigations have demonstrated latitudinal gradients in microbial community richness, with markedly reduced diversity at high latitudes [26–28]. The functional consequences of these variations are unknown, since microbial phylogeny and function are not well correlated. Our results demonstrate that microbial community function varies systematically across latitudinal gradients. The lower summed hydrolysis rates and the more limited spectrum of substrates enzymatically accessible to microbial communities at high latitudes coincides with a reduction in community richness (Figs. 1 and 2).

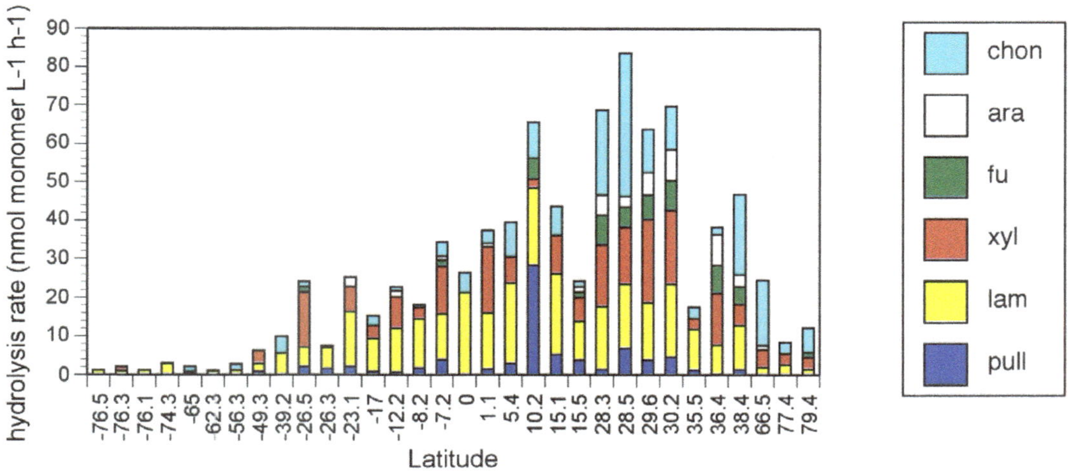

Figure 1. Summed enzymatic hydrolysis rates in surface water at each station plotted against latitude (south latitudes shown with negative numbers). Bar height shows the sum of the maximum enzymatic hydrolysis rate of each substrate at each station. All stations were visited once, with the exception of Station J (79°N); values shown for Station J are averages from 4 visits. (See Fig. 4 and Table S1 for data from each sampling time at Station J.) Pullulan hydrolysis is shown in blue, laminarin in yellow, xylan in red, fucoidan in green, arabinoglactan in white, and chondroitin sulfate in aqua. Hydrolysis rates and standard deviations for all substrates and stations are in Table S1.

Differences in the extent to which genes for polysaccharide hydrolases are expressed may also contribute to differences in enzyme activities among microbial communities, but the genetic diversity of polysaccharide hydrolases even among well-studied organisms is barely beginning to be explored [32], complicating efforts for a concerted genetic investigation. One study suggests that a high diversity of hydrolases is potentially available to microbial communities in the North Atlantic [14], but the extent and the conditions under which this genetic potential might be expressed are still matters of speculation. A search of the CAMERA database (http://camera.calit2.net/), for example, yielded numerous gene sequences related to an alpha-L-fucosidase of *Pseudoalteromonas atlantica* T6c (http://img.jgi.doe.gov/cgi-bin/pub/main.cgi). These sequences were found at 60 different Global Ocean Survey (GOS) sites (available through CAMERA), spanning latitudes of 32°S to 45°N, despite the fact that this enzyme activity was not detected in many of our samples from the same range of latitudes (Fig. 1). Likewise, a search (via CAMERA) of the two marine sites in the Antarctic Aquatic Metagenome produced sequences from Newcomb Bay (66°S) closely matching pullulanases from fully-sequenced marine bacteria (http://blast.ncbi.nlm.nih.gov/Blast.cgi), although pullulanase activity was not detectable at any of our sites at latitudes higher than 49°S or 38°N (Fig. 1). Limited geographical overlap between the current CAMERA database and our samples presently preclude a more detailed comparison among sites that would provide insight into the extent to which microbial communities vary in their genetic response to environmental parameters.

Patterns of microbial community composition at a given location are temporally repeatable [33–37]. Results from our investigation suggest that patterns of hydrolytic activities are also consistent over multi-year timescales. One high-latitude station

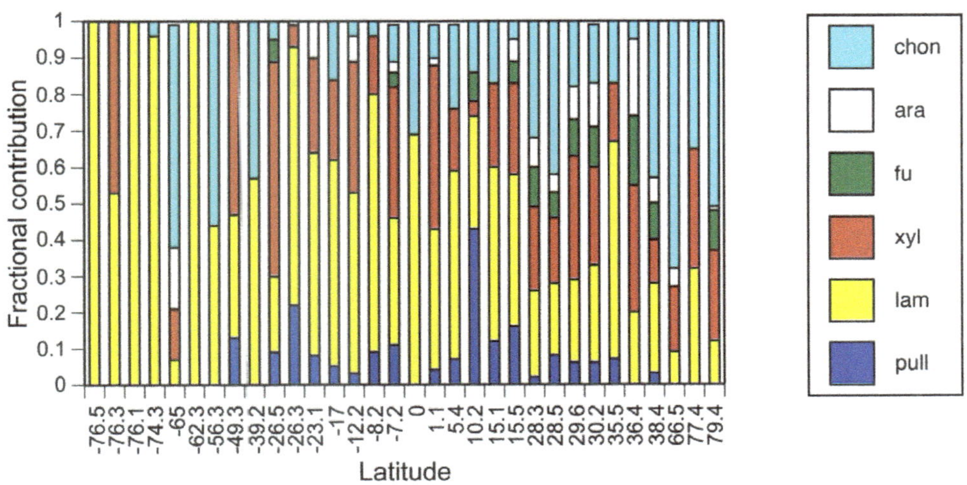

Figure 2. Proportionate contribution of each enzyme activity to summed hydrolysis rates, normalized to 100%. Station latitude and color key for enzyme identity are as in Fig. 2.

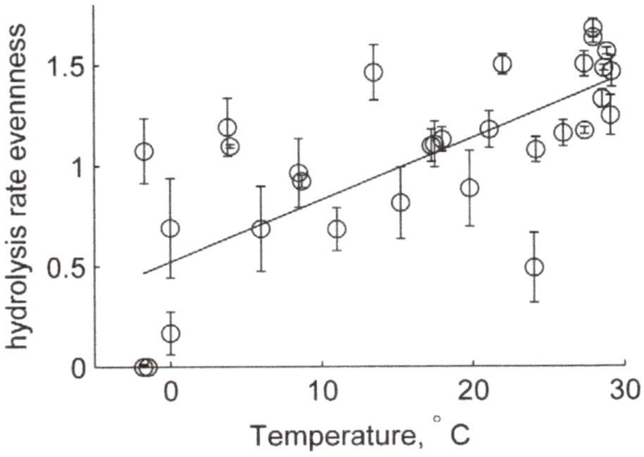

Figure 3. Evenness of enzymatic hydrolysis rates as a function of temperature. Average values (as shown in Figs. 1 and 2) were used for Station J, the only station that was sampled more than once. Line fit: $r^2 = 0.6154$, $n = 26$, $p < 0.001$.

(79°N, Stn. J; Table 1) sampled four times over the course of 10 years consistently showed hydrolytic activity dominated by chondroitinase (>50% of total activity), with similar patterns and levels of total activity (Table S1; Fig. 4). Likewise, two pairs of stations close to one another in the Gulf of Mexico (GOM1 and GOM11, 28°N; also GOM072 and GOM073, 28.2–30.2°N; Table 1) sampled 6 years apart showed the same broad substrate spectrum (nearly unique among sample locations; Fig. 2) and high levels of total activity (Fig. 1).

The patterns in enzymatic activities observed here provide insight into the potential of microbial communities to access components of the DOC pool, rather than a snapshot of enzyme activities expressed at the time of sampling, since our enzyme incubations lasted several days to two weeks. These experiments demonstrate the potential of a microbial community to access specific substrates over timescales sufficiently long to allow for cellular growth as well as for enzyme induction. The observation that many substrates remained unhydrolyzed over these timescales suggests that some microbial communities collectively lack the

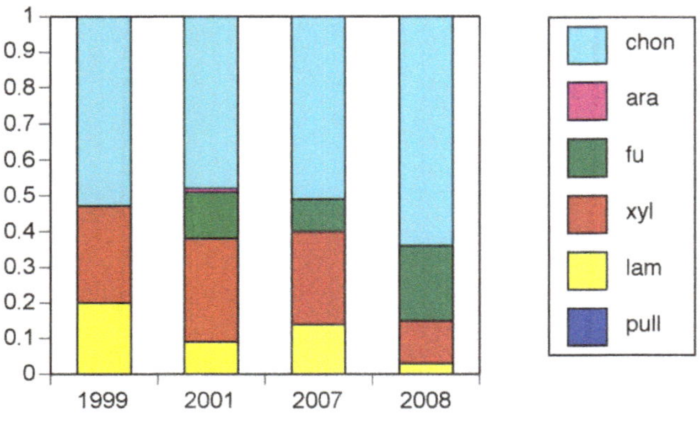

Figure 4. Relative contributions to summed hydrolysis rates for four visits to Station J, Svalbard. Each enzyme activity is shown in a different color; summed rates are normalized to 1.0. Pullulan hydrolysis is shown in blue, laminarin in yellow, xylan in red, fucoidan in green, arabinoglactan in white, and chondroitin sulfate in aqua. All rates plus standard deviations are listed in Table S1.

capabilities (enzymes, inducers, or organisms) to access specific substrates [38].

This information is unique, since no other currently available analytical method tests the abilities of microbial communities to access specific polysaccharide structures. Likewise, the concentrations of specific polysaccharides in the ocean cannot be measured with currently-available techniques [19], precluding direct measurement of polysaccharide production or concentration. Comparable data cannot be obtained by measuring surface water DOC concentrations at diverse locations, because DOC concentrations are a function of DOC production, DOC degradation, and water mass history [39] (the radiocarbon age of bulk DOC is ca. 6000 years [40].) Measurements of DOC concentrations in the surface ocean therefore cannot differentiate unambiguously between changes in the metabolic capabilities of heterotrophic microbes, changes in DOC production, and differences in water mass history. Moreover, attempts to constrain production and consumption terms for DOC across broad latitudinal gradients are greatly complicated by the multitude of sources and varying bioavailabilities of different components of the DOC pool [6], as well as the paucity of data from high latitude environments [9,41]. A measure of the spectrum of enzyme activities capable of hydrolyzing a class of biomolecules constituting the largest identified component of the DOC pool thus yields insight into processes otherwise not amenable to quantification.

The broader spectrum of enzyme activities observed at temperate and tropical sites indicates that those microbial communities can access a wider range of substrates than their high-latitude counterparts. The fact that this trend was discernable in our global data set despite variations in season, levels of productivity, oceanic province, and a host of other factors, is remarkable. The functional factors controlling the breadth of microbial community metabolic capabilities remain to be determined. However, the pattern observed here matches the decrease in bacterial species richness observed at high latitudes [26–27]. The reason for that trend is not well understood [42]; bacterial richness correlates to a similar extent with water temperature and latitude [26]. A recent model investigation of global patterns of phytoplankton diversity, however, points at the magnitude of seasonal variability in environments and at relative rates of organism dispersal as key factors controlling latitudinal diversity gradients for phytoplankton [43]. Similar factors may control diversity gradients of heterotrophic bacteria.

Projected changes in ocean environments over the coming decades driven by global warming [44–45] may increase the rates and widen the spectrum of enzyme activities due to changes in microbial community composition that are facilitated by changes in ocean temperature, circulation, or biogeochemical parameters [46]. Particularly in the Arctic, where rapid temperature increases are projected for the near future, the input of terrestrially-derived DOC into the Arctic basin may be greatly accelerated due to melting of permafrost and increased runoff [47]. Currently, much of this organic matter is buried or exported to the North Atlantic [41,48–49]. In the future, more organic matter may reach the Arctic Ocean, and if the range of complex substrates available to heterotrophic microbial communities is broader in a future Arctic Ocean, a larger fraction of it may be remineralized there.

Kirchman et al. [46] predict that climate change in polar waters may also lead to changes in food web structure resulting in greater transfer of carbon from phytoplankton to DOC to bacteria, and thus to the respiration of a larger fraction of marine primary production to CO_2. These processes would be greatly facilitated if future polar microbial communities can use a wider range of enzymatic tools, as their temperate to tropical counterparts currently do. Since DOC reactivity is a function of the metabolic capabilities of microbial communities as well as intrinsic chemical characteristics, predictive understanding of the global carbon cycle will require further investigation of the ways in which these metabolic capabilities are likely to change in response to changing climate.

Materials and Methods

Sample collection, substrate incubation, and sample analysis

Surface water was collected at each site (Table 1; Fig. S1), by Niskin or Go-Flo bottles; surface water at Stns. J and AB was collected by bucket. Water samples were dispensed into replicate vials. Each vial received a single fluorescently labeled polysaccharide [50] as a substrate, at a concentration of 3.5 µmol monomer equivalent L^{-1}. Triplicate vials were incubated at *in situ* temperature. Controls were autoclaved or poisoned with mercuric chloride before substrate addition. Subsamples from each vial were filtered through 0.2 µm pore-sized filters, and were stored frozen prior to analysis. Changes in substrate molecular weight with time were quantified using gel permeation chromatography, with Sephadex G-50 and G-75 columns linked in series, and fluorescence detection (excitation and emission wavelengths of 490 and 530 nm, respectively), as previously described [51]. Hydrolysis rates were calculated from changes in molecular weight distribution of the polysaccharides, as described previously [50–51]. No specific permits were required for the collection of water from the ocean as described above. The field studies did not involve any endangered or protected species.

Statistical analyses

Statistical comparisons were carried out using two-tailed Student's t-test. The evenness of hydrolysis rates (the contribution of each activity to summed activities) was calculated using Shannon's entropy [52], $H = -\sum_{i=1}^{6} f_i \ln f_i$, where f_i represents the hydrolysis rate of the ith substrate expressed relative to the sum of all hydrolysis rates for that station. As previously discussed [53], H is maximized ($H \approx 1.79$) when enzymatic hydrolysis rates of all substrates are equal, and minimized ($H = 0$) when only one hydrolysis rate was measurable. Error bars in Fig. 3 represent the standard deviation of the ensemble results from a Monte Carlo error simulation, as described in [53].

Acknowledgments

We thank the captains, crews, and scientific parties of R/V *Farm*, R/V *Cape Hatteras*, R/V *Cape Henlopen*, R/V *N.B. Palmer*, R/V *Hugh Sharp*, and R/V *Pelican* for help with sampling, and A.J. Baldwin for her assistance with shipboard experiments.

Author Contributions

Conceived and designed the experiments: CA WHJ. Performed the experiments: CA ADS KZ SG WHJ. Analyzed the data: CA ADS KZ SG. Wrote the paper: CA. Contributed to writing the manuscript: ADS KZ SG WHJ. Performed statistical analyses: ADS KZ.

References

1. Hedges JI (2002) Why dissolved organics matter. In: Hansell DA, Carlson CA, eds. Biogeochemistry of Marine Dissolved Organic Matter. San Diego: Academic Press. pp 1–33.
2. Azam F (1998) Microbial control of oceanic carbon flux: The plot thickens. Science 280: 694–696.
3. Hansell DA, Carlson C (2002) Biogeochemistry of Dissolved Organic Matter. San Diego: Academic Press.
4. McCarren J, Becker JW, Repeta DJ, Shi Y, Young CR, et al. (2010) Microbial community transcriptomes reveal microbes and metabolic pathways associated with dissolved organic matter turnover in the sea. PNAS.
5. Carlson CA (2002) Production and removal processes. In: Hansell DA, Carlson CA, eds. Biogeochemistry of Dissolved Organic Matter. San Diego: Academic Press. pp 91–139.
6. Jiao N, Herndl GJ, Hansell DA, Benner R, Kattner G, et al. (2010) Microbial production of recalcitrant dissolved organic matter: long-term carbon storage in the global ocean. Nature Rev Microbiol 8: 593–599.
7. Weiss MS, Abele U, Weckesser J, Welte W, Schiltz E, et al. (1991) Molecular architecture and electrostatic properties of a bacterial porin. Science 254: 1627–1630.
8. Benner R, Pakulski JD, McCarthy M, Hedges JI, Hatcher PG (1992) Bulk chemical characteristics of dissolved organic matter in the ocean. Science 255: 1561–1564.
9. Benner R (2002) Chemical composition and reactivity. In: Hansell DA, Carlson CA, eds. Biogeochemistry of Marine Dissolved Organic Matter. New York: Academic Press. pp 59–90.
10. DeLong EF, Pace N (2001) Environmental diversity of bacteria and archea. Syst Biol 50: 470–478.
11. Hoppe H-G (1983) Significance of exoenzymatic activities in the ecology of brackish water: measurements by means of methylumbelliferyl-substrates. Mar Ecol Prog Ser 11: 299–308.
12. Arnosti C (2011) Microbial extracellular enzymes and the marine carbon cycle, In: Carlson C, Giovannoni S, eds. (2011) Annual Review of Marine Science, vol 3. Palo Alto: Annual Reviews. pp 401–425.
13. Bauer M, Kube M, Telling H, Richter M, Lombardot T, et al. (2006) Whole genome analysis of the marine Bacteroidetes 'Gramella forsetii' reveals adaptations to degradation of polymeric organic matter. Environmental Microbiology 8: 2201–2213.
14. Elifantz H, Waidner LA, Michelou VK, Cottrell MT, Kirchman DL (2008) Diversity and abundance of glycosyl hydrolase family 5 in the North Atlantic Ocean. FEMS Microbiol Ecol 63: 316–327.
15. Weiner RM, Taylor LE, II, Henrissat B, Hauser L, Land M, et al. (2008) Complete genome sequence of the complex carbohydrate-degrading marine bacterium, Saccharophagus degradans Strain 2-40T. PLOS Genetics 5(5): e1000087.
16. Gilbert J, Field D, Swift P, Simon T, Cummings D, et al. (2010) Taxonomic and functional diversity of microbes at a temperate coastal site: A 'multi-omic' study of seasonal and diel temporal variation. PLoS ONE 5(11): e15545.
17. Cantarel BL, Coutinho PM, Rancurel C, Bernard T, Lombard V, et al. (2008) The carbohydrate-active enzyme database (CAZy): an expert resource for glycogenomics. Nucleic Acids Res 37: 233–38.
18. Arrieta JM, Herndl GJ (2002) Changes in bacterial β-glucosidase diversity during a coastal phytoplankton bloom. Limnol Oceanogr 47: 594–599.
19. Hedges JI, Eglinton G, Hatcher PG, Kirchman DL, Arnosti C, et al. (2000) The Molecularly-uncharacterized component (MUC) of nonliving organic matter in natural environments. Organic Geochem 31: 945–958.
20. Painter TJ (1983) Algal Polysaccharides. In: Aspinall GO, ed. The Polysaccharides. New York: Academic Press. pp 195–285.
21. Glockner FO, Kube M, Bauer M, Teeling H, Lombardot T, et al. (2003) Complete genome sequence of the marine planctomycete Pirellula sp. strain 1. Proc Nat Acad Sci 100: 8298–8303.
22. Alderkamp A-C, van Rijssel M, Bolhuis H (2007) Characterization of marine bacteria and the activity of their enzyme systems involved in degradation of the algal storage glucan laminarin. FEMS Microbiol Ecol 59: 108–117.
23. Arnosti C, Durkin S, Jeffrey WH (2005) Patterns of extracellular enzyme activities among pelagic marine microbial communities: implications for cycling of dissolved organic carbon. Aq Microb Ecol 38: 135–145.
24. Baldwin AJ, Moss JA, Pakulski JD, Catala P, Joux F, et al. (2005) Microbial diversity in a Pacific Ocean transect from the Arctic to Antarctic circles. Aq Microb Ecol 41: 91–102.
25. Martiny JBH, Bohannan BJM, Brown JH, Colwell RK, Fuhrman JA, et al. (2006) Microbial biogeography: putting microorganisms on the map. Nature Microb Rev 6: 102–112.

26. Fuhrman JA, Steele JA, Hewson I, Schwalbach MS, Brown MV, et al. (2008) A latitudinal diversity gradient in planktonic marine bacteria. Proc Nat Acad Sci 105(22): 7774–7778.
27. Pommier T, Canback B, Riemann L, Bostrom KH, Simu K, et al. (2007) Global patterns of diversity and community structure in marine bacterioplankton. Molecular Ecol 16: 867–880.
28. Wietz M, Gram L, Jorgensen B, Schramm A (2010) Latitudinal patterns in the abundance of major marine bacterioplankton groups. Aq Microb Ecol 61: 179–189.
29. Arnosti C (2008) Functional differences between Arctic sedimentary and seawater microbial communities: Contrasts in microbial hydrolysis of complex substrates. FEMS Microb Ecol 66: 343–351.
30. Bano N, Hollibaugh JT (2002) Phylogenetic composition of bacterioplankton assemblages from the Arctic Ocean. Appl Environ Microbiol 68: 505–518.
31. Malmstrom RR, Straza TRA, Cottrell MT, Kirchman DL (2007) Diversity, abundance, and biomass production of bacterial groups in the western Arctic Ocean. Aq Microb Ecol 47: 45–55.
32. Hess M, Sczyrba A, Egan R, Kim T-W, Chokhawala H, et al. (2011) Metagenomic discovery of biomass-degading genes and genomes from cow rumen. Science 331: 463–467.
33. Murray AE, Preston CM, Massana R, Taylor LT, Blakis A, et al. (1998) Seasonal and spatial variability of bacteria and archael assemblages in the coastal waters near Anvers Island, Antarctica. Appl Environ Microbiol 64: 2585–2595.
34. Morris RM, Vergin KL, Cho J-C, Rappe M, Carlson CA, et al. (2005) Temporal and spatial response of bacterioplankton lineages to annual convective overturn at the Bermuda Atlantic Time-series Study site. Limnol Oceanogr 50: 1687–1696.
35. Kan J, Crump BC, Wang K, Chen F (2006) Bacterioplankton communities in Cheasapeake Bay: Predictable or random assemblages. Limnol Oceanogr 51: 2157–2169.
36. Fuhrman JA, Steele JA (2008) Community structure of marine bacterioplankton: patterns, networks, and relationships to function. Aq Microb Ecol 53: 69–81.
37. Fuhrman JA, Hewson I, Schwalbach MS, Steele JA, Brown MV, et al. (2006) Annually reoccurring bacterial communities are predictable from ocean conditions. Proc Nat Acad Sci USA 103: 13104–13109.
38. Arnosti C (2004) Speed bumps and barricades in the carbon cycle: Substrate structural effects on carbon cycling. Mar Chem 92: 263–273.
39. Goldberg SJ, Carlson CA, Bock B, Nelson NB, Siegel DA (2010) Meridional variability in dissolved organic matter stocks and diagenetic state within the euphotic and mesopelagic zone of the North Atlantic subtropical gyre. Mar Chem 119: 9–21.
40. Bauer JE, Williams PM, Druffel ERM (1992) ^{14}C activity of dissolved organic carbon fractions in the north-central Pacific and Sargasso Sea. Nature 357: 667–670.
41. Amon RMW (2004) The role of dissolved organic matter for the organic carbon cycle in the Arctic Ocean. In: Stein R, Macdonald RW, eds. The organic carbon cycle in the Arctic Ocean. Berlin: Springer. pp 83–100.
42. Fuhrman JA (2009) Microbial community structure and its functional implications. Nature 459: 193–199.
43. Barton AD, Dutkiewicz S, Flierl G, Bragg J, Follows MJ (2010) Patterns of diversity in marine phytoplankton. Science 327: 1509–1511.
44. Behrenfeld MJ, O'Malley RT, Siegel DA, McClain CR, Sarmiento JL, et al. (2006) Climate-driven trends in contemporary ocean productivity. Nature 444: 752–755.
45. Vezina AF, Hoegh-Guldberg O (2008) Effects of ocean acidification on marine ecosystems: Introduction. Mar Ecol Prog Ser 373: 199–201.
46. Kirchman DL, Moran XAG, Ducklow H (2009) Microbial growth in the polar oceans—role of temperature and potential impact of climate change. Nature Rev Microbiol 7: 451–459.
47. Stocker TF, Raible CC (2005) Water cycle shifts gear. Nature 434: 830–832.
48. Stein R, Macdonald RW (2004) Organic carbon budget: Arctic Ocean vs. Global Ocean. In: Stein R, Macdonald RW, eds. The organic carbon cycle in the Arctic Ocean. Berlin: Springer. pp 315–323.
49. Benner R, Louchouarn P, Amon RMW (2005) Terrigenous dissolved organic matter in the Arctic Ocean and its transport to surface and deep waters of the North Atlantic. Global Biogeochem Cycles 19: GB2025. doi:2010.1029/2004GB002398.
50. Arnosti C (1996) A new method for measuring polysaccharide hydrolysis rates in marine environments. Org Geochem 25: 105–115.
51. Arnosti C (2003) Fluorescent derivatization of polysaccharides and carbohydrate-containing biopolymers for measurement of enzyme activities in complex media. J Chromatog B 793: 181–191.
52. Legendre P, Legendre L (1998) Numerical Ecology. New York: Elsevier. 853 p.
53. Steen AD, Ziervogel K, Arnosti C (2010) Comparison of multivariate microbial datasets with the Shannon Index: an example using enzyme activities from diverse marine environments. Org Geochem 41: 1019–1021.

11

Rapid Response of Hydrological Loss of DOC to Water Table Drawdown and Warming in Zoige Peatland: Results from a Mesocosm Experiment

Xue-Dong Lou[1,2], Sheng-Qiang Zhai[1], Bing Kang[2], Ya-Lin Hu[3], Li-Le Hu[1]*

1 Chinese Research Academy of Environmental Sciences, Beijing, China, **2** College of Life Sciences, Northwest Agriculture & Forestry University, Yangling, Shaanxi, China, **3** Institute of Applied Ecology, Chinese Academy of Sciences, Shenyang, China

Abstract

A large portion of the global carbon pool is stored in peatlands, which are sensitive to a changing environment conditions. The hydrological loss of dissolved organic carbon (DOC) is believed to play a key role in determining the carbon balance in peatlands. Zoige peatland, the largest peat store in China, is experiencing climatic warming and drying as well as experiencing severe artificial drainage. Using a fully crossed factorial design, we experimentally manipulated temperature and controlled the water tables in large mesocosms containing intact peat monoliths. Specifically, we determined the impact of warming and water table position on the hydrological loss of DOC, the exported amounts, concentrations and qualities of DOC, and the discharge volume in Zoige peatland. Our results revealed that of the water table position had a greater impact on DOC export than the warming treatment, which showed no interactive effects with the water table treatment. Both DOC concentration and discharge volume were significantly increased when water table drawdown, while only the DOC concentration was significantly promoted by warming treatment. Annual DOC export was increased by 69% and 102% when the water table, controlled at 0 cm, was experimentally lowered by −10 cm and −20 cm. Increases in colored and aromatic constituents of DOC (measured by $Abs_{254\ nm}$, $SUVA_{254\ nm}$, $Abs_{400\ nm}$, and $SUVA_{400\ nm}$) were observed under the lower water tables and at the higher peat temperature. Our results provide an indication of the potential impacts of climatic change and anthropogenic drainage on the carbon cycle and/or water storage in a peatland and simultaneously imply the likelihood of potential damage to downstream ecosystems. Furthermore, our results highlight the need for local protection and sustainable development, as well as suggest that more research is required to better understand the impacts of climatic change and artificial disturbances on peatland degradation.

Editor: Shiping Wang, Institute of Tibetan Plateau Research, China

Funding: The authors received funding from the Nature Science Foundation of China under Grant Nos. 41103041 and 41271318 to support this study. The funders had no role in study design, data collection and analysis, decision to publish, or preparation of the manuscript.

Competing Interests: All authors have declared that no competing interests exist.

* Email: hulile@craes.org.cn

Introduction

Generally, peat-accumulating wetlands provide waterlogged conditions where carbon accumulation is encouraged [1], and therefore have huge carbon storage potential. However, there is increasing concern that carbon storage in peatlands is unstable and may be susceptible to water table drawdown and higher temperatures over the next two centuries due to projected climatic change [2–6]. Furthermore, the water table in peatlands may also be significantly lowered by drainage resulting from human activities [7,8]. As the largest highland wetland in the world [9] and the largest peat storage area in China, the Zoige alpine wetland serves as a natural barrier and prevents desertification in Northwest China, extending farther toward Southeast China, and is very sensitive to climate change [10]. The peatland in Zoige is also the major water source of the world's largest plateau reserve (i.e., Three-Rivers Source Nature Reserve), supplying water for the three most important rivers in East Asia (i.e. the Yellow, Yangtze, and Lancang rivers) [11]. The Zoige wetland is particularly closely associated with the ecological security of the Yellow River drainage basin [12] because it provides about 40% of the total flow of the Yellow River [13]. Zoige peatland covers an estimated area of 0.5 million hectares and accounts for 47.53% of the total organic carbon reserves in Chinese peatland. Thus, it accounts for the highest organic carbon accumulation of any peatland in China [14].

Unfortunately, due to climate warming, artificial drainage for pastures, and peat exploitation since the 1970s, Zoige peatland has suffered extensive biodiversity loss and ecosystem degradation, including severe peat deterioration [9]. The Zoige wetland has decreased by 30% in the past 30 years due to water table drawdown [15], and artificial drainage has been regarded to be the most important cause of Zoige wetland (including peatland) degradation [9]. Previous studies suggested that the carbon cycle in peatland could change rapidly with climate change [16–18] and that is sensitive to water table [4,19–21]. Therefore, climate warming and a lowered of the water table could potentially create a carbon storage and ecosystem stability crisis in Zoige peatland.

Dissolved organic carbon (DOC) is the most active and sensitive indictor in the carbon cycle [22], and connects the biogeochemical cycle from terrestrial to aquatic ecosystems [23]. Hydrological losses of aquatic carbon can be of significant concern when determining carbon storage in peatlands [24] and may be increasing [25,26]. Among the aquatic constituents of peatlands, DOC is generally considered to have the largest aquatic carbon flux [27,28]. The peculiar water–peat interaction system and strong hydrological connectivity in peatlands ensures that the export of DOC from peatland to downstream plays a key role in the regional redistribution of terrestrial carbon [29] and the carbon balance [30]. Furthermore, the transfer of carbon from terrestrial peatland to fluvial downstream locations has a large influence on the water quality in aquatic ecosystems [31,32]. Previous studies have warned that larger amounts of DOC feeding into downstream locations could increase the levels of aquatic organic acids, decrease the buffering ability of the water, and attenuate the penetration of visible and ultraviolet (UV) light due to changes in the water color [33]. This is likely to cause damage to the sustainable and stable development of aquatic ecosystems, such as their net primary productivity [34] and production of bacteria [2,35]. A large body of literature has reported changes in the color or aromatic components of water in peatlands that has occurred in recent years [11,31,36,37]. $SUVA_{254\ nm}$ was a useful parameter for determining the aromatic characteristic of DOC [38], and absorbance at 400 nm was used as a measure of the color composition [36] and could further indicate changes in DOC composition when combined with specific absorbance [39]. Therefore, DOC is likely to be an important part of the carbon cycle linking peatland and downstream ecosystems, although it is not the only pathway of carbon loss from an upland peatland.

The amount of DOC exported from peatlands is believed to depend on interactions between discharged water through peatland and the production and consumption of DOC within the peatland [4]. However, it has also been reported to increase with a higher discharge [25,40,41] without any effect on DOC concentration. Climate change can regulate the import and export of DOC [42,43], mainly by controlling the most important environmental factors (i.e., temperature and the water table) affecting the peatland carbon cycle. A high water table and low soil temperatures are believed to be major reasons for the low decay rates, which could restrict the production of DOC compounds [44–46]. However, previous observations have indicated that DOC concentrations in peat could be either elevated [47–49] or lowered [4,37,50,51] with a decline in the water level, which could be contributed to the complicated mechanisms and processes involved in the production, consumption, and transport of DOC in peat along with inevitable site-specific characteristics [52]. Similarly, high temperatures can not only improve DOC production through enhanced phenol oxidase activity but also increase the consumption of DOC [25,43]. Thus, it is difficult to determine DOC concentrations in specific regions without performing practical experiments. Moreover, some studies have observed significant changes in water color and aromatic content with shifting water tables and soil temperatures at a range of sites [28,36,49,52–54]. Many previous studies have produced inconsistent results regarding the effects of changes to the water table and/or warming on aquatic DOC release, with both factors able to impact DOC concentrations, the amount of discharge, or both, in a confounding way. Specifically, the response of these variables in peatlands could depend on the length of the observation period. For example, the response of DOC production to drought conditions in the year of drought may differ from that a few years after the drought [55,56], and in a Tibetan alpine meadow experiment, the response of the aboveground environment to warming treatments in the third year was found to be different from the trend of the first two years [57]. Therefore, our observations in the year immediately after a controlled experiment are helpful for understanding how DOC export might react to climate change and anthropogenic drainage.

Zoige peatland is known to be undergoing a warming and drying climate trend [46], and severe artificial drainage [58]. Several studies in Zoige recently have reported that changes of temperature and/or water table could cause effect on the emissions of CH4 and CO2 [23,26,58,59], and Luo et al. [60] has noticed that DOC could response to experimental warming and grazing. However, there is currently knowledge of the potential response of hydrological DOC loss to the variation of temperature and water table in Zoige peatland. Furthermore, most previous studies on DOC have been conducted countries other than China, particularly in Europe and North America. Therefore, investigating regarding the response of DOC export to warming and water table treatments could provide insight into the impact and mechanisms of climate warming and artificial drainage on the regional carbon budget of Zoige peatland, as well as provide guidance for the local protection and restoration of this deteriorating natural environment. Thus, we undertook a mesocosm experiment to investigate how the hydrological loss of DOC would respond to climatic warming and artificial drainage. The specific objectives were to determine whether the export quantity and concentration, as well as the qualities of DOC and the discharge of flow water, could respond significantly to water table and temperature manipulations. In terms of potential changes of DOC export, the study provides evidence of possible changes to the carbon cycle and storage under the impact of climate change or artificial disturbance and provides evidence for the need to protect and further restore the Zoige peatland.

Methods

Field Site

The peat columns used for mesocosms were collected from Zoige peatland in Hongyuan County, Sichuan Province, on the northeastern margin of the Qinghai–Tibet Plateau (32.76°N, 102.5°E), with a mean altitude of about 3,500 m. Peat was extracted in the area for energy production until 2003, which has left a peat layer of approximately 2 m deep and created severe long-term water shortages [61]. The vegetation community mainly consists of *Carex muliensis* (relative coverage of 41%) and *Kobresia setchwanensis* (39%), as well as a small number of scattered *Potentilla anserina* (15%) and *Plantago depressa* (11%). The topography and vegetation characteristics of the study area are shown in Figure S1. During the period 2002–2011, the site experienced a mean annual temperature and precipitation level of 2.27°C and 700 mm year^{-1}, respectively. During that period, the mean temperature and precipitation from May to October were 8.26°C and 596.34 mm, respectively (data obtained from the China Meteorological Data Sharing Service System at http://cdc.cma.gov.cn/home.do). This study was conducted from May to October in 2012, when the mean temperature and precipitation were 8.65°C and 808 mm, respectively. Therefore, the site experienced higher rainfall and higher temperatures than the average of the previous 10 years.

The study was carried out on the private land of Mr. Jiang in Hongyuan County. Please contact the author first if further information is required. No further permits were required for the locations/activities in the study, and our work did not involve any endangered or protected species.

Mesocosm Experiment

All peat columns were extracted intact from the source plot in December 2011, when the peat was totally frozen and easy to move and reset. Frozen peat cores (cuboid-shaped, with intact vegetation and peat structure) with a surface area of 1 m^2 and a depth of 50–66 cm were carefully placed into stainless-steel barrels with only an open top. We used perforated stainless steel (diameter 9.0 cm) as a pocket sand filter (gravel particle size <4 mm), passing water through its inlet to maintain a near-natural infiltration rate. The perforated stainless-steel filter was buried into the peat column and connected by a drainage system to 5-L tanks in the closed bottom used to store the discharge [62]. The drainage system was connected to a manostat device with a similar pocket sand filter in the interface to lessen the peat outflow. Eighteen mesocosms were constructed for the manipulation of temperature and water table levels (three water table levels, two temperature, and three replicates, n = 18) in a crossed factorial experiment that commenced in May 2012. Positions of the water table level were controlled by hanger loops of the drainage system and set to 0 cm (W0), −10 cm (W1), and −20 cm (W2). They were calibrated using engraved rulers placed adjacent to the bottom of the steel barrels (i.e., the height of the water table was equal to the depth of the peat column plus the observed value). Warming treatment was achieved by using open top chambers (OTCs) during the snow-free period following Walker et al. [63], with 0.43-m-high polycarbonate solar panels placed outside of the mesocosms instead of infrared lamps. Actually, OTCs realize warming mainly through reducing both wind-speed and air convection and increasing incoming solar radiation [64,65]. It can be confirmed by results of previous studies [63,65,66]. During the first growing season, we observed an overall temperature increase of 1.35°C (on an annual basis) was observed for the peat with a −10 cm water table in the warming mesocosms (T1) compared to the ambient mesocosms (T0) (Figure S2). The details on the experiment design in the study were shown in Figure S4.

To closely monitor the output–input water budget in the mesocosms, water discharged from the mesocosms, rainfall, and recharge water were measured using a gauge at least once a week, and more frequently for the first two measurements when rainfall occurred. Water in each mesocosm was mainly supplied by natural precipitation and supplemented by water pumped from a nearby drainage ditch to maintain the preset water table level when necessary. As the drainage ditch extended from the same continuous *C. muliensis* peatland, thus this supplementary water had a similar attributes to the water at the field site. We buried four HOBOPro data-loggers to record peat temperature at −10 cm depth in the mesocosms: two in warming mesocosms with a water table level of 0 cm and two in control mesocosms with a water table level of −20 cm. Monthly weather data from the Hongyuan County weather station were collected for reference.

Sample Analysis

During the study period, discharged water was collected every month for DOC analysis during the growing season (May–October) in 2012. Water samples collected from the manostat tanks were mixed well before sampling, stored in sterile containers (volume 100 ml), and then filtrated through a syringe microfilter (0.45 μm) as preparation for further testing. The DOC concentration was equivalent to total carbon (TC) minus dissolved inorganic carbon (DIC), and both were determined directly using a TOC/TN analyzer (Multi N/C3100TOC/TN; Analytik Jena, Germany). TC was measured by wet combustion, and DIC was measured after sample acidification by 10% H_3PO_4 as proposed by Guo et al. [67]. The water budget data and the measured DOC concentrations were used to estimate DOC export by Method 3 proposed by Walling and Webb [68]. The UV absorbances of filtered water samples at wavelengths of 254 nm and 400 nm were determined using a UV-visible spectrophotometer (UV-2600; Shimadzu, Kyoto, Japan). UV absorption characteristics of DOC are generally measured to obtain information regarding changes in the composition of DOC compounds. We thus determined the characteristics of DOC composition by means of four related measurements of specific- and UV absorption (i.e., $Abs_{254\ nm}$, $SUVA_{254\ nm}$, $Abs_{400\ nm}$, and $SUVA_{400\ nm}$).

Statistical Analysis

Statistical analysis was done using a three-way ANOVA, including the effect of interactions between the time variable (month) and the two treatments on the monthly changes of DOC. Then a sequential full model of two-way repeated-measures ANOVA and main effect analysis and a Sidak post hoc comparison of means test were successively conducted to determine the effects of two treatments. All of these analyses were conducted after testing for essential homogeneity of variance ($p>$ 0.05, meaning that variances were homogenous; see Table. 1). Further correlation and regression analyses were conducted to determine the relationships of the monthly mean DOC concentrations and DOC export values with the corresponding peat temperature and precipitation. We also conducted linear regression analysis to determine what proportion of the two treatments and discharged volumes respectively. All statistical analyses were performed using SPSS 17.0 (SPSS Inc., Chicago, IL, USA).

Results

Microclimate in Mesocosms and Its Correlation with DOC

Soil Temperature in Peat and Precipitation. The mean monthly temperatures in the mesocosms showed that peat temperature (from May to October) was significantly higher on average in warmed (11.95 ± 4.03°C) than in the controlled mesocosms (10.60 ± 4.25°C, $p = 0.003$; n = 164; Figure S2). The mean monthly precipitation measured with the rain gauge in the mesocosms was 139.14 mm (139.14 L m^{-2}) during the growing season, which was very similar to the value of 134.92 mm recorded by the meteorological station in Hongyuan County. In general, precipitation in all mesocosms was nearly identical regardless of the possible changes in discharge and evapotranspiration among mesocosms located at the same site. A mean number of 17.4 rainy days per month occurred from May to September. Specifically, October only had five rainy days, and June and July each had twenty-one rainy days. The average discharge in the mesocosms was 281.19 L m^{-2} $year^{-1}$, which accounted for 32% of the water input (precipitation and water recharge) and 34% of the rainfall. In contrast, the annual precipitation of temperate biomes (e.g. temperate forest) in China ranges from 400 to 650 mm, i.e., less than the average precipitation (834.84 mm for 6 months) at Zoige. This suggests an abundance of precipitation in Zoige peatland, which is necessary to maintain its year-round spongy condition. It was found that the effect of warming on discharge was nonsignificant in the experiment ($p>0.05$; Table 1). Meanwhile, the interactive effect of warming and water table on discharge was also insignificant ($p>0.05$; Table 1).

Correlation of Peat Temperature and Precipitation with DOC. One year study provides limited perspective on the interannual patterns of DOC hydrological export. However, we obtained extra information by the correlation analysis between the two main microclimatic factors and the mean monthly concen-

Table 1. *P*-values of a two-way ANOVA and Levene's test for the effects of the water table level, temperature, and their interactions on the amount of annual DOC export, DOC concentration, absorbance and specific absorbance, and water discharge.

Treatment	DOC		Discharge	Absorbances and specific absorbances			
	Export	Concentration	Discharge	$Abs_{254\ nm}$	$SUVA_{254}$	$Abs_{400\ nm}$	$SUVA_{400}$
Water table	<0.001**	<0.001**	0.037*	0.001**	0.005**	0.003**	0.008**
Temperature	0.075	0.012**	0.764	0.007**	0.008**	0.010**	0.018**
Water table × Temperature	0.734	0.735	0.553	0.689	0.077	0.439	0.431
Levene's Test	0.179	0.318	0.235	0.520	0.318	0.616	0.110

*indicates a significant difference ($p<0.05$, n = 18)
**indicates a highly significant difference ($p<0.01$, n = 18).

tration and export of DOC, see Figure S3. Monthly DOC export was positively correlated with peat temperature ($R^2 = 0.4984$, $p<0.01$; n = 24) and precipitation ($R^2 = 0.8982$, $p<0.01$; n = 6), while the mean monthly DOC concentration was significantly correlated with peat temperature ($R^2 = 0.4025$, $p<0.01$; n = 24), but had a nonsignificant correlation with precipitation ($R^2 = 0.3046$, $p = 0.128$; n = 6).

Annual Export of DOC

Neither the water table × warming nor the month × two controlling-factor interactions were statistically significant. Thus, we examined the single effect of water table and temperature manipulation on the measured variables in the study (i.e. DOC export, concentration, and quality, and the water budget).

We found a difference between the experimental warming treatment and water table treatment in terms of the annual amount of DOC exported. During the period of the study, the manipulation of water table depth in the mesocosms significantly influenced DOC export ($p<0.001$; Figure 1). The export of DOC displayed an upward trend with decreasing water levels. DOC export was 5.76 ± 0.63 g C m^{-2} $year^{-1}$ when the water table was at 0 cm, significantly lower than the levels of 9.75 ± 0.84 g C m^{-2} $year^{-1}$ (−10 cm water-level; $p = 0.004$) and 11.65 ± 1.68 g C m^{-2} $year^{-1}$ (−20 cm water-level; $p<0.001$) respectively. Although all values were within the range (5–40 C m^{-2} $year^{-1}$) found in the natural peatland [3], the results indicated that DOC export would increase by 69% and 102% annually if the water table at 0 cm was lowered by 10 cm and 20 cm, respectively. In contrast, no significant effect on DOC export was observed in the warming treatments when the peat temperature was raised by 1.35°C (9.81±3.32 g C m^{-2} $year^{-1}$ vs. 8.30±1.82 g C m^{-2} $year^{-1}$, $p = 0.059$) throughout the growing season. Previous studies have demonstrated that DOC loss during the nongrowing season is similar between controlled and manipulated sites [41]. Therefore, our estimation of DOC export within the growing season did not suggest an alteration in its overall tendency throughout the year.

Eighty-seven percent of the variability in DOC annual export was explained by the combination of the water table level, temperature, and discharge. Among these variables, the level of the water table was the most important predictor, explaining more than 68% of the variation in DOC annual export. Furthermore, discharge was found to be significantly affected only by the water table treatment ($p = 0.028$, n = 18; Figure 2). The discharge

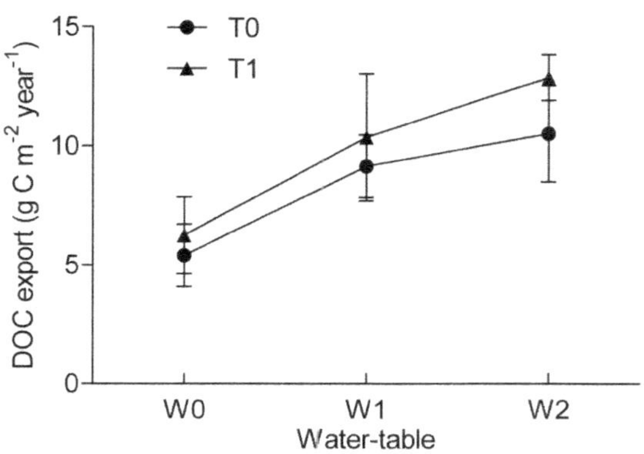

Figure 1. Effect of water table levels and temperature on DOC annual export. Data are means ± standard error. T0 and T1 correspond to ambient temperature and warming temperature, respectively, and W0, W1, and W2 indicate water table depths of 0 cm, −10 cm, and −20 cm, respectively.

volume was 265.08 ± 1.88 L $year^{-1}$ when the water table was at 0 cm, slightly lower than 293.58.04 ± 6.71 L $year^{-1}$ (−10 cm water-level; $p = 0.063$) and significantly lower than 294.92 ± 10.84 L $year^{-1}$ (−20 cm water-level; $p = 0.050$). In contrast, the discharge in the warming treatment was nonsignificantly smaller (283.11 ± 17.66 L vs. 279.28 ± 29.28 L). Correlation analysis also showed that water table levels were significantly negatively correlated with discharge ($R^2 = 0.31$, $p = 0.008$; n = 18), while experimental warming was almost irrelevant to discharge ($R^2 = 0.004$, $p = 0.398$; n = 18).

DOC Concentrations

DOC concentrations in discharged water varied significantly between the warmed and ambient temperature treatments as well as among the three water table treatments, despite the differences in their effects on DOC export. As shown in Figure 2, the DOC concentration was lower (23.18±5.76 mg L^{-1}) when the position of the water table level was at 0 cm, significantly lower than 32.81±2.88 mg L^{-1} (−10 cm water-level; $p = 0.028$) and 38.92±4.98 mg L^{-1} (−20 cm water-level; $p = 0.001$). Similarly, DOC concentration was higher in the warmed mesocosms (34.90±5.25 mg L^{-1}, $p = 0.005$) than in the ambient temperature mesocosms (28.48±8.13 mg L^{-1}). In addition, almost 78% of the variation in DOC concentrations was explained by water table and temperature treatments, and 61% of these could be contributed to the water table treatment alone.

Qualities of DOC

There were similar trends for the effects of experimental warming and water table level on absorbance (at wavelengths of 254 nm and 400 nm) and specific absorbance ($SUVA_{254\ nm}$ and $SUVA_{400\ nm}$) of DOC in the filtered discharge. These four measures of the quality of DOC were all significantly higher under the warming treatment ($p = 0.004$, $p = 0.016$, $p = 0.004$, and $p = 0.015$, respectively; Table 1, Figure 3) than in the control. Similarly, values of the four measures for a lower water table were significantly higher than those observed at higher water table level ($p<0.001$, $p = 0.010$, $p<0.001$, $p = 0.006$, respectively; Table 1). We assessed the impact of the three positions of the water table level on the four DOC quality measures using a multiple comparison analysis as shown in Figure 3. Therefore, the results above showed that the water table and warming treatments clearly led to several changes in both the quality and absolute DOC concentrations of DOC, indicating a higher aromatic content and changes in the color of downstream water.

Discussion

Effect of Water Table Treatment on Export, Concentration, and Qualities of DOC

Water table manipulation had significant effects on the annual amount of DOC exported, DOC concentration, and the water discharge. Lower water tables were often accompanied by higher DOC exports and concentrations, and a larger discharge volume. These effects were significant when the water table at 0 cm was lowered to −10 cm and −20 cm. This supports the observed variation in the export of DOC, possibly due to site-specific characteristics, and is mainly derived from both fluctuations in the DOC concentration in runoff water and the quantity of water discharged, which disagrees with several previous studies [4,26,69]. Peatland has its own specific features such as plant community construction [70] and hydro-topographical characteristics [67]. Meanwhile, the export of DOC varies with catchment properties and hydrogeologic setting [71], such as precipitation, evapotranspiration [72] and annual runoff [73]. Therefore, it may have disparate performances in DOC production and export which may result in difficulty to draw a universal conclusion [67]. Actually, most reported studies on the response of DOC to water table were done in Europe [74] or North America [4,69]. In this study, when the water table at 0 cm was artificially lowered to −10 cm and −20 cm, DOC export increased by 69% and 102%, respectively. This suggests that an estimated extra 18.4 × 10^9 g C and 27.2 × 10^9 g C, respectively, would be transported downstream during the growing season in Zoige. Our ranges of estimated DOC exports with a water table level of −10 cm and −20 cm (9.75–11.65 g C m^{-2} $year^{-1}$) were similar to those (8.4–11.3 C m^{-2} $year^{-1}$) observed in Quebec, Canada, in two growing seasons following a water table drawdown [41]. This could result in a substantial DOC loading into downstream ecosystems, potentially altering the physical and chemical characteristics of aquatic ecosystems, such as acidity, light penetration, and metal and nutrient availability [75]. Such indirect effects on aquatic ecosystems may be the most severe consequence of the elevated DOC export from peatlands. Increases in DOC export may also indicate shifts in the carbon budget, suggesting either a decrease in carbon uptake and storage, or an increase in the turnover of organic carbon [39].

As expected, DOC concentrations were elevated when the water level declined, which agreed with the results of several previous studies [47–49], but disagreed with some other published observations [4,37,50,51]. These inconsistencies reflect the complicated mechanisms and processes involved in the production, consumption, and transport of DOC in peat [52]. Initially, water

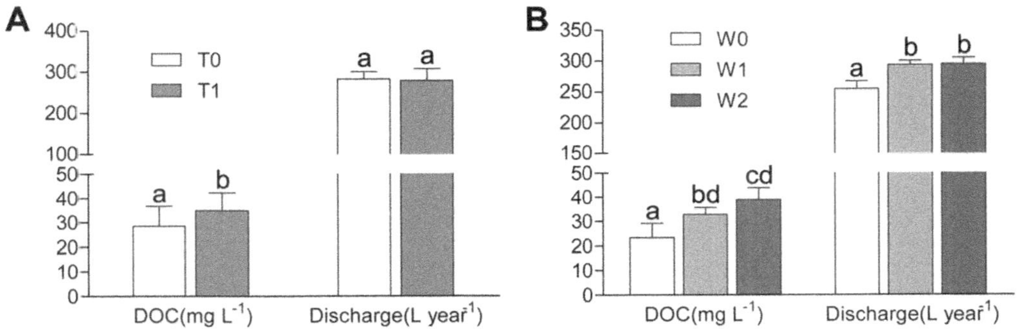

Figure 2. Variations in DOC concentrations and discharge volumes under different treatments. Data are means ± standard error. Same letter superscripts denote insignificant differences among the three water table levels from post hoc tests.

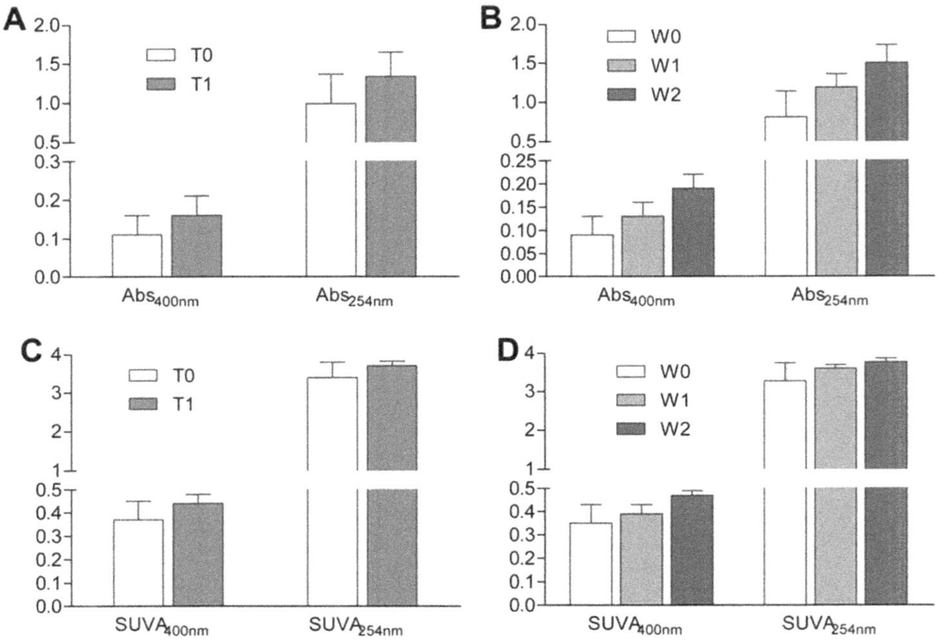

Figure 3. Effects of water table levels and temperature treatments on $Abs_{254\ nm}$, $Abs_{400\ nm}$, $SUVA_{254\ nm}$, and $SUVA_{400\ nm}$. Data are means ± standard error. Same letter superscripts denote nonsignificant differences among the three water table levels from post hoc tests.

table drawdown, which is closely associated with peat moisture content, can promote aerobic respiration [76,77] and release the activity of degrading hydrolase enzymes in peat, which is supported by reported changes in the specific absorbance of DOC [78]. Biomass is often related to water table, larger biomass being associated with lower water table [79]. Besides, water table could also affect decomposition indirectly through changes in plant community composition [80] or reduce productivity and even cause death in wetland species as the water table is lowered too far [79], thus it can't be determined water table effects on decomposition through changes in plant community or microbial activity within the results of our study. These two mechanisms described above lead to an increase in DOC production, which can then be flushed out from stagnant peat horizons [41] during rainfall. Furthermore, studies have shown that this could lead to peat subsidence and a lower porosity when the water level declines [42,81], likely resulting in slower interflow and a longer residence time for water moving through peat. This may also contribute to more DOC compounds from peat transferring into flowing water. There were 17.4 rainy days every month from May to September 2012 in Zoige, with a maximum of 21 in each of June and July. According to Harrison et al. [82], this high rainfall rate may have potential effect on DOC concentration in water by promoting DOC release from peat soil. Moreover, the water discharge also increased when the level of the water table was lowered in the mesocosm experiment. This suggests a decreased capacity for water storage, which might exacerbate severe water loss and peat erosion if artificial water drainage continues in Zoige peatland. Higher DOC concentrations could also create problems for the stable and sustainable development of peatland ecosystems because it would alter aquatic habitats through its effect on pH and various biological activities (e.g., transportation of nutrients). The weak correlation between precipitation and DOC concentration also indicates that precipitation could promote the export of DOC with a limited enhancement of the DOC concentration. It also indicates that the elevated DOC in runoff water was the main contributor to the high discharge rather than the high concentration observed under higher rainfall.

The changes in DOC concentrations resulting from the water table treatment were accompanied by changes in UV absorbance characteristics. Our results showed that the aromatic and colored components of DOC increased when the water table was low, suggesting a possible increase in peat degradation. The colored components of DOC, as measured by absorbance at 400 nm and $SUVA_{400\ nm}$, increased sharply when the water table at 0 cm level was lowered to −20 cm, but showed only a slight decline when it was lowered to −10 cm. The increase in DOC aromaticity, measured by absorbance at 254 nm and $SUVA_{254\ nm}$, implied that more aromatic DOC substances should occur in discharge water following a decline in the water table level. Recent studies of Zoige peatland have demonstrated that aromaticity has substantially increased in sites that have experienced a long period of aerobic oxidation and water loss [11]. This is in accordance with our observation of a higher aromatic content when the water table was lower because more peat would be exposed to the air, resulting in increased aerobic respiration [76]. This would also support a reduction in the colored components and aromatic content when the water table rises, which has also been observed elsewhere [36,49,53]. Consequently, DOC in runoff consists of more colored and aromatic components when the water table was drawdown, making it less accessible to microbes within the fluvial ecosystem [54]. Finally, as our results suggested, it would lead to carrying more DOC compounds downstream.

Effect of Warming on Export, Concentration, and Qualities of DOC

Experimental warming at a rate of 1.35°C·year^{-1} in peat had a significant effect on DOC concentration but limited effects on DOC export and discharge. DOC concentrations in warming mesocosms were higher than in normal mesocosms. However, the observed variability of DOC concentrations generally had a limited impact on the hydrological export of DOC which is explained by the relevance of the discharge volume to DOC exports and displayed a nonsignificant correlation with warming in our experiment. Some studies have observed high DOC concentrations at higher temperatures that increased the amount of DOC exported for as long as 12 years under natural warming conditions [25], whereas others have reported a decrease in DOC exports due to the lower discharge following a temperature increase of 1.6–4.1°C [4]. In the study, warming mesocosms have significantly higher DOC concentration but with insignificant lower discharge, which may result from its specific feature (such as evapotranspiration) [72] in Zoige that differs from anywhere else. Therefore, we assumed that temperature variation may have a complex influence on DOC export that is probably associated with both the rate of warming and the temporal scale. It is important to consider both present and future climate change when investigating the effects of experimental warming on peatland carbon turnover.

Temperature is the main factor influencing bacterial metabolism and the rate of decomposition of organic materials, and it also affects DOC dynamics in ecosystems [83]. High temperatures can not increase DOC production through enhanced phenol oxidase activity, but it also can increase the consumption of DOC [25,43]. Thus, determining the DOC concentration in specific regions is difficult without conducting practical experiments. Warming can decrease plant species richness but increase aboveground net primary production [84], thus it may influence inputs of carbon into peatland and lead to unstable DOC concentrations. Our correlation analysis in our experiment indicated that the monthly mean DOC concentration in the discharge increased with temperature. This suggests that the higher temperature increased the DOC concentration because the enhanced decomposition exceeded the gain in DOC consumption in the first year following a rise in temperature. Walker et al. [63] suggested that warming in OTC experiments increased the height and cover of deciduous shrubs and graminoids and decreased species diversity and evenness, implying that the increased DOC concentration we observed might also be attributable to the changes in plant primary productivity. However, temperature had a weak effect on the water discharge during the whole growing season, indicating a relatively stable discharge volume independent of the warming treatment, and differing slightly from the observations of Bridgham et al. [62]. However, this is understandable given the relatively plentiful precipitation at the study site as well as the relatively small change in peat temperature.

Similarly, all of the colored components displayed an upward trend with warming in the study, which was contrary to the results obtained in a laboratory experiment by Tang et al. [52]. But one recent study showed that warming could cause a shift in the composition of bacterial communities in the surface (1–3 cm) and middle layers (9–11 cm) of peat [85], which supported our observations of greater colored and aromatic content under warmer conditions. Therefore, our results indicated that the rising temperature could influence the composition of peat (especially color and aromatic content). Furthermore, peat degradation might arise following climatic warming according to the results, probably due to potential shifts in the function and structure of microbial communities in the peatland, a hypothesis that requires further investigation.

Conclusion

We investigated the response of the hydrological export of DOC in Zoige peatland to changes in the water table level and temperature. Our one-year study provides a basis for understanding the rapid response of the carbon budget in Zoige peatland to climate change and/or artificial drainage, as well as the potential damage to downstream ecosystems, particularly in the Yellow River.

The differences between the water table and peat-temperature treatments implied that future short-duration water table drawdown events could have a greater impact than rising temperature on the export of DOC. In this study, water table effects DOC concentration and export as well as discharge, while temperature treatment only causes obvious effect on DOC concentration. It probably derives from that water table drawdown influenced temperature patterns in the decomposing litter [80] in the crossed factorial experiment. Meanwhile, the experimental warming may also not be high enough for changing the amount of DOC export, as warming could also increase evapotranspiration and therefore decrease discharge [62], which can be known from the relatively smaller discharge in warming mesocosm in the study. Thus, it supported the view that the influence of the water table or water content on peatland ecosystems (such as DOC loss in the study) is stronger than in variations in other environmental conditions, such as temperature [79]. The two experimental water table positions (−10 m and −20 m) resulted in increases in the annual export of DOC by 69% and 102%, respectively, through both a higher DOC concentration and larger discharge volume, indicating the potential release of both carbon and water from peatland after the level of the water table is lowered. The temperature treatment resulted in clear changes in the DOC concentration but had a limited effect on DOC export, probably indicating a shift in the turnover rate of organic carbon in peat because temperature is the main factor affecting bacterial metabolism and the rate of decomposition of organic materials. The nonsignificant effect of warming on DOC export and the notably positive relationship between mean monthly DOC export and peat temperature, which resulted from the shortage of recorded data for all mesocosms, appear contradictory, and suggest that further studies should be careful to consider these issues. Variable water levels and temperatures changed the absorbances of DOC in the year immediately after this experiment. This result suggests the varied nature or qualities of DOC that might influence fluvial systems, and also warns that using absorbance records as a proxy for DOC concentrations when studied in peatland should be done with caution.

Therefore, our observations in the first year immediately after the controlled experiment were helpful for understanding how the carbon budget might react to climate change and anthropogenic interference (i.e., drainage). This mesocosm experiment also provides useful information for local protection and sustainable development in Zoige peatland. Additional experiments and observations are required to achieve a comprehensive understanding of the carbon cycle (such as DOC fluxes) in peatlands facing changing environmental conditions.

Supporting Information

Figure S1 Topography and vegetation characteristics of the study area. (A) Drainage ditch. (B) Vegetation community growing in shallow water. (C) Scattered vegetation surrounded by surface water. All photographs were taken during May 2012 in Hongyuan County located in Zoige peatland.

Figure S2 Peat temperature at −10 cm depth and precipitation during the growing season in 2012.

Figure S3 Correlation analysis of temperature and precipitation with DOC. (A) Results of the correlation analysis between peat temperature recorded in four mesocosms and corresponding mean monthly export and concentration of DOC (DOC concentration: $y = 1.4573x + 9.9416$, $R^2 = 0.4025$, $p<0.01$, n = 24; DOC export: $y = 0.2043x - 0.9177$, $R^2 = 0.4984$, $p<0.01$, n = 24). (B) Results of the correlation analysis between mean monthly precipitation and mean monthly export and concentration of DOC in all mesocosms (DOC concentration: $y = 0.0379x + 26.409$, $R^2 = 0.3046$, $p = 0.128$, n = 6; DOC export: $y = 0.013x - 0.2987$, $R^2 = 0.8982$, $p< 0.01$, n = 6).

Figure S4 The schematic drawing of the mesocosm in the study. The references of the number are shown as below: 1. Polycarbonate solar panels; 2. Water intake system; 3. Drainage system; 4. Water storage barrel; 201. Observation tube of water-level; 202. Sand filter pocket of inlet; 301. Sand filter pocket of outlet; 302. High pressure valves; 303. Water pipe; 304. Hanger loop; 305. Observation rule.

Acknowledgments

The authors wish to acknowledge the assistance of Prof. Zhang Fen Chun at CRAES (Chinese Research Academy of Environmental Sciences) for his expert revision of the manuscript. The authors are also grateful to two anonymous reviewers for their valuable comments on the manuscript.

Author Contributions

Conceived and designed the experiments: LLH BK. Performed the experiments: XDL SQZ. Analyzed the data: XDL BK YLH. Contributed reagents/materials/analysis tools: XDL SQZ BK LLH. Wrote the paper: XDL LLH.

References

1. Aselmann I, Crutzen PJ (1989) Global distribution of natural fresh-water wetlands and rice paddies, their net primary productivity seasonality and possible methane emissions. Journal of Atmospheric Chemistry 8: 307–358.
2. Hobbie JE (1992) Microbial control of dissolved organic-carbon in lakes - research for the Future. Hydrobiologia 229: 169–180.
3. Moore TR, Roulet NT, Waddington JM (1998) Uncertainty in predicting the effect of climatic change on the carbon cycling of Canadian peatlands. Climatic Change 40: 229–245.
4. Pastor J, Solin J, Bridgham SD, Updegraff K, Harth C, et al. (2003) Global warming and the export of dissolved organic carbon from boreal peatlands. Oikos 100: 380–386.
5. Turunen J (2008) Development of Finnish peatland area and carbon storage 1950–2000. Helsinki, FINLANDE: Finnish Environment Institute. 16 p.
6. Yu Z, Beilman DW, Jones MC (2013) Sensitivity of northern peatland carbon dynamics to Holocene climate change. Carbon Cycling in Northern Peatlands: American Geophysical Union. pp. 55–69.
7. Moore TR, Dalva M (1993) The influence of temperature and water table position on carbon dioxide and methane emissions from laboratory columns of peatland soils. Journal of Soil Science 44: 651–664.
8. Price JS (2003) Role and character of seasonal peat soil deformation on the hydrology of undisturbed and cutover peatlands. Water Resources Research 39: 1241.
9. Xiang S, Guo R, Wu N, Sun S (2009) Current status and future prospects of Zoige Marsh in Eastern Qinghai-Tibet Plateau. Ecological Engineering 35: 553–562.
10. Shi C-c, Tu J (2009) Remote Sensing Monitory Study on Land Desertification in Ruoergai Plateau of Sichuan Province during 40 Years. Southwest China Journal of Agricultural Sciences 6: 035 (in Chinese).
11. Guo X, Du W, Wang X, Yang Z (2013) Degradation and structure change of humic acids corresponding to water decline in Zoige peatland, Qinghai-Tibet Plateau. Sci Total Environ 445–446: 231–236.
12. Zhang X, Lv X, Gu H (2005) To analyze threats, to describe present conservation situation and to provide management advices of the Ruoergai marshes. Wetland Sci 3: 292–297(in Chinese).
13. SAFS (Sichuan Academy of Forest Science) (2006) Scientific investigation report on Zoige marsh. Chengdu: Sichuan Science and Technology Press(in Chinese).
14. Wang M, Liu Z, Ma X, Wang G (2012) Division of organic carbon reserves of peatlands in China. Wetland Sci 10: 156–163(in Chinese).
15. Gao J (2006) Degradation factor analysis and solutions of Ruoergai Wetland in Sichuan. Sichuan Environ 25: 48–53(in Chinese).
16. Gorham E (1991) Northern peatlands- role in the carbon-cycle and probable responses to climatic warming Ecological Applications 1: 182–195.
17. Trettin CC, Laiho R, Minkkinen K, Laine J (2006) Influence of climate change factors on carbon dynamics in northern forested peatlands. Canadian Journal of Soil Science 86: 269–280.
18. Bridgham SD, Pastor J, Dewey B, Weltzin JF, Updegraff K (2008) Rapid carbon response of peatlands to climate change. Ecology 89: 3041–3048.
19. Hogg EH, Lieffers VJ, Wein RW (1992) Potential carbon losses from peat profiles - effects of temperature, drought cycles, and fire. Ecological Applications 2: 298–306.
20. Freeman C, Lock MA, Reynolds B (1993) Fluxes of CO2, CH4 and N2O from a Welsh peatland following simulation of water-table draw-down - potential feedback to climatic-change. Biogeochemistry 19: 51–60.
21. Bohrer G, Chen H, Wu N, Wang Y, Zhu D, et al. (2013) Inter-Annual Variations of Methane Emission from an Open Fen on the Qinghai-Tibetan Plateau: A Three-Year Study. PLoS ONE 8: e53878.
22. Evans CD, Chapman PJ, Clark JM, Monteith DT, Cresser MS (2006) Alternative explanations for rising dissolved organic carbon export from organic soils. Global Change Biology 12: 2044–2053.
23. Zhang G, Tian J, Jiang NA, Guo X, Wang Y, et al. (2008) Methanogen community in Zoige wetland of Tibetan plateau and phenotypic characterization of a dominant uncultured methanogen cluster ZC-I. Environmental microbiology 10: 1850–1860.
24. Billett MF, Palmer SM, Hope D, Deacon C, Storeton-West R, et al. (2004) Linking land-atmosphere-stream carbon fluxes in a lowland peatland system. Global Biogeochemical Cycles 18: GB1024.
25. Freeman C, Evans CD, Monteith DT, Reynolds B, Fenner N (2001) Export of organic carbon from peat soils. Nature 412: 785–785.
26. Chen H, Yao S, Wu N, Wang Y, Luo P, et al. (2008) Determinants influencing seasonal variations of methane emissions from alpine wetlands in Zoige Plateau and their implications. Journal of Geophysical Research: Atmospheres 113: D12303.
27. Limpens J, Berendse F, Blodau C, Canadell JG, Freeman C, et al. (2008) Peatlands and the carbon cycle: from local processes to global implications – a synthesis. Biogeosciences 5: 1475–1491.
28. Dinsmore KJ, Billett MF, Dyson KE (2013) Temperature and precipitation drive temporal variability in aquatic carbon and GHG concentrations and fluxes in a peatland catchment. Global Change Biology 19: 2133–2148.
29. Aerts R, De Caluwe H (1999) Nitrogen deposition effects on carbon dioxide and methane emissions from temperate peatland soils. Oikos 84: 44–54.
30. Arnosti C, Holmer M (2003) Carbon cycling in a continental margin sediment: contrasts between organic matter characteristics and remineralization rates and pathways. Estuarine Coastal and Shelf Science 58: 197–208.
31. Wallage ZE, Holden J (2010) Spatial and temporal variability in the relationship between water colour and dissolved organic carbon in blanket peat pore waters. Science of The Total Environment 408: 6235–6242.
32. Chin W-C, Lennon JT, Hamilton SK, Muscarella ME, Grandy AS, et al. (2013) A Source of Terrestrial Organic Carbon to Investigate the Browning of Aquatic Ecosystems. PLoS ONE 8: e75771.
33. Evans CD, Monteith DT, Cooper DM (2005) Long-term increases in surface water dissolved organic carbon: Observations, possible causes and environmental impacts. Environmental Pollution 137: 55–71.
34. Carpenter SR, Pace ML (1997) Dystrophy and eutrophy in lake ecosystems: Implications of fluctuating inputs. Oikos 78: 3–14.
35. Wetzel RG (1992) Gradient-dominated ecosystems - sources and regulatory functions of dissolved organic-matter in fresh-water ecosystems. Hydrobiologia 229: 181–198.
36. Wallage ZE, Holden J, McDonald AT (2006) Drain blocking: An effective treatment for reducing dissolved organic carbon loss and water discolouration in a drained peatland. Science of the Total Environment 367: 811–821.
37. Grayson R, Holden J (2012) Continuous measurement of spectrophotometric absorbance in peatland streamwater in northern England: implications for understanding fluvial carbon fluxes. Hydrological Processes 26: 27–39.
38. Weishaar JL, Aiken GR, Bergamaschi BA, Fram MS, Fujii R, et al. (2003) Evaluation of specific ultraviolet absorbance as an indicator of the chemical composition and reactivity of dissolved organic carbon. Environmental Science & Technology 37: 4702–4708.
39. Worrall F, Armstrong A, Adamson JK (2007) The effects of burning and sheep-grazing on water table depth and soil water quality in a upland peat. Journal of Hydrology 339: 1–14.
40. Fraser CJD, Roulet NT, Moore TR (2001) Hydrology and dissolved organic carbon biogeochemistry in an ombrotrophic bog. Hydrological Processes 15: 3151–3166.
41. Strack M, Waddington JM, Bourbonniere RA, Buckton EL, Shaw K, et al. (2008) Effect of water table drawdown on peatland dissolved organic carbon export and dynamics. Hydrological Processes 22: 3373–3385.

42. Sommer M (2006) Influence of soil pattern on matter transport in and from terrestrial biogeosystems—A new concept for landscape pedology. Geoderma 133: 107–123.
43. Briggs J, Large DJ, Snape C, Drage T, Whittles D, et al. (2007) Influence of climate and hydrology on carbon in an early Miocene peatland. Earth and Planetary Science Letters 253: 445–454.
44. Strack M, Waddington JM, Tuittila ES (2004) Effect of water table drawdown on northern peatland methane dynamics: Implications for climate change. Global Biogeochemical Cycles 18: GB4003.
45. Scanlon D, Moore T (2000) Carbon dioxide production from peatland soil profiles: The influence of temperature, oxic/anoxic conditions and substrate. Soil Science 165: 153–160.
46. Morris PJ, Belyea LR, Baird AJ (2011) Ecohydrological feedbacks in peatland development: A theoretical modelling study. Journal of Ecology 99: 1190–1201.
47. Dai Y, Luo Y, Wang C, Shen Y, Ma Z, et al. (2010) Climate variation and abrupt change in wetland of Zoig Plateau during 1961 and 2008. Journal of Glaciology and Geocryology 32: 35–42(in Chinese).
48. Jager DF, Wilmking M, Kukkonen JVK (2009) The influence of summer seasonal extremes on dissolved organic carbon export from a boreal peatland catchment: Evidence from one dry and one wet growing season. Science of The Total Environment 407: 1373–1382.
49. Blodau C, Siems M (2012) Drainage-induced forest growth alters belowground carbon biogeochemistry in the Mer Bleue bog, Canada. Biogeochemistry 107: 107–123.
50. Sapek A, Sapek B, Chrzanowski S, Urbaniak M (2009) Nutrient mobilisation and losses related to the groundwater level in low peat soils. International Journal of Environment and Pollution 37: 398–408.
51. Ellis T, Hill PW, Fenner N, Williams GG, Godbold D, et al. (2009) The interactive effects of elevated carbon dioxide and water table draw-down on carbon cycling in a Welsh ombrotrophic bog. Ecological Engineering 35: 978–986.
52. Tang R, Clark JM, Bond T, Graham N, Hughes D, et al. (2013) Assessment of potential climate change impacts on peatland dissolved organic carbon release and drinking water treatment from laboratory experiments. Environmental Pollution 173: 270–277.
53. Watts CD, Naden PS, Machell J, Banks J (2001) Long term variation in water colour from Yorkshire catchments. Science of The Total Environment 278: 57–72.
54. Wilson L, Wilson J, Holden J, Johnstone I, Armstrong A, et al. (2011) Ditch blocking, water chemistry and organic carbon flux: Evidence that blanket bog restoration reduces erosion and fluvial carbon loss. Science of the Total Environment 409: 2010–2018.
55. Mitchell GN (1990) Natural discoloration of freshwater: Chemical composition and environmental genesis. Progress in Physical Geography 14: 317–334.
56. Mitchell G, McDonald AT (1992) Discolouration of water by peat following induced drought and rainfall simulation. Water Research 26: 321–326.
57. Li G, Liu Y, Frelich LE, Sun S (2011) Experimental warming induces degradation of a Tibetan alpine meadow through trophic interactions. Journal of Applied Ecology 48: 659–667.
58. Chen H, Wu N, Wang Y, Zhu D, Zhu Qa, et al. (2013) Inter-Annual Variations of Methane Emission from an Open Fen on the Qinghai-Tibetan Plateau: A Three-Year Study. PLoS ONE 8: e53878.
59. Yanbin H, Yanfen W, Xurong M, Xiangzhong H, Xiaoyong C, et al. (2008) CO2H2O and energy exchange of an Inner Mongolia steppe ecosystem during a dry and wet year. Acta Oecologica-international Journal Of Ecology 33: 133–143.
60. Luo C, Xu G, Wang Y, Wang S, Lin X, et al. (2009) Effects of grazing and experimental warming on DOC concentrations in the soil solution on the Qinghai-Tibet plateau. Soil Biology and Biochemistry 41: 2493–2500.
61. Zhang XH, Liu HY, Baker C, Graham S (2012) Restoration approaches used for degraded peatlands in Ruoergai (Zoige), Tibetan Plateau, China, for sustainable land management. Ecological Engineering 38: 86–92.
62. Bridgham SD, Pastor J, Updegraff K, Malterer TJ, Johnson K, et al. (1999) Ecosystem control over temperature and energy flux in Northern peatlands. Ecological Applications 9: 1345–1358.
63. Walker MD, Wahren CH, Hollister RD, Henry GH, Ahlquist LE, et al. (2006) Plant community responses to experimental warming across the tundra biome. Proc Natl Acad Sci U S A 103: 1342–1346.
64. Debevec EM, MacLean JrSF (1993) Design of greenhouses for the manipulation of temperature in tundra plant communities. Arctic and Alpine Research: 56–62.
65. Turetsky M, Treat C, Waldrop M, Waddington J, Harden J, et al. (2008) Short-term response of methane fluxes and methanogen activity to water table and soil warming manipulations in an Alaskan peatland. Journal of Geophysical Research: Biogeosciences (2005–2012) 113.
66. Chivers M, Turetsky M, Waddington J, Harden J, McGuire A (2009) Effects of experimental water table and temperature manipulations on ecosystem CO2 fluxes in an Alaskan rich fen. Ecosystems 12: 1329–1342.
67. Guo Y, Wan Z, Liu D (2010) Dynamics of dissolved organic carbon in the mires in the Sanjiang Plain, Northeast China. Journal of Environmental Sciences 22: 84–90.
68. Walling DE, Webb BW (1981) The reliability of suspended sediment load data: IAHS Publication.
69. Clair TA, Arp P, Moore TR, Dalva M, Meng FR (2002) Gaseous carbon dioxide and methane, as well as dissolved organic carbon losses from a small temperate wetland under a changing climate. Environmental Pollution 116, Supplement 1: S143–S148.
70. Weltzin JF, Pastor J, Harth C, Bridgham SD, Updegraff K, et al. (2000) Response of bog and fen plant communities to warming and water-table manipulations. Ecology 81: 3464–3478.
71. Fraser C, Roulet N, Moore T (2001) Hydrology and dissolved organic carbon biogeochemistry in an ombrotrophic bog. Hydrological Processes 15: 3151–3166.
72. Moore T (1989) Dynamics of dissolved organic carbon in forested and disturbed catchments, Westland, New Zealand: 1. Maimai. Water Resources Research 25: 1321–1330.
73. Urban N, Bayley S, Eisenreich S (1989) Export of dissolved organic carbon and acidity from peatlands. Water Resources Research 25: 1619–1628.
74. Worrall F, Reed M, Warburton J, Burt T (2003) Carbon budget for a British upland peat catchment. Science of The Total Environment 312: 133–146.
75. Steinberg (2003) Ecology of humic substances in freshwaters: determinants from geochemistry to ecological niches. Berlin: Springer.
76. Clymo RS (1984) The limits to peat bog growth. Philosophical Transactions of the Royal Society of London B, Biological Sciences 303: 605–654.
77. Mars H, Wassen MJ, Peeters WHM (1996) The effect of drainage and management on peat chemistry and nutrient deficiency in the former Jegrznia-floodplain (NE-Poland). Vegetatio 126: 59–72.
78. Freeman C, Ostle N, Kang H (2001) An enzymic 'latch' on a global carbon store. Nature 409: 149–149.
79. Moore PD (2002) The future of cool temperate bogs. Environmental Conservation 29: 3–20.
80. Strakova P, Penttilä T, Laine J, Laiho R (2012) Disentangling direct and indirect effects of water table drawdown on above-and belowground plant litter decomposition: consequences for accumulation of organic matter in boreal peatlands. Global Change Biology 18: 322–335.
81. Whittington PN, Price JS (2006) The effects of water table draw-down (as a surrogate for climate change) on the hydrology of a fen peatland, Canada. Hydrological Processes 20: 3589–3600.
82. Harrison AF, Taylor K, Scott A, Poskitt J, Benham D, et al. (2008) Potential effects of climate change on DOC release from three different soil types on the Northern Pennines UK: examination using field manipulation experiments. Global Change Biology 14: 687–702.
83. Froberg M, Berggren D, Bergkvist B, Bryant C, Mulder J (2006) Concentration and fluxes of dissolved organic carbon (DOC) in three norway spruce stands along a climatic gradient in sweden. Biogeochemistry 77: 1–23.
84. Lin X, Zhang Z, Wang S, Hu Y, Xu G, et al. (2011) Response of ecosystem respiration to warming and grazing during the growing seasons in the alpine meadow on the Tibetan plateau. Agricultural and Forest Meteorology 151: 792–802.
85. Kim SY, Freeman C, Fenner N, Kang H (2012) Functional and structural responses of bacterial and methanogen communities to 3-year warming incubation in different depths of peat mire. Applied Soil Ecology 57: 23–30.

Carbon Footprint of Telemedicine Solutions - Unexplored Opportunity for Reducing Carbon Emissions in the Health Sector

Åsa Holmner[1]*, Kristie L. Ebi[2,4], Lutfan Lazuardi[3], Maria Nilsson[4]

1 Department of Radiation Sciences, Umeå University, Umeå, Sweden, **2** ClimAdapt, LLC, Seattle, Washington, United States of America, **3** Department of Public Health, Faculty of Medicine, Gadjah Mada University, Yogyakarta, Indonesia, **4** Department of public health and clinical medicine, epidemiology and global health, Umeå University, Umeå, Sweden

Abstract

Background: The healthcare sector is a significant contributor to global carbon emissions, in part due to extensive travelling by patients and health workers.

Objectives: To evaluate the potential of telemedicine services based on videoconferencing technology to reduce travelling and thus carbon emissions in the healthcare sector.

Methods: A life cycle inventory was performed to evaluate the carbon reduction potential of telemedicine activities beyond a reduction in travel related emissions. The study included two rehabilitation units at Umeå University Hospital in Sweden. Carbon emissions generated during telemedicine appointments were compared with care-as-usual scenarios. Upper and lower bound emissions scenarios were created based on different teleconferencing solutions and thresholds for when telemedicine becomes favorable were estimated. Sensitivity analyses were performed to pinpoint the most important contributors to emissions for different set-ups and use cases.

Results: Replacing physical visits with telemedicine appointments resulted in a significant 40–70 times decrease in carbon emissions. Factors such as meeting duration, bandwidth and use rates influence emissions to various extents. According to the lower bound scenario, telemedicine becomes a greener choice at a distance of a few kilometers when the alternative is transport by car.

Conclusions: Telemedicine is a potent carbon reduction strategy in the health sector. But to contribute significantly to climate change mitigation, a paradigm shift might be required where telemedicine is regarded as an essential component of ordinary health care activities and not only considered to be a service to the few who lack access to care due to geography, isolation or other constraints.

Editor: Igor Linkov, US Army Engineer Research and Development Center, United States of America

Funding: This work was partly undertaken within the Umeå Centre for Global Health Research, with support from FAS, the Swedish Council for Working Life and Social Research (grant no. 2006-1512). The work was also supported in part by funding from the Swedish International Development Cooperation Agency (SIDA). The funders had no role in study design, data collection and analysis, decision to publish, or preparation of the manuscript.

Competing Interests: ClimAdapt, LLC is a sole proprietorship consulting company working on health and global change issues. Clients are predominantly international organizations, such as WHO and UNEP, and national institutions, such as Health Canada. Dr. Kristie Ebi conducts research on the impacts of and adaptation to climate change, including on extreme events, thermal stress, foodborne safety and security, and vectorborne diseases. Her work focuses on understanding sources of vulnerability and designing adaptation policies and measures to reduce the risks of climate change in a multi-stressor environment. Hence, the authors can confirm that this does not alter their adherence to all PLOS ONE policies on sharing data and materials.

* Email: Asa.Holmner-Rocklov@vll.se

Introduction

The health care sector is facing a series of major challenges. In developed countries, the health care sector is committed to improving the provision of care, while at the same reducing costs. In developing countries, the health sector is facing the challenge of meeting the fundamental right of all citizens to adequate health according to the WHO and UN goals of universal health coverage and healthy life expectancy. Challenges range from aging populations with an increasing prevalence of lifestyle related disease, to changing disease patterns because of development patterns and global environmental changes [1,2]. At the same time, the health care sector contributes significantly to climate change [3,4], which means that if the sector does not introduce climate friendly policies and practices, it paradoxically will continue to contribute directly and indirectly to negative health impacts through its emissions of CO_2 and other greenhouse gases. To meet the growing demand of health care resources without furthering climate change, future health services must be built on sustainable and low-carbon systems and work models.

Information and communication technology (ICT) has been suggested as one solution for reducing the carbon footprint of many sectors, including the health care sector [5]. Telemedicine, which is defined as the use of ICT to provide health services across distance, time, or other barriers, has been suggested to be a potent tool to reduce emissions from travel that, according to the UK National Health Services (NHS), represent as much as 18% of the total carbon footprint of the UK health sector [4]. Telemedicine covers a broad range of technologies and activities that have been applied to many different clinical disciplines, such as radiology, pathology, dermatology, rehabilitation, and chronic disease management. Large-scale clinical studies are often lacking, but in general telemedicine studies show high patient satisfaction and acceptance [6,7], although evidence on cost effectiveness is still relatively weak [8]. Regarding clinical outcomes, some telemedicine applications, such as specialist rehabilitation using videoconferencing, have shown to be comparable to or even more efficient than traditional interventions [9]. In our experience, this can be explained by the increased access to specialists that makes it possible to commit to a more intense rehabilitation regime that would have been feasible if the patient had been required to travel to the clinic for each appointment.

Attempts have been made to evaluate telemedicine programs from a climate mitigation perspective by estimating the potential reduction in tailpipe emissions [10–14]. However, to disregard the impacts from the technology used to facilitate telemedicine can be seriously misleading and it is important to take into account the cost of manufacturing, using and discarding the equipment, to evaluate the actual impact of different virtual meeting solutions [15,16].

To assess the carbon costs and benefits of telemedicine, we performed a simplified life cycle inventory for telemedicine activities within two rehabilitation units at the Umeå University Hospital in Northern Sweden. These two units have provided evidence that rehabilitation using telemedicine, i.e., telerehabilitation, is a cost efficient and well-accepted alternative to traditional care in Northern Sweden (personal communication). Net emissions for the telemedicine appointments in these two cases were compared to care-as-usual scenarios constructed using the patient's place of residence. Thresholds for when the telemedicine work model becomes favorable were estimated and sensitivity analyses performed to provide insight into (1) what factors contribute most to carbon emission in the telemedicine work model, (2) what influence technological set-ups and different use scenarios have on net carbon emission and (3) what would be the impact of greener transport options on the carbon reduction potential. These are the first such assessments of telemedicine activities and have the potential to contribute important information for the development of telemedicine guidelines in sustainable health care practice.

Methods

Ethics statement

This study is based on data from two clinical units within the University Hospital of Umeå. The study is based on aggregated data and patient information used in the study was anonymized and de-identified prior to analysis. It was thus judged by the Regional Ethical Review Board in Umeå that no ethical approval was needed.

Data capture

To assess the carbon footprint of telerehabilitation at the University Hospital of Northern Sweden we analyzed data from two units offering specialist rehabilitation using telemedicine. These are the rehabilitation unit of the hand and plastic surgery section, which is part of the centre for reconstructive surgery, and the speech therapy clinic, which is a county clinic belonging to the ear, nose and throat clinic. The reason for including these two clinics in our study was twofold; telemedicine has become a well integrated part of the clinical activities in these units and there are extensive records of their telemedicine appointments available for analysis. Telemedicine appointments were compared with care-as-usual scenarios that require the patient travel to the hospital for a face-to-face visit. Patients enrolled in the telemedicine programs are primarily inhabitants of Västerbotten County, but the hospital also provides specialized health care to the northern care region (approximately the northern half of Sweden), as well as to some patients from other counties. Travel distances are estimated as the distance from the town closest to the patient's place of residence to Umeå where the university hospital is located. One-way distances vary from less than one km to 700 km although the hand rehabilitation unit only registers travel distances above 2.5 km (one way). Car is the transport option assessed in our scenarios as most patients, particularly in the speech therapy clinic, utilize car or subsidized taxi services to reach the hospital. Patients from other counties may travel by other means, such as airplane or bus, but this is not accounted for in our study. All calculations were performed using Microsoft Excel.

Hand- and plastic surgery section

The study included all patients involved in the telerehabilitation program from January to December 2012. Appointments included follow-ups, interventions, consultations, and assessments of various conditions, such as amputations of one or more fingers, osteoarthritis, flexor tendon injuries, radius fractures, finger fractures, and ligament injuries. There were 238 telemedicine appointments during this period that thus avoided travel to Umea. Of these, 81 were conducted in the patient's home using a PC or tablet computer and 157 at the closest primary health centre using standard videoconferencing equipment. Travel to appointments made at the primary health centre is not accounted for. Based on the patient's places of residence, an accumulated travel distance of 82,310 km was avoided during the study. Data on the exact length of individual appointments are lacking, but they typically varied from 10 to 50 minutes with an average of 25 minutes (verbal communication).

Speech therapy unit

The study included data for all patients involved in a telemedicine project conducted in 2005–2006 at the speech therapy unit. Today, telemedicine is well integrated into the clinical activities; however, readily accessible statistics on the patients' place of residence (used to calculate travel-related emissions for comparison) was only available from the time of the project. Telemedicine treatment was delivered to patients of all ages for conditions including aphasia, dysarthria, and dyslexia. 481 therapy sessions using telemedicine were performed either in the patients' home or at the closest primary health centre, although details about exact location were not available. Data on the length of individual appointments were not available but varied from 30 to 40 minutes, with an average of 35 minutes (verbal communication). Based on the patient's place of residence, an accumulated travel distance of 154,842 km was avoided during the study.

Life cycle assessment

A Life Cycle Assessment (LCA) is defined as a complete ecological assessment of all energy, material, and waste flows of a product, and their impact on the environment. This "cradle to grave" evaluation begins with the design of the product and ends with the disposal or recycling of material and components (end-of-life). Typically, LCAs are performed in several steps, starting with a life cycle inventory to determine the raw materials and energy used and the emissions that occur during the life cycle of a product. This inventory is followed by a life cycle impact assessment to estimate the impacts of these emissions and raw material depletions on, for example, human health or certain ecosystems. If the focus of the study is on a single waste product it can be specified as a stream lined LCA.

This study adopts the form of a simplified, stream lined life cycle inventory, with the aim to evaluate the most important aspects of telemedicine with respect to CO_2 emissions. We chose one hour as the functional unit in the telemedicine scenario. This is a typical meeting duration and is, in addition, a convenient entity to work with considering that the units W and kWh are applied in our calculations.

The study builds on results from life cycle inventories on hardware and software required to connect the patient and the specialist, called mediated meeting solutions or videoconferencing solutions. We largely adopted the strategy described in Ong et al. [17] that takes into account end-point devices, such as computers [18], monitors, cameras, local area network (LAN) components [19] and video codecs used to compress and decompress digital video signals, as well as the costs for Internet traffic [19], although technically, data is transmitted using the dedicated hospital network, Sjunet. The method was modified to fit our device set-ups (see Figure 1 and Table 1) and assumptions on life length and use rates of the equipment. To make the protocol more generic, life cycle and operating carbon costs are given in kWh to account for the carbon footprint of energy production itself, which may differ significantly between countries and different stages of the product life cycle.

The total carbon cost of a virtual meeting was estimated by adding i) the cumulative emissions generated by all equipment during the use-phase (from energy consumption) and ii) emissions generated during design, manufacturing, disposal and recycling of the equipment, whenever these data were available. To obtain an hourly carbon cost of a mediated meeting (functional unit), the cost of operating the equipment for one hour was added to the carbon emissions generated throughout all other life cycle phases amortized over the whole life length of the equipment. In energy terms this can be expressed as;

$$E_h = \sum_i (E_o + E_{emb}/T_{life}) \quad (1)$$

where E_h is the total hourly energy cost in kWh (functional unit) for a videoconferencing set-up consisting of a specific combination of end-point devices or set of devices (i). E_o is the hourly energy consumption of the end-point device during operation, E_{emb} is the accumulated embodied energy for all other life cycle phases available (see Table 1) and T_{life} is the expected number of hours in use. A generalized conversion factor of 0.6 kCO_2e/kWh was used to convert these results to carbon dioxide equivalents, based on the methods developed by Ong and Malmodin [17,20], because the processes of designing, manufacturing, using, and disposing of the technology may differ across locations and countries. It is thus likely that we overestimate the carbon cost of the use phase because northern Sweden has access to environmentally friendly electricity in the form of hydropower and wind power. Life cycle data on computers, monitors, and other end-point devices are summarized in Table 1. Data on the Internet operating expenditures (opex) and embodied energy for different bandwidths are summarized in Table 2.

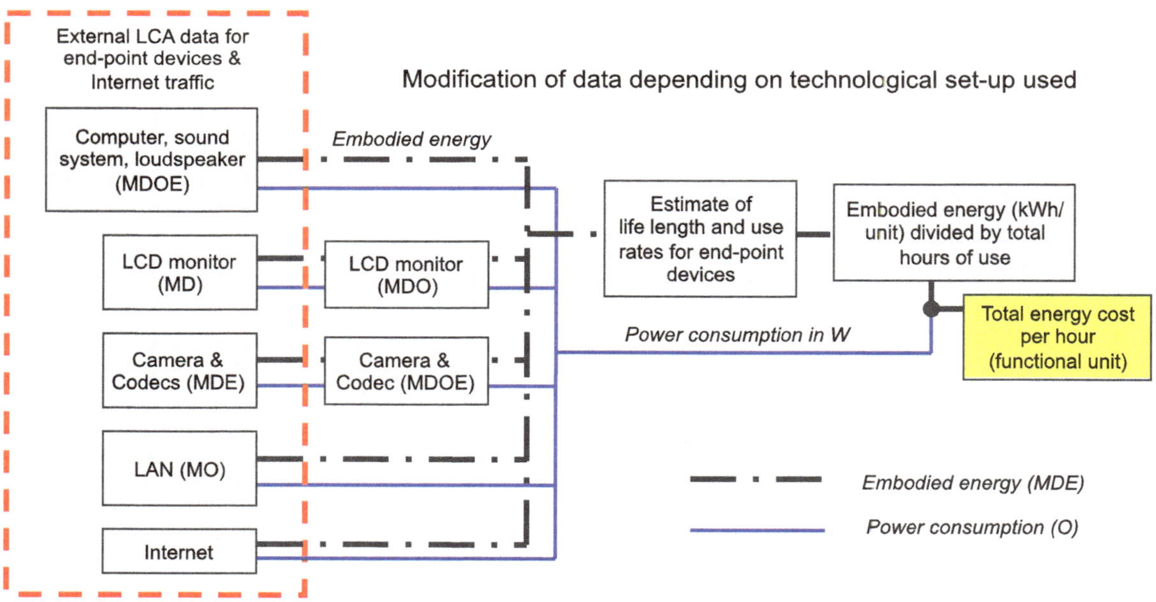

Figure 1. Methods summary. The study is based on the method used by Ong et al. that builds on existing LCA data for end-point devices used in videoconferencing and emission estimates for Internet traffic [17]. Completeness of the LCA data varies between devices, but includes energy costs and/or carbon emissions generated during manufacturing (M), distribution (D), operation (O) and end-of-life stages (E). Emissions data for MDE are provided in, or has been converted to, energy equivalents (kWh/unit) and is called embodied energy. Data for the videoconferencing solution (monitors, camera and video codecs) were modified to better fit our technological set-up. To obtain the hourly carbon cost of telerehabilitation in $kgCO_2e$, we divided the embodied energy with estimates of the life length and use rates of all equipment, and applied a conversion factor of 0.6 kg CO_2e/kWh [17,20]. See also Equation 1.

Table 1. Life cycle and operating costs of end point devices.

	Power consumption (W)	*Embodied energy (kWh/unit)^*	*LCA phases included**
PC			
Desktop	150	583	M D O E
Laptop	40	378	M D O E
Camera	9.5	33	M D O E
Sound system	4.1	104	M O
Microphone	2.5	52	M O
Monitor			
46" NEC LCD	188	145¤	M D O
Videoconference			
Camera + codec (Cisco SX20)	40	134#	M D O E
Local Area Network (LAN) end-points	20	278	M O

* M = manufacturing, D = distribution, O = operation, E = end-of-life (disposal & recycle)
^In the original reference, some values were converted from CO_2e to MJ using a conversion factor of 0.6 kg CO_2e/kWh [17,20]. Data were further converted from MJ to kWh to better fit our calculations.
#Embodied energy of the Cisco SX20 camera and codec was estimated from the embodied energy of camera and entry level video codec used in Ong et al. [17].
¤Data on the embodied energy of the NEC LCD monitor was based on an active screen size of 1018×572.7 mm and calculated from data provided by Ong et al. [17].

To enable comparison with emissions generated during a traditional physical meeting, we accounted for the life cycle carbon costs of travelling. The estimates were primarily based on the research by Lenzen et al. [21] that addresses tailpipe emissions as well as energy consumption and/or carbon emissions generated during other life cycle phases of transport, such as manufacturing of cars, fuels, and road infrastructure. According to Lenzen, the total carbon cost of a car is 0.86 $kgCO_2e/km$. To account for improved energy efficiency in more modern set-ups, we assessed more recent data by Leduc et al. [22] that accounts for life cycle phases related to automobiles and fuels, exclusively. This study generated a total of 0.25–0.27 $kgCO_2e/km$ for the most commonly purchased petrol and diesel cars in Europe in 2010; we utilized the value of 0.26 $kgCO_2e/km$ in our study. The EU emissions target of 0.130 kg CO_2 for automobiles in 2015 is also addressed in the sensitivity analysis.

System identification

Telemedicine appointments were conducted with different technological set-ups depending on location of the treatment; at home or at the closest primary health centre. For home treatments, the patient and the specialist are both assumed to be using a standard desktop PC with additional web camera, microphone, and loud speakers. The patient's computer is assumed to be used for 780 hours in total before disposal, based on a use rate of 5 hours per week for three years. This is a very rough estimate because the users vary significantly in age and computer habit; therefore, use rates are further addressed in the sensitivity analysis. The PC of the specialist is used for 7,300 hours before disposal based on an average use rate of 5 hours per day and a life expectancy of four years (data from the investment database). The bandwidth is set to 512 kbps, which is considered to be an acceptable lower limit for treatment provided in the home.

For treatment provided in the primary health centre, the patient is assumed to be utilizing a Cisco TelePresence SX20 with a 46" LCD monitor from NEC (or equivalent) that is used for 360 hours before disposal, based on an average use of 60 hours per year in six years. This estimate is derived from the average use-rate of videoconferencing equipment in the county council, which is

Table 2. Estimates of the Internet opex and embodied energy.

	Maximum estimates *	
Bandwidth	**Networking energy opex (kWh)**	**Embodied energy cost (kW)**
20 Mbit/s	32	29.3
10 Mbit/s	16	14.6
4 Mbit/s	6.4	5.6
2 Mbit/s	3.2	2.9
1 Mbit/s	1.6	2.2
768 kbit/s	1.2	1.5
512 kbit/s	0.8	1.1

*For simplicity and to avoid underestimating the cost of Internet traffic, the network operating expenditure (opex) and embodied energy (in kWh) are based on the maximum estimate of the operating energy intensity (3.61 kWh/GB) and embodied energy (3.33 kWh/GB) of the Internet. To calculate the opex in kWh, the bandwidth was converted to bytes per hour (or MB/h).

120 hours per year, with the number decreased to better fit the rural primary health centers. The specialist is using the same Cisco system but the use rate is assumed to be 240 hours per year for six years on average, i.e., 1440 hours. This estimate is based on the average use rate of videoconferencing equipment, with the number increased to better fit the two clinics as they are more frequent users of telemedicine than the average department. The LAN is expected to be operating for 14,600 hours derived from 10 hours per day for four years. The data transfer rate is estimated as 4Mbps, slightly exceeding the bandwidth typically used for telerehabilitation. The power consumption and embodied energy of computers [18], monitors, and audio/video peripherals are based on literature values for equivalent products, including the LAN end-point devices.

Access to information on the technological set-ups used for individual appointments was only available for the hand and plastic surgery unit. Therefore, we estimated a realistic upper and lower emission limit based on the following scenarios. These scenarios were applied to both clinics.

Upper bound scenario: The upper bound calculations were based on the scenario described above for treatment in the primary health centre, but with the addition of a second 46" screen in the videoconference room of the specialist.

Lower bound scenario: The lower bound calculations were based on the scenario described above for treatments provided in the patients' home; two standard desktop PCs with use rates of 780 hours (patient) and 7,300 hours (specialist) and a bandwidth of 512 kbps. Regarding carbon costs of Internet traffic, the same assumptions apply as for the upper bound scenario to avoid underestimating the carbon costs. The patients home LAN is likely to be of significantly lower complexity than the hospital LAN and both power consumption and embodied energy costs are likely to be significantly lower. Nonetheless, we chose to apply the same estimates as the higher bound scenario, with the only exception that the use rate applied to the patient scenario is 780 hours, similar to the other end-point devices.

Results

Carbon cost comparisons

For the hand and plastic surgery clinic, the carbon cost of the 238 telemedicine appointments during 2012 was 602 $kgCO_2e$ based on our baseline assumptions. This corresponds to an average of 1.4–2.8% of the carbon costs of travelling to/from the clinic by car or subsidized taxi services, based on a total avoided travel distance of 82,310 km for study patients. Based on the upper and lower bound scenarios, a telerehabilitation visit in the hand and plastic surgery clinic generated 0.4–0.9% and 3.2–6.4% of the expected carbon costs for a traditional face-to-face appointment, respectively. Similar numbers were obtained in the speech therapy clinic. In summary, the telerehabilitation activities of the two clinics resulted in a cut in carbon emissions by 15–250 times for the telemedicine work model compared to traditional care. Data on the carbon costs of the different scenarios are summarized in Table 3 and in Figure 2.

Based on the upper and lower bound scenarios defined in this paper, a one hour telemedicine appointment was estimated to generate 1.86 and 8.43 $kgCO_2e$, respectively. Consequently, telerehabilitation is carbon cost-effective once there is a need for the patient to travel at least 3.6 km by car for a one-hour appointment using the Lenzen estimate [21] and 7.2 km based on the Leduc estimate [22]. Corresponding values for the upper bound videoconference scenario are 16 km and 32 km, respectively. In reality, appointments are often shorter than one hour. For the care model described in this paper, these distances may well be reduced by 50%.

Sensitivity analysis

Sensitivity analyses were performed to gain a better understanding of how different choices in technological set-up and usage of videoconference equipment affect the magnitude of carbon emission for telerehabilitation. Sensitivity analyses were also performed to address LCA data on passenger transport. Results from the sensitivity analyses are summarized below and in Figure 3.

It is clear from the upper and lower bound scenarios that the choice of hardware affects net emission rates. Based on the upper and lower bound scenarios, a one-hour meeting emits 8.43 $kgCO_2e$ and 1.85 $kgCO_2e$, respectively. Thus, a standard desktop solution produces approximately 20% of the carbon equivalents of the Cisco SX20 based on our assumptions. The difference would be even more significant for dedicated desktop videoconferencing solutions, such as the Cisco EX60 that has a significantly lower power consumption in the use phase, or when compared to more complex tele-presence systems or multipart meeting solutions. Meeting duration is another of the dominant factors influencing the carbon emissions of telerehabilitation. A two-hour meeting would essentially produce twice the emission of a one-hour meeting if the use rates are kept constant and within reasonable limits. Another important factor is bandwidth given that data transfer is one of the most energy consuming processes of videoconferencing. For the upper bound scenario, data transfer contributes 87% of the total emissions and the corresponding emissions for the lower bound scenario are approximately 50%. Because the energy consumption of the Internet is calculated in kWh/GB, there is a linear increase in carbon emission with an increase in bandwidth, at least theoretically. Hence, for a one-hour meeting using the otherwise fixed lower bound set-up, the net carbon emissions would increase by more than 200% when increasing the bandwidth from 4 to 10 Mbps. A decrease to 1 Mbps would yield a corresponding 4-times reduction in energy consumption for data transfer and a 3-times reduction in total carbon emissions.

The use rates and expected life of the equipment also will influence the hourly energy and carbon cost of telerehabilitation given that the embodied energy is amortized over the lifetime of the equipment. The use rate of the specialist applied to the lower bound scenario is estimated at 5 hours per day for four years on average, i.e., 7,300 hours in total. A ten-time reduction in use rate would increase the net carbon emission by 20% for a one-hour meeting. However, if the use rates were to increase ten times the corresponding reduction in carbon emission would only be about 2%. Thus, the contribution from the embodied energy to the hourly emission is most influential on equipment used infrequently with the current lower bound estimates.

Finally, the Lenzen [21] reference used to assess the carbon emissions from patient transport is relatively old (1999) and using the newer reference [22] is unrealistic because all cars used are unlikely to be newer models. Moreover, this reference did not account for additional emissions generated by building and maintaining the road infrastructure, for example. Hence, sensitivity analysis was used to address the impact of lower tailpipe emissions and other life cycle contributions to emissions due to utilization of more modern automobiles. If we account for an annual 5% reduction of all life cycle phases of a private car based on an average emission of 0.86 kgCO2e/km according to Lenzen et al., the net emissions of all face-to-face meetings in the hand rehabilitation clinic in 2012 would be 22,950 $kgCO_2e$, which still

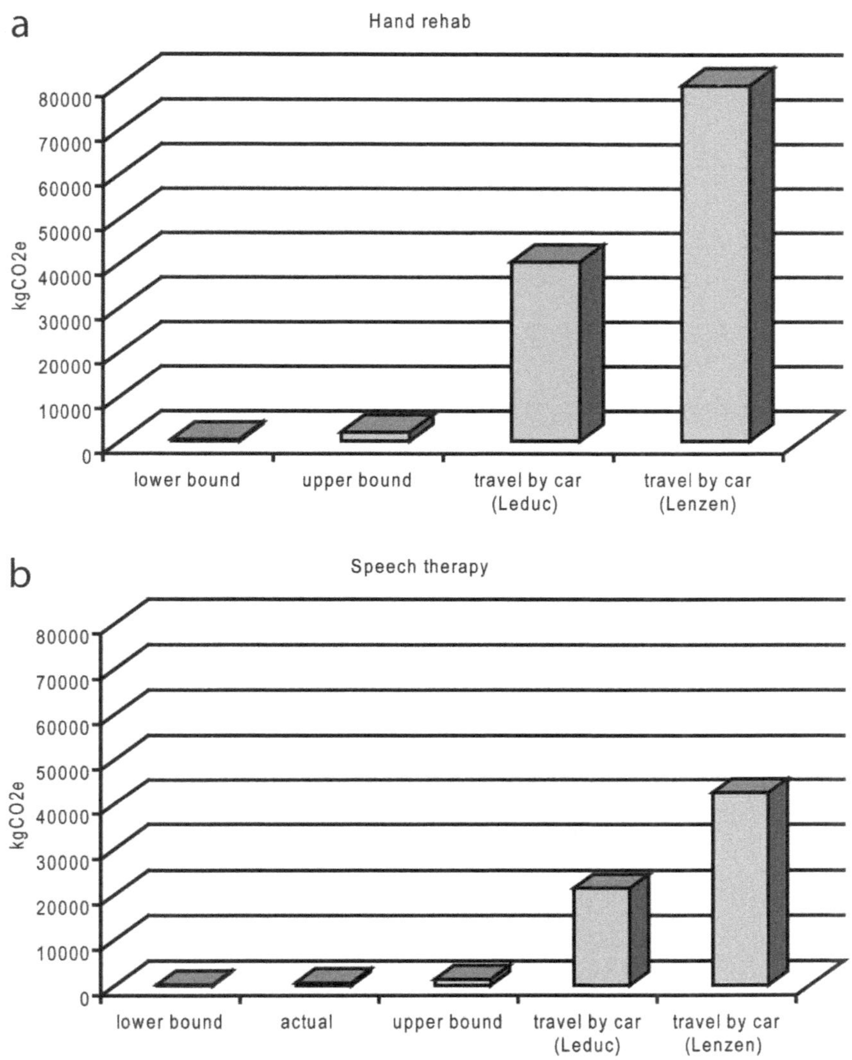

Figure 2. Summary of results. Net emissions for the two scenarios. Hand and plastic surgery section (a), and speech therapy unit (b). Information on the technological set-up used for individual appointments was only available in the hand and plastic surgery section (actual). Therefore, we applied the upper and lower bound scenarios in the speech therapy clinic.

corresponds to 38 times the emissions of the telerehabilitation scenario. If we perform the same calculations using the EU target for automobiles in 2015 of 0.130 $kgCO_2e$/km, emissions are reduced to 6,419 $kgCO_2e$, which are ten times the carbon emissions with respect to telerehabilitation in the hand rehabilitation clinic. These calculations do not take into account corresponding reductions in life cycle costs of the videoconferencing equipments.

Discussion

The carbon footprint of telemedicine services has been assessed by estimating the reduction in carbon emission due to reduced need for transportation [10–14]. However, the environmental impact of ICT is complex and to look at travel savings alone is misleading[5]. By performing a simplified LCA we accounted for the carbon cost of telemedicine and not just the potential to reduce tailpipe emissions. This is a new contribution to the scientific literature. Our analysis reveals that factors such as choice of teleconferencing solutions, duration of the appointment, capacity of the Internet connection, and use rates of the technology contribute to emissions to various degrees. These results are highly policy relevant. The outcome stresses the benefits of using telemedicine for short meetings and implies that the choice of bandwidth should be based on the clinical need rather than the access to highest possible Internet capacity. At higher bandwidths, data transfer is the main contributor to emissions of telerehabilitation, reaching 87% of total emissions in our upper bound scenario. It is also clear that traditional videoconferencing solutions emit more carbon than desktop solutions per hour of usage, and that use rates should be considered, particularly for equipment used infrequently. Careful planning is thus needed on the local level to make the best use of a videoconference system, which could be a very poor investment if wrongly placed or when in low demand due to the rapid growth of high-quality desktop and mobile videoconferencing solutions.

Based on our results, the magnitude of the carbon reduction *per appointment* is extensive and clearly indicates that up-scaling the use of telemedicine could have a large impact on the over-all

Table 3. Accumulated life cycle carbon costs of telemedicine versus face-to-face meetings.

	Speech therapy (481 visits)		Hand rehabilitation (238 visits)	
	kWh	$kgCO_2e$*	kWh	$kgCO_2e$
Telemedicine				
Authentic conditions^			1004	602
Lower bound	741	409	305	183
Upper bound	3307	1984	818	1364
Face-to-face visit				
Travel by car (Lenzen)		79 909		42 472
Travel by car (Leduc)		40 258		21 400
Emissions with respect to car travel (Lenzen/Leduc)				
Authentic conditions			1.4%/2.8%	
Lower bound	0.6%/1,0%		0.4%/0.9%	
Upper bound	2.5%/4,9%		3.2%/6.4%	

*Based on a conversion factor of 0.6 CO2e/kWh

^Based on 157 connections to primary health centres using videoconferencing solutions and 81 connections to the patients home using desktop solutions. Such detailed data were not available in the speech therapy section.

carbon footprint of the health sector. The hand- and plastic surgery clinic reduced the carbon emissions per appointment by more than 70 times without having to make major financial investments. The yearly monetary cost of a videoconferencing equipment of this standard is approximately 1100 EUR or 1500 USD. Further, when taking into account trends towards greener transports, telerehabilitation is the most climate-smart work model based on our sensitivity analyses.

These results are from a rural perspective and it is reasonable to expect that the carbon reduction potential of this work model could be significantly smaller in urban environments where patients have shorter distances to travel to receive specialist rehabilitation. However, carbon emissions from a one-hour meeting using a desktop solution are exceeded by the emissions from a car driving as little as a few kilometers based on our baseline assumptions and car queues and traffic jams will significantly increase the emission rates by any type of motor vehicle [23]. Thus, telemedicine also might be promising in cities, particularly those in regions that suffer from poor air quality and experience the majority of all traffic accidents. This new way of thinking about telemedicine and virtual meetings in general could thus be very relevant for city planning and future megacities; reducing risks while saving time and being climate friendly.

When evaluating the potential of telemedicine as climate mitigation strategy, it is important to consider its full potential rather than only assessing the effects from single visits or individual telemedicine programs. Traditional telemedicine activities reduce the carbon costs by only a small fraction in relation to the over all burden of travel, according to a study in the Grampaign region of UK [13]. In this study, telemedicine was estimated to reduce net travel by as little as 0.1%. This brings us to the question whether telemedicine should only be used to serve the needs of those who lack access to health care by virtue of geography, isolation, or other constraints, or if telemedicine can be accepted as an essential component of any ordinary health care activities? For the sake of the environment, we strongly support the latter. To gain true insight into the potential impact of telemedicine on health systems and environment, we anticipate that larger scale studies are needed from the viewpoint of telemedicine as a fully integrated part of any medical or health care activity. Only when becoming part of mainstream medical care and health care can telemedicine reach its full potential as climate mitigation strategy. Some steps in this direction have been taken by the Västerbotten County, which is a geographically widespread and sparsely populated County (population of 260,000) in northern Sweden. Virtual meetings have been highly prioritized for more than a decade and telemedicine is already an integrated part of many clinical units, supporting activities as diverse as surgical planning and follow-ups, specialist rehabilitation, tele-pathology, radiation treatment planning, collaborative care planning and chronic disease management. Consequently, the number of logged videoconferencing hours increased on average 30% per year for the past five years, reaching almost 20,000 hours in 2012 according to the county council statistics. For sake of debate, this corresponds to a reduction of several tons of CO_2e had all these meetings been applied to clientele and activities similar to the ones addressed in our study. When applying this work model in new clinical contexts it is, however, crucial to take into account possible trade-offs between environmental benefits and clinical outcomes. Intuitively, a clinician is unlikely to adapt a more environmentally friendly solution when there is the potential for any negative impact on the patient or on the quality of services. However, intuitive choices might not accurately reflect actual outcomes. Further, promoting sustainability means the adjustments to current practice should be part of standard evaluation. To support this complex decision process, we suggest continuing this work by performing life cycle impact assessments to estimate the impact on climate change and potential health co-benefits from reductions in CO_2 emissions.

The method applied in this paper has its limitations; there are very few studies available on this topic and the LCA data on videoconferencing peripherals and transport are sometimes rough estimates and not always completely up to date. It is thus important to keep in mind that the aim of our study was to describe the central aspects of emissions generated by the system under investigation to guide future studies and policy development

There are a lot of global initiatives focusing on reducing the impacts from travel, for example by implementing regulations to reduce tailpipe greenhouse gas emissions, improving vehicle technology, and introducing lower-carbon fuels. At the same time, the rapid development of the ICT sector is leading to more

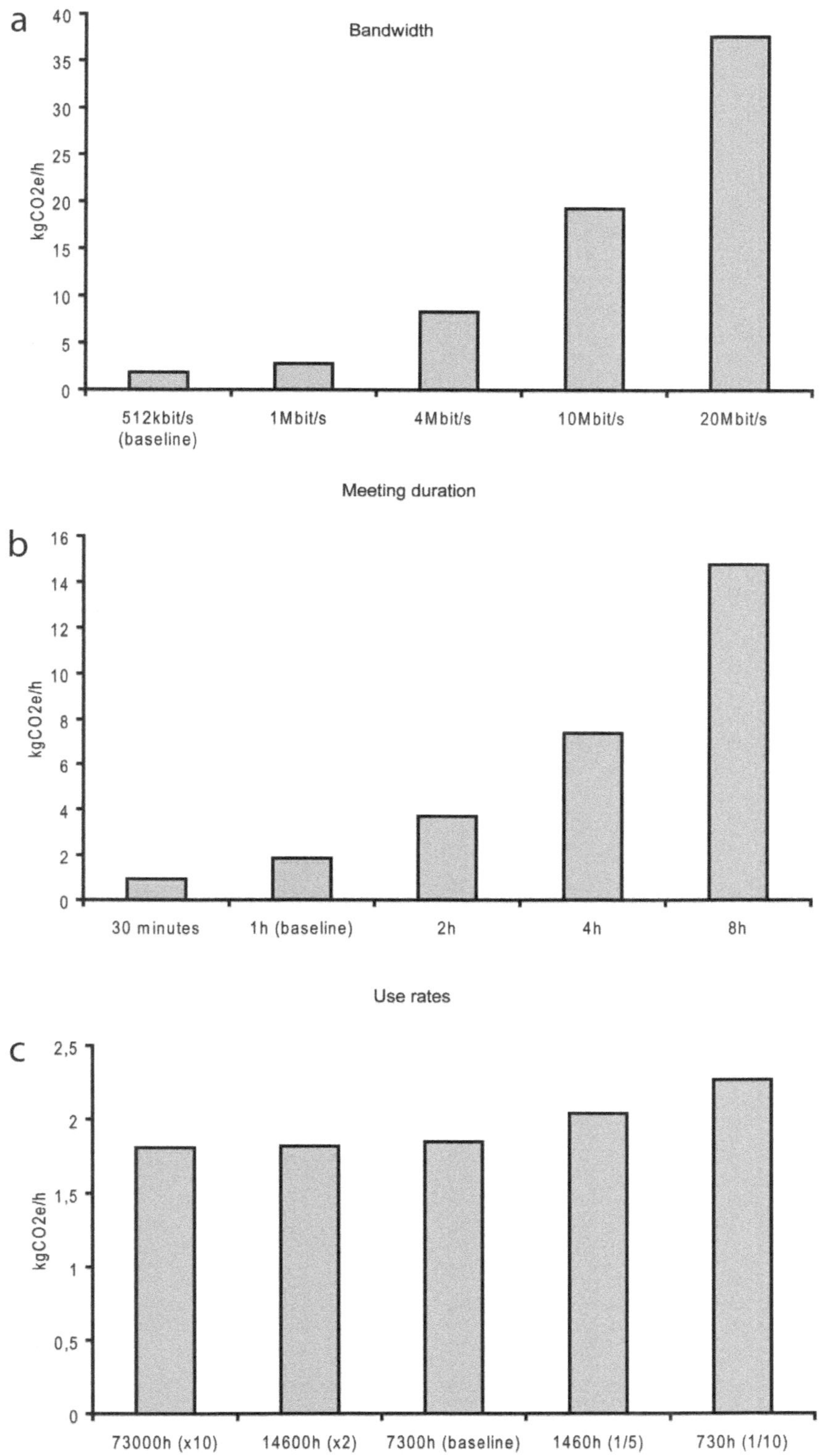

Figure 3. Sensitivity analyses. The analyses are based on the lower bound emissions scenario that emits 1.85 $kgCO_2e$ during a one hour meeting. All other factors were kept constant while changing the bandwidth (a), meeting duration (b) or use rates (c), respectively.

integrated and energy efficient local area networks and less energy demanding PCs and videoconferencing equipments. In contrast to this positive development, the automobile industry is growing steadily, and there are tendencies towards people investing in larger screen sizes and more advanced videoconference systems. These initiatives have the potential to significantly shift the anticipated net reduction in emissions from choosing virtual meeting solutions over face-to-face visits, in either direction. A way to address this in future studies is to adapt methods for calculating time-corrected life cycle emissions intensity for all scenarios by taking into account future development in all relevant sectors, such as ICT and transport [24,25].

It is a common perception that telemedicine can cut costs, extend health services to remote areas, maintain or improve clinical outcomes, and save time for patients and health workers. Some scientific evidence challenges these perceptions, particularly with respect to cost efficiency [8] but for many telemedicine services, these expectations are unquestionably met [9,26]. Nonetheless, there is a striking delay in the up scaling and implementation of successful telemedicine work models [27,28]. There is a lack of studies investigating the impacts of such work models on routine care [29] and a lack of clear organization and economic policies and guidelines [30,31]; such policies require empirical evidence and dissemination of results to relevant stakeholders and decision makers. It is our hope that the empirical evidence of the environmental benefits of telemedicine generates additional momentum and accelerates the implementation of successful telemedicine systems and work models in health systems globally, for the benefit of patients, health care providers, and the planet.

Conclusions

This study shows that telemedicine generates far fewer carbon emissions than do traditional care models where patients and health workers are expected to travel by car to appointments. Telemedicine is thus judged to be a potent climate change mitigation strategy, not just in rural areas but also in urban environments where additional co-benefits might be even greater if few people use public or active transport. Although previous research indicates that current telemedicine programs reduce travel-related emissions to a limited extent [14], we conclude that implementing telemedicine more broadly could make a significant contribution to reducing the greenhouse gas emissions driving climate change. This requires a paradigm shift in health care and proactive efforts from health care decision makers. This also calls for empirical evidence on the clinical, economical and environmental benefits of telemedicine. Research is needed on a larger scale to evaluate the current and future impact of different telemedicine solutions on carbon emissions, from the viewpoint of telemedicine as a well-accepted and fully integrated part of any health care activity.

Acknowledgments

The authors wish to thank Agneta Carlsson and Käte Alrutz for supplying the data and for generous support throughout the study.

Author Contributions

Conceived and designed the experiments: ÅH. Performed the experiments: ÅH. Analyzed the data: ÅH KE LL MN. Wrote the paper: ÅH KE LL MN.

References

1. Woodward A, Smith KR, Campbell-Lendrum D, Chadee DD, Honda Y, et al. (2014) Climate change and health: on the latest IPCC report. Lancet 383: 1185–1189.
2. McMichael AJ (2013) Globalization, Climate Change, and Human Health. New England Journal of Medicine 368: 1335–1343.
3. Pencheon D, Rissel CE, Hadfield G, Madden DL (2009) Health sector leadership in mitigating climate change: experience from the UK and NSW. N S W Public Health Bull 20: 173–176.
4. NHS (2009) Saving Carbon, Improving Health, NHS carbon reduction strategy for England. Available: http://www.sduhealth.org.uk/documents/publications/1237308334_qylG_saving_carbon,_improving_health_nhs_carbon_reducti.pdf. Accessed 2013 September 29.
5. Holmner A, Rocklov J, Ng N, Nilsson M (2012) Climate change and eHealth: a promising strategy for health sector mitigation and adaptation. Glob Health Action 5.
6. Johansson T, Wild C (2011) Telerehabilitation in stroke care - a systematic review. Journal of Telemedicine and Telecare 17: 1–6.
7. Rozenblum R, Donze J, Hockey PM, Guzdar E, Labuzetta MA, et al. (2013) The impact of medical informatics on patient satisfaction: A USA-based literature review. International Journal of Medical Informatics 82: 141–158.
8. Mistry H (2012) Systematic review of studies of the cost-effectiveness of telemedicine and telecare. Changes in the economic evidence over twenty years. Journal of Telemedicine and Telecare 18: 1–6.
9. Kairy D, Lehoux P, Vincent C, Visintin M (2009) A systematic review of clinical outcomes, clinical process, healthcare utilization and costs associated with telerehabilitation. Disability and Rehabilitation 31: 427–447.
10. Yellowlees PM, Chorba K, Parish MB, Wynn-Jones H, Nafiz N (2010) Telemedicine Can Make Healthcare Greener. Telemedicine Journal and E-Health 16: 230–233.
11. Connor A, Lillywhite R, Cooke MW (2011) The carbon footprints of home and in-center maintenance hemodialysis in the United Kingdom. Hemodialysis International 15: 39–51.
12. Lewis D, Tranter G, Axford AT (2009) Use of videoconferencing in Wales to reduce carbon dioxide emissions, travel costs and time. Journal of Telemedicine and Telecare 15: 137–138.
13. Wootton R, Tait A, Croft A (2010) Environmental aspects of health care in the Grampian NHS region and the place of telehealth. Journal of Telemedicine and Telecare 16: 215–220.
14. Wootton R, Bahaadinbeigy K, Hailey D (2011) Estimating travel reduction associated with the use of telemedicine by patients and healthcare professionals: proposal for quantitative synthesis in a systematic review. Bmc Health Services Research 11: 185.
15. Borggren C, Moberg A, Rasanen M, Finnveden G (2013) Business meetings at a distance - decreasing greenhouse gas emissions and cumulative energy demand? Journal of Cleaner Production 41: 126–139.
16. Arnfalk P, Kogg B (2003) Service transformation - managing a shift from business travel to virtual meetings. Journal of Cleaner Production 11: 859–872.
17. Ong D, Moors T, Sivaraman V (2012) Complete life-cycle assessment of the energy/CO2 costs of videoconferencing vs face-to-face meetings. 2012 Ieee Online Conference on Green Communications, p50–55.
18. Fujitsu (2011) White paper - Life Cycle Assessment and Product Carbon Footprint - Fujitsu ESPRIMO E9900 Desktop PC. Available: http://globalsp.ts.fujitsu.com/dmsp/Publications/public/wp-LCA-PCF-ESPRIMO-E9900.pdf. Accessed 2013 September 25.
19. Raghavan B, Ma J (2011) The energy and emergy of the internet. Proceedings of the 10th ACM Workshop on Hot Topics in Networks. Cambridge, Massachusetts: ACM. 1–6.
20. Malmodin J, Moberg A, Lunden D, Finnveden G, Lovehagen N (2010) Greenhouse Gas Emissions and Operational Electricity Use in the ICT and Entertainment & Media Sectors. Journal of Industrial Ecology 14: 770–779.
21. Lenzen M (1999) Total requirements of energy and greenhouse gases for Australian transport. Transportation Research Part D-Transport and Environment 4: 265–290.
22. Leduc G, Mongelli I, Uihlein A, Nemry F (2010) How can our cars become less polluting? An assessment of the environmental improvement potential of cars. Transport Policy 17: 409–419.
23. Barkenbus JN (2010) Eco-driving: An overlooked climate change initiative. Energy Policy 38: 762–769.
24. Kendall A (2012) Time-adjusted global warming potentials for LCA and carbon footprints. International Journal of Life Cycle Assessment 17: 1042–1049.
25. Kendall A, Price L (2012) Incorporating Time-Corrected Life Cycle Greenhouse Gas Emissions in Vehicle Regulations. Environmental Science & Technology 46: 2557–2563.
26. Hilty DM, Ferrer DC, Parish MB, Johnston B, Callahan EJ, et al. (2013) The Effectiveness of Telemental Health: A 2013 Review. Telemedicine and E-Health 19: 444–454.
27. Zapka J, Simpson K, Hiott L, Langston L, Fakhry S, et al. (2013) A mixed methods descriptive investigation of readiness to change in rural hospitals participating in a tele-critical care intervention. Bmc Health Services Research 13:33.
28. Christensen MC, Remler D (2009) Information and Communications Technology in US Health Care: Why Is Adoption So Slow and Is Slower Better? Journal of Health Politics Policy and Law 34: 1011–1034.

29. Schmidt S, Grimm A (2009) Health service research of telemedicine applications. Bundesgesundheitsblatt-Gesundheitsforschung-Gesundheitsschutz 52: 270–278.
30. Koch S (2006) Home telehealth - Current state and future trends. International Journal of Medical Informatics 75: 565–576.
31. Broens TH, Huis in't Veld RM, Vollenbroek-Hutten MM, Hermens HJ, van Halteren AT, et al. (2007) Determinants of successful telemedicine implementations: a literature study. J Telemed Telecare 13: 303–309.

Spatial Distribution of Soil Organic Carbon and Its Influencing Factors in Desert Grasslands of the Hexi Corridor, Northwest China

Min Wang[1,2]*, Yongzhong Su[1], Xiao Yang[1]

1 Linze Inland River Basin Research Station, Chinese Ecosystem Network Research, Cold and Arid Regions Environmental and Engineering Research Institute, Chinese Academy of Sciences, Lanzhou, Gansu, China, 2 University of Chinese Academy of Sciences, Beijing, China

Abstract

Knowledge of the distribution patterns of soil organic carbon (SOC) and factors that influence these patterns is crucial for understanding the carbon cycle. The objectives of this study were to determine the spatial distribution pattern of soil organic carbon density (SOCD) and the controlling factors in arid desert grasslands of northwest China. The above- and belowground biomass and SOCD in 260 soil profiles from 52 sites over 2.7×10^4 km^2 were investigated. Combined with a satellite-based dataset of an enhanced vegetation index during 2011–2012 and climatic factors at different sites, the relationships between SOCD and biotic and abiotic factors were identified. The results indicated that the mean SOCD was 1.20 (SD:+/− 0.85), 1.73 (SD:+/− 1.20), and 2.69 (SD:+/− 1.91) kg m^{-2} at soil depths of 0–30 cm, 0–50 cm, and 0–100 cm, respectively, which was smaller than other estimates in temperate grassland, steppe, and desert-grassland ecosystems. The spatial distribution of SOCD gradually decreased from the southeast to the northwest, corresponding to the precipitation gradient. SOCD increased significantly with vegetation biomass, annual precipitation, soil moisture, clay and silt content, and decreased with mean annual temperature and sand content. The correlation between BGB and SOCD was closer than the correlation between AGB and SOCD. Variables could together explain about 69.8%, 74.4%, and 78.9% of total variation in SOCD at 0–30 cm, 0–50 cm, and 0–100 cm, respectively. In addition, we found that mean annual temperature is more important than other abiotic factors in determining SOCD in arid desert grasslands in our study area. The information obtained in this study provides a basis for accurately estimating SOC stocks and assessing carbon (C) sequestration potential in the desert grasslands of northwest China.

Editor: Ben Bond-Lamberty, DOE Pacific Northwest National Laboratory, United States of America

Funding: The Strategic Priority Research Program - Climate Change: Carbon Budget and Relevant Issues of the Chinese Academy of Sciences, grant number XDA05050406-3. (2) National Natural Science Foundation of China (91125022). The funders had no role in study design, data collection and analysis, decision to publish, or preparation of the manuscript.

Competing Interests: The authors have declared that no competing interests exist.

* E-mail: wmin85@126.com

Introduction

Soil plays a crucial role in the global carbon cycle by linking carbon transformation with the pedosphere, biosphere, and atmosphere. Therefore, minor changes in the soil carbon pool will greatly affect the alteration of atmospheric CO_2 concentration and have potential feedbacks to climate change [1–4]. SOC storage in temperate grasslands is heavily studied [5–14], while there is little research examining SOC storage in arid regions such as desert-grassland or desert-steppe. Arid regions cover about 47.2% of the earth's land area, with soils containing nearly 241 Pg of soil organic carbon, which is about 40 times more than what was added into the atmosphere through anthropogenic activities [15]. Additionally, soils in these regions are fragile and may experience degradation, desertification, wind erosion, and over-grazing. Small changes in soil conditions can modify the original balance of soil carbon cycle, increase the C loss from soil, and release more greenhouse gases into the atmosphere. Therefore, SOC storage in the desert-grassland ecosystem is a critical component of global C cycle and has a considerable effect on reducing the rate of enrichment of atmospheric CO_2.

In northwest China, desert-grasslands are widely distributed (near 6.5×10^7 hm^2); however, the SOC storage here has not been widely studied. Among the limited estimates of SOC storage in desert-grassland in northwest China [16–18], large differences were found, potentially due to different data sources or approaches. Data from the Second National Soil Survey were usually used for these previous estimates [16–18], but few soil profiles were sampled from the grasslands in northwest China, and these soil profiles lacked data on bulk density and gravel fractions [19]. Regarding different data approaches, previous studies usually calculated SOC stock using average SOC density (SOCD); however, this approach could be constrained by limited soil profiles and large soil heterogeneity. Accordingly, satellite-based approaches will be useful to scale up site-level observations to regional-scale estimates [19].

SOC storage in grasslands is closely correlated with biological, climatic, and edaphic factors. SOC storage exhibits a balance between C inputs from organic material, and C losses through decomposition and mineralization [20–23]. In particular, this balance depends on climatic conditions [18,24–26]. Precipitation and temperature determine the vegetation types, size of plant productivity, and the speed of microbial degradation of soil organic matter. In addition to climate, soil texture plays an important role, which results in an increasing clay content and decreasing C outputs through its stabilizing effect on SOC [5,26,27].

The Hexi Corridor, one characteristically arid area in northwest China, with annual precipitation ranging from about 200 mm in the east to less than 50 mm in the west, represents a desert-grassland ecosystem. Desert-grasslands here always contain transition zones between desert and grassland or between desert and oases, and play a crucial role in maintaining a stable ecological environment and productivity. The desert grassland ecosystem in the Hexi Corridor has unique features, such as limited precipitation, low vegetation cover, coarse soil particles, large gravel content, and highly intensified wind erosion. Accordingly, the vegetation composition varies more than other desert-grasslands. For example, desert steppe in Inner Mongolia is mainly composed of gramineous plants and shrubs with deep roots [28], whereas small shrubs or subshrubs with shallow roots dominate the desert grassland in the Hexi Corridor [29], which results in small biomass productivities and low soil C inputs. These characteristics could lead to greater differences in SOC storage compared with other typical grasslands, but few studies have focused on this desert grassland. The working hypothesis for our study was that the spatial distribution of SOC will show clear relationships with unique climate and vegetation biomass, as well as specific soil conditions. Throughout our field investigations, our research objectives were: (1) to identify SOC density and its spatial distribution characteristics; and (2) to analyze the influence of biotic, climatic, and edaphic factors on SOC density and its distribution in an arid desert grassland. Well water conditions and soil particle composition can promote plant growth and soil organic carbon fixation, and high temperature can accelerate the decomposition of soil organic carbon. For these reasons we hypothesized that SOCD would increase with the increase of moisture and soil clay content, and decrease with the rise of temperature.

Materials and Methods

Ethics Statements

The location of field studies is not privately-owned or protected in any way, so no specific permission was required. All field studies in the desert-grassland were undertaken with support from Linze Inland River Basin Research Station, Cold and Arid Regions Environmental and Engineering Research Institute, Chinese Academy of Sciences. The field studies did not involve endangered or protected species.

Study area

The study area is found in the central region of the Hexi Corridor in Gansu province, northwestern China (spanning from 101°42′36″ to 97°45′36″E and 40°31′12″to 38°8′26″N, elevation ranging from 1200 m to 1500 m); the study region has an area of $\sim 2.7 \times 10^4$ km^2. The mean annual precipitation varied from 250 mm to 50 mm from the southeast to the northwest, and 70–80% of the rainfall occurs between June and August. The annual mean temperature varies from 5 to 9°C. The main soil types are Calcic-Orthic Aridosols according to Chinese Soil Taxonomy, which is equivalent to the Aridosols and Entisols of the USDA soil taxonomy classification (Group of Chinese Soil Taxonomy, Institute of Soil Science, Chinese Academy of Sciences, 2001). Soil thickness ranges from 0.2 m to 1.5 m, and most soils contain a large amount of gravel (in 0.5–6 cm), especially below the 30 cm soil layer. The vegetation population structure in desert grassland is relatively simple, and the main plant species is composed of small shrubs and sub-shrubs including *Asterothamnus centraliasiaticus*, *Reaumuria songarica*, *Salsola passerina*, *Sympegma regelii* and some ephemeral plant species such as *Suaeda glauca*, *Bassia dasyphylla* and *Artemisia scoparia*. The research areas have been subjected to grazing prohibition since 2000.

Sampling sites

A total of 52 sites were selected (Figure 1). Each site contains five plots (the area of a square plot is 1 m×1 m for herbaceous plants and small semi-shrubs or 5 m×5 m for shrubs). A total of 260 soil profiles were sampled (i.e., five profiles at each site) in August of 2011–2012. Based on the pre- investigation of vegetation community types in the study area,we set 52 sampling points. These sampling points basically cover the main vegetation communities of the area (*Reaumuria songarica* community, *Salsola passerine* community, *Sympegma regelii* community, and *Asterothamnus centraliasiaticus* community), and can represent the community characteristics, biomass and soil organic carbon contents and other information of desert grassland. On the other hand, we chose areas with no animal dung, no trace of vegetation were eaten, no trace of trampling, and within the barbed wire enclosure as sampling points. That can ensure selected sample sites without human grazing interference.

Soil sampling, analysis, and biomass survey

At each sampling plot, three soil pits were randomly excavated and mixed into a composite sample at seven depths of 0–5, 5–10, 10–20, 20–30, 30–50, 50–70, and 70–100 cm. Bulk density samples for each depth interval were obtained using a cutting ring (volume of 100 cm^3). Soil moisture (SM) was measured gravimetrically after 24 h desiccation at 105°C. Bulk density was also calculated as the ratio of the oven-dry soil weight to the cutting ring volume. Soil samples were air-dried, hand-picked to remove plant residues, visible soil organisms, and stones, and weighed. The air-dried samples were then sieved through a screen with 2 mm openings, and gravels (>2 mm) were weighed. The weight percent of gravel to soil was obtained. A proportion of samples that passed through the 2 mm sieve were finely ground to pass through a 0.10 mm sieve and analyzed for soil organic carbon

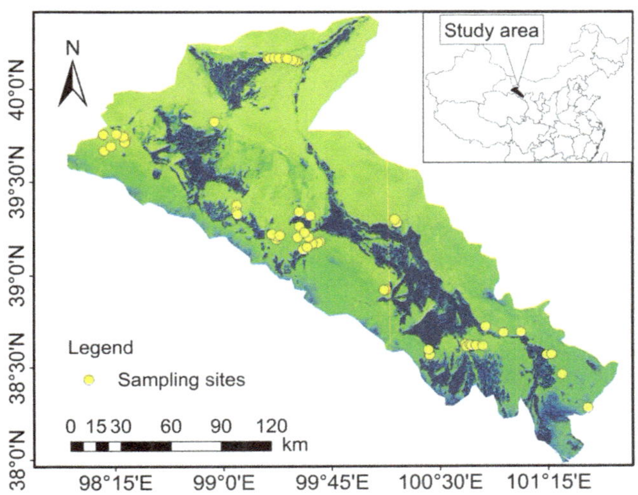

Figure 1. Spatial distribution of sampling sites in the Hexi Corridor.

by the $K_2Cr_2O_7$-H_2SO_4 oxidation method developed by Walkley-Black [30]. A subsample was then analyzed for soil texture by the wet sieve method [31].

Aboveground biomass (AGB) and belowground biomass (BGB) were harvested at 260 plots. At sites with herbaceous plants, which were always present at a low plant density (1 m×1 m), we excavated the total plants to obtain aboveground and belowground biomass. At sites with shrubs (5 m×5 m), one or several of the dominant plant species in the plot were chosen according to the crown breadth proportion (large: length × width ≥ 50 cm×50 cm; medium: 20 cm×20 cm≤length × width <50 cm×50 cm, and small: length × width <20 cm×20 cm); the whole plants were then dug out in accordance with the crown breadth survey to estimate above and belowground biomass. In the laboratory, the roots samples were soaked in water and cleaned of residual soil using a 0.5 mm sieve. Biomass samples were oven-dried at 65°C to a constant weight and weighed to the nearest 0.01 g.

MODIS data and climate information

The MODIS-EVI data used in this study were obtained from the United States Geological Survey at a spatial resolution of 250 m×250 m and 16-day intervals for the period 2011 to 2012 (http://LPDAAC.usgs.gov). Monthly maximum EVI composites were generated using the Maximum Value Composition method proposed by Holben [32] from 2011 to 2012. The EVI data used were the average of monthly EVI during the growing season from July to August.

Climate data, such as mean annual air temperature (MAT) and annual precipitation (AP), were separated from the climate database of the China monthly ground weather dataset during 2011–2012 (http://cdc.cma.gov.cn). These data were spatially interpolated from the records of 25 climatic stations located throughout the Hexi Corridor.

SOC evaluation

SOC densities for each soil profile at 0–30 cm, 0–50 cm, and 0–100 cm depth intervals were calculated:

$$SOCD = \sum_{h=1}^{n} H_h \times BD_h \times SOC_h \times (1 - C_h)/100 \qquad (1)$$

where SOC density (SOCD) in kg m^{-2}, soil thickness (H_h) in cm, bulk density (BD_h) in g cm^{-3}, SOC (SOC_h) in g kg^{-1}, and volume percentage of the fraction >2 mm (C_h) at layer h were used.

To investigate the spatial distribution of SOCD, we established the relationship between SOCD and MODIS-EVI for three soil depth intervals (Table 1), which was based on the linear relationship between AGB-EVI (Figure 2A) and AGB-SOCD (Figure 2B, and equation for AGB-$SOCD_{0\text{-}50\ cm}$ and AGB-$SOCD_{0\text{-}100\ cm}$ was $SOCD_{0\text{-}50\ cm} = 0.014AGB+0.091$, $R^2 = 0.41$, $SOCD_{0\text{-}100\ cm} = 0.023AGB+0.0039$, $R^2 = 0.44$, respectively).

Using the regression equations (Table 1), each pixel of EVI was converted to SOCD. We then obtained the spatial distribution of SOCD for different soil layers (Figure 3A–C). The spatial distribution of SOCD was performed in ArcGIS, version 9.3 (ESRI, RedLands, California).

Statistical analysis

We used simple linear regression to analyze the relationship between dependent and independent variables (AGB-EVI, SOCD-AGB, SOCD-EVI, SOCD-BGB, and SOCD-TB,

Table 1. Relationships between soil organic carbon density and enhanced vegetation index at three soil depth intervals (0–30 cm, 0–50 cm, and 0–100 cm).

Soil depth	Equation	R^2	P	RMSE	SSE	F-statistic
0–30 cm	$SOCD_{0\text{-}30\ cm} = 18.87\ EVI - 1.05$	0.62	<0.001	0.52	13.97	82.77
0–50 cm	$SOCD_{0\text{-}50\ cm} = 25.64\ EVI - 1.33$	0.58	<0.001	0.70	30.76	69.98
0–100 cm	$SOCD_{0\text{-}30\ cm} = 37.12\ EVI - 1.75$	0.48	<0.001	1.35	95.32	47.69

Notes: RMSE, root-mean-square error; SSE, sum of squares for error.

respectively). Additionally, to evaluate integrative effects of MAT, AP, SM, and soil texture on SOCD, a general linear model (GLM) was employed. Ordinary least squares regression was used to fit these models. Residuals from the linear models were examined for normality (by Shapiro-Wilk test), independence (by Durbin-Watson test), and linearity (by plotting the residuals after fitting linear regression between dependent variable and independent variable) and data met the assumption of normality, independence and linearity. Therefore, linear regressions and GLM models were effective and meaningful for these analyses. Statistical analyses were conducted with R version 3.0.2 package (http://www.R-project.org). We also analyzed the correlations between SOCD and environmental factors (MAT, AP, SM, Silt, Clay, and Sand) using nonlinear regression (by Levenberg-Marquardt and Universal Global Optimization analyses). We constructed exponential equations to describe relationships between SOCD and MAT, AP, SM, soil texture. Nonlinear regression analyses were performed using 1stOpt software, version 1.5 (First Optimization, 7D-Soft High Technology Inc., Xian, China).

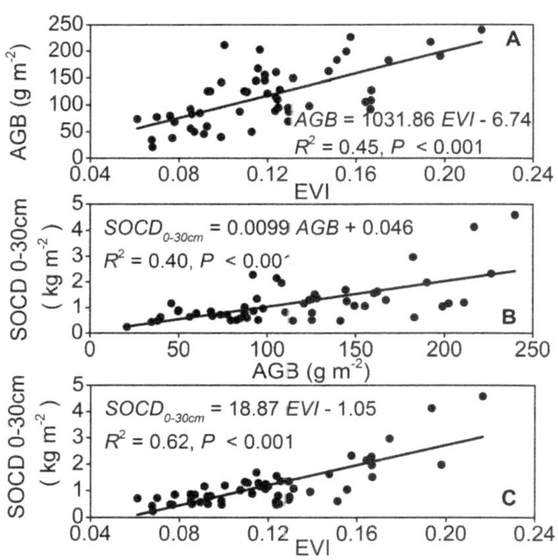

Figure 2. Relationships between above-ground biomass, enhanced vegetation index, and soil organic carbon density at depth interval of 0–30 cm. *Notes:* AGB, above-ground biomass; EVI, enhanced vegetation index; SOCD, soil organic carbon density.

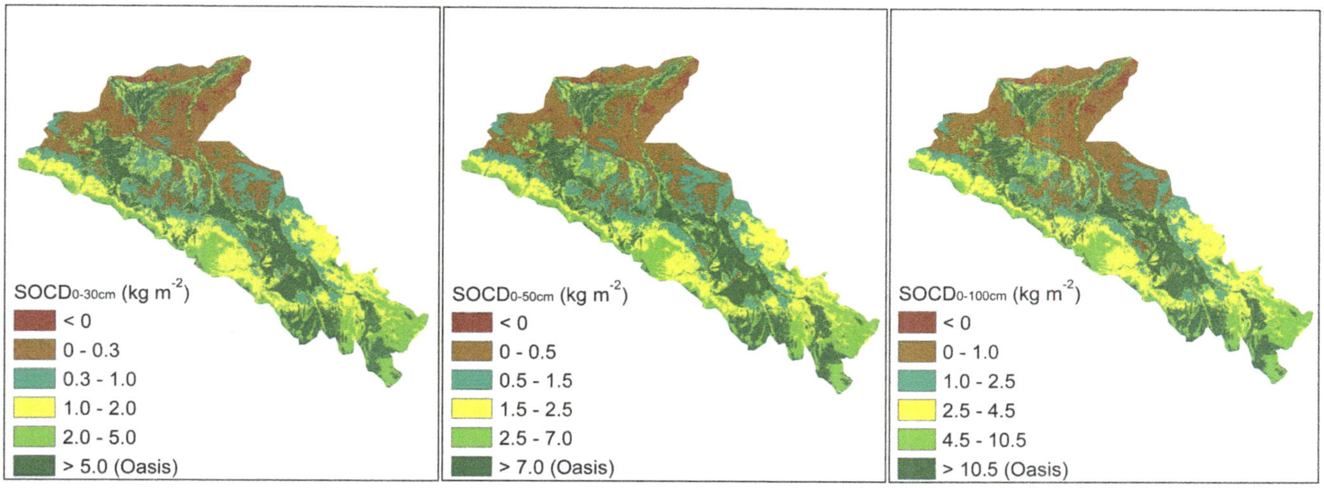

Figure 3. Spatial distributions of soil organic carbon density at different soil layers (0–30 cm, 0–50 cm, and 0–100 cm). *Notes:* SOCD, soil organic carbon density.

Results

SOC stocks and spatial distribution

The statistical description of SOCD across 52 sites at soil depths of 0–30 cm, 0–50 cm, and 0–100 cm is shown in Table 2. As shown, SOCD exhibited large variations among the three soil depths, ranging from 0.24–4.58 kg m^{-2} for 30 cm in depth, 0.35–6.32 kg m^{-2} for 50 cm, and 0.59–9.57 kg m^{-2} for 100 cm, respectively. The corresponding average SOC densities were 1.20, 1.73, and 2.69 kg m^{-2}, respectively. The SOC content in 0–30 cm interval accounted for almost 45% of total SOC in the top 1 m of soil.

The density of SOC decreased from the southeast to the northwest (Figure 3A–C), which corresponds to the precipitation gradient.

Effects of biomass and environmental factors on SOCD

The significant positive relationships between SOCD and AGB, BGB, and TB at different soil depths were characterized by linear functions (Figure 4, $P<0.001$). The R^2 values of regression functions between TB and SOCD were the highest of three biological variables. The R^2 values of regression functions between BGB and SOCD were higher than that between AGB and SOCD and thus showed the closer correlation between BGB and SOCD compared with the correlation between AGB and SOCD (Figure 4, A–C for BGB, D–F for AGB) and exhibited the crucial influence of BGB on SOC distribution (Figure 4G–I). The association of SOC content with BGB and TB was closest in the top soil and decreased at deeper intervals (Figure 4D and G). Nevertheless, the relationships between AGB and SOCD exhibited only minimal differences at three soil depths (Figure 4A–C).

We established regression relationships between SOCD and various environmental factors, such as MAT, AP, SM, silt, clay, and sand content (Figure 5). In the top 0–30 cm interval, SOCD decreased markedly with MAT (Figure 5A) and sand content (Figure 5P). On the contrary, SOCD increased significantly with an increase SM (Figure 5G). Furthermore, SOCD was positively related with silt (Figure 5J) and clay content (Figure 5M) as well as AP and SM. The regression curves indicated that the closest relationship between SOCD and environmental variables in the top 30 cm soil was with MAT (Figure 5A, $R^2=0.79$, $P<0.001$), followed by SM (Figure 5G, $R^2=0.67$, $P<0.001$). Additionally, similar relationships between SOCD and environmental factors were also observed in other soil intervals (Figure 5B, E, H, K, N, and Q for 0–50 cm; Figure 5C, F, I, L, O, and R for 0–100 cm). The R^2 value of fitted curves for associations of SOCD with MAT, AP, SM, and sand content declined with soil depth, but an increasing trend was found for association of SOCD with silt and clay content (Figure 5).

According to the above-described relationship between SOCD and environmental factors, we chose five variables (MAT, SM, clay, silt, and sand content, and soil moisture as a measure of water availability) to establish a GLM model. The results suggested that environmental factors explained 69.81%, 74.41%, and 78.87% of the overall variation of SOCD at the soil intervals of 0–30 cm, 0–50 cm, and 0–100 cm, respectively (Table 3).

MAT was the most important parameter for SOCD at 0–30 cm and 0–50 cm (and accounted for 35.53% and 33.58% of

Table 2. Statistics of soil organic carbon density at three soil depth intervals (0–30 cm, 0–50 cm, and 0–100 cm).

SOCD	N	Mean (kg m^{-2})	Std.D. (kg m^{-2})	Min (kg m^{-2})	Median (kg m^{-2})	Max (kg m^{-2})
$SOCD_{0-30\ cm}$	52	1.20	0.85	0.24	0.99	4.58
$SOCD_{0-50\ cm}$	52	1.73	1.20	0.35	1.44	6.32
$SOCD_{0-100\ cm}$	52	2.69	1.91	0.59	2.16	9.57

Notes: SOCD, soil organic carbon density; N, number of samples; Std.D., standard Deviation.

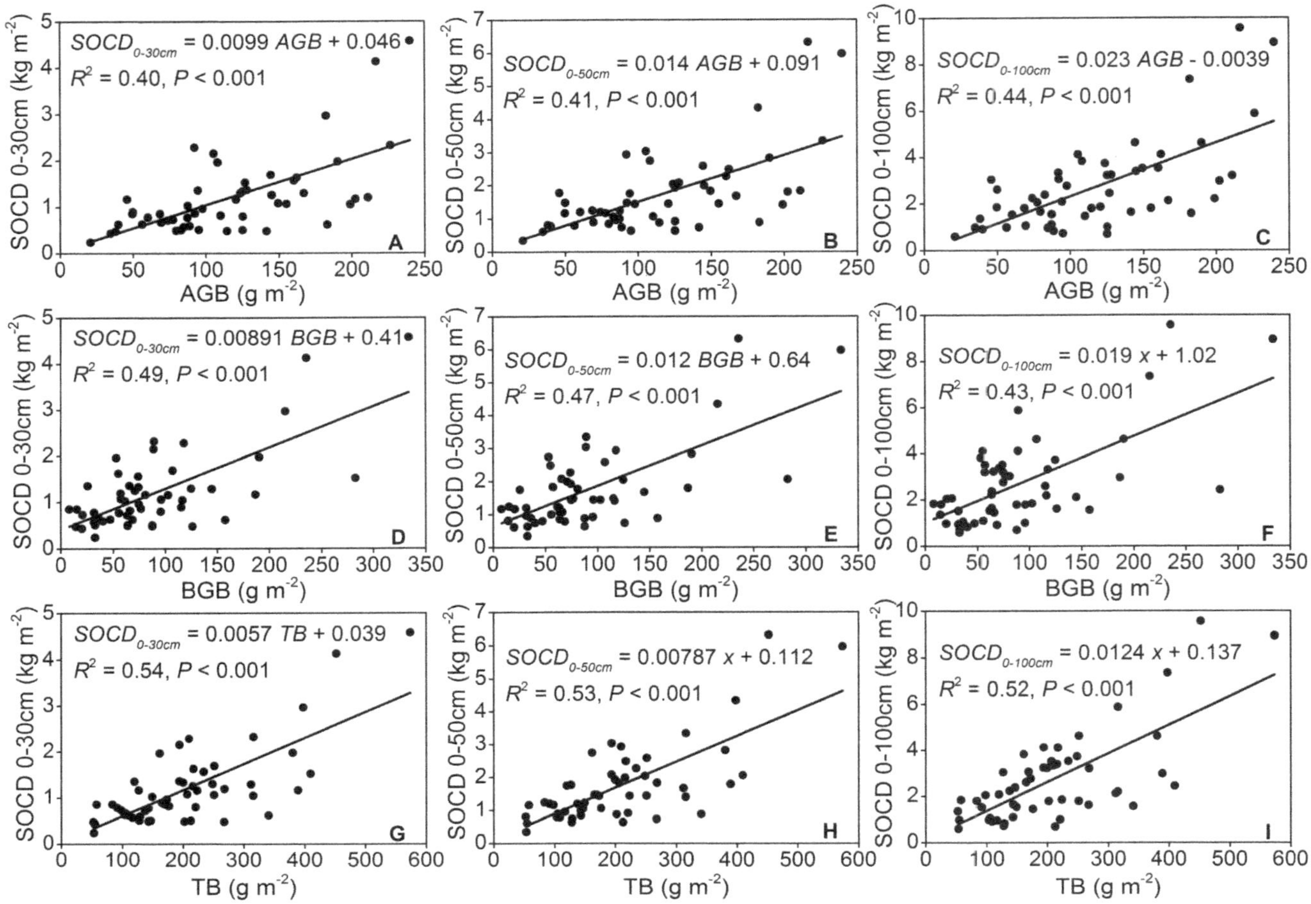

Figure 4. Relationships between soil organic carbon density and biomass at different depth intervals. *Notes*: AGB, above-ground biomass; BGB, below-ground biomass; TB, total biomass; SOCD, soil organic carbon density.

variation), whereas SM was the most important parameter for SOCD at 0–100 cm (where it accounted for 23.61% of variation). The proportion of variances explained by both factors decreased with an increase of soil depth. Soil texture variables (clay, silt, and sand content) explained 7.41%, 15.71%, and 35.95% of the variance at 0–30 cm, 0–50 cm, and 0–100 cm, respectively. In contrast to MAT and SM, the proportion of variance explained by soil texture markedly increased along with soil depth. The proportion of variance explained by sand content increased more rapidly than those explained by the other two variables, and instead silt content become the most important textural variable at soil depth of 0–100 cm (accounted for 13.60% of variation).

Discussion

SOC storage estimation

We summarized previous estimations on SOC storage on different vegetation types at global and regional scales in Table 4. In this study, the mean SOCD of 260 soil profiles at a soil depth of 0–100 cm in the desert grassland of the Hexi Corridor was 2.69 kg m^{-2}. Our results were generally lower than the global mean SOCD (10.8 kg m^{-2}), the average SOCD in China (7.8 kg m^{-2}), and other records based on vegetation types (temperate desert, steppe, and grassland). This difference is most likely due to differences in climate and soil conditions, which play a critical role in determining vegetation types and biomass productivity. The drier local climate, together with a greater gravel content and thinner soil layer thickness compared to other temperate steppe, grassland, and desert-grassland regions lead to lower vegetation production and limited SOC inputs [29].

Relationship between SOCD and biomass

SOC concentrations are closely linked to biotic processes, such as biomass production, decomposition, and the placement of aboveground litter and root litter in and on the soil [33]. Through regression analyses, we detected a higher coefficient of R^2 for the linear functions between SOCD and BGB compared with the association between SOCD and AGB at the three soil depth intervals. This finding indicated that BGB was most likely the main resource of C inputs and a dominant biological factor on the determination of SOCD. Belowground roots can provide abundant and stable organic material into soil and enhance SOC density. However, due to intense wind erosion in our research area, large amounts of litter fall were blown away by wind, which resulted in only a small amount of aboveground organic matter entering the soil.

Both BGB and TB exhibited the highest correlation in the top 30 cm soil, and this correlation decreased at deeper intervals likely due to high organic matter inputs at surface soil [26,34]. Large gravel content at depths below 30 cm make it hard for roots to extend to deep soil layers, so most of root biomass is concentrated in the upper 0–30 cm soil interval (almost 97%) [29]. According to

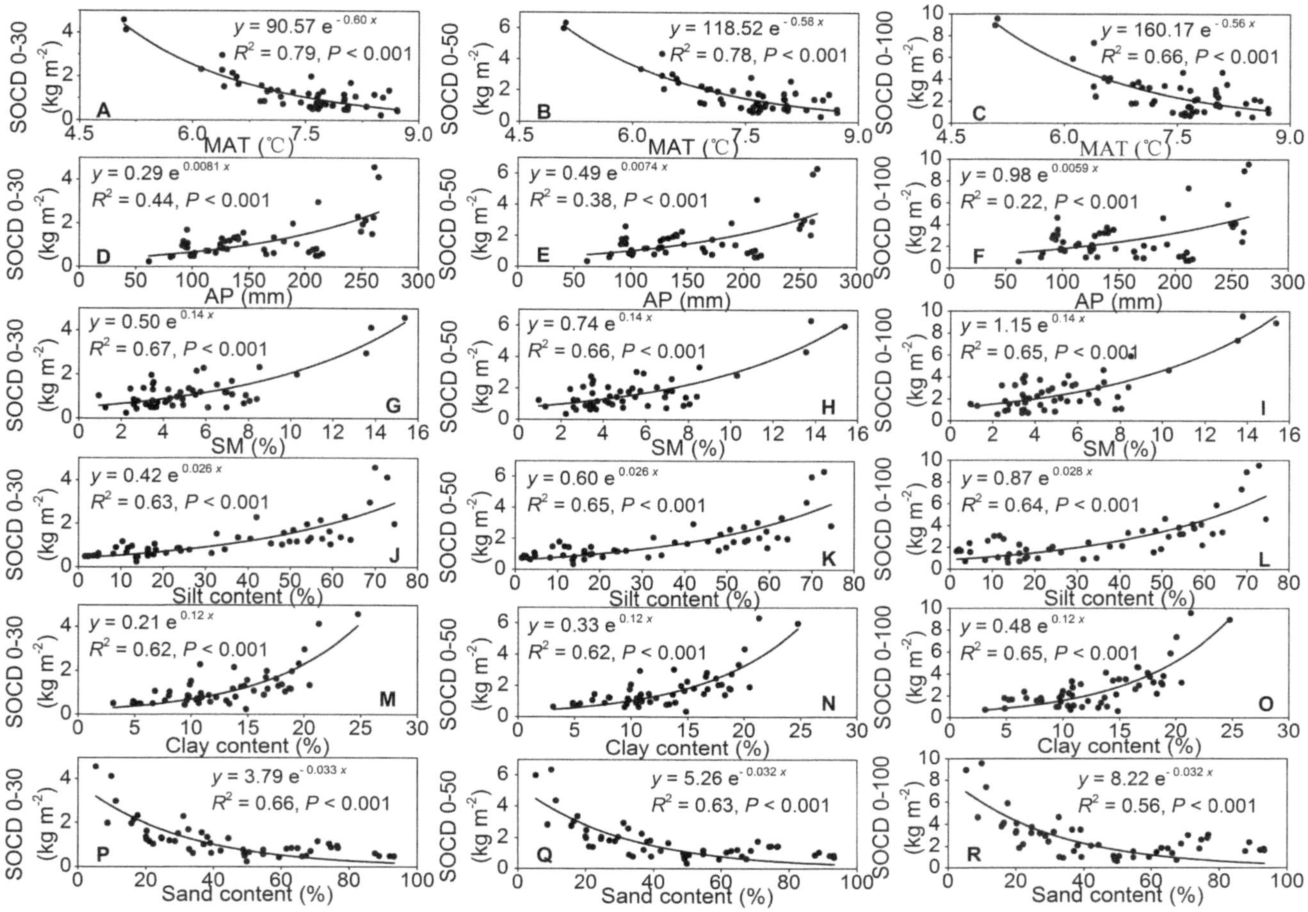

Figure 5. Relationships between soil organic carbon density and environmental factors at different depth intervals. *Notes*: SOCD, soil organic carbon density; MAT, mean annual temperature; AP, annual precipitation; SM, soil moisture.

Table 3. Integrative effects of mean annual temperature, soil moisture and soil texture (clay, silt, and sand) on soil organic carbon density at three soil depth intervals (0–30 cm, 0–50 cm, and 0–100 cm).

Source	0–30 cm				0–50 cm				0–100 cm			
	df	MS	SS%	P	df	MS	SS%	P	df	MS	SS%	P
MAT	1	4.45***	35.53	−0.47	1	8.70***	33.58	−0.66	1	15.13***	19.31	−0.87
SM	1	3.37***	26.87	0.10	1	6.51***	25.12	0.14	1	18.49***	23.61	0.24
Clay	1	0.33*	2.60	0.026	1	1.22**	4.69	0.051	1	7.04***	8.99	0.12
Silt	1	0.52*	4.14	0.013	1	1.89**	7.28	0.024	1	10.46***	13.36	0.057
Sand	1	0.08	0.67	0.004	1	0.97*	3.74	0.014	1	10.65***	13.60	0.047
Residuals	46	0.08	30.19		46	0.14	25.59		46	0.36***	21.13	
Intercept				3.22				3.78				2.25

Notes:
***P<0.001;
**P<0.01;
*P<0.05.
df, degree of freedom; MS, mean squares; SS%, proportion of variances explained by the variable; P, parameters of best-fitted GLM equations; MAT, mean annual temperature; SM, soil moisture.
The results were obtained from general linear model analysis.

Table 4. Comparisons of soil organic carbon density in desert grasslands in the Hexi Corridor with previous estimates.

Research types	Soil organic carbon density (kg m^{-2})				Reference
	0–30 (cm)	0–50 (cm)	0–100 (cm)	Actual depth (cm)	
Global mean	–	–	10.8	–	[23]
Global cool temperate desert	–	–	9.7	–	
Global cool temperate steppe	–	–	13.3	–	
Global temperate grassland	–	–	11.7	–	[26]
Global desert	–	–	6.2	–	
Average of China	3.7	–	7.8	–	[18]
Temperate grassland of Inner Mongolia	4	5.19	6.68	–	[28]
Desert steppe of Inner Mongolia	2.3	3.1	4.01	–	
				–	
Temperate typical steppe	–	–	12.3	–	[40]
Temperate deserted steppe	–	–	8.7	–	
Temperate desert	–	–	6.2	–	
Temperate steppe-desert	–	–	8	–	[10]
Temperate desert-steppe	–	–	8.7	–	
Temperate desert	–	–	6.2	–	
Desert	–	–	4.39	–	[41]
Desert steppe	–	–	7.09	–	
Temperate typical steppe	–	–	–	8	[16]
Temperate deserted steppe	–	–	–	2.8	
Temperate desert	–	–	–	2.2	
Desert grasslands in the Hexi Corridor	1.2	1.73	2.69	–	This study

Notes: "–" mean not measured.

the shallow distribution of roots, the majority of C inputs from roots are concentrated at soil depths of 0–30 cm; additionally, the topsoil is the first layer that directly receives C inputs from aboveground biomass.

Relationship between SOCD and climate and soil environmental factors

Under natural conditions, the distribution of SOC was controlled by climate, vegetation, parent material, and soil texture [3,5,23,26,35]. Our studies observed SOC distribution was positively associated with precipitation and clay content and negatively correlated with temperature and sand content and indicated that the strength of relationships between SOCD and environmental variables decreased with the increase of soil depth.

As shown in Figure 5, SOCD decreased markedly with MAT, likely due to the accelerated mineralization of soil organic matter with increasing temperature [3,24,26]. On the other hand, increasing temperature in arid regions results in a considerable decline in water use efficiency by increasing evapotranspiration can lead to low biomass production and low SOC density [15]. Additionally, the effect of MAT on SOCD in each soil layer was stronger than that for AP and edaphic factors. This result underlines the importance of MAT as a predicator for the SOC density in desert grasslands.

In arid ecosystems, precipitation and soil moisture constrain plant production and decomposition [26]. Our results revealed a substantial increase in the SOCD in desert grasslands with an increase in AP and SM. These results imply that water availability is a powerful parameter for assessment of SOCD. Water is the limiting factor for plant production in desert grasslands, as a small increase of water could significantly stimulate bio-productivity and thus contribute to the accumulation of SOCD [26,36]. Meanwhile, higher precipitation and soil moisture will affect SOC sequestration mainly through higher soil acidity and lower base saturation at the exchange sites, which would reduce the litter decomposition rate. In addition to the abovementioned results, the relationship between SOCD and SM was stronger than that between SOCD and AP. Precipitation in arid desert-grassland has difficulty entering the soil through infiltration due to low vegetation coverage, less litter content, and coarse soil texture, and consequently, soil moisture can represent the actual water content and play a more important role in model construction between SOCD and environmental factors.

In our study, we observed that the increased SOCD was positively correlated with the accumulation of silt and clay content but negatively correlated with sand content. Finely textured soils with appropriate clay and silt content can increase physical and hydrological protection of SOC by inhibiting decomposition through stabilizing SOC, and increasing residence time to decrease C leaching [3,5,26,37,38]. Moreover, an increase in clay and silt content could enhance the formation of aggregates, which have two advantages for improving SOC content: (1) by improving the soil quality and capacity to efficiently retain water [3,5,39] and (2) by mitigating wind erosion [27]. These two advantages could stimulate plant productivity and thus result in additional C inputs.

The GLM analysis suggested that climate factors were more important in determining the distribution of SOC than soil texture, but the effect decreased gradually with an increase in soil depth. In contrast, the impact of soil texture on SOC distribution in surface soils was not obvious but increased alongside soil depth, which indicates that soil texture plays a critical role in SOC distribution at deeper soil layers. Among all climatic and edaphic parameters, MAT had the highest contribution to explain the distribution of SOC in surface soils (0–30 cm and 0–50 cm). Considering the vast majority of SOC accumulates in the surface soil [26], MAT plays a decisive role in the spatial distribution of SOC in the arid desert- grassland ecosystem in the Hexi Corridor. That may be because biomass production is strictly limited by water in arid desert- grasslands, and the rise of temperature can promote the decomposition of SOC, and conversely promote the accumulation of SOC.

Author Contributions

Conceived and designed the experiments: MW. Performed the experiments: MW YS XY. Analyzed the data: MW YS. Contributed reagents/materials/analysis tools: MW. Wrote the paper: MW.

References

1. Davidson EA, Janssens IA (2006) Temperature sensitivity of soil carbon decomposition and feedbacks to climate change. Nature 440: 165–173.
2. Albaladejo J, Ortiz R, Garcia-Franco N, Navarro AR, Almagro M, et al. (2013) Land use and climate change impacts on soil organic carbon stocks in semi-arid Spain. Journal of Soils and Sediments 13: 265–277.
3. Schimel DS, Braswell BH, Holland EA, McKeown R, Ojima DS, et al. (1994) Climatic, edaphic, and biotic controls over storage and turnover of carbon in soils. Global biogeochemical cycles 8: 279–293.
4. Lal R (2004) Soil carbon sequestration impacts on global climate change and food security. Science 304: 1623–1627.
5. Yang YH, Fang JY, Tang YH, Ji CJ, Zheng CY, et al. (2008) Storage, patterns and controls of soil organic carbon in the Tibetan grasslands. Global Change Biology 14: 1592–1599.
6. Conant R, Paustian K (2002) Spatial variability of soil organic carbon in grasslands: implications for detecting change at different scales. Environmental Pollution 116: S127–S135.
7. Fan JW, Zhong HP, Harris W, Yu GR, Wang SQ, et al. (2008) Carbon storage in the grasslands of China based on field measurements of above-and below-ground biomass. Climatic Change 86: 375–396.
8. Wang GX, Qian J, Cheng GD, Lai YM (2002) Soil organic carbon pool of grassland soils on the Qinghai-Tibetan Plateau and its global implication. Science of the Total Environment 291: 207–217.
9. Han X, He N, Wu L, Wang Y (2008) Storage and Dynamics of Carbon and Nitrogen in Soil after Grazing Exclusion in Grasslands of Northern China. Journal of Environmental Quality 37: 663–668.
10. Ni J (2002) Carbon storage in grasslands of China. Journal of Arid Environments 50: 205–218.
11. Parton W, Scurlock J, Ojima D, Gilmanov T, Scholes R, et al. (1993) Observations and modeling of biomass and soil organic matter dynamics for the grassland biome worldwide. Global biogeochemical cycles 7: 785–809.
12. Piao SL, Fang JY, Zhou LM, Tan K, Tao S (2007) Changes in biomass carbon stocks in China's grasslands between 1982 and 1999. Global biogeochemical cycles 21 Doi: 10.1029/2005gb002634.
13. Qibin W, Linghao L, Xianhua L (1998) Spatial heterogeneity of soil organic carbon and total nitrogen in Xilin River basin grassland, Inner Mongolia. Acta Phytoecologica Sinica 22: 409–414.
14. Yang YH, Fang JY, Guo DL, Ji CJ, Ma WH (2010) Vertical patterns of soil carbon, nitrogen and carbon: nitrogen stoichiometry in Tibetan grasslands. Biogeosciences Discussions 7: 1–24.
15. Lal R (2004) Carbon sequestration in dryland ecosystems. Environ Manage 33: 528–544.
16. Wu HB, Guo ZT, Peng CH (2003) Distribution and storage of soil organic carbon in China. Global biogeochemical cycles 17 Doi: 10.1029/2001gb001844.
17. Wang SQ, Tian HQ, Liu JY, Pan SF (2003) Pattern and change of soil organic carbon storage in China: 1960s–1980s. Tellus B 55: 416–427.
18. Yang YH, Mohammat A, Feng JM, Zhou R, Fang JY (2007) Storage, patterns and environmental controls of soil organic carbon in China. Biogeochemistry 84: 131–141.
19. Fang JY, Yang YH, Ma WH, Mohammat A, Shen HH (2010) Ecosystem carbon stocks and their changes in China's grasslands. Science China Life sciences 53: 757–765.
20. Schlesinger WH, Andrews JA (2000) Soil respiration and the global carbon cycle. Biogeochemistry 48: 7–20.
21. Bastida F, Moreno JL, Hernandez T, Garcia C (2007) The long-term effects of the management of a forest soil on its carbon content, microbial biomass and activity under a semi-arid climate. Applied Soil Ecology 37: 53–62.
22. Johnston CA, Groffman P, Breshears DD, Cardon ZG, Currie W, et al. (2004) Carbon cycling in soil. Frontiers in Ecology and the Environment 2: 522–528.
23. Post WM, Emanuel WR, Zinke PJ, Stangenberger AG (1982) Soil carbon pools and world life zones. Nature 298: 156–159.
24. Burke IC, Yonker CM, Parton WJ, Cole CV, Flach K, et al. (1989) Texture, climate, and cultivation effects on soil organic matter content in US grassland soils. Soil Sci Soc Am J 53: 800–805.
25. Canadell JG, Kirschbaum MUF, Kurz WA, Sanz M-J, Schlamadinger B, et al. (2007) Factoring out natural and indirect human effects on terrestrial carbon sources and sinks. Environmental Science & Policy 10: 370–384.
26. Jobbágy EG, Jackson RB (2000) The vertical distribution of soil organic carbon and its relation to climate and vegetation. Ecological Applications 10: 423–436.
27. Su YZ, Wang XF, Yang R, Lee J (2010) Effects of sandy desertified land rehabilitation on soil carbon sequestration and aggregation in an arid region in China. Journal of Environmental Management 91: 2109–2116.
28. Ma WH (2006) Carbon Storage in the Temperate Grassland of Inner Mongolia. Beijing: Peking University.
29. Wang M, Su YZ, Yang R, Yang X (2013) Allocation patterns of above- and below-ground biomass in desert grassland in the middle reaches of Heihe River, Gansu Province, China. Chinese Journal of Plant Ecology 37: 209–219 (in Chinese).
30. Nelson DW, Sommers LE (1982) Total carbon, organic carbon, and organic matter. In: Methods of soil analysis Part 2 Chemical and microbiological properties: 539–579.
31. Chaudhari SK, Singh R, Kundu DK (2008) Rapid textural analysis for saline and alkaline soils with different physical and chemical properties. Soil Sci Soc Am J 72: 431–441.
32. Holben BN (1986) Characteristics of maximum-value composite images from temporal AVHRR data. International Journal of Remote Sensing 7: 1417–1434.
33. Lal R, Kimble J (2001) Soil erosion and carbon dynamics on grazing land. The potential of US grazing lands to sequester carbon and mitigate the greenhouse effect: Lewis Publishers, Boca Raton, Florida: 231–247.
34. Li Z, Zhao QG (2001) Organic carbon content and distribution in soils under different land uses in tropical and subtropical China. Plant and Soil 231: 175–185.
35. Wang SQ, Huang M, Shao XM, Mickler RA, Li KR, et al. (2004) Vertical distribution of soil organic carbon in China. Environmental Management 33: 200–209.
36. Callesen I, Liski J, Raulund-Rasmussen K, Olsson MT, Tau-Strand L, et al. (2003) Soil carbon stores in Nordic well-drained forest soils - relationships with climate and texture class. Global Change Biology 9: 358–370.
37. Paul EA (1984) Dynamics of organic matter in soils. Plant and Soil 76: 275–285.
38. Wynn JG, Bird MI, Vellen L, Grand-Clement E, Carter J, et al. (2006) Continental-scale measurement of the soil organic carbon pool with climatic, edaphic, and biotic controls. Global biogeochemical cycles 20 Doi: 10.1029/2005gb002576.
39. Schimel DS, Parton WJ (1986) Microclimatic controls of nitrogen mineralization and nitrification in shortgrass steppe soils. Plant and Soil 93: 347–357.
40. Zinke PJ, Stangenberger AG, Post WM, Emanuel WR, Olson JS (1984) Worldwide organic soil carbon and nitrogen data. Oak Ridge National Lab., TN (USA) .
41. Liu WJ, Chen SY, Qin X, Baumann F, Scholten T, et al. (2012) Storage, patterns, and control of soil organic carbon and nitrogen in the northeastern margin of the Qinghai–Tibetan Plateau. Environmental Research Letters 7 Doi: 10.1088/1748-9326/7/3/035401.

14

Combining XCO_2 Measurements Derived from SCIAMACHY and GOSAT for Potentially Generating Global CO_2 Maps with High Spatiotemporal Resolution

Tianxing Wang*, Jiancheng Shi, Yingying Jing, Tianjie Zhao, Dabin Ji, Chuan Xiong

State Key Laboratory of Remote Sensing Science, Institute of Remote Sensing and Digital Earth, Chinese Academy of Sciences. Beijing, China

Abstract

Global warming induced by atmospheric CO_2 has attracted increasing attention of researchers all over the world. Although space-based technology provides the ability to map atmospheric CO_2 globally, the number of valid CO_2 measurements is generally limited for certain instruments owing to the presence of clouds, which in turn constrain the studies of global CO_2 sources and sinks. Thus, it is a potentially promising work to combine the currently available CO_2 measurements. In this study, a strategy for fusing SCIAMACHY and GOSAT CO_2 measurements is proposed by fully considering the CO_2 global bias, averaging kernel, and spatiotemporal variations as well as the CO_2 retrieval errors. Based on this method, a global CO_2 map with certain UTC time can also be generated by employing the pattern of the CO_2 daily cycle reflected by Carbon Tracker (CT) data. The results reveal that relative to GOSAT, the global spatial coverage of the combined CO_2 map increased by 41.3% and 47.7% on a daily and monthly scale, respectively, and even higher when compared with that relative to SCIAMACHY. The findings in this paper prove the effectiveness of the combination method in supporting the generation of global full-coverage XCO_2 maps with higher temporal and spatial sampling by jointly using these two space-based XCO_2 datasets.

Editor: Juan A. Añel, University of Oxford, United Kingdom

Funding: The work described in this paper has been jointly supported by project of "Climate Change: Carbon Budget and Related Issues" (Grant nr. XDA05040402) from Chinese Academy of Sciences (CAS), the CAS/SAFEA International Partnership Program for Creative Research Teams (Grant nr. KZZD-EW-TZ-09) and National Natural Science foundation of China (Grant nr. 41301177). The funders had no role in study design, data collection and analysis, decision to publish, or preparation of the manuscript.

Competing Interests: The authors have declared that no competing interests exist.

* Email: wangtx@radi.ac.cn

Introduction

In recent years, global warming caused by emission of CO_2 has attracted considerable attention from the public. During the past decade, although tremendous efforts have been made toward improving the understandings of the mechanism between CO_2 increase in the atmosphere and global warming, some uncertainties still exist in the spatiotemporal characteristics of CO_2 sinks/sources on regional and global scales due to the lack of high-density measurements of such variables with good accuracy [1,2]. To date, the estimates of CO_2 flux from inverse methods rely mainly on ground-based measurements [3,4]. Although providing highly accurate atmospheric CO_2 records, the traditional ground-based networks intrinsically suffer from sparse spatial coverage [2,5]. Satellite-based measurements with various spatial and temporal resolutions provide a unique opportunity to accurately map atmospheric CO_2 in both daytime and nighttime over large areas, thus having the potential to bridge this gap. As a result, various satellite-based platforms have been equipped in recent years for deriving the CO_2 concentrations.

Generally, methods for retrieving CO_2 from space can be grouped into two categories: (1) inferring CO_2 concentrations by measuring shortwave infrared (SWIR) reflected solar radiation around 1.6 and 2.0 μm with sufficient spectral resolution. This includes the Greenhouse gases Observing SATellite (GOSAT), operating since 2009 [6], the Scanning Imaging Absorption spectrometer for Atmospheric CartograpHY (SCIAMACHY), in orbit since 2002 [7], and the second Orbiting Carbon Observatory (OCO-2), which, as a rebuild of OCO [8,9], is planned to be launched in July 2014. In addition, CarbonSat will also be scheduled to be launched in 2018 (http://www.iup.uni-bremen.de/carbonsat/). These measurements have a nearly uniform sensitivity to CO_2 from the surface up through the middle troposphere, and thus are frequently used to derive the column-average dry air mole fraction of atmosphere CO_2 (XCO_2) during the daytime; (2) retrieving CO_2 concentrations by interpreting the recorded spectra of the Earth-atmosphere system in thermal infrared (TIR) bands (around 15 μm). Instruments that work in such a way include AIRS [10,11], IASI [12,13], and FTS (Band 4) of GOSAT [6]. These measurements bring the advantage that they can detect CO_2 during both day and night time, while the lack of sensitivity in the lower troposphere makes them inappropriate to estimate CO_2 near the surface where the largest signals of CO_2 sources and sinks occur [1]. The complementarities of these platforms allow us to combine the SWIR and TIR measurements for obtaining enhanced understanding of CO_2 spatiotemporal variations globally. Since XCO_2 is much less affected by vertical transport of CO_2, it is particularly useful for investigation of CO_2 sources and sinks using inversion modeling [14,15]. On the other hand, the spatial and temporal variations in XCO_2 are even smaller than that in the surface CO_2; therefore, unprecedented measurement precision and accuracy are highly required for such column measurements [16–19]. SCIAMACHY (operation stopped in April 2012) and GOSAT are two typical instruments that can be used to derive XCO_2 from space, and a variety of retrieval algorithms have been developed for SCIAMACHY [1,20–27] and GOSAT [2,4,5,28–30] with eyes on improving XCO_2 retrieval accuracy to a great extent. At present, a number of XCO_2 products have been released. These will definitely enhance our understanding of the global carbon cycle.

Unfortunately, almost all typical instruments currently used to derive atmospheric CO_2 concentration are working in the infrared spectral range (less than 16 μm). Thus, except for the instrument's observation mode (for example, GOSAT observes in lattice points), the spatial coverage of the derived CO_2 is severely restricted by the presence of clouds. In addition, the lower signal-to-noise level over ice/snow covered surfaces and ocean for SWIR instruments (e.g., SCIAMACHY) also contributes to the CO_2 sparse coverage. For instance, it has been pointed out that only about 10% of GOSAT data can be used for retrieval of XCO_2 due to the cloud contaminations [4]. The amount of CO_2 measurements will be even smaller if additional screening criteria such as quality of spectral fit, aerosol loadings, etc. are further applied. Although the amount of remaining CO_2 measurements from certain space-based instruments may largely surpass that of ground-based sites, it is still not sufficient enough for accurately quantifying the spatiotemporal distribution of CO_2 over the global scale. As a result, it is greatly desired to jointly use these available CO_2 measurements derived from various space-based data. Recently, a novel method has been proposed for combining CO_2 values from seven different algorithms, and a new Level-2 CO_2 database (EMMA) from one algorithm is composed according to the median of monthly average of seven CO_2 products in each 10°×10° latitude/longitude grid box [31]. In fact, this method cannot increase the number of CO_2 observations but chooses a product with moderate oscillation among the available products. Despite the usefulness of the XCO_2 measurements (Level 2) in their own right, further spatiotemporal analysis for interpreting their scientific merit is essentially necessary due to the retrieval uncertainties and sparse coverage of such Level-2 observations [32]. For this point, many works have attempted to generate global full-coverage (i.e., Level 3) maps from XCO_2 values derived from single satellite observations using a geospatial statistics approach [32–34]. However, as reflected in these studies (for instance, Fig. 1 in the work of [33]), a compromise has to be made between the interpolated accuracy and the spatiotemporal resolution of Level-3 product because of the limited amount of Level-2 XCO_2 observations being used. For this point, instead of using Level-2 XCO_2 from a single dataset (e.g., GOSAT or OCO-2) as performed in the existing literature, we attempt to explore the potential of combining two CO_2 datasets (GOSAT and SCIAMACHY) in assisting in global Level-3 generation, aiming to: (1) propose a general strategy for combining (fusing) various CO_2 datasets with different instruments, algorithms, averaging kernels, etc.; and 2) increase the number of daily CO_2 points (utilized in Level-3 map interpolations) through the combination of two datasets, so that potentially improved Level-3 maps with higher accuracy and shorter time scale can be generated. The better the interpretation of the satellite-based CO_2 observations one can make, the higher the resolution (both temporal and spatial) of the generated global CO_2 maps.

Datasets

For GOSAT, the Fourier transform spectrometer (FTS) on GOSAT is the fundamental unit to retrieve atmospheric CO_2 and CH_4. It observes sunlight reflected from the earth's surface, and light emitted from the atmosphere and the surface. It is composed of three narrow bands in the SWIR region (0.76, 1.6, and 2.0 μm) and a wide TIR band (5.5–14.3 μm) at a spectral and spatial resolution of 0.2 cm^{-1} and 10.5 km, respectively [35]. Specifically, four CO_2 products from GOSAT have currently been released to the public: University of Leicester product [9,36], the RemoTeC product [28], NIES GOSAT product [35] and the product generated by NASA's Atmospheric CO_2 Observations from Space (ACOS) team (hereafter called ACOS product) [2,30]. The difference between some of the above mentioned products with various versions have been investigated in a recent study [37]. In the present paper, the ACOS product of 2009–2010 with version v2.9 has been employed.

SCIAMACHY was successfully launched on board Environmental Satellite (ENVISAT) in 2002 (unfortunately ceased in April 2012), which is a detector elements satellite spectrometer covering the spectral range 0.24–2.38 μm with a moderate spectral resolution of about 0.2–1.6 nm, and spatial resolution at nadir of 60×30 km [7]. It has eight spectral channels, with 1024 individual detector diodes for each band, observing the spectral regions 0.24–1.75 μm (band 1–6), 1.94–2.04 μm (band 7), and 2.26–2.38 μm (band 8) simultaneously in nadir and limb and solar and lunar occultation viewing geometries [22]. As mentioned in Section 1, till today, a number of CO_2-retrieval algorithms have been developed for SCIAMACHY. The IUP/IFE of University of Bremen has released two XCO_2 products, i.e., WFM-DOAS product [21,22] and the Bremen Optimal Estimation DOAS (BESD) product [1,26]. In this study, the BESD product with the versions of v02.00.08 for 2009–2010 is used.

In addition, CO_2 profiles of CT [38] are also collected here to allow the data mentioned above to be properly fused. CT is a NOAA data assimilation system, which provides the 3D profiles of CO_2 mole fractions in the atmosphere over the globe. For this study, CT data with version CT2011 is collected. This dataset provides global CO_2 profiles with 3°×2° latitude/longitude grid and 3 hours temporal resolution (a total 8 times from 01 to 22 in UTC) spanning the time period from January 2000 to December 2010. The CT dataset is used here mainly to assist in adjusting and time-shifting of the two CO_2 products being combined.

Methodologies

For combining the different space-based CO_2 measurements, three steps are adapted in this study. First, taking the global ground measurements of CO_2 as reference, remove the bias of the individual CO_2 retrievals for ensuring the accuracy of the fused CO_2 product; then make some adjustment for both the ACOS and BESD products, so that they can be physically comparable and thus combined; finally fuse the ACOS and BESD CO_2 products considering their retrieval uncertainties, spatial scales, differences in averaging kernels and overpass times, etc.

3.1 Global bias corrections

Removal of any global bias of the retrieved CO_2 when compared with the ground *in situ* measurements is essential before performing joint use. Many researches [4,25] frequently pointed out that CO_2 retrievals from GOSAT are low biased with different levels due to the uncertainties in pressure, radiometric calibration, line shape model, cloud and aerosol scattering, etc.

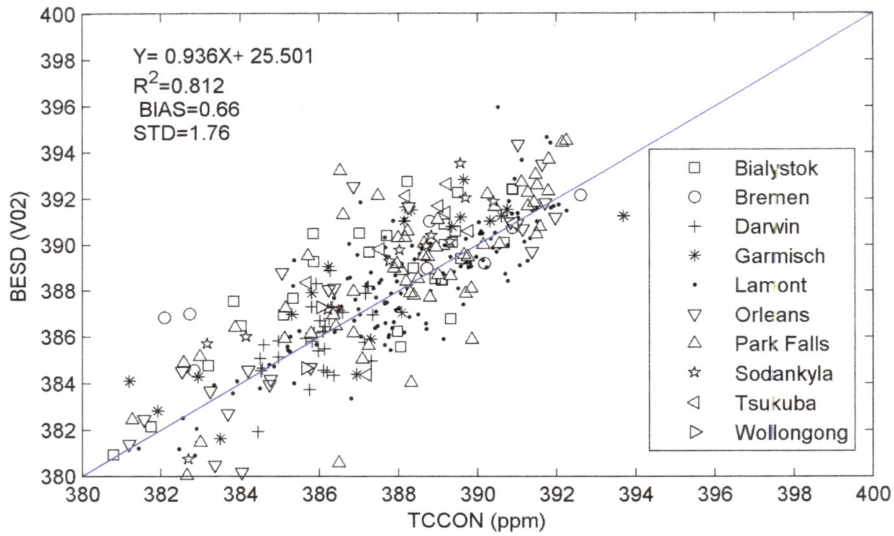

Figure 1. Validation of the BESD products against *in situ* TCCON CO_2 measurements over globe for 2009–2010.

Fortunately, a recent study has proposed a method for evaluating systematic errors in CO_2 and showed that the new version of ACOS product (v2.9) has a low global bias (<0.5 ppm) [39]. Thus, there is no global bias correction for the ACOS product being conducted here, but only the ACOS retrievals that pass the filter of table B1 in the work of [38] and marked as "good" in the quality flag are used. For the BESD product, we select Total Carbon Column Observing Network (TCCON) [15] measurements for 2009–2010 as the ground truth to determine its global bias. Specifically, BESD retrievals within ±2.5° and ±2.5° latitude/longitude box centered at each TCCON site and the mean FTS value (within ±1 h time window of satellite overpass time) are

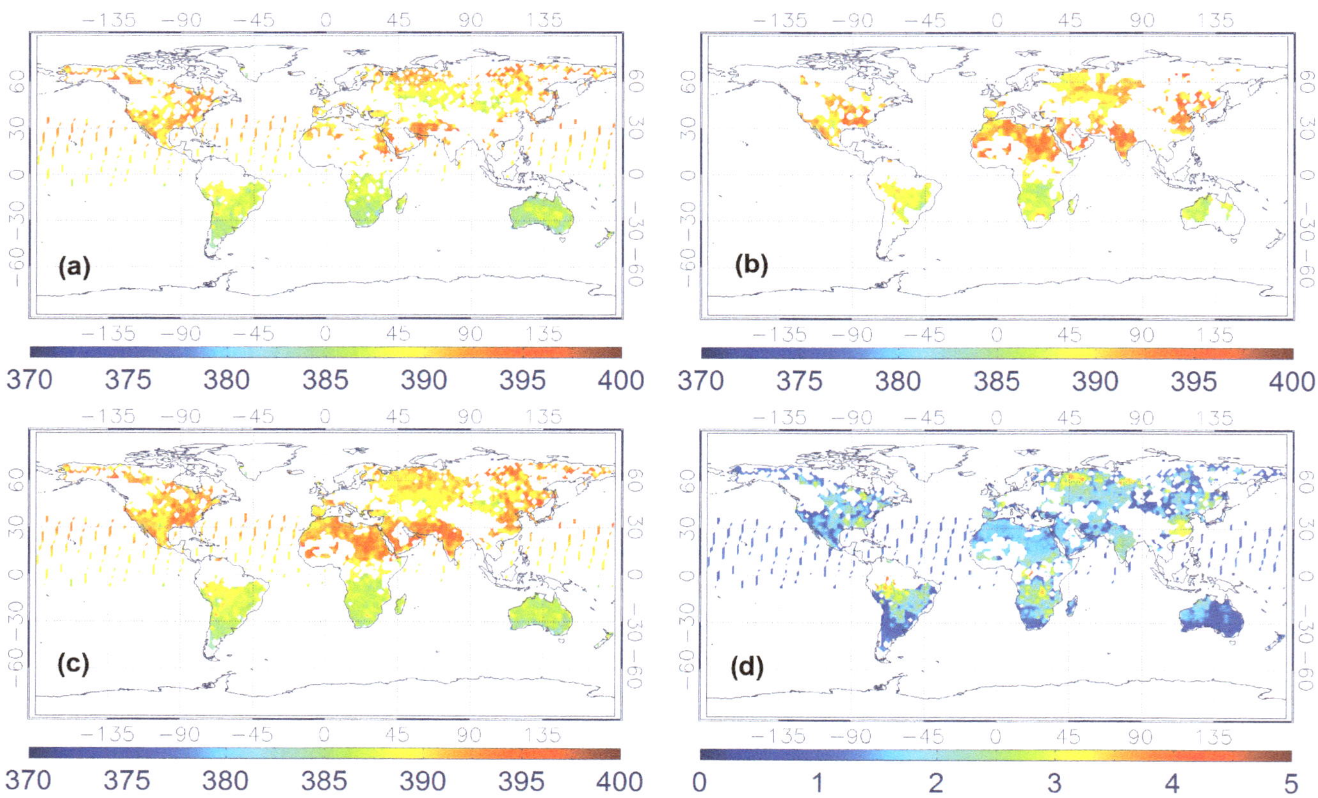

Figure 2. XCO_2 monthly mean maps in May of 2010. ((a) ACOS XCO_2, (b) BESD XCO_2, (c) combined product, and (d) XCO_2 uncertainties of the combined product).

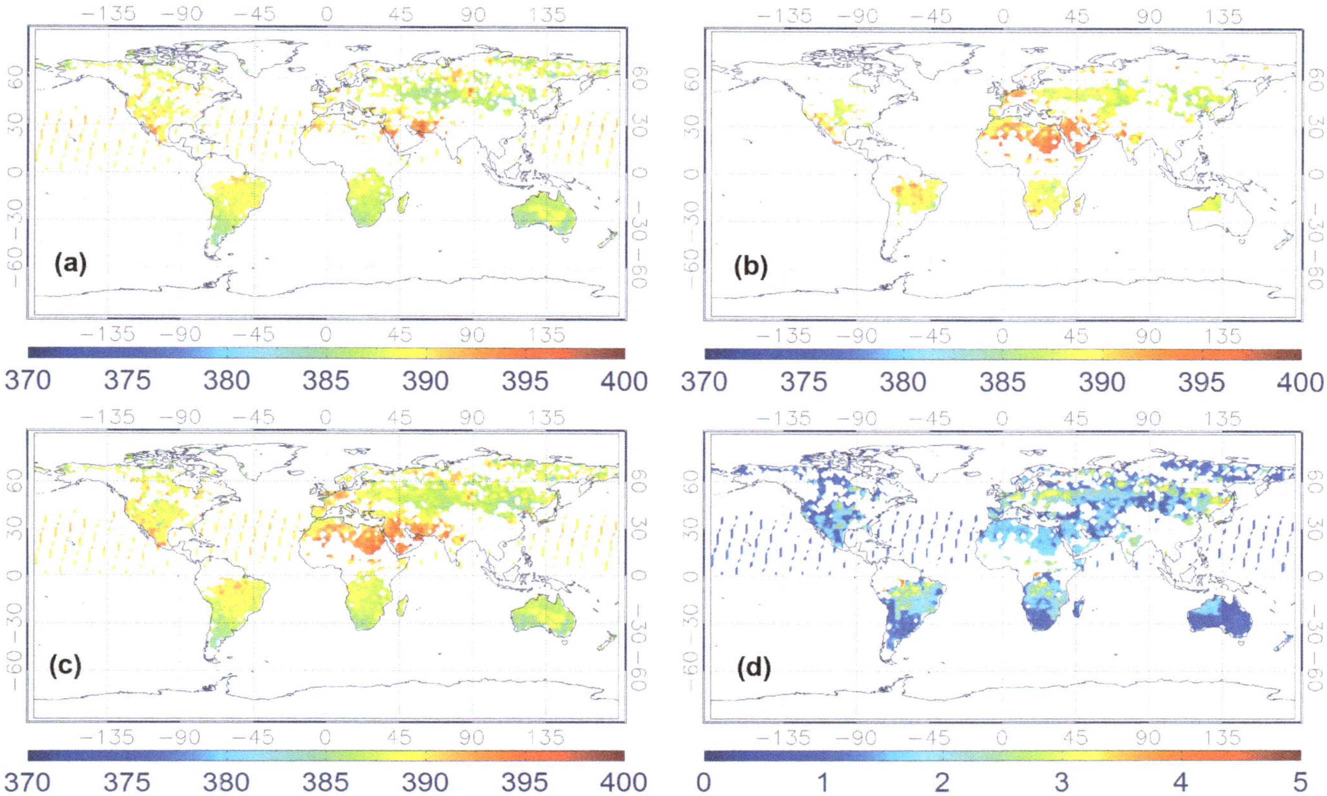

Figure 3. XCO_2 monthly mean maps in June of 2010. ((a) ACOS XCO_2, (b) BESD XCO_2, (c) combined product, and (d) XCO_2 uncertainties of the combined product).

extracted and compared (totally ten TCCON sites are utilized). The coincidence criteria mentioned above ultimately yield a total of 338 pairs of CO_2 measurements. The comparison result is shown in Fig. 1.

3.2 Retrieval adjustments

As pointed out by most researchers, it is not reasonable to directly compare or use two XCO_2 measurements. A suitable way to do that is to take the a priori profiles and variations in averaging kernel into account during the comparison [26,40]. To tackle the a priori issue, after correcting their global biases, both BESD and ACOS products are adjusted for a common a priori profile, which we assume to be the CT profile interpolated at the middle of the two overpass times (Equation (1)). Specifically, the a priori CO_2 profile of both the ACOS and BESD are first interpolated or extrapolated to the level of the CT CO_2 profile according to their pressure layers. After interpolation, the a priori profiles for both ACOS and BESD have the same dimension as the CT profile. Here the reason we take the CT profile at the middle of the two overpass times is that the time difference for GOSAT (1:00 pm) and SCIAMACHY (10:00 am) is relative large (3 hours), if we take one satellite time as reference, the induced error would be large for the other satellite measurements considering the CO_2 natural diurnal variation. So a middle time between these two satellite overpass times is selected for minimizing the CO_2 uncertainties during the adjustment.

$$XCO_2\_adj = XCO_2\_ret + (h^T I - a)(xCT - xa) \tag{1}$$

Here, XCO_2_adj is the adjusted XCO_2 for ACOS or BESD; XCO_2_ret corresponds to retrieved XCO_2 of ACOS or BESD; a is the column-averaging kernel (row vector) of ACOS or BESD; h is pressure-weighting function (column vector); I is an identity matrix; xCT and xa (column vectors) are the common CT CO_2 profile and the corresponding a priori CO_2 profile for ACOS or BESD, respectively.

While it is not trivial to accurately consider the smoothing error without an estimate of the true atmospheric variability which is generally not readily available for most cases [39]. Fortunately, some works revealed that the smoothing error is generally small [26,39]. Consequently, for the remainder of this paper, only the adjustment in Equation (1) is applied for both the ACOS and BESD CO_2 products (after bias corrections).

3.3 Combination and time shifting

Based on the processes described above, the world is divided into a number of 0.5°×0.5°latitude/longitude grid box (totally 720×360). For each grid cell, Equation (2) is used to combine the corresponding CO_2 measurements within that grid.

$$XCO_2\_Fued = \sum_{i=1}^{m} \left(XCO_2\_i \times \frac{1 - Uncert\_ratio^i}{\sum_{i=1}^{m} (1 - Uncert\_ratio^i)} \right) \tag{2}$$

where XCO_2_Fued is the combined XCO_2; m is the total number of space-based CO_2 retrievals (ACOS and/or BESD) within a certain grid; XCO_2_i is the ith XCO_2 retrieval in a grid for which

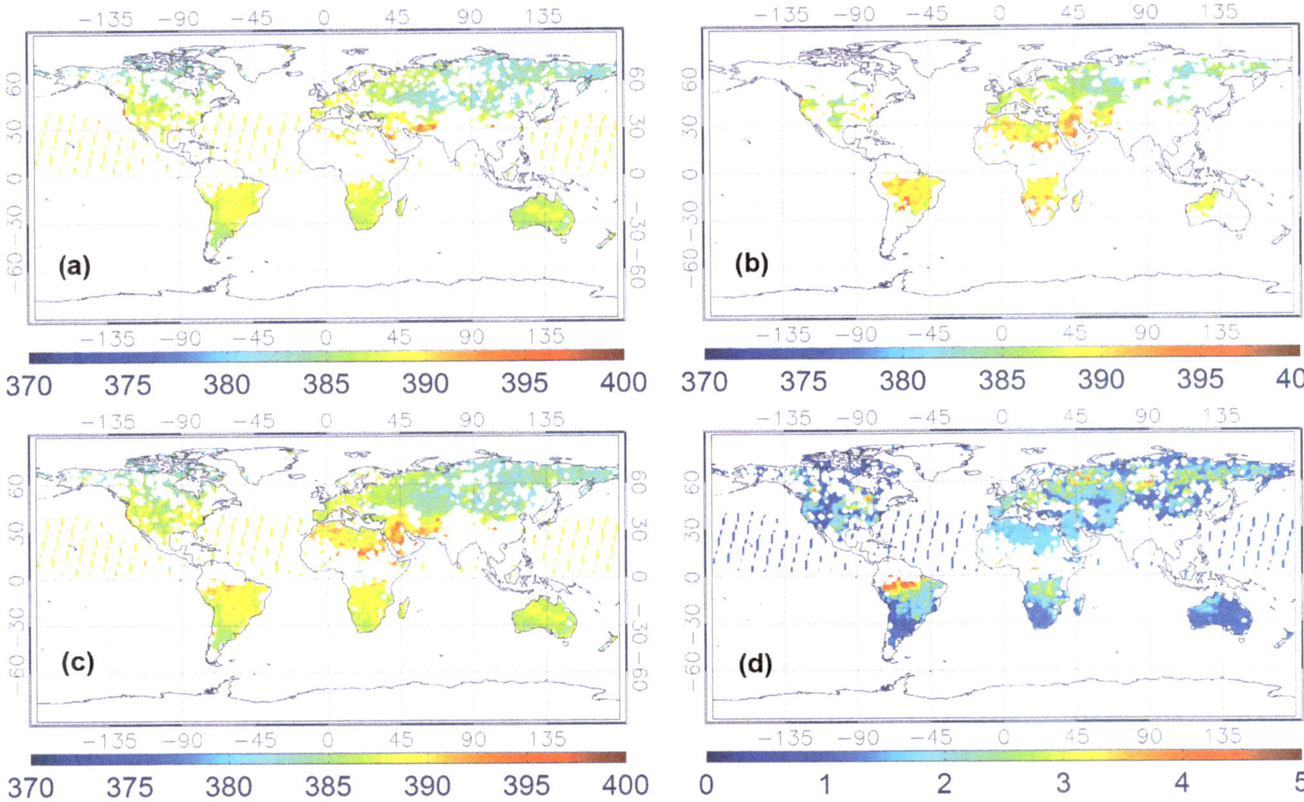

Figure 4. XCO_2 monthly mean maps in July of 2010. ((a) ACOS XCO_2, (b) BESD XCO_2, (c) combined product, and (d) XCO_2 uncertainties of the combined product).

the global bias and Equation (1) are supposed to be applied; $Uncert\_ratio^i$ is the ratio of uncertainty of the ith XCO_2 retrieval to its XCO_2 value.

Please note that since different CO_2 retrievals have distinct overpass times, it is necessary to unify them to avoid uncertainties induced from the time discrepancy before fusion. To this end, a method for considering the CO_2 shifting along time has been developed (Equation (3)). First, designate a specific time or select one overpass time as reference, then transfer CO_2 measurements at various overpass times to that of the reference time by interpolating the CT CO_2 at temporal scale. Here, it should be pointed out that despite the CO_2 absolute values of CT not being accurate enough, the daily cycle pattern of atmospheric CO_2 it reflects is assumed to be correct.

$$XCO_2\_ref = \frac{\omega^T X\_ref^{CT}}{\omega^T X\_t^{CT}} \times XCO_2\_t \qquad (3)$$

Here, XCO_2_ref is the transformed XCO_2 (ACOS or BESD) at the reference time; XCO_2_t is the retrieved XCO_2 from ACOS or BESD at overpass time t; X_ref^{CT} and X_t^{CT} are CO_2 profiles of CT at times of reference and t, respectively; ω is the pressure-weighting vector (column vector).

Based on the time-shifting strategy proposed here, a global CO_2 map at any specific time can be theoretically produced by employing the pattern of the CO_2 daily cycle reflected by CT data. For instance, we can unify all XCO_2 retrievals being combined with various overpass times to that of UTC = 1.

Results

Evaluation analysis showed that the global bias for the BESD product is generally small. In this study, the bias of the BESD product is corrected by subtracting 0.6 ppm from all XCO_2 values according to the results in Fig. 1. Although the systematic bias of the XCO_2 retrievals is removed, it is supposed that the error characteristics (random error) within the data are still unchanged. The bias-corrected XCO_2 retrievals of both ACOS and BEDS are used as fundamental data for the combination algorithm.

By applying the series of processes shown in Section 3, daily, weekly, as well as monthly maps of combined XCO_2 for 2009 and 2010 are generated. Here, as an example, only four maps (from May to August) of monthly mean XCO_2 of 2010 are shown here (Fig. 2–Fig. 5). In addition, the total XCO_2 uncertainties of the combined product which mainly depend on the uncertainties of the original ACOS or BESD XCO_2 retrievals are also illustrated.

From Fig. 2–Fig. 5, it is not difficult to observe that the combined data realize the physical complementary of the two products in terms of spatial coverage. The number of valid CO_2 measurements in the fused product is the union of the CO_2 data from both the ACOS and BESD at the same geographical location. In addition, the combined XCO_2 demonstrates similar spatiotemporal characteristics with that of ACOS and BESD over the globe, which implies that all processes associated with the combination do not distort the essential information of the original XCO_2 products (ACOS or BESD). Similar findings can also be observed in the daily mean and weekly mean XCO_2 maps. To quantitatively investigate the improvement of fused XCO_2 in spatial coverage, the fractional coverage of all three variables

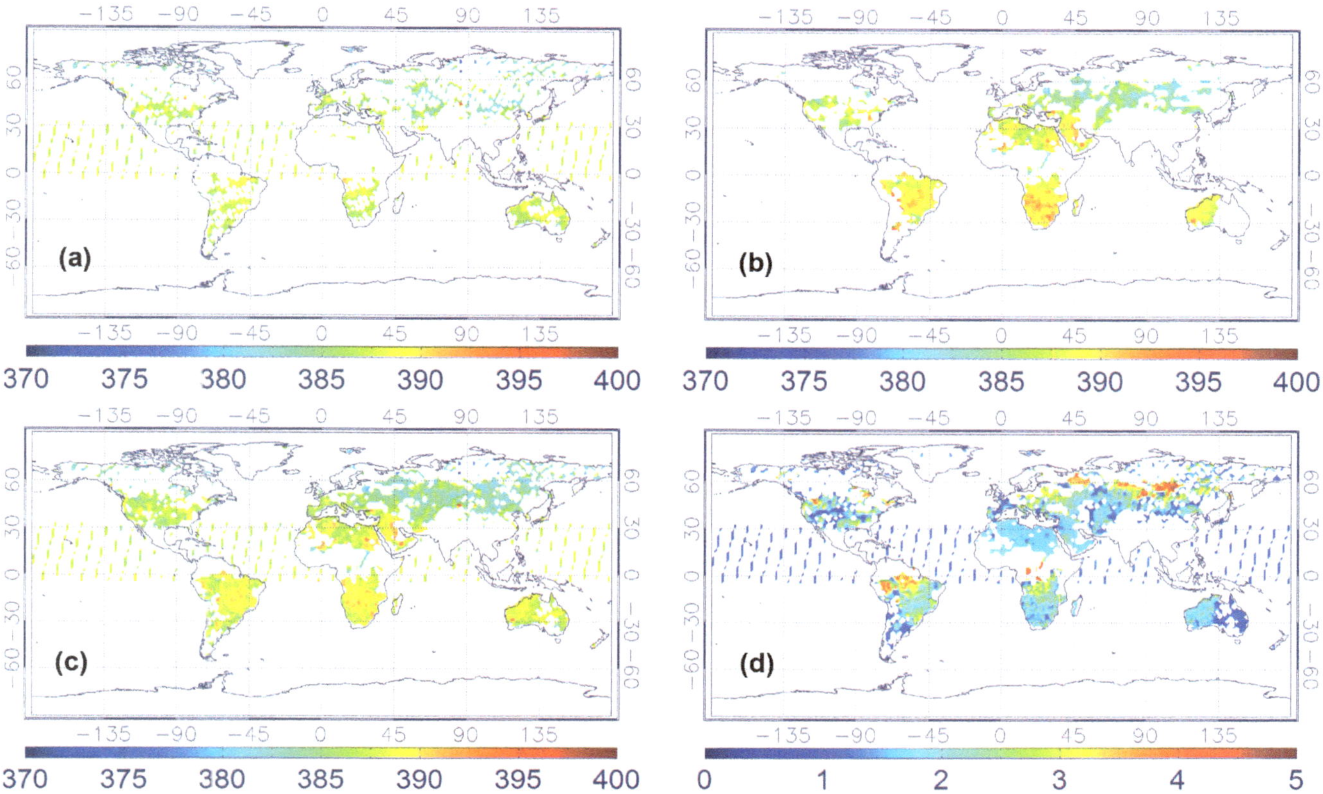

Figure 5. XCO_2 monthly mean maps in August of 2010. ((a) ACOS XCO_2, (b) BESD XCO_2, (c) combined product, and (d) XCO_2 uncertainties of the combined product).

(ACOS, BESD, and combined XCO_2) on both daily and monthly scales is calculated (Fig. 6). From Fig. 6, it can be seen that the average global coverage of ACOS and BESD is around 0.46% and 0.21%, respectively, on a daily scale. The monthly mean coverage of such products accounts for about 5.70% and 3.75%, respectively. While spatial coverage of combined XCO_2 can reach up to 0.65% and 8.42% on daily and monthly scales, respectively, it accounts for increments of 41.3% and 47.7% on the daily and

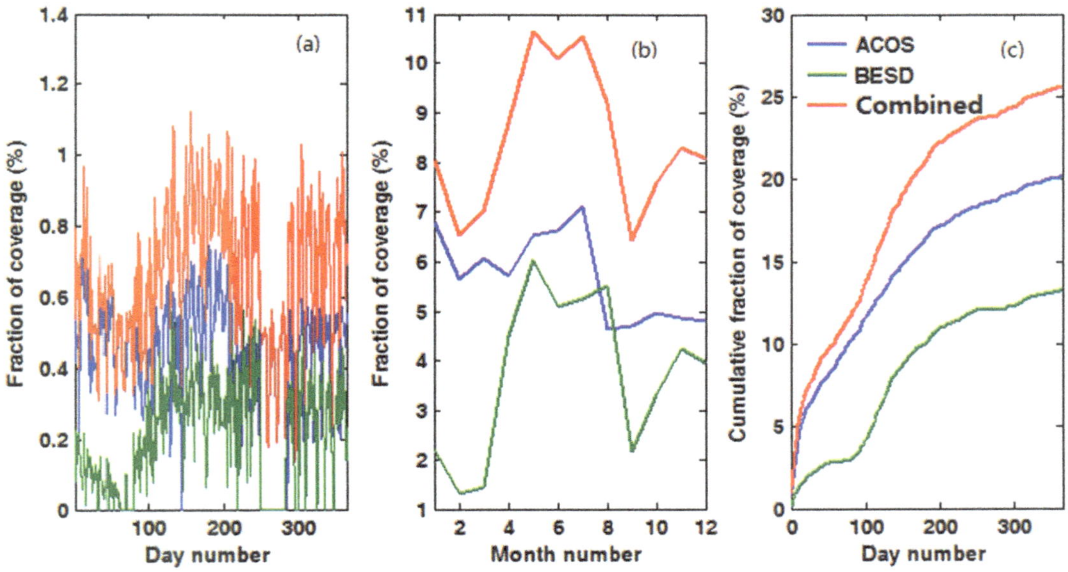

Figure 6. XCO_2 fraction of coverage of ACOS, BESD, and combined products. (a) Daily coverage. (b) Monthly coverage. (c) Cumulative coverage.

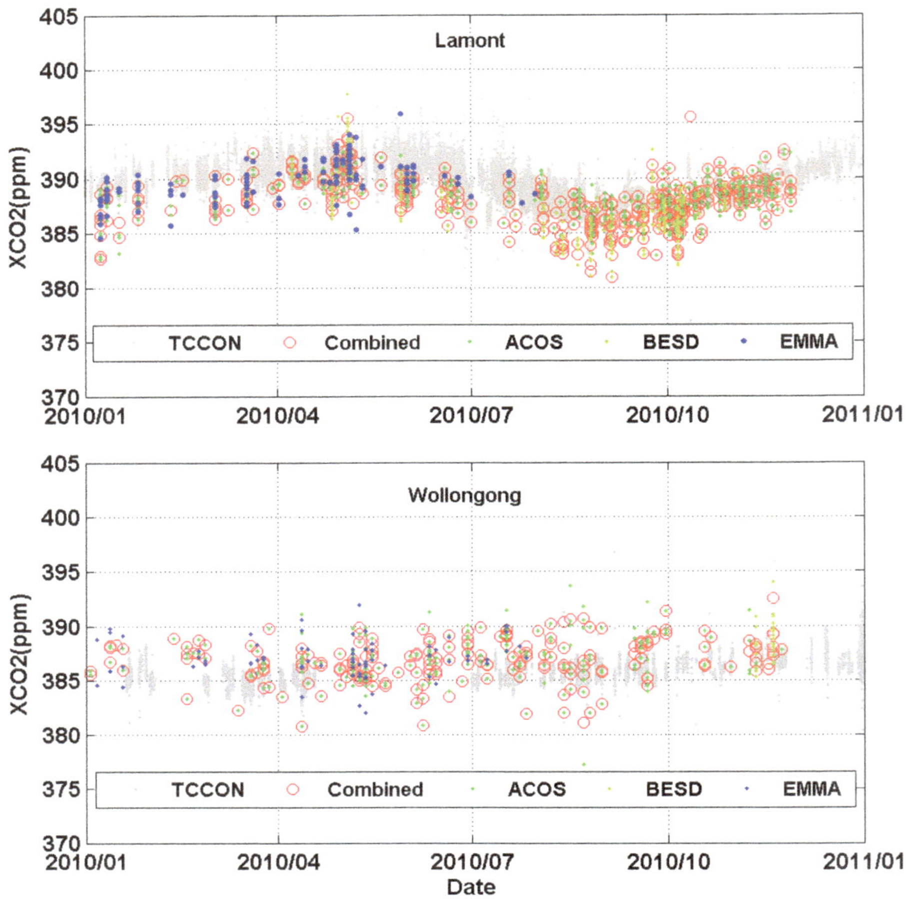

Figure 7. Comparison of XCO_2 measurements from TCCON, ACOS, BESD, EMMA, and our new combination method over Wollongong and Lamont sites (distance<0.25 degree, temporal difference<1 hour).

monthly scales with respect to that of GOSAT and it is even higher relative to the coverage of SCIAMACHY. Likewise, the cumulative fraction of coverage of the combined XCO_2 has risen to 25% when compared with 20% and 13% for ACOS and BESD, respectively. The increase in the XCO_2 spatial coverage indicates the potential advantage of the combined XCO_2 observations in generating global Level-3 XCO_2 maps when compared with any single dataset by providing more satellite-based XCO_2 retrievals used for optimal interpolating.

For evaluating the performance of our combination strategy, the combined XCO_2 values are compared with that retrieved from ACOS and BESD as well as XCO_2 in the EMMA database at two TCCON sites (Fig. 7). The results reveal that the XCO_2 values from the combination method show generally consistent variation in time with TCCON measurements except for a small overall bias (especially for the Lamont site). On the whole, the new combined XCO_2 product shows good consistency with the EMMA data, and they are comparable in terms of CO_2 magnitude, while the combined XCO_2 are shown with a longer time period, which is in line with the satellite observations, and possess more data points even over the same period.

Discussions and Conclusions

Despite the fact that space-based measurements can provide a unique opportunity to map atmospheric CO_2 over large areas, the number of valid CO_2 measurements from a single space-based instrument is generally limited for a certain day over a specific region due to the presence of clouds. In addition, although these Level-2 XCO_2 retrievals themselves are very important for inversion modeling of surface carbon sources/sinks, further comprehensive analysis by investigating the spatiotemporal full-coverage XCO_2 (Level 3) distribution is needed for interpreting their significant scientific merit [32]. While the limited satellite observations restrict the generation of Level-3 XCO_2 maps with high spatial and temporal resolutions when only a single satellite-based XCO_2 dataset is considered. This is our main motivation in this paper.

In this study, a strategy for combining SCIAMACHY and GOSAT CO_2 measurements has been proposed by fully accounting for the CO_2 global bias, differences in averaging kernels and overpass times, and the Level-2 retrieval errors of the CO_2 measurements being used. The results indicated that the average global coverage of both ACOS and BESD is less than 0.5% on a daily scale, and less than 6% on a monthly scale. While spatial coverage of combined XCO_2 can reach up to 0.65% and 8.42% on daily and monthly scales, respectively, the comparison analysis reveals that the combined XCO_2 product is consistent with TCCON and EMMA in both temporal variation and magnitude except for a small bias when compared with the TCCON measurements. All these findings herein prove the effectiveness of the combination method in supporting generation global full-coverage XCO_2 maps with higher temporal and spatial sampling by jointly using two space-based XCO_2 datasets. Similar to the existing studies (e.g. [32–34]), although these combined XCO_2 are not intended to be used in inverse modeling studies, they deliver a key complement for such research, and can be deemed as an independent dataset for comparison with model predictions. Similar to the existing study [31], an improved fusion approach (based on multiple XCO_2 datasets) to create Level-2 XCO_2 measurements that can be directly used for inverse modeling is also attempted and will be presented in another paper.

A last point that needs to be addressed is that although we employed CO_2 data of GOSAT and SCIAMACHY in this study, the proposed strategies are not restricted to such data. As a general strategy, it can be refined and adapted to further combine other XCO_2 products, such as OCO-2, CarbonSat, etc. in the future, and even to be applied to the fusion of other trace gases, such as O_3, CH_4.

Acknowledgments

The authors would like to thank the SCIAMACHY team at University of Bremen IUP/IFE as well as the ACOS scientific teams for providing us the CO_2 products. The authors also thank the anonymous reviewers for their helpful and valuable comments to improve this work.

Author Contributions

Conceived and designed the experiments: TW JS YJ TZ DJ CX. Performed the experiments: TW JS YJ TZ DJ CX. Analyzed the data: TW JS YJ TZ DJ CX. Contributed reagents/materials/analysis tools: TW JS YJ TZ DJ CX. Wrote the paper: TW JS YJ TZ DJ CX.

References

1. Reuter M, Bovensmann H, Buchwitz M, Burrows JP (2010) A method for improved SCIAMACHY CO_2 retrieval in the presence of optically thin clouds. Atmos. Meas. Tech 3: 209–232.
2. O'Dell CW, Connor B, Bösch H, O'Brien D, Frankenberg C, et al. (2012) The ACOS CO_2 retrieval algorithm – Part 1: Description and validation against synthetic observations. Atmos Meas Tech 5(1): 99–121.
3. Baker DF, Law RM, Gurney KR, Rayner P, Peylin P, et al (2006) TransCom 3 inversion intercomparison: Impact of transport model errors on the interannual variability of regional CO_2 fluxes, 1988–2003. Global Biogeochem Cy 20: GB1002, doi:10.1029/2004GB002439.
4. Morino I, Uchino O, Inoue M, Yoshida Y, Yokota T, et al. (2011) Preliminary validation of column-averaged volume mixing ratios of carbon dioxide and methane retrieved from GOSAT short-wavelength infrared spectra. Atmos Meas Tech 4: 1061–1076.
5. Butz A, Hasekamp OP, Frankenberg C, Aben I (2009) Retrievals of atmospheric CO_2 from simulated space-borne measurements of backscattered near-infrared sunlight: accounting for aerosol effects. Appl Opt 18: 3322–3336.
6. Kuze A, Suto H, Nakajima M, Hamazaki T (2009) Thermal and near infrared sensor for carbon observation Fourier-transform spectrometer on the Greenhouse Gases Observing Satellite for greenhouse gases monitoring, Appl Opt 35: 6716–6733.
7. Bovensmann H, Burrows JP, Buchwitz M, Frerick J, Noel S, et al (1999) SCIAMACHY –Mission objectives and measurement modes. J Atmos Sci 56: 127–150.
8. Crisp D, Atlas RM, Breon FM, Brown LR, Burrows JP, et al (2004) The Orbiting Carbon Observatory (OCO) mission. Adv Space Res 34: 700–709.
9. Boesch H, Baker D, Connor B, Crisp D, Miller C (2011) Global Characterization of CO_2 Column Retrievals from Shortwave-Infrared Satellite Observations of the Orbiting Carbon Observatory-2 Mission. Remote Sens 3: 270–304.
10. Aumann HH, Chahine MT, Gautier C, Goldberg MD, Kalnay E, et al. (2003) AIRS/AMSU/HSB on the Aqua Mission: Design, science objectives, data products, and processing systems. IEEE Trans Geosci Remote Sens 41: 253–264.
11. Chahine M, Barnet C, Olsen ET, Chen L, Maddy E (2005) On the determination of atmospheric minor gases by the method of vanishing partial derivatives with application to CO_2. Geophys Res Lett 32: L22803, doi:10.1029/2005GL024165.
12. Phulpin T, Cayla F, Chalon G, Diebel D, Schlussel P (2002) IASI on board Metop: Project status and scientific preparation. Proceedings of the 12th International TOVS Study Conference. Lorne, Vi ctoria, Australia.
13. Turquety S, Hadji-Lazaro J, Clerbaux C, Hauglustaine DA, Clough SA, et al. (2004) Operational trace gas retrieval algorithm for the Infrared Atmospheric Sounding Interferometer. J Geophys Res 109: D21301, doi:10.1029/2004JD004821.
14. Yang Z, Washenfelder RA, Keppel-Aleks G, Krakauer NY, Randerson JT, et al. (2007) New constraints on Northern Hemisphere growing season net flux. Geophys Res Lett 34: L12807. doi:10.1029/2007GL029742.
15. Wunch D, Toon GC, Blavier JFL, Washenfelder R, Notholt J, et al. (2011) The Total Carbon Column Observing Network. Philos T Roy Soc A 369: 2087–2112.
16. Rayner PJ, O'Brien DM (2011) The utility of remotely sensed CO_2 concentration data in surface source inversions. Geophys Res Lett 28: 175–178.
17. Houweling S, Breon FM, Aben I, Rodenbeck C, Gloor M, et al. (2004) Inverse modeling of CO_2 sources and sinks using satellite data: a synthetic intercomparison of measurement techniques and their performance as a function of space and time. Atmos Chem Phys 4: 523–538.
18. Olsen SC, Randerson JT (2004) Differences between surface and column atmospheric CO_2 and implications for carbon cycle research. J Geophys Res 109; D02301. doi:10.1029/2003JD003968.
19. Miller CE, Crisp D, DeCola PL, Olsen SC, Randerson JT, et al. (2007) Precision requirements for space-based XCO_2 data. J Geophys Res 112: D10314, doi:10.1029/2006JD007659.
20. Buchwitz M, Beek R, Burrows JP, Bovensmann H, Warneke T, et al. (2005) Atmospheric methane and carbon dioxide from SCIAMACHY satellite data: initial comparison with chemistry and transport models, Atmos Chem Phys 5: 941–962.
21. Buchwitz M, Beek R, Nol S, Burrows JP, Bovensmann H, et al. (2005) Carbon monoxide, methane and carbon dioxide columns retrieved from SCIAMACHY by WFM-DOAS: year 2003 initial data set. Atmos Chem Phys 5: 3313–3329.
22. Buchwitz M, Rozanov VV, Burrows JP (2000) A near-infrared optimized DOAS method for the fast global retrieval of atmospheric CH_4, CO, CO_2, H_2O, and N_2O total column amounts from SCIAMACHY Envisat-1 nadir radiances. J Geophys Res 105: 15231–15245, doi:10.1029/2000JD900191.
23. Houweling S, Hartmann W, Aben I, Schrijver H, Skidmore J, et al. (2005) Evidence of systematic errors in SCIAMACHY-observed CO_2 due to aerosols. Atmos Chem Phys 5: 3003–3013.
24. Barkley MP, Frieβ U, Monks PS (2006) Measuring atmospheric CO_2 from space using Full Spectral Initiation (FSI) WFM-DOAS, Atmos Chem Phys 6: 3517–3534.
25. Schneising O, Buchwitz M, Burrows JP, Bovensmann H, Reuter M, et al. (2008) Three years of greenhouse gas column-averaged dry air mole fractions retrieved from satellite – Part 1: Carbon dioxide. Atmos Chem Phys 8: 3827–3853.
26. Reuter M, Bovensmann H, Buchwitz M, Burrows JP, Connor BJ, et al. (2011) Retrieval of atmospheric CO_2 with enhanced accuracy and precision from SCIAMACHY: Validation with FTS measurements and com-parison with model results. J Geophys Res 116: D04301, doi:10.1029/2010JD015047.
27. Bösch H, Toon GC, Sen B, Washenfelder RA, Wennberg PO, et al. (2006) Space-based near-infrared CO_2 measurements: Testing the Orbiting Carbon Observatory retrieval algorithm and validation concept using SCIAMACHY observations over Park Falls, Wisconsin. J Geophys Res 111: D23302, doi:10.1029/2006JD007080.
28. Butz A, Guerlet S, Hasekamp O, Schepers D, Galli A, et al. (2011) Toward accurate CO_2 and CH_4 observations from GOSAT. Geophys Res Lett 14: L14812. DOI:10.1029/2011GL047888.
29. Yoshida Y, Ota Y, Eguchi N, Kikuchi N, Nobuta K, et al. (2011) Retrieval algorithm for CO_2 and CH_4 column abundances from short-wavelength infrared spectral observations by the Greenhouse Gases Observing Satellite. Atmos Meas Tech 4: 717–734.
30. Crisp D, Fisher B, O'Dell C, Frankenberg C, Basilio R, et al. (2012)The ACOS CO_2 retrieval algorithm - Part II: Global XCO_2 data characterization. Atmos Meas Tech 5: 687–707.
31. Reuter M, Bösch H, Bovensmann H, Bril A, Buchwitz M, et al. (2013) A joint effort to deliver satellite retrieved atmospheric CO_2 concentrations for surface flux inversions: the ensemble median algorithm EMMA. Atmos Chem Phys 13: 1771–1780.
32. Hammerling DM, Michalak AM, O'Dell C, Kawa SR (2012) Global CO_2 distributions over land from the Greenhouse Gases Observing Satellite(GOSAT). Geophys Res Lett 39: L08804, doi:10.1029/2012GL051203.
33. Hammerling DM, Michalak AM, Kawa SR (2012) Mapping of CO_2 at high spatiotemporal resolution using satellite observations: Global distributions from OCO-2. J Geophys Res 117: D06306, doi:10.1029/2011JD017015.
34. Zeng Z, Lei L, Hou S, Ru F, Guan X, et al. (2014) A Regional Gap-Filling Method Based on Spatiotemporal Variogram Model of CO_2 Columns. IEEE Trans Geosci Remote Sens 6: 3594–3603.
35. Yokota T, Yoshida Y, Eguchi N, Ota Y, Tanaka T, et al. (2009) Global Concentrations of CO_2 and CH_4 Retrieved from GOSAT: First Preliminary Results. SOLA 5: 160–163.
36. Parker R, Boesch H, Cogan A, Fraser A, Feng L, et al. (2011) Methane observations from the Greenhouse Gases Observing SATellite: Comparison to

ground-based TCCON data and model calculations. Geophys Res Lett 15: L15807, doi:10.1029/2011GL047871.
37. Wang TX, Shi JC, Jing YY, Xie YH (2013) Investigation of the consistency of atmospheric CO_2 retrievals from different space-based sensors: Intercomparison and spatio-temporal analysis, Chin Sci Bull, 33: 4161–4170.
38. Peters W, Jacobson AR, Sweeney C, Andrews AE, Conway TJ, et al. (2007) An atmospheric perspective on North American carbon dioxide exchange: Carbon Tracker. Proc Natl Acad Sci U.S.A. 48: 18925–18930.
39. Wunch D, Wennberg PO, Toon GC, Connor BJ, Fisher B, et al. (2011) A method for evaluating bias in global measurements of CO_2 total columns from space. Atmos Chem Phys 11: 12317–12337.
40. Rodgers CD (2000) Inverse Methods for Atmospheric Sounding: Theory and Practice. World Scientific Publishing Co. Ltd.193 p.

15

Predicting Protein Phenotypes Based on Protein-Protein Interaction Network

Lele Hu[1,2]☯, Tao Huang[3,4]☯, Xiao-Jun Liu[5]*, Yu-Dong Cai[1,6]*

1 Institute of Systems Biology, Shanghai University, Shanghai, China, **2** Department of Chemistry, College of Sciences, Shanghai University, Shanghai, China, **3** Key Laboratory of Systems Biology, Shanghai Institutes for Biological Sciences, Chinese Academy of Sciences, Shanghai, China, **4** Shanghai Center for Bioinformation Technology, Shanghai, China, **5** College of Animal Science and Technology, Shihezi University, Shihezi City, Xinjiang, China, **6** Centre for Computational Systems Biology, Fudan University, Shanghai, China

Abstract

Background: Identifying associated phenotypes of proteins is a challenge of the modern genetics since the multifactorial trait often results from contributions of many proteins. Besides the high-through phenotype assays, the computational methods are alternative ways to identify the phenotypes of proteins.

Methodology/Principal Findings: Here, we proposed a new method for predicting protein phenotypes in yeast based on protein-protein interaction network. Instead of only the most likely phenotype, a series of possible phenotypes for the query protein were generated and ranked acording to the tethering potential score. As a result, the first order prediction accuracy of our method achieved 65.4% evaluated by Jackknife test of 1,267 proteins in budding yeast, much higher than the success rate (15.4%) of a random guess. And the likelihood of the first 3 predicted phenotypes including all the real phenotypes of the proteins was 70.6%.

Conclusions/Significance: The candidate phenotypes predicted by our method provided useful clues for the further validation. In addition, the method can be easily applied to the prediction of protein associated phenotypes in other organisms.

Editor: Hitoshi Okazawa, Tokyo Medical and Dental University, Japan

Funding: The authors have no support or funding to report.

Competing Interests: The authors have declared that no competing interests exist.

* E-mail: xiaojun.liu@ed.ac.uk (XJL); cai_yud@yahoo.com.cn (YDC)

☯ These authors contributed equally to this work.

Introduction

Identifying phenotypes of proteins is a central challenge of the modern genetics in post-genome era. The study on phenotypes always involves many major diseases, such as HIV [1,2,3,4,5], different kinds of cancers [6,7,8,9], chronic liver diseases [10], Gaucher disease [11]. The high-throughput phenotype assays [12,13] combining with gene perturbation technology [14,15] provide fast identification for gene active in a response [16]. For example, yeast mutant strain collections has become increasingly used to identify the phenotypes [17]. However, these assays are often trapped in the high false negative rates [18]. On the other hand, the study on phenotypes is highly complex for the multifactorial trait often results from contributions of many proteins. Consequently, using experimental approaches alone is insufficient, and the computational methods should be applied for the identification of protein phenotypes [18].

In principle, there are two kinds of computational methods: the sequence-based methods and network-based methods. A sequence-based method is often designed on a benchmark dataset, sequence features such as amino acid composition [19], pseudo amino acid [20] (PseAAC), are used to represent the data (e.g. protein sequence), then a prediction model can be built according to the machine learning algorithm (e.g. nearest neighbor algorithm). In the past decade, a series of predictors have been designed for phenotype prediction. For example, Resch W *et al.* used a neural network model to identify the phenotype of HIV type 1 from loop 3 sequences [21]. Pillai S *et al.* proposed a classifier based on support vector machine for V3 phenotype prediction [22]. Recently, Onuki R *et al.* also employed a support vector machine method for predicting phenotype from genotype data [23]. With the ever-increasing build-up of high-throughput techniques, biological data acquisition has never increased more rapidly. More and more biological networks, such as gene-regulatory networks and metabolic networks are constructed from multi data sources (e.g. microarrays, literature mining, and protein-protein interaction). Consequently, many network-based methods are proposed to contribute to various aspects of biology, including phenotype prediction. For instance, Keleta C *et al.* implemented the prediction of the 16 different growth phenotypes in E.*coli* based on regulated metabolic networks [24]. McGary KL *et al.* demonstrate that the loss-of function *Saccharomyces cerevisiae* phenotypes are predictable in the functional gene network, and the proposed network-based method succeeded in the identification of yeast orthologs of human disease genes.

In this research, we presented a new network-based method for predicting budding yeast protein phenotypes. Unlike previous methods, our method can rank the possible phenotypes associated with the query protein and shows a more comprehensive view of the protein's biological effects. With the results, we also demonstrated that using protein-protein network is effective for predicting protein phenotypes. Owing to many protein-protein network of other organisms are available, we suggest that this method will be widely applied.

Materials and Methods

Data Set

Because of the complexity of phenotype research, we selected the budding yeast *Saccharomyces cerevisiae* (a well studied model organism [25,26]) as a model system. The protein data used here was taken from CYGD [27] (the MIPS Comprehensive Yeast Genome Database, ftp://ftpmips.gsf.de/yeast/), which dedicated to information on the molecular structure and functional network of the budding yeast. Among the 6,732 proteins of the yeast proteome, only those with both sequence and phenotypic annotations were selected. Thus we obtained 1,460 such proteins belonging to 11 phenotypic categories (see Table S1). The number of proteins in each category was listed in the Table 1, from which we can easily find that the total number of proteins (2,397) in 11 phenotypic categories is much larger than the total number of proteins (1,460). That is because many proteins exhibit more than one phenotype and this is the reason why we developed this method to predict the possible phenotypes with ranked scores, rather than only one predicted phenotype like previous tools.

The yeast protein-protein interaction (PPI) network used here was retrieved from STRING [28] (http://string.embl.de/), whose primary mission is to provide researchers with both physical (direct) and functional (indirect) interactions. For each species, a PPI network is constructed by integrating huge information derived from numerous sources such as experimental repositories, computational methods, and text-mining methods. In the functional protein association network, the interaction unit consists of two nodes (proteins) and an edge between them. The interaction confidence score is used as the edge weight to represent the likelihood that a predicted association exists between two nodes. Weight confidence limits are as follows: low confidence −15% (or better), medium confidence −40%, high confidence −70%, highest confidence −90%. In this research, we chose the highest confidence limit −90% to obtain reliable yeast PPI network (see Table S2), which contains 32,513 functional linkages among 4,209 yeast proteins.

Among the 1,460 proteins with phenotypic annotations, 1,267 proteins could be mapped to the yeast PPI network downloaded from STRING. Thus, the nodes in the network could be grouped into two kinds: those with phenotypic information, others without phenotypic information. Here, we called the protein with phenotypic annotation in the PPI network "seed protein", and the dataset consisting of 1,267 seed proteins "seed set", which were then used to test the network-based method.

Table 1. Breakdown of 1,460 budding yeast proteins according to their 11 phenotypes.

Number	Phenotype category	Number of proteins
1	Conditional phenotypes	536
2	Cell cycle defects	271
3	Mating and sporulation defects	198
4	Auxotrophies, carbon and nitrogen utilization defects	266
5	Cell morphology and organelle mutants	534
6	Stress response defects	147
7	Carbohydrate and lipid biosynthesis defects	46
8	Nucleic acid metabolism defects	218
9	Sensitivity to amino acid analogs and other drugs	124
10	Sensitivity to antibiotics	43
11	Sensitivity to immunosuppressants	14

See the texts of the paper for further explanation.

The availability of using the PPI network to predict protein phenotypes

In the functional network, PPI contains both physical (direct) and functional (indirect) interactions. Physically interacting proteins exist in the same complex, while functional interacting proteins tend to participate in the same pathway or cellular process. Here, we investigated the relationships between complex/pathway and phenotype to explain the availability of using the PPI network to predict protein phenotypes. In order to analyze the relationship conveniently, we selected the proteins with single phenotype. The complex annotation of proteins was also downloaded from CYGD [27], and the pathway annotation of proteins was retrieved from KEGG [29] (Kyoto Encyclopedia of Genes and Genomes) (see Table S3 and Table S4). Totally, these proteins belonged to 733 complexes and 86 pathways. Each protein was coded by the vectors:

$$\begin{cases} V_{complex} = [v_1, v_2, \ldots, v_i, \ldots v_{733}] \\ V_{pathway} = [v_1, v_2, \ldots, v_i, \ldots v_{86}] \end{cases} \quad (1)$$

where $v_i = 1$ if the protein belonged to the *i-th* complex/pathway, otherwise $v_i = 0$. Then *m-th* phenotype can be represented by the protein complex/pathway information as the vector:

$$\overline{V_m} = \frac{\sum_{1}^{n} V_j}{n} \quad (2)$$

where n is the number of proteins that had the *m-th* phenotype. The similarity between any two phenotypes was calculated as:

$$C = \frac{\overline{V_p} \cdot \overline{V_q}}{||\overline{V_p}|| \cdot ||\overline{V_q}||} (p = 1,2,\ldots,11; q = 1,2,\ldots,11; p \neq q) \quad (3)$$

where $\overline{V_p} \cdot \overline{V_q}$ is the vectors' inner product, $||\overline{V}||$ is the module of vector. Generally, two phenotypes are difficult to discriminated from each other using the complex/pathway if the value of the similarity of them is larger than 0.5. Using the protein complex information, the distribution of the similarities of 11 phenotypes was shown in **Figure 1**. Clearly, all the 55 similarities are smaller than 0.5. Because the proteins with the phenotype of sensitivity to immunosuppressants lacked the pathway annotation, the similarities of other 10 phenotypes were calculated using the protein pathway information. The distribution of the similarities of 10 phenotypes was shown in **Figure 2**, where two thirds of the

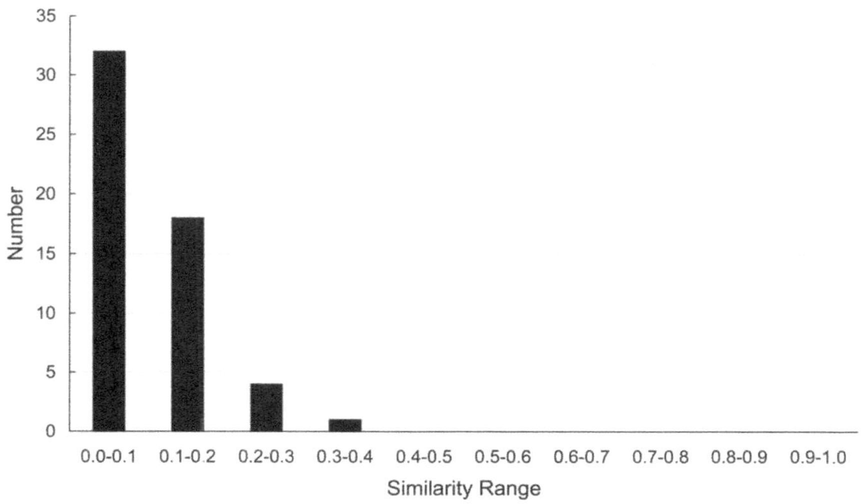

Figure 1. The distribution of the similarities of 11 phenotypes that were represented by protein complex information.

similarities are smaller than 0.5. The results indicate the phenotypes can be classified by using protein complex/pathway information. Therefore, protein phenotypes can be predicted by using the functional PPI network.

Network-based Method

In the PPI network, when we were to predict the phenotypes of a node (protein), just like the weighted vote, not only the number of its neighbor nodes, but also the strengths of interactions (i.e., the edge weights) were considered by the method. The phenotypic categories of each protein in the network can be predicted as following.

First, let us consider the PPI network consisting of n proteins $\{P_1,P_2,...,P_n\}$, in which seed proteins belonged to 11 phenotypic categories ($\mathbb{T}=[T_1,T_2,...,T_{11}]$), where T_1 represents the "Conditional phenotypes" category, T_2 the "Cell cycle defects", T_3 the "Mating and sporulation defects", and so forth (cf. **Table 1**). And the phenotypes of the i-th protein in the network can be denoted by

$$T(P_i)=[t_{i,1},t_{i,2},...,t_{i,j},...,t_{i,11}]^{\mathbf{T}}(i=1,2,...,n;j=1,2,...,11) \quad (4)$$

where

$$t_{i,j}=\begin{cases}1, & \text{if } P_i \text{ belongsto } j\text{-th phenotype}\\ 0, & \text{otherwise}\end{cases} \quad (5)$$

Towards a query protein P_k, its interaction weights with m seed proteins can be defined as follows

$$W(P_k)=[w_{k,1},w_{k,2},...,w_{k,i},...,w_{k,m}]^{\mathbf{T}}(i=1,2,...m) \quad (6)$$

where $w_{k,i}$ is the interaction weight (confidence score [28]) between P_k and the i-th protein in the seed set. If there is no edge between them, $w_{k,i}=0$. Since we did not consider the self-interaction of protein, $w_{k,i}=0$ when $k=i$. Subsequently, we

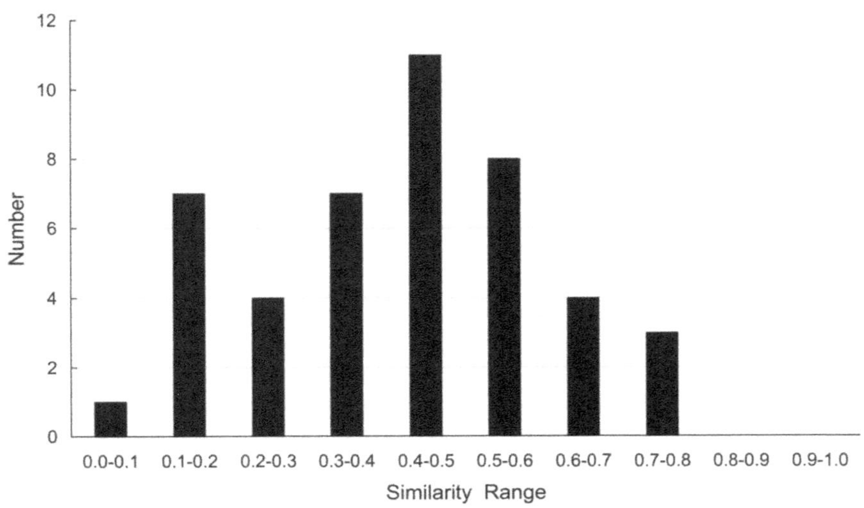

Figure 2. The distribution of the similarities of 10 phenotypes that were represented by protein pathway information.

proposed a new concept called "tethering potential" of protein P_k to the j-th phenotype to reflect the potential of protein P_k belonging to the j-th phenotype, which can be calculated as follows

$$S(P_k \Rightarrow j) = \sum_{i=1}^{m} w_{k,i} t_{i,j} \qquad (j = 1,2,...,11) \tag{7}$$

From this equation, we know that the proteins in seed set without association with the query protein do not contribute to the score of $S(P_k \Rightarrow j)$. Thus the tethering potential of protein P_k to the j-th phenotype can be also described as the sum of interaction weights of it with neighbor proteins of the j-th phenotype in seed set. Obviously, the larger the value of $S(P_k \Rightarrow j)$ is, the more likely the protein P_k belongs to the j-th phenotypic category. Therefore, the most likely phenotype of the query protein P_k can be predicted to belong to the μ-th phenotypic category as follows

$$\mu = \mathbf{arg\ max}_j \{ S(P_k \Rightarrow j) \mid j = 1,2,...,11 \} \tag{8}$$

where μ stands for the argument of j that maximizes the value of $S(P_k \Rightarrow j)$. However, many proteins in yeast give rise to more than one phenotype; the prediction result with only the most likely candidate phenotype is insufficient. In view of this, to make the method able to handle the proteins with multiple phenotypes and benefit biologists with more flexible information in prioritizing candidate phenotypes, we introduced a 11-D (dimensional) vector to reflect the likelihood that the query protein may give rise to each of the 11 phenotypes, which can be formulated as follows

$$\mathbf{D}^{\downarrow} \{ S(P_k \Rightarrow j) \mid j = 1,2,...,11 \} = \mathbf{V} = \begin{bmatrix} \mu_1 \\ \mu_2 \\ \vdots \\ \mu_j \\ \vdots \\ \mu_{10} \\ \mu_{11} \end{bmatrix} \tag{9}$$

where $\mathbf{D}^{\downarrow}$ is a descending operator to sort the 11 scores of $S(P_k \Rightarrow j)$ in descending order. Hence, we have $\mu_1 \geq \mu_2 \geq \cdots \geq \mu_j \geq \cdots \geq \mu_{11}$. Accordingly, if $\mu_1 = S(P_k \Rightarrow 1)$, $\mu_2 = S(P_k \Rightarrow 7)$, $\mu_3 = S(P_k \Rightarrow 5)$, ..., then that the query protein P_k gives rise to the 1st phenotype (Conditional phenotypes) will have the maximum likelihood, that P_k gives rise to the 7th phenotype (Carbohydrate and lipid biosynthesis defects) will have the second maximum likelihood, that P_k gives rise to the 5th phenotype (Cell morphology and organelle mutants) will have the third maximum likelihood, and so forth (cf. **Table 1**). In rare cases, when more than one element of the vector in Eq.6 has the same value, the order will be randomly sorted. Based on the descending order of Eq.6, the predicted results are respectively called the 1st-order predicted result, the 2nd-order predicted result, the 3rd-order predicted result, and so forth.

Jackknife Cross-validation and Evaluation

In statistical prediction, three cross-validation methods are often used to examine the prediction quality: subsampling (K-fold) test, independent dataset test and jackknife test [30]. Among the three methods, jackknife test is regarded as the most objective as discussed in Chou's work [31,32] and has been used more and more frequently to test and evaluate various predictors [33,34,35,36,37,38,39,40,41,42]. In this research, the jackknife cross-validation was also applied to test the network-based method. During the validation, each protein in the seed set is in turn knocked out as a query protein sample, and the remaining proteins of the seed set in the PPI network are used for prediction by the network-based method. Thus, the *i-th* order prediction accuracy Q_i can be calculated as follows

$$Q_i = \frac{\sum_{j=1}^{11} M_{ij}}{N} (i = 1,2,...,11) \tag{10}$$

Where M_{ij} is the number of correctly predicted proteins of the *j-th* phenotypic category in the seed set, and N is the total number of proteins in the seed set. Finally, the 11-order prediction accuracies are obtained to evaluate the network-based method. The large Q_i with a small i and the small Q_i with a large i imply a good performance of the method.

The average number of phenotypes that each protein in the network exhibits can be calculated as follows

$$N = \sum_{i=1}^{11} Q_i \tag{11}$$

Therefore, another evaluation for the network-based method was proposed as the likelihood that the first *r-order* predicted results include all the phenotypes of proteins, which can be calculated as follows

$$P_r = \frac{\sum_{i=1}^{r} Q_r}{\sum_{i=1}^{11} Q_i} \tag{12}$$

A large P_r accompanied with a small r also implies a good performance of the method for the protein phenotype prediction.

Prediction

Besides the seed proteins, there are also 2,942 proteins in the PPI network. The tethering potential of such protein to the each phenotype can be calculated according to Eq. (7) and then ranked in descending order. In this manner, the phenotypes of these proteins can be predicted by the network-based method.

Results and Discussion

Performance of Network-based method

Through leave-one-out cross-validation, the overall 11-order success rates by the network-based method on the aforementioned 1,267 seed proteins are listed in **Table 2**. As we can see from the table, the most likely (first-order) prediction accuracy is 65.4%, and the least likely (last-order) one is 3.39%. The former minus the latter equals 61%. Based on the prediction criteria, the bigger the difference value is, the better the method performs. According to **Table 2**, a downward-slope curve is

Table 2. The leave-one-out cross-validation (Jackknife test) success rates by a random guess and the network-based method.

Most likely category							
	Order	1	2	3	4	5	6
Random Guess	Accuracy (%)	15.5	15.5	15.5	15.5	15.5	15.5
Network-based Method		65.4	34.1	20.7	13.3	8.76	6.47
Least likely category							
	Order	7	8	9	10	11	
Random Guess	Accuracy (%)	15.5	15.5	15.5	15.5	15.5	
Network-based Method		5.84	5.21	3.47	3.39	3.39	

drawn in the **Figure 3**, showing that higher-order phenotype prediction is better than the lower-order one. This is the exact phenomenon that we want to see, and it may imply that the predicted phenotypic categories of proteins are well arranged by the method according to the prediction criteria.

The average number of phenotypes that each seed protein has is 1.7 according to Eq. (11). The chance that a random guess of a protein phenotype will succeed is 1.7/11 = 15.4%, much lower than the first order prediction success rate. As is shown in the **Table 2**, the first 3 prediction accuracies are larger than the success rates of random guess. And the likelihood of the first 3-order predicted results including the phenotypic categories of the proteins in seed set is 70.6% according to the Eq. (12). These results may imply that our method performs well in the prediction of protein phenotypes in budding yeast.

In genetics, mutations that cause the same phenotype are inferred to functionally associated, and vice versa [18]. Phenotype is a multifactorial trait that often results from the contribution of many proteins. Because the interacting proteins are often in the same complex or pathway, it is rational to expect that interacting proteins often share the common phenotypes. For example, the interactions of seed protein YBR039W with the other seed proteins are listed in **Table 3**. The complex information about those proteins is retrieved from CYGD [27]. We can easily see that protein YBR039W and its neighbors YBL099W, YDL004W, YDR298C, YLR295C, YML081C-A, YPL078C, YPL271W are members of the same F0/F1 ATP synthase (complex V) complex. Additionally, proteins YDR298C and YPL078C are also members of complex in [43], and protein YPR024W is component of Yme1 protease complex. And these proteins share the common phenotype auxotrophies, carbon and nitrogen utilization defects. Therefore, when protein YBR039W is predicted as a test sample by the method, the first candidate phenotype will be assigned its real phenotype. For another example, the interactions of seed protein YDL028C with the other seed proteins are listed in **Table 4**. The information of pathways that yeast proteins participate in is retrieved from Kyoto Encyclopedia of Genes and Genomes [44] (KEGG). Except proteins YDR168W, YKL042W, YPL209C with no pathway annotation, proteins YDL028C, YBL084C, YGL116W, YGR113W, YGR188C, YIL106W, YKL022C, YMR055C, YOR026W involve in the same pathway sce04111 (Cell cycle in budding yeast). The loss-of-function of any one of these 9 proteins likely disrupts the mitotic cell cycle progression and lead to cell cycle defects. Based on the interactions listed in the table, we can arrange the first, second candidate phenotype of protein YDL028C as the cell cycle defects, cell morphology and organelle mutants respectively according to the prediction criteria. The correct phenotype predictions of proteins YBR039W and YDL028C support the hypothesis that the functional associated proteins often share the same phenotypes. Therefore, the protein phenotypes can be predicted from the phenotypes of its interacting proteins by the method.

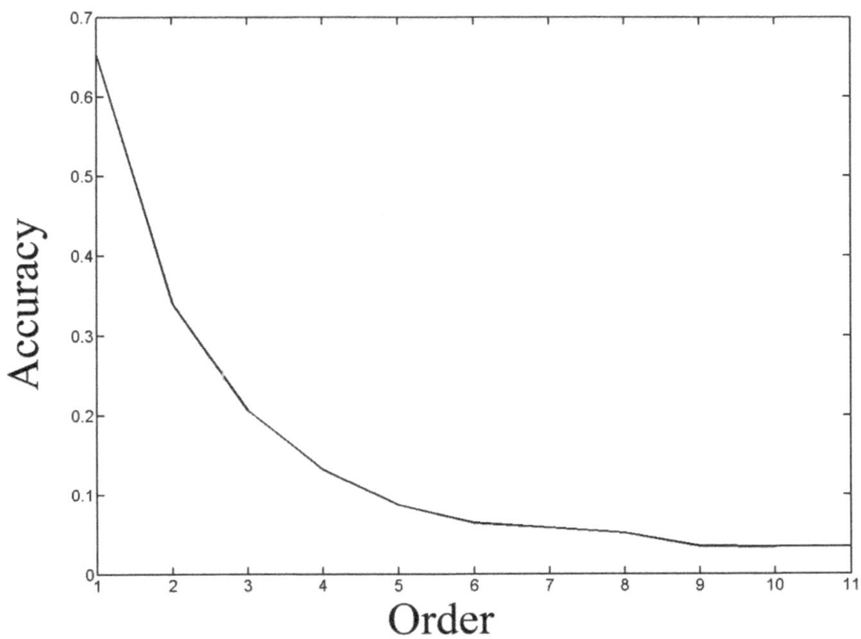

Figure 3. A downward-slope curve to show the relations among the different order prediction accuracies.

Table 3. Interactions of protein YBR039W with its neighbor proteins.

Protein A	Phenotype	Complex	Protein B	Phenotype	Complex	Weight
YBR039W	P1	C1	YBL099W	P1	C1	999
YBR039W	P1	C1	YDL004W	P1	C1	999
YBR039W	P1	C1	YDR298C	P1	C1; C2	999
YBR039W	P1	C1	YLR295C	P1	C1	917
YBR039W	P1	C1	YML081C-A	P1; P2	C1	934
YBR039W	P1	C1	YPL078C	P1	C1; C2	999
YBR039W	P1	C1	YPL271W	P1	C1	997
YBR039W	P1	C1	YPR024W	P1; P2; P3	C3	986

C1 represents F0/F1 ATP synthase (complex V), C2 represents Complex in study [38], C3 represents Yme1 protease complex, P1 represents Auxotrophies, carbon and nitrogen utilization defects, P2 represents Cell morphology and organelle mutants, P3 represents Conditional phenotypes.

Protein phenotype prediction with inactivating its interacting protein

Here, we discuss the robustness of our method by applying the method to the proteins whose interacting proteins are inactivated. First, we chose a protein and took away one of its interacting proteins from the PPI network. Then the phenotype of the protein was predicted by the method based on the broken PPI network. In this way, the phenotypes of 6 proteins were predicted, as shown in **Table 5**. The phenotypes predicted from the unbroken network and the recent phenotype studies focusing on these proteins are also listed in **Table 5**. We found that the phenotypes predicted from the broken network were different from the phenotypes predicted from the unbroken network, while the proteins were verified to have these new phenotypes predicted from broken network in the recent studies. For example, with protein YOR196C in the network, the 1st order predicted phenotype of protein YER178W by the method is "auxotrophies, carbon and nitrogen utilization defects", which is the same as the annotation from CYGD [27]. After inactivating protein YOR196C, the phenotype of protein YER178W is predicted as the "conditional phenotypes". In the study [45], protein YER178W was reported to have the phenotype-"Heat sensitivity: increased", which is one kind of "conditional phenotypes" according to the phenotype classification in CYGD. In the table, the new phenotypes of other proteins predicted from the broken network can also be supported by the literatures [13,46,47,48,49,50]. The examples listed in the table indicate that our method may provide new phenotypes for proteins and serve as a complementary tool for the existing resources.

Application and improvement

As is discussed above, the first 3-order predicted results (approximately double the average number of phenotypes 1.7) can be considered as the candidate phenotypes of the proteins concerned by the biologists. Genetic experiments can focus on these candidate phenotypes of the proteins, which may accelerate the research progress and decrease the cost. At least, the last three predicted phenotypes can be excluded because the last 3-order prediction accuracies are lower than 5% (See **Table 2**).

The effectiveness of the functional network for predicting phenotypes of proteins in yeast suggests the possibility of application to other species. The method is based on the functional protein association network. Besides an abundance of such networks in STRING [28] (Version 8.0 of STRING covered 630 networks of different organisms), the PPI networks can also derived from worm PPI database [51], fly database [52], human

Table 4. Interactions of protein YDL028C with its neighbor proteins.

Protein A	Phenotype	Pathway	Protein B	Phenotype	Pathway	Weight
YDL028C	P4; P5	sce04111	YBL084C	P4; P6	sce04111; sce04113; sce04120	929
YDL028C	P4; P5	sce04111	YDR168W	P4	no annotation	999
YDL028C	P4; P5	sce04111	YGL116W	P4	sce04111; sce04113; sce04120	956
YDL028C	P4; P5	sce04111	YGR113W	P4; P5	sce04111	999
YDL028C	P4; P5	sce04111	YGR188C	P5	sce04111; sce04113	999
YDL028C	P4; P5	sce04111	YIL106W	P4; P5	sce04111	988
YDL028C	P4; P5	sce04111	YKL022C	P4; P7	sce04111; sce04113; sce04120	929
YDL028C	P4; P5	sce04111	YKL042W	P4	no annotation	990
YDL028C	P4; P5	sce04111	YMR055C	P4	sce04111	984
YDL028C	P4; P5	sce04111	YOR026W	P4; P7	sce04111	978
YDL028C	P4; P5	sce04111	YPL209C	P4; P5; P7	no annotation	984

P4 represents Cell cycle defects, P5 represents Cell morphology and organelle mutants, P6 represents Nucleic acid metabolism defects, P7 represents Conditional phenotypes, Sce04111 represents cell cycle pathway in budding yeast.

Table 5. Phenotypes of proteins predicted by our method with/without inactivating its interacting protein.

Protein	Phenotype from CYGD [27]	Phenotype predicted by our method without inactivating the interacting protein	Inactivated interacting protein	Phenotype predicted by our method with inactivating the interacting protein	Phenotype from literatures
YER178W	Auxotrophies, carbon and nitrogen utilization defects	Auxotrophies, carbon and nitrogen utilization defects	YOR196C	Conditional phenotypes	Heat sensitivity: increased [45]
YML035C	Conditional phenotypes	Conditional phenotypes	YDR226W	Cell morphology and organelle mutants	Toxin resistance: increased [46]
YMR198W	Cell cycle defects	Cell cycle defects	YPR141C	Cell morphology and organelle mutants	Bud morphology: abnormal [47]
YOR254C	Conditional phenotypes Cell cycle defects Mating and sporulation defects	Conditional phenotypes	YKL073W	Cell morphology and organelle mutants	Mitochondrial morphology: abnormal [48] Telomere length: increased [49]
YDL198C	Conditional phenotypes	Conditional phenotypes	YPL240C	Auxotrophies, carbon and nitrogen utilization defects	Utilization of nitrogen source: absent [50] Utilization of carbon source: decreased [13]
YPR166C	Auxotrophies, carbon and nitrogen utilization defects Cell morphology and organelle mutants	Auxotrophies, carbon and nitrogen utilization defects	YHR147C	Conditional phenotypes	Heat sensitivity: increased [45]

PPI database [53,54,55], and so on. When in possession a series of proteins with known phenotypes, one can predict the possible phenotypes of other proteins in the networks. Therefore, the method can be easily applied to the prediction of protein phenotypes in other organisms, especially model organisms.

The performance of our method can be improved if the following problems are solved. First, increase the quality of PPI network and exclude the false positive interaction; currently we used high confidence score cutoff to filter the network (See section Data Set). Second, proteins in the same complex or pathway may exert opposite effects on a phenotype, playing as actors or repressors [18]. If the network can discriminate the positive or negative regulation, our method can be modified and the performance will be improved. Third, the performance of the network-based method depends on the number of seed proteins. This problem can be solved in future when the phenotypes of more proteins are investigated. In summary, identification of protein phenotypes is an extremely complicated work and there is a long way to go.

Conclusion

In this research, we proposed a multi-target model [40] to predict phenotypes of proteins in budding yeast based on the protein-protein network. Because some proteins can give rise to more than one phenotype, rather than the most likely phenotype, a series of candidate phenotypes are predicted for each protein. With the performance of the method, it is anticipated that the promising approach may serve as a useful tool for annotating the phenotypes for uncharacterized protein sequences.

Supporting Information

Table S1 The 1,460 proteins with both sequence and phenotype information retrieved from CYGD (the Comprehensive Yeast Genome Database) (Guldener U, Munsterkotter M, Kastenmuller G, Strack N, van Helden J, et al. (2005) CYGD: the Comprehensive Yeast Genome Database. Nucleic acids research 33: D364-368.). The corresponding phenotype of the phenotype number can be found in Table 1.

Table S2 Yeast Protein-Protein Interaction Network. The protein-protein interaction were downloaded from STRING (Search Tool for the Retrieval of Interacting Genes/Proteins) (Jensen LJ, Kuhn M, Stark M, Chaffron S, Creevey C, et al. (2009) STRING 8–a global view on proteins and their functional interactions in 630 organisms. Nucleic acids research 37: D412-416.) with the highest confidence, i.e., the confidence scores are not less than 900.

Table S3 The proteins and the complexes they belong to in yeast. The information was retrieved from CYGD (the Comprehensive Yeast Genome Database) (Guldener U, Munsterkotter M, Kastenmuller G, Strack N, van Helden J, et al. (2005) CYGD: the Comprehensive Yeast Genome Database. Nucleic acids research 33: D364-368.).

Table S4 The proteins and the pathways they belong to in yeast. The information was retrieved from KEGG (Kyoto Encyclopedia of Genes and Genomes) (Kanehisa M, Goto S, Hattori M, Aoki-Kinoshita KF, Itoh M, et al. (2006) From genomics to chemical genomics: new developments in KEGG. Nucleic Acids Res 34: D354-357.).

Acknowledgments

We thank CYGD and STRING for supply data to support not-for-profit research efforts.

Author Contributions

Conceived and designed the experiments: LH YDC. Performed the experiments: LH TH. Analyzed the data: LH XJL YDC. Contributed reagents/materials/analysis tools: LH XJL. Wrote the paper: LH TH YDC.

References

1. Van Houtte M, Picchio G, Van Der Borght K, Pattery T, Lecocq P, et al. (2009) A comparison of HIV-1 drug susceptibility as provided by conventional phenotyping and by a phenotype prediction tool based on viral genotype. Journal of medical virology 81: 1702–1709.
2. Vasil'ev AV, Kazennova EV, Bobkova MR (2009) [Prediction of phenotype R5/X4 of HIV-1 variants circulating in Russia, by using computer methods]. Voprosy virusologii 54: 17–21.
3. Xu S, Huang X, Xu H, Zhang C (2007) Improved prediction of coreceptor usage and phenotype of HIV-1 based on combined features of V3 loop sequence using random forest. Journal of microbiology 45: 441–446.
4. Vermeiren H, Van Craenenbroeck E, Alen P, Bacheler L, Picchio G, et al. (2007) Prediction of HIV-1 drug susceptibility phenotype from the viral genotype using linear regression modeling. Journal of virological methods 145: 47–55.
5. Foulkes AS, De GV (2002) Characterizing the relationship between HIV-1 genotype and phenotype: prediction-based classification. Biometrics 58: 145–156.
6. Lin TY, Chang JT, Wang HM, Chan SH, Chiu CC, et al. (2010) Proteomics of the Radioresistant Phenotype in Head-and-Neck Cancer: Gp96 as a Novel Prediction Marker and Sensitizing Target for Radiotherapy. International journal of radiation oncology, biology, physics 78: 246–256.
7. Bathen TF, Jensen LR, Sitter B, Fjosne HE, Halgunset J, et al. (2007) MR-determined metabolic phenotype of breast cancer in prediction of lymphatic spread, grade, and hormone status. Breast cancer research and treatment 104: 181–189.
8. Lakhani SR, Reis-Filho JS, Fulford L, Penault-Llorca F, van der Vijver M, et al. (2005) Prediction of BRCA1 status in patients with breast cancer using estrogen receptor and basal phenotype. Clinical cancer research: an official journal of the American Association for Cancer Research 11: 5175–5180.
9. Dwyer T, Stankovich JM, Blizzard L, FitzGerald LM, Dickinson JL, et al. (2004) Does the addition of information on genotype improve prediction of the risk of melanoma and nonmelanoma skin cancer beyond that obtained from skin phenotype? American journal of epidemiology 159: 826–833.
10. Piruzyan LA, Korshunov IB, Morozova NV, Pyn'ko NE, Radkevich LA (2004) Prediction of chronic liver diseases on the basis of the N-acetyltransferase 2 phenotype. Doklady Biochemistry and biophysics 395: 84–87.
11. Whitfield PD, Nelson P, Sharp PC, Bindloss CA, Dean C, et al. (2002) Correlation among genotype, phenotype, and biochemical markers in Gaucher disease: implications for the prediction of disease severity. Molecular genetics and metabolism 75: 46–55.
12. Drees BL, Thorsson V, Carter GW, Rives AW, Raymond MZ, et al. (2005) Derivation of genetic interaction networks from quantitative phenotype data. Genome biology 6: R38.
13. Dudley AM, Janse DM, Tanay A, Shamir R, Church GM (2005) A global view of pleiotropy and phenotypically derived gene function in yeast. Molecular systems biology 1: 2005 0001.
14. Fire A, Xu S, Montgomery MK, Kostas SA, Driver SE, et al. (1998) Potent and specific genetic interference by double-stranded RNA in Caenorhabditis elegans. Nature 391: 806–811.
15. Winzeler EA, Liang H, Shoemaker DD, Davis RW (2000) Functional analysis of the yeast genome by precise deletion and parallel phenotypic characterization. Novartis Foundation symposium 229: 105–109; discussion 109-111.
16. Carter GW, Prinz S, Neou C, Shelby JP, Marzolf B, et al. (2007) Prediction of phenotype and gene expression for combinations of mutations. Molecular systems biology 3: 96.
17. Scherens B, Goffeau A (2004) The uses of genome-wide yeast mutant collections. Genome biology 5: 229.
18. McGary KL, Lee I, Marcotte EM (2007) Broad network-based predictability of Saccharomyces cerevisiae gene loss-of-function phenotypes. Genome biology 8: R258.
19. Cedano J, Aloy P, Perez-Pons JA, Querol E (1997) Relation between amino acid composition and cellular location of proteins. Journal of molecular biology 266: 594–600.
20. Chou KC (2001) Prediction of protein cellular attributes using pseudo-amino acid composition. Proteins 43: 246–255.
21. Resch W, Hoffman N, Swanstrom R (2001) Improved success of phenotype prediction of the human immunodeficiency virus type 1 from envelope variable loop 3 sequence using neural networks. Virology 288: 51–62.
22. Pillai S, Good B, Richman D, Corbeil J (2003) A new perspective on V3 phenotype prediction. AIDS research and human retroviruses 19: 145–149.
23. Onuki R, Shibuya T, Kanehisa M (2010) New kernel methods for phenotype prediction from genotype data. Genome informatics International Conference on Genome Informatics 22: 132–141.
24. Kaleta C, Centler F, di Fenizio PS, Dittrich P (2008) Phenotype prediction in regulated metabolic networks. BMC systems biology 2: 37.
25. Gimeno CJ, Ljungdahl PO, Styles CA, Fink GR (1992) Unipolar cell divisions in the yeast S. cerevisiae lead to filamentous growth: regulation by starvation and RAS. Cell 68: 1077–1090.
26. Lengeler KB, Davidson RC, D'Souza C, Harashima T, Shen WC, et al. (2000) Signal transduction cascades regulating fungal development and virulence. Microbiology and molecular biology reviews: MMBR 64: 746–785.
27. Guldener U, Munsterkotter M, Kastenmuller G, Strack N, van Helden J, et al. (2005) CYGD: the Comprehensive Yeast Genome Database. Nucleic acids research 33: D364–368.
28. Jensen LJ, Kuhn M, Stark M, Chaffron S, Creevey C, et al. (2009) STRING 8–a global view on proteins and their functional interactions in 630 organisms. Nucleic acids research 37: D412–416.
29. Kanehisa M, Goto S, Hattori M, Aoki-Kinoshita KF, Itoh M, et al. (2006) From genomics to chemical genomics: new developments in KEGG. Nucleic acids research 34: D354–357.
30. Chou KC, Zhang CT (1995) Prediction of protein structural classes. Critical reviews in biochemistry and molecular biology 30: 275–349.
31. Chou KC, Shen HB (2008) Cell-PLoc: a package of Web servers for predicting subcellular localization of proteins in various organisms. Nature protocols 3: 153–162.
32. Chou KC, Shen HB (2007) Recent progress in protein subcellular location prediction. Analytical biochemistry 370: 1–16.
33. Cai Y, He J, Li X, Lu L, Yang X, et al. (2009) A novel computational approach to predict transcription factor DNA binding preference. Journal of proteome research 8: 999–1003.
34. Afjehi-Sadat L, Lubec G (2007) Identification of enzymes and activity from two-dimensional gel electrophoresis. Nature protocols 2: 2318–2324.
35. Chen C, Chen L, Zou X, Cai P (2009) Prediction of protein secondary structure content by using the concept of Chou's pseudo amino acid composition and support vector machine. Protein and peptide letters 16: 27–31.
36. Zeng YH, Guo YZ, Xiao RQ, Yang L, Yu LZ, et al. (2009) Using the augmented Chou's pseudo amino acid composition for predicting protein submitochondria locations based on auto covariance approach. Journal of theoretical biology 259: 366–372.
37. Ding H, Luo L, Lin H (2009) Prediction of cell wall lytic enzymes using Chou's amphiphilic pseudo amino acid composition. Protein and peptide letters 16: 351–355.
38. Zhou XB, Chen C, Li ZC, Zou XY (2007) Using Chou's amphiphilic pseudo-amino acid composition and support vector machine for prediction of enzyme subfamily classes. Journal of theoretical biology 248: 546–551.
39. Huang T, Tu K, Shyr Y, Wei CC, Xie L, et al. (2008) The prediction of interferon treatment effects based on time series microarray gene expression profiles. J Transl Med 6: 44.
40. Huang T, Cui W, Hu L, Feng K, Li YX, et al. (2009) Prediction of pharmacological and xenobiotic responses to drugs based on time course gene expression profiles. PLoS ONE 4: e8126.
41. Huang T, Shi XH, Wang P, He Z, Feng KY, et al. (2010) Analysis and prediction of the metabolic stability of proteins based on their sequential features, subcellular locations and interaction networks. PLoS ONE 5: e10972.
42. Huang T, Wang P, Ye ZQ, Xu H, He Z, et al. (2010) Prediction of Deleterious Non-Synonymous SNPs Based on Protein Interaction Network and Hybrid Properties. PLoS ONE 5: e11900.
43. Ho Y, Gruhler A, Heilbut A, Bader GD, Moore L, et al. (2002) Systematic identification of protein complexes in Saccharomyces cerevisiae by mass spectrometry. Nature 415: 180–183.
44. Kanehisa M, Goto S (2000) KEGG: kyoto encyclopedia of genes and genomes. Nucleic acids research 28: 27–30.
45. Sinha H, David L, Pascon RC, Clauder-Munster S, Krishnakumar S, et al. (2008) Sequential elimination of major-effect contributors identifies additional quantitative trait loci conditioning high-temperature growth in yeast. Genetics 180: 1661–1670.
46. Morton CO, Dos Santos SC, Coote P (2007) An amphibian-derived, cationic, alpha-helical antimicrobial peptide kills yeast by caspase-independent but AIF-dependent programmed cell death. Molecular microbiology 65: 494–507.
47. Watanabe M, Watanabe D, Nogami S, Morishita S, Ohya Y (2009) Comprehensive and quantitative analysis of yeast deletion mutants defective in apical and isotropic bud growth. Current genetics 55: 365–380.
48. Altmann K, Westermann B (2005) Role of essential genes in mitochondrial morphogenesis in Saccharomyces cerevisiae. Molecular biology of the cell 16: 5410–5417.
49. Ungar L, Yosef N, Sela Y, Sharan R, Ruppin E, et al. (2009) A genome-wide screen for essential yeast genes that affect telomere length maintenance. Nucleic acids research 37: 3840–3849.
50. Cai H, Kauffman S, Naider F, Becker JM (2006) Genomewide screen reveals a wide regulatory network for di/tripeptide utilization in Saccharomyces cerevisiae. Genetics 172: 1459–1476.
51. Li S, Armstrong CM, Bertin N, Ge H, Milstein S, et al. (2004) A map of the interactome network of the metazoan C. elegans. Science 303: 540–543.
52. Giot L, Bader JS, Brouwer C, Chaudhuri A, Kuang B, et al. (2003) A protein interaction map of Drosophila melanogaster. Science 302: 1727–1736.
53. Rual JF, Venkatesan K, Hao T, Hirozane-Kishikawa T, Dricot A, et al. (2005) Towards a proteome-scale map of the human protein-protein interaction network. Nature 437: 1173–1178.
54. Lehner B, Fraser AG (2004) A first-draft human protein-interaction map. Genome biology 5: R63.
55. Stelzl U, Worm U, Lalowski M, Haenig C, Brembeck FH, et al. (2005) A human protein-protein interaction network: a resource for annotating the proteome. Cell 122: 957–968.

16

Climate Change Affects Winter Chill for Temperate Fruit and Nut Trees

Eike Luedeling[1]*, Evan H. Girvetz[2], Mikhail A. Semenov[3], Patrick H. Brown[4]

1 World Agroforestry Centre (ICRAF), Nairobi, Kenya, **2** The Nature Conservancy, Seattle, Washington, United States of America, **3** Rothamsted Research, Harpenden, United Kingdom, **4** Department of Plant Sciences, University of California Davis, Davis, California, United States of America

Abstract

Background: Temperate fruit and nut trees require adequate winter chill to produce economically viable yields. Global warming has the potential to reduce available winter chill and greatly impact crop yields.

Methodology/Principal Findings: We estimated winter chill for two past (1975 and 2000) and 18 future scenarios (mid and end 21st century; 3 Global Climate Models [GCMs]; 3 greenhouse gas emissions [GHG] scenarios). For 4,293 weather stations around the world and GCM projections, Safe Winter Chill (SWC), the amount of winter chill that is exceeded in 90% of all years, was estimated for all scenarios using the "Dynamic Model" and interpolated globally. We found that SWC ranged between 0 and about 170 Chill Portions (CP) for all climate scenarios, but that the global distribution varied across scenarios. Warm regions are likely to experience severe reductions in available winter chill, potentially threatening production there. In contrast, SWC in most temperate growing regions is likely to remain relatively unchanged, and cold regions may even see an increase in SWC. Climate change impacts on SWC differed quantitatively among GCMs and GHG scenarios, with the highest GHG leading to losses up to 40 CP in warm regions, compared to 20 CP for the lowest GHG.

Conclusions/Significance: The extent of projected changes in winter chill in many major growing regions of fruits and nuts indicates that growers of these commodities will likely experience problems in the future. Mitigation of climate change through reductions in greenhouse gas emissions can help reduce the impacts, however, adaption to changes will have to occur. To better prepare for likely impacts of climate change, efforts should be undertaken to breed tree cultivars for lower chilling requirements, to develop tools to cope with insufficient winter chill, and to better understand the temperature responses of tree crops.

Editor: Anna Traveset, Institut Mediterrani d'Estudis Avançats (CSIC/UIB), Spain

Funding: Rothamsted Research receives grant-aided support from the Biotechnology and Biological Sciences Research Council of the UK (BBSRC). The funders had no role in study design, data collection and analysis, decision to publish, or preparation of the manuscript.

Competing Interests: The authors have declared that no competing interests exist.

* E-mail: e.luedeling@cgiar.org

Introduction

Commercially successful cultivation of many fruit and nut trees requires the fulfillment of a winter chilling requirement, which is specific for every tree cultivar [1,2,3,4]. In order to avoid frost damage of sensitive tissue in the cold winters of their regions of origin, trees from temperate or cold climates evolved a period of dormancy during the cold season. After a certain duration of cold conditions (chilling), endodormancy is broken and the tree is ready to resume growth in spring. Chilling requirements vary substantially between species and cultivars from different parts of the world and commercial production of temperate tree crops requires selecting appropriate cultivars for the climatic conditions of the planned production site.

Climate change is likely to affect future winter chill and could have a major impact on the US$ 93 billion global fruit and nut industry (only species with chilling requirements, production statistics for 2005 from ref. [5], currencies converted into 2005 US$ according to ref. [6]). Temperatures are expected to increase in most parts of the world, with minimum temperatures rising most rapidly. This development may compromise the ability of many growers of temperate fruits and nuts to successfully produce the same array of crops as in the past. Climate change effects on winter chill have recently been analyzed for California [7,8], Germany [9] and high-mountain oases in Oman [10]. While conditions in Germany were relatively stable during the 20th century, winter chill was found to have declined in California and Oman, and this process was expected to continue in the future. The differences between these studies indicate that different growing regions may be differentially impacted, but to date, no estimates are available at a global scale to indicate which regions will maintain adequate winter chill for temperate fruits and nuts in the future. This study aims to fill this knowledge gap and to provide important information needed to evaluate the future viability of fruit and nut growing regions around the world.

Several models have been developed for quantifying winter chill, e.g. the Chilling Hours Model [11], the Utah Model [12] and the Dynamic Model [13,14]. These models differ greatly in their sensitivity to climate change [15], making the choice of the model a crucial determinant of the predicted extent of climate change effects on winter chill. When using different models for similar climate change scenarios, the Chilling Hours Model and the Utah

Model tend to show much stronger decreases in winter chill than the Dynamic Model, especially in warm growing regions [15]. Using the former two models probably overestimates winter chill losses, because several studies have shown the Dynamic Model to be more accurate, especially in subtropical climates [16,17,18]. One more study found it to be equal to the Utah Model in Spain [19], and another one reported failure of all models on the Tropical island of Réunion [20]. Calculating ratios between winter chill estimates with different models at warm locations shows large differences, with strong variation and strong temperature dependence [21]. For colder regions, however, such ratios tend to be fairly similar and much less variable [21]. Consequently, the Dynamic Model can be used as a proxy of winter chill in both warm and cold growing regions and, among the common winter chill models, is the one most suitable for a global analysis. The Chilling Hours and Utah Models may produce reasonably accurate results in cold regions, but are not applicable for warmer parts of the world, where their use would produce misleading overestimates of likely impacts of climate change [8]. In this study, we therefore only use the Dynamic Model (for equations, see [21]), which quantifies winter chill in Chill Portions (CP).

We quantified winter chill for the entire terrestrial globe using climate scenarios based on observed daily weather from 4293 weather stations around the world and climate projections from three Global Climate Models (GCMs). Based on this analysis we calculated the safe winter chill (SWC) metric [8], which quantifies the amount of winter chill that is exceeded in 90% of years. This metric is meaningful to fruit and nut producers, because failure to meet chilling requirements in more than 10% of years is likely to render production uneconomical. This analysis identifies important fruit and nut producing areas in the world where SWC has already decreased and is projected to decrease further, and

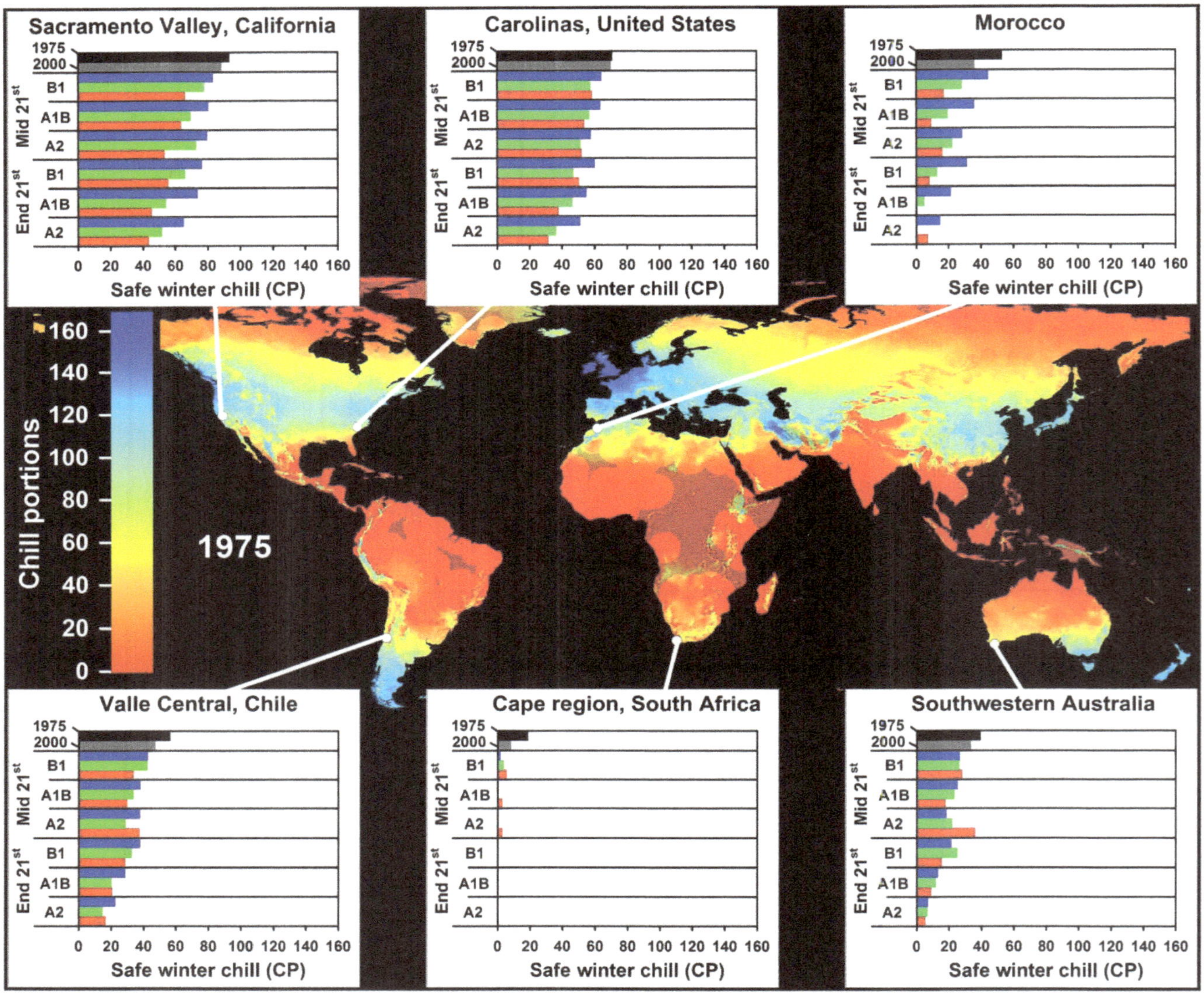

Figure 1. Modeled Safe Winter Chill around the year 1975 (large map), as well as site-specific estimates of Safe Winter Chill for six growing regions and for 20 climate scenarios, representing four points in time (1975, 2000, mid and end 21st century). Future projections include three greenhouse gas emissions scenarios (B1, A1B and A2) and three Global Climate Models (CSIRO - green bars; HADCM3 - blue bars; and MIROC - red bars). Areas that are more than 5° away from the closest weather station are shaded, because interpolated results are unreliable.

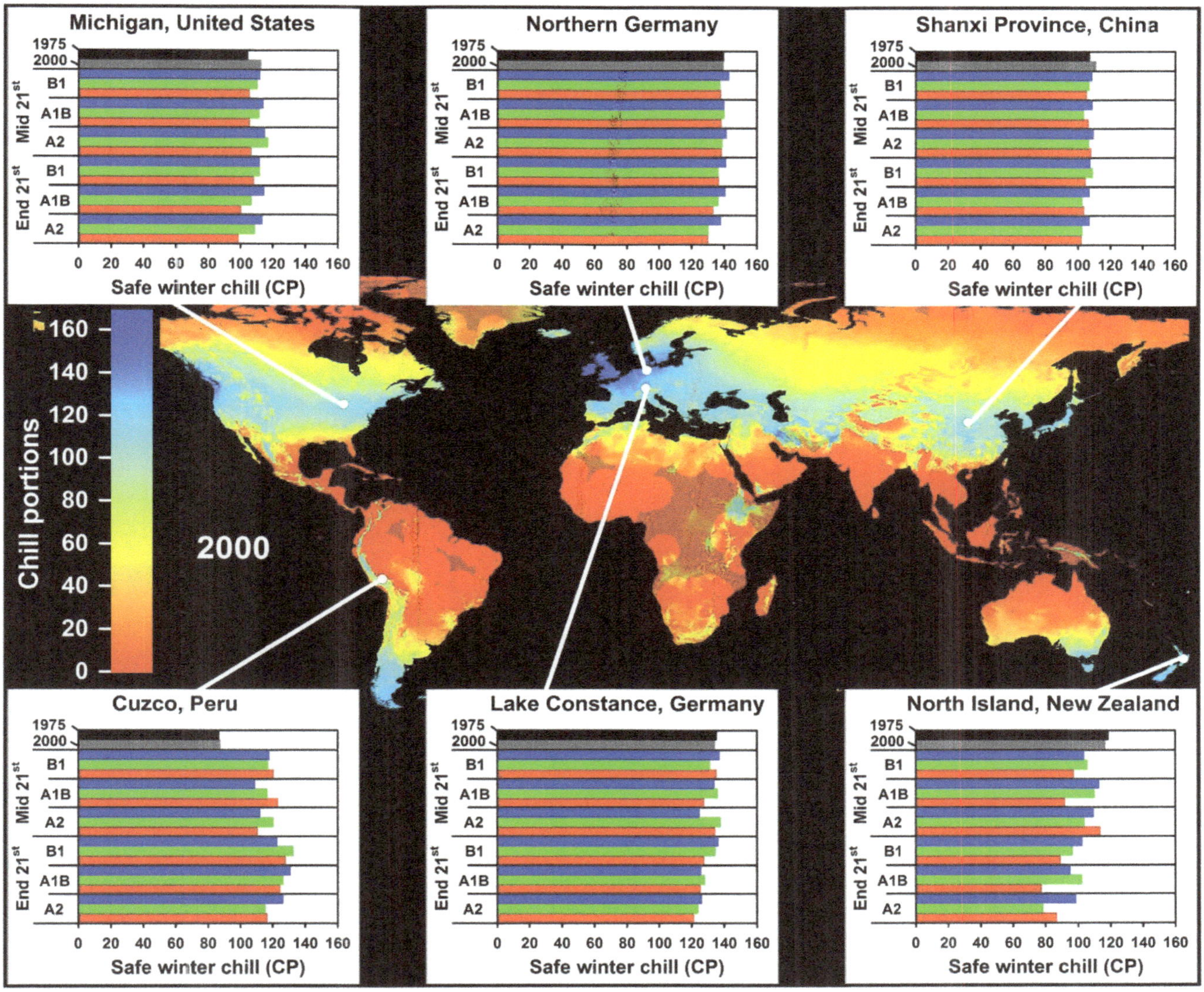

Figure 2. Modeled Safe Winter Chill around the year 2000 (large map), as well as site-specific estimates of Safe Winter Chill for six growing regions and for 20 climate scenarios, representing four points in time (1975, 2000, mid and end 21st century). Future projections include three greenhouse gas emissions scenarios (B1, A1B and A2) and three Global Climate Models (CSIRO - green bars; HADCM3 - blue bars; and MIROC - red bars). Areas that are more than 5° away from the closest weather station are shaded, because interpolated results are unreliable.

provides a comprehensive assessment of the magnitude of changes to winter chill that will likely occur.

Daily weather records for all weather stations were obtained from the National Climatic Data Center of the United States [22], subjected to a data quality filter, and used to calibrate a weather generator [23,24]. Daily weather records were then generated for 20 climate scenarios. Two scenarios represented typical climatic conditions in 1975 and 2000. Eighteen future scenarios were generated by extracting future projections from datasets assembled in the ClimateWizard tool [25]. These included statistically downscaled projections with three GCMs (MIROC3.2 (medres), UKMO-HadCM3 and CSIRO-Mk3.0; referred to as MIROC, HADCM3 and CSIRO in the following) and for three IPCC greenhouse gas emissions scenarios (B1 - global curbing of emissions over the 21st century; A1B - emissions leveling off at mid 21st century; and A2 - continually increasing rate of greenhouse gas emissions). Projections were made for the middle (2040–2059) and end (2080–2099) of the 21st century. Idealized daily temperature curves were used for converting daily to hourly weather records and allow calculation of winter chill. For each scenario, 101 years of weather records were generated and the 10% quantile of the resulting distribution of annual winter chill interpreted as Safe Winter Chill. For each scenario, SWC from all stations was spatially interpolated, and winter chill for all scenarios was extracted from the resulting layers for 24 important growing regions around the world.

Results

In all climate scenarios, estimates of Safe Winter Chill ranged from 0 CP in tropical and very cold regions to about 170 CP in maritime temperate climates of Northwestern Europe (Figs. 1–4).

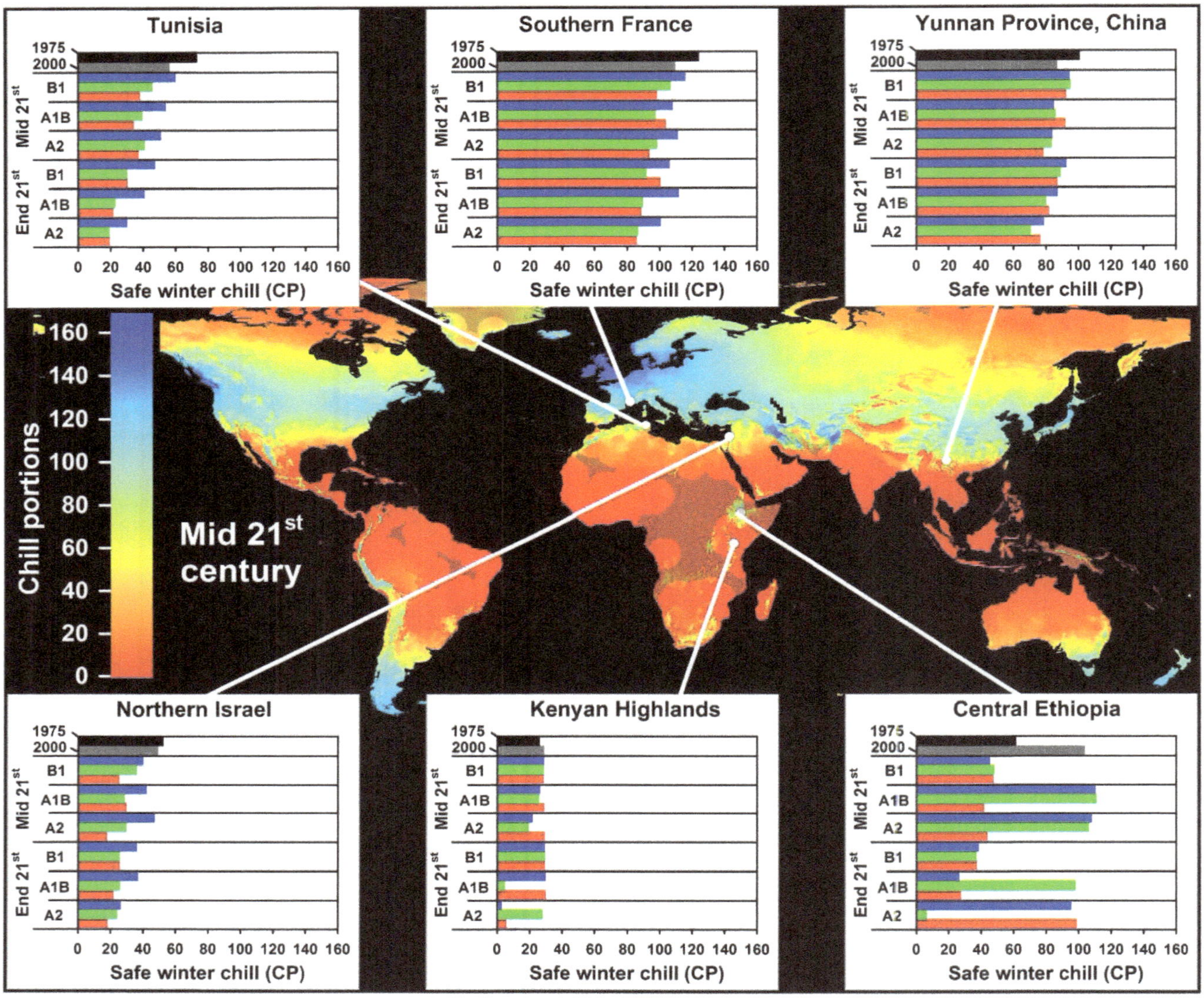

Figure 3. Modeled Safe Winter Chill around the middle of the 21st century averaged over three greenhouse gas emissions scenarios and three Global Climate Models (large map), as well as site-specific estimates of Safe Winter Chill for six growing regions and for 20 climate scenarios, representing four points in time (1975, 2000, mid and end 21st century). Future projections include three greenhouse gas emissions scenarios (B1, A1B and A2) and three Global Climate Models (CSIRO - green bars; HADCM3 - blue bars; and MIROC - red bars). Areas that are more than 5° away from the closest weather station are shaded, because interpolated results are unreliable.

While the overall range of the winter chill distribution did not change much across all scenarios, our results show changes in the global distribution of winter chill, as well as site-specific trends. Because the Dynamic Model does not consider freezing temperatures to be effective for chilling, reduced incidence of frost tends to increase the number of Chill Portions in cold regions. This process is reflected in increasing Safe Winter Chill in cold regions (Fig. 5), which may affect fruit growing regions in Canada, Southern Scandinavia and Eastern Europe (Fig. 4). Decreases are projected for warmer regions, in particular around the Mediterranean Sea (Fig. 6) and in Southwestern North America (Fig. 7), where losses up to 40 CP are expected by the end of the 21st century (Fig. 5). Many warm growing regions are projected to lose most of their winter chill, with South Africa, Southern Australia and Northern Africa particularly affected (Figs. 6 and 8).

Trends in site-specific projections for different growing regions varied substantially (site diagrams in Figs. 1–4). Most warm growing regions of temperate fruits and nuts are expected to experience decreasing winter chill, regardless of the emissions scenario or climate model used. The Sacramento Valley in California, the Southeastern United States, Chile's Valle Central, Yunnan Province in China, as well as South and Southwestern Australia are all projected to lose winter chill. This will likely require growers to transition to different species or cultivars than are grown today or to develop management practices that can help overcome shortages in winter chill. The highest losses relative to current winter chill levels occurred in Morocco, Tunisia, Israel, in the Cape region of South Africa and, for some GCMs, in the highlands of Kenya and Ethiopia. In these regions, climate change is likely to severely challenge current production systems, some of

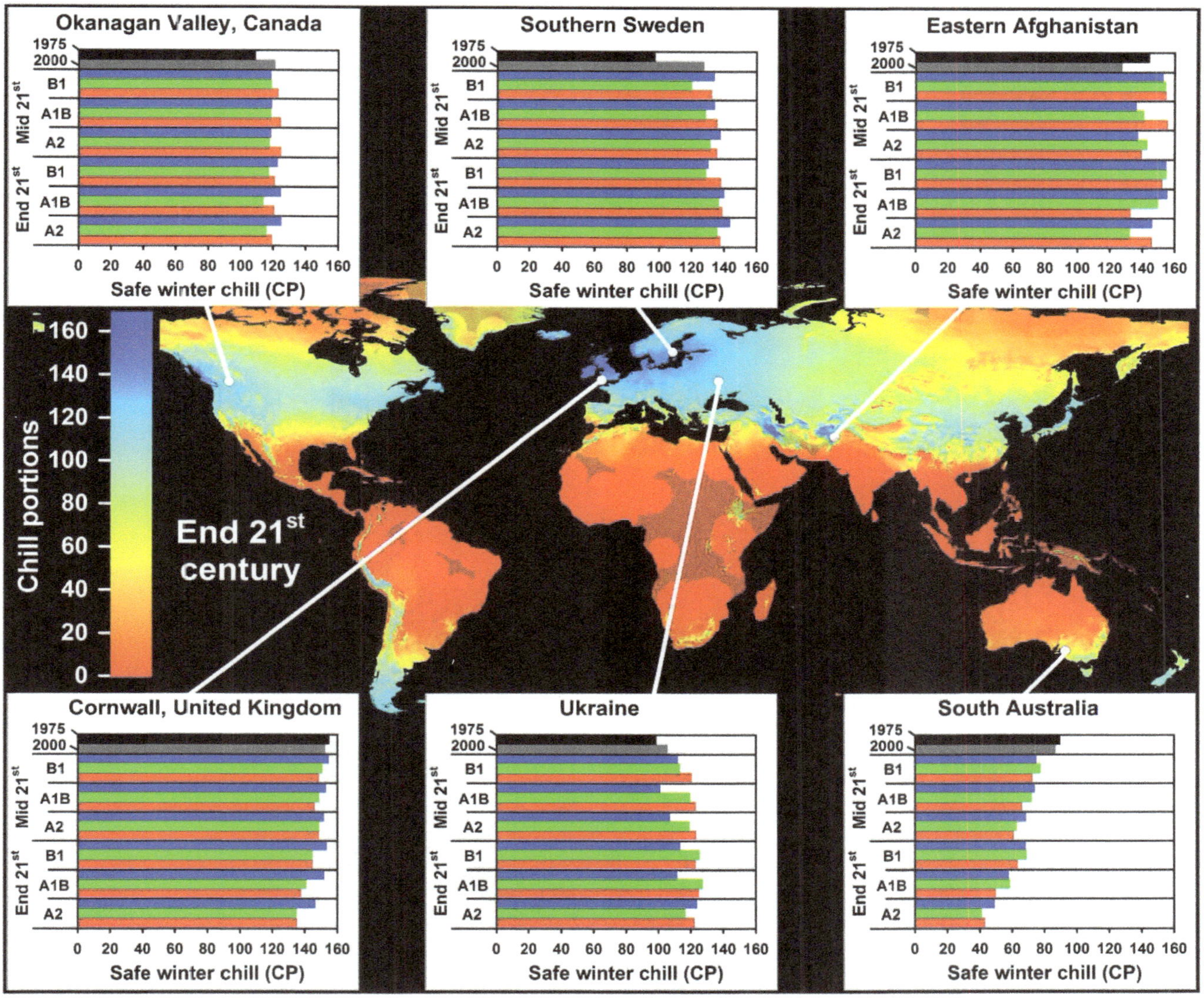

Figure 4. Modeled Safe Winter Chill around the end of the 21st century averaged over three greenhouse gas emissions scenarios and three Global Climate Models (large map), as well as site-specific estimates of Safe Winter Chill for six growing regions and for 20 climate scenarios, representing four points in time (1975, 2000, mid and end 21st century). Future projections include three greenhouse gas emissions scenarios (B1, A1B and A2) and three Global Climate Models (CSIRO - green bars; HADCM3 - blue bars; and MIROC - red bars). Areas that are more than 5° away from the closest weather station are shaded, because interpolated results are unreliable.

which already rely on cultural measures such as rest-breaking chemicals and artificial defoliation.

Cool regions are less likely to experience decreasing winter chill. Growing regions in Germany, the United Kingdom, the Midwestern United States (Fig. 7), Northern China and Central Asia are projected to see little change in SWC levels. Southern France (Fig. 6) and New Zealand (Fig. 8) may experience slight but likely insignificant losses. The coldest current growing regions (e.g. the Okanagan Valley in Canada, Southern Sweden and Eastern Europe) are expected to see more winter chill in the future. Whether these changes will require growers to adapt is currently unclear and is likely to depend more on the effects of summer warming than on winter temperatures.

In addition to the time period analyzed, the amplitude of expected changes also depended on the greenhouse gas emissions scenario. The A2 scenario consistently projected the greatest changes in winter chill, followed by the A1B scenario and the B1 scenario. If emissions are curbed to levels assumed in the B1 scenario, few growing regions are likely to see decreases by more than 20 CP by the end of the 21st century (Fig. 9). If business-as-usual emissions continue (A2 scenario), many subtropical regions will see chilling declines up to 40 CP, which can be expected to disrupt production systems.

The choice of the climate model also influenced model results. The MIROC model produced the greatest changes, followed by HADCM3 and CSIRO (Fig. 10). Within the same time period and emissions scenario, mean absolute differences between winter chill levels projected by CSIRO and HADCM3 were always smaller than for comparisons of either model with MIROC projections (Table 1). By the end of the 21st century, mean absolute differences between all modeled scenarios and historic SWC levels for 1975 were between 12.8 and 29.0 CP, on average

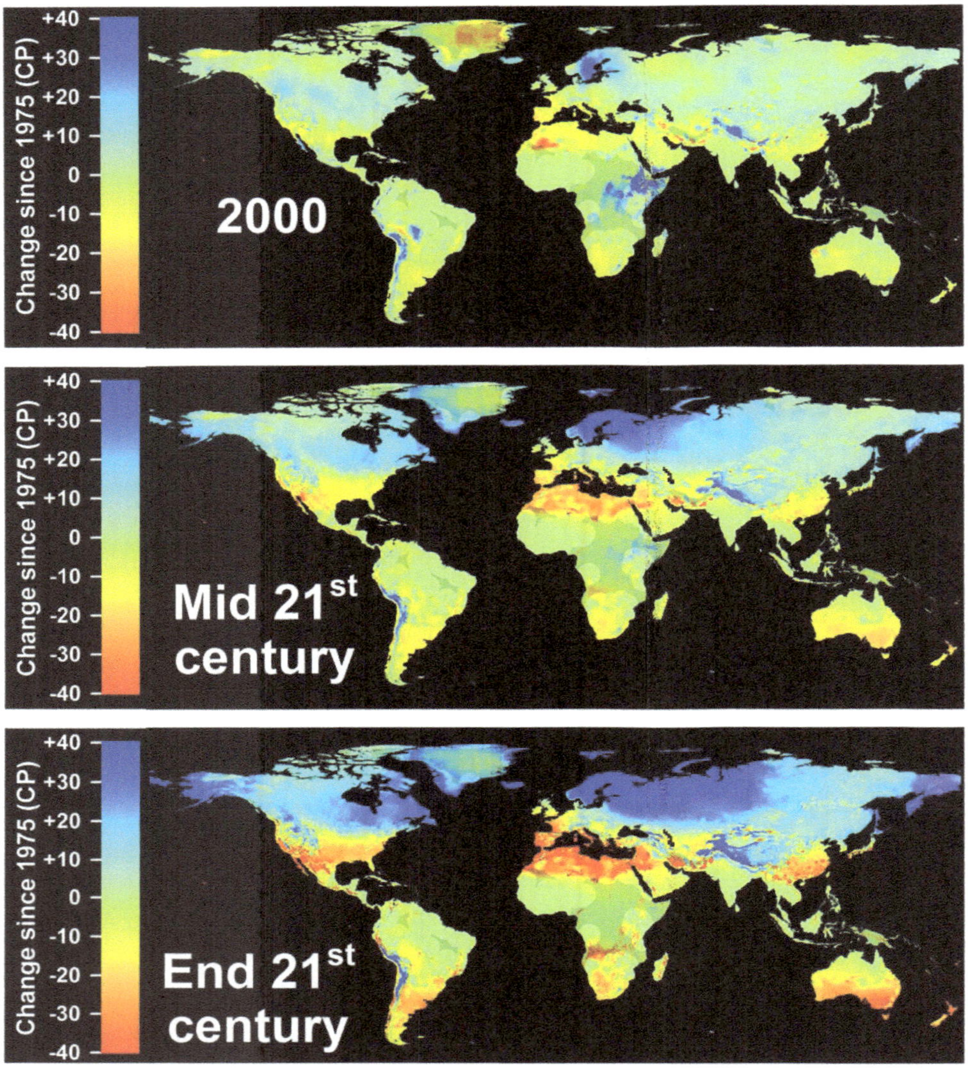

Figure 5. Modeled and projected losses in Safe Winter Chill compared to 1975 for the year 2000 (top), the middle of the 21st century (middle), and the end of the 21st century (bottom). For each point in time, results are averaged over three greenhouse gas emissions scenarios and three Global Climate Models. Areas that are more than 5° away from the closest weather station are shaded, because interpolated results are unreliable.

over all grid cells. These levels of change indicate that vegetation that relies on winter dormancy will experience very different temperature cues in the future than it does now.

Discussion

Our projections indicate that most warm growing regions will experience severe declines in Safe Winter Chill over the course of the 21st century. In contrast to this, cool regions may not see much change, because reductions in winter chill due to warming are compensated for by chilling gains caused by less frequent frost. For cold growing regions, these opposing trends play out to result in more winter chill in response to warming.

Among these changes, reduced winter chill is likely to have the most severe consequences for fruit production. Lack of winter chill can delay or prevent flowering, lead to staggered bloom, and cause various forms of anomalous growth [26,27]. Anecdotal evidence of this has been reported from various growing regions but seldom found its way into the literature.

Increases in winter chill in cold areas are less likely to lead to disruptions in fruit production, but even there, a mismatch between the chilling requirements of common species and cultivars and available winter chill could cause some problems for fruit and nut growers. In all areas, however, even in those where no changes in winter chill are projected, other manifestations of climate change are also likely to affect fruit and nut production. Plantations may be impacted by changes in rainfall, changes to summer and spring heat or increases in pest pressure due to faster reproduction of ectothermic pest organisms [28]. Due to all these additional effects of climate change, we do not attempt to predict future yields, but focus on pointing out potential problems due to lack of winter chill.

Even assuming no additional changes due to climate change, modeling the effect of changes in winter chill on crop yields is not

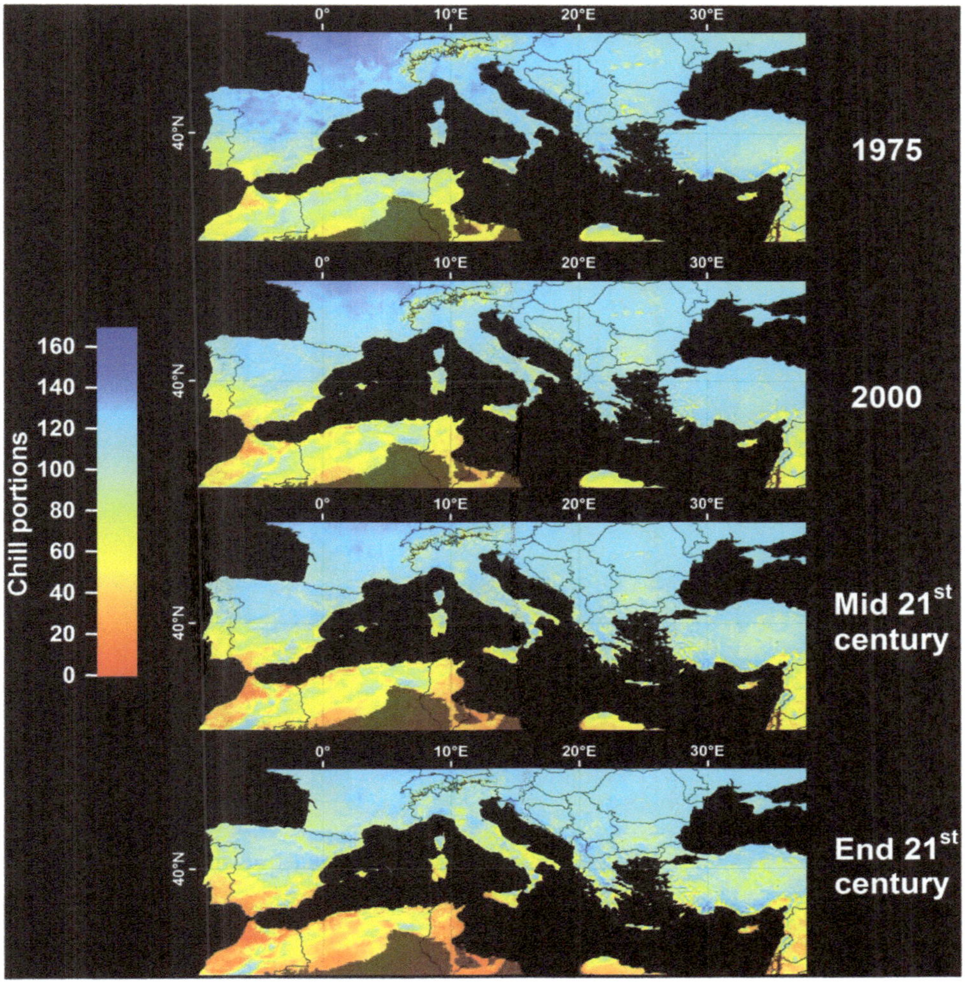

Figure 6. Modeled and projected Safe Winter Chill in the Mediterranean region, for 1975, 2000, the middle of the 21st century (middle), and the end of the 21st century (bottom). For each point in time, results are averaged over three greenhouse gas emissions scenarios and three Global Climate Models. Areas that are more than 5° away from the closest weather station, and areas with mean annual temperatures >20 or <°0C are shaded.

possible at present. In spite of two centuries of research on winter chill, it is still unclear what happens during chilling accumulation, and how exactly this process is influenced by ambient temperatures. There may also be other environmental factors, such as relative humidity, photoperiod or the Red/Far Red light ratio [29], that impact the breaking of dormancy but are not recognized in any dormancy models for tree crops. All chilling models, including the Dynamic Model, were developed to assist fruit and nut growers in selecting appropriate species and cultivars, rather than describing a biological process with scientific accuracy. They are all empirically derived rather than based on a functional understanding of the dormancy process.

Lack of knowledge about the chilling requirements of most species and cultivars, in appropriate units, also precludes detailed projections of future yields or suitable ranges. Most growers and researchers have used Chilling Hours to quantify chilling requirements, but this model has been shown to perform poorly in warm regions [16,17,18,19,20] and to be very sensitive to climate change [15]. Estimates of chilling requirements, when given in Chilling Hours, must also be adjusted before they are useful in a different location or in a warmer climate [21]. Existing lists of species-specific chilling requirements are thus of limited value for estimating future ranges of cultivars and species. Estimates in Chill Portions are less widely available, and we are not aware of a comprehensive list that compiles them. Research efforts are needed to close the pertinent knowledge gaps and allow quantitative projections of the effect that changes in winter chill will have on fruit and nut production.

As much as precise quantitative projections are impossible, changes to available winter chill and summer heat will likely change the suitable ranges of many tree crops, and it seems likely that many growing regions will become unsuitable for the cultivars that are currently produced. However, whether or not tree crops will actually be moved to cooler climates will depend on many factors, such as availability of land and critical infrastructure, land tenure and competition with other crops. For example, the ecological niche of many fruits and nuts in the Western United States is likely to move north, from California's Central Valley towards Northern California, Oregon and Washington. These new potentially suitable areas have adverse topography, poorer soils, and limited water availability compared to the Central Valley, making the economic viability of production there

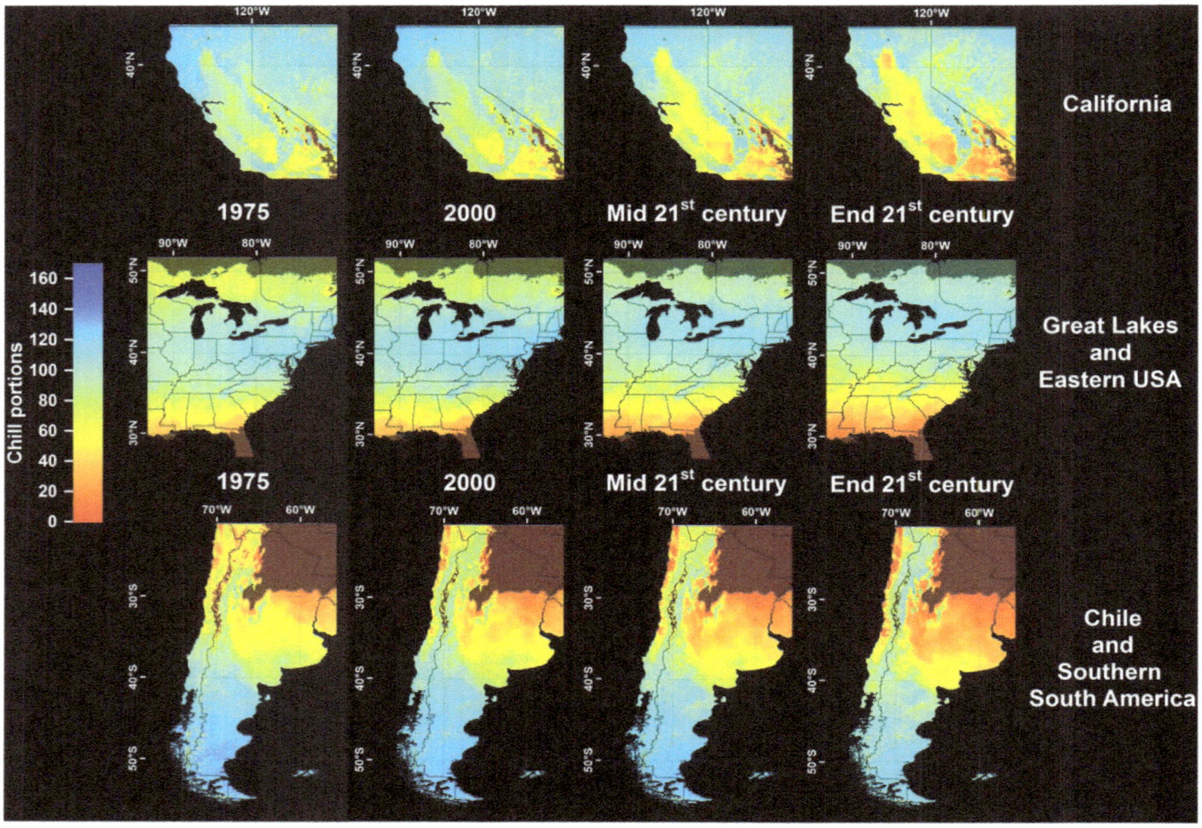

Figure 7. Modeled and projected Safe Winter Chill in California, the Eastern United States and Southern South America, for 1975, 2000, the middle of the 21st century (middle), and the end of the 21st century (bottom). For each point in time, results are averaged over three greenhouse gas emissions scenarios and three Global Climate Models. Areas that are more than 5° away from the closest weather station, and areas with mean annual temperatures >20 or <°0C are shaded.

questionable. Similarly, other regions, such as parts of Scandinavia, Canada and Siberia, that could potentially become more suitable for these tree crops, may be limited by cold winters, lack of summer heat, or adverse photoperiodic conditions. Production potential of tree crops under many of these novel conditions expected in the future has not been studied sufficiently to be discussed here.

To a certain extent, adaptation to changes in winter chill will be possible. Agrochemicals have been developed to artificially break a tree's dormancy during the later stages of chilling accumulation [30,31], irrigation and shading may influence orchard microclimates favorably [32], and other cultural practices, such as artificial defoliation [33], also have potential for reducing chilling requirements. Inclusion of low chilling requirements as an explicit target in breeding programs is likely to produce cultivars that will remain suitable in a warmer future. It will be necessary, however, to intensify efforts to develop such adaptation strategies. In particular breeding programs for low-chilling cultivars, which can take decades to produce useable results, need more attention and more resources.

The economic cost of climate change incurred by fruit and nut growers could be substantial. Many businesses may be confronted with the decision to either abandon their production or adapt as well as possible to altered climatic conditions. Applying adaptation treatments could be economically unviable. And even if crops move towards more suitable climatic zones, most small orchard operations lack the capital to move their production to a different area, potentially impacting many livelihoods.

Natural plant communities respond to similar temperature cues as fruit and nut trees and will likely be affected as well by changes in the amount of available winter chill. It seems very likely that the projected decreases in winter chill in the Subtropics, but also the increases in the colder regions, will affect local natural vegetation, potentially impacting the suitable areas for many plants. Rapid climate-driven changes in plant communities at the local scale have already been reported [34]. Most studies that have investigated climate change effects on large numbers of plant species have found that most species showed advances in spring phenology, indicating that the impact of reduced chilling is compensated, in most cases, by increases in spring heat [35,36,37]. However, the same studies include a sizeable number of species, which show an opposite trend - delayed spring phenology in response to increases in temperature. For meadow and steppe vegetation on the Tibetan Plateau, Yu et al. [38] have recently shown a clear correlation of such a delay with increases in winter temperature, implicating lack of winter chill as the cause of the delay.

Conclusion

Even under the most conservative emissions scenario examined here the projected substantial decreases in winter chill would negatively impact productivity of current cultivars and viability of fruit and nut industries in warm growing regions. Chemical, mechanical and physical methods to compensate for a loss of chilling are available but add significantly to production costs. A

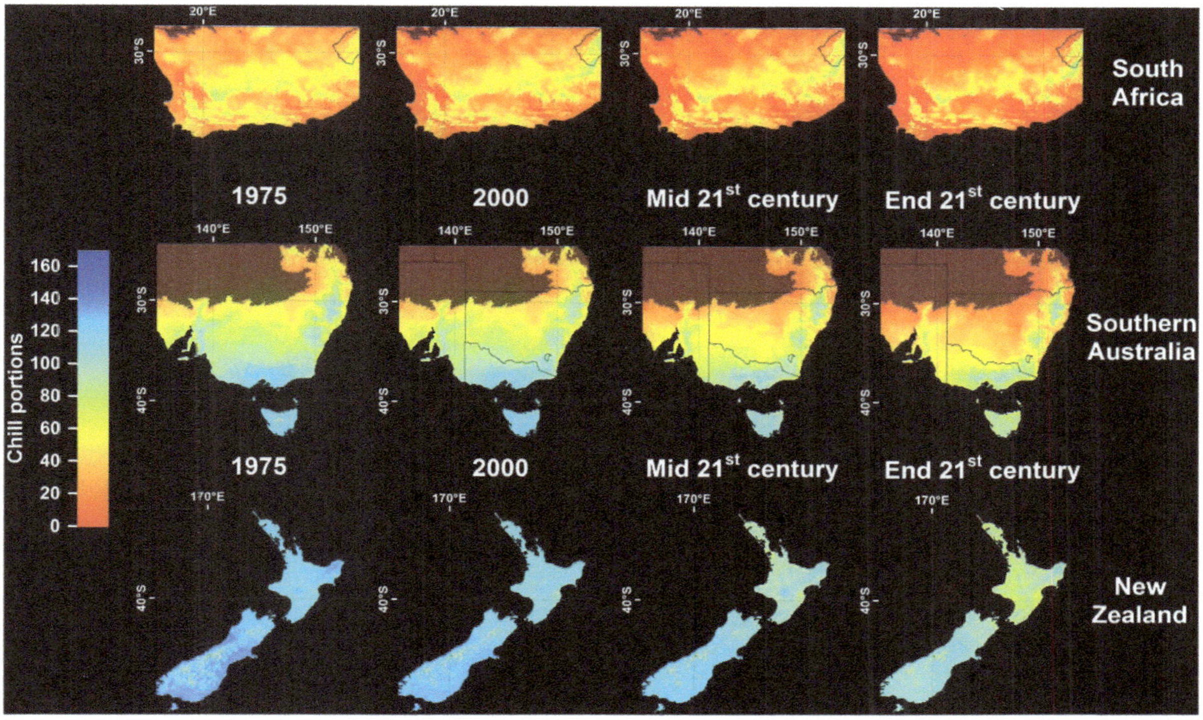

Figure 8. Modeled and projected Safe Winter Chill in South Africa, Southern Australia and New Zealand, for 1975, 2000, the middle of the 21st century (middle), and the end of the 21st century (bottom). For each point in time, results are averaged over three greenhouse gas emissions scenarios and three Global Climate Models. Areas that are more than 5° away from the closest weather station, and areas with mean annual temperatures >20 or <°0C are shaded.

shift to new production areas may be possible but would incur substantial infrastructure costs. Moreover, production viability depends on many critical factors in addition to climatic suitability, which may not be available in areas with suitable future climates. The most viable approach to adapt to climate change in deciduous fruit and nut species is through the development of new cultivars that are more productive under lower chill conditions. This will require investment in efforts to understand the biological basis for chilling, to develop better models that relate environmental cues with yield and phenology, as well as renewed emphasis on breeding programs.

Materials and Methods

Weather data

Daily temperature and rainfall records were downloaded for all 11,361 available weather stations at the National Climatic Data Center [22]. This dataset was filtered, removing all stations that had less than 5000 daily records between 1973 and 2002, and excluding all stations with more than 25% of daily minimum or maximum temperatures or 50% of daily rainfall data missing. For the remaining 5078 weather stations, all available temperature and precipitation records were used to calculate site parameters for use in the LARS-WG stochastic weather generator [23,24]. A stochastic weather generator (WG) is a model which, after calibration of site parameters with observed weather at that site, is capable of simulating synthetic time-series of daily weather that are statistically similar to observed weather [39]. By altering the site parameters of the WG using changes in climate predicted from a global climate model (GCM), it is possible to generate synthetic daily weather for the future. WGs are extensively used as a computationally inexpensive tool to produce daily site-specific climate scenarios for impact assessments of climate change [40,41,42,43]. Because the generation of daily weather is dependent on the site-specific duration of wet and dry spells, the modeling procedure required precipitation records in addition to temperatures.

Following Luedeling et al. [8], for each station we evaluated the entire weather record for all days between 1973 and 2002, calculating separate linear regression equations between time (in years) and the minimum temperature, maximum temperature and precipitation for each month of the year. We then used these regression equations to develop climate scenarios representing typical climatic conditions in 1975 and 2000. These scenarios do not represent actually observed temperatures and precipitation in these years, but rather typical conditions at these times that are representative of long-term trends over the calibration period (1973–2002). The climate scenarios contain the mean deviation of monthly minimum and maximum temperatures and precipitation from the mean of the calibration period for each station.

Using the same method, we evaluated the weather record for 18 future scenarios, based on projections by three Global Climate Models (GCMs; MIROC3.2 (medres), UKMO-HadCM3 and CSIRO-Mk3.0) that had been statistically downscaled to a 0.5 degree resolution using the CRU TS 2.0 data set to calibrate the downscaling (R. Neilson, unpublished data). Using the Climate Wizard tool (www.climatewizard.org; ref. [25]), we extracted projected minimum and maximum temperatures and precipitation for three greenhouse gas emissions scenarios of the IPCC Special Report on Emissions Scenarios [44]: the A2 scenario (continually increasing rate of greenhouse gas emissions), the A1B scenario (emissions leveling off at mid 21st century) and the B1 scenario

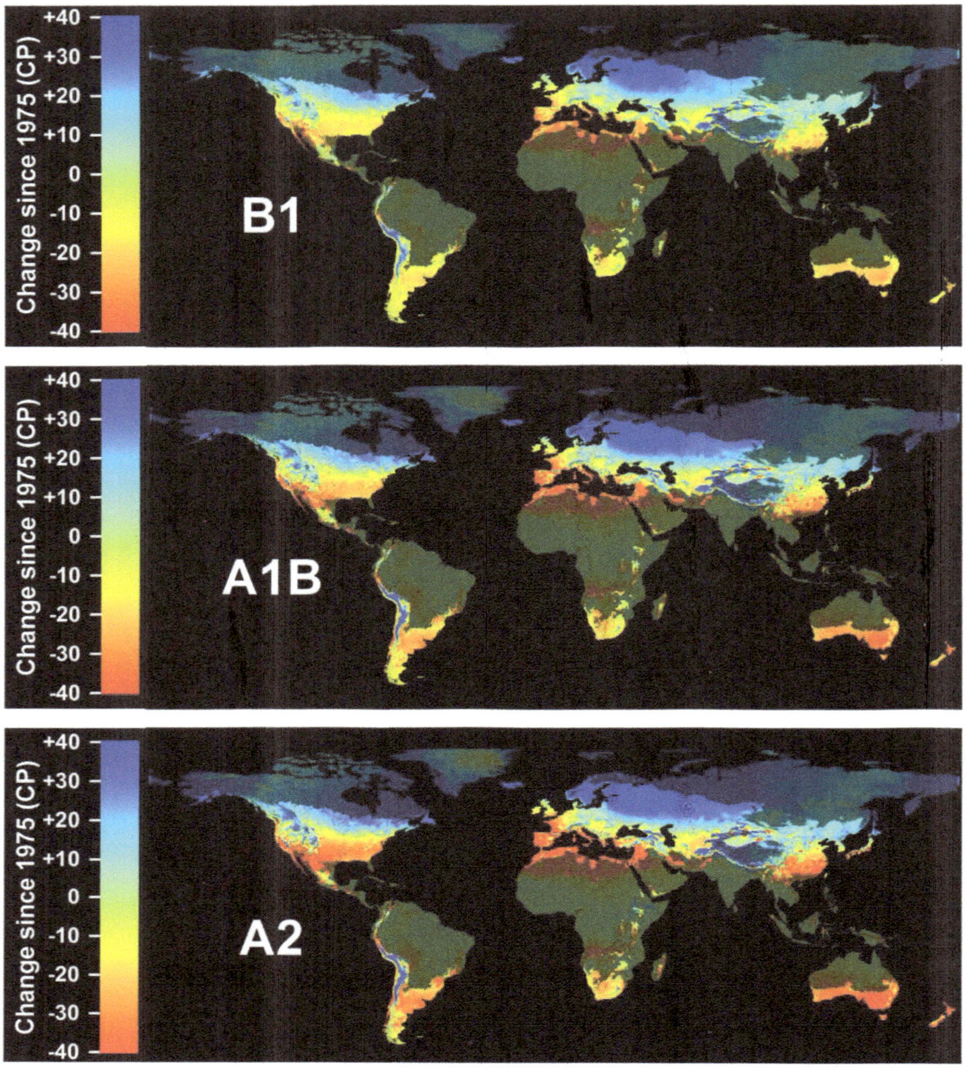

Figure 9. Projected losses in Safe Winter Chill at the end of the 21st century compared to 1975, for three greenhouse gas emissions scenarios: B1 (top), A1B (middle) and A2 (bottom). For each scenario, results are averaged over projections from three Global Climate Models. Areas that are more than 5° away from the closest weather station are shaded, because interpolated results are unreliable.

(global curbing of emissions over the 21st century). For each weather station, emissions scenario and GCM, mean monthly anomalies of minimum and maximum temperatures and precipitation relative to the calibration period were obtained for two periods of time: 2040–59 and 2080–2099, representing conditions at mid and end 21st century. For several weather stations, mostly along coast lines, no GCM projection data were available. These stations were excluded, bringing the total number of weather stations down to 4293.

For each station and for each of the 2 past and 18 future scenarios, we then generated 101 years of synthetic daily weather data using a command line version of the LARS-WG weather generator [24]. Synthetic daily maximum and minimum temperatures were converted into hourly temperatures using the idealized temperature curve proposed by Linvill [9,45]. Linvill's equations, which use a sine curve for daytime temperatures and a logarithmic decline curve for nighttime cooling, require sunset and sunrise hours, as well as daylength, as input parameters. These data were generated using equations by Spencer [46] and Almorox et al. [47]. Resulting from these processing steps were 101 years of synthetic hourly temperature for each weather station, representing typical weather conditions for each climate scenario.

Winter chill

Based on the generated hourly temperature, we calculated winter chill for 100 winters for each weather station and climate scenario, with start and end dates of the winter season set to October 1st and May 1st, respectively, for stations in the Northern Hemisphere, and to April 1st and November 1st, respectively, for stations in the Southern Hemisphere. In a global analysis, effective times of winter chill accumulation often deviate from these dates, depending on local temperature curves. Since the Dynamic Model contains a self-regulating mechanism, which only allows accumulation of Chill Portions during times with appropriate temperatures, this variability should not affect the accuracy of our results.

For the resulting distribution over 100 winters, we then calculated Safe Winter Chill (SWC), the 10% quantile of the distribution [8]. This metric is more meaningful for growers than

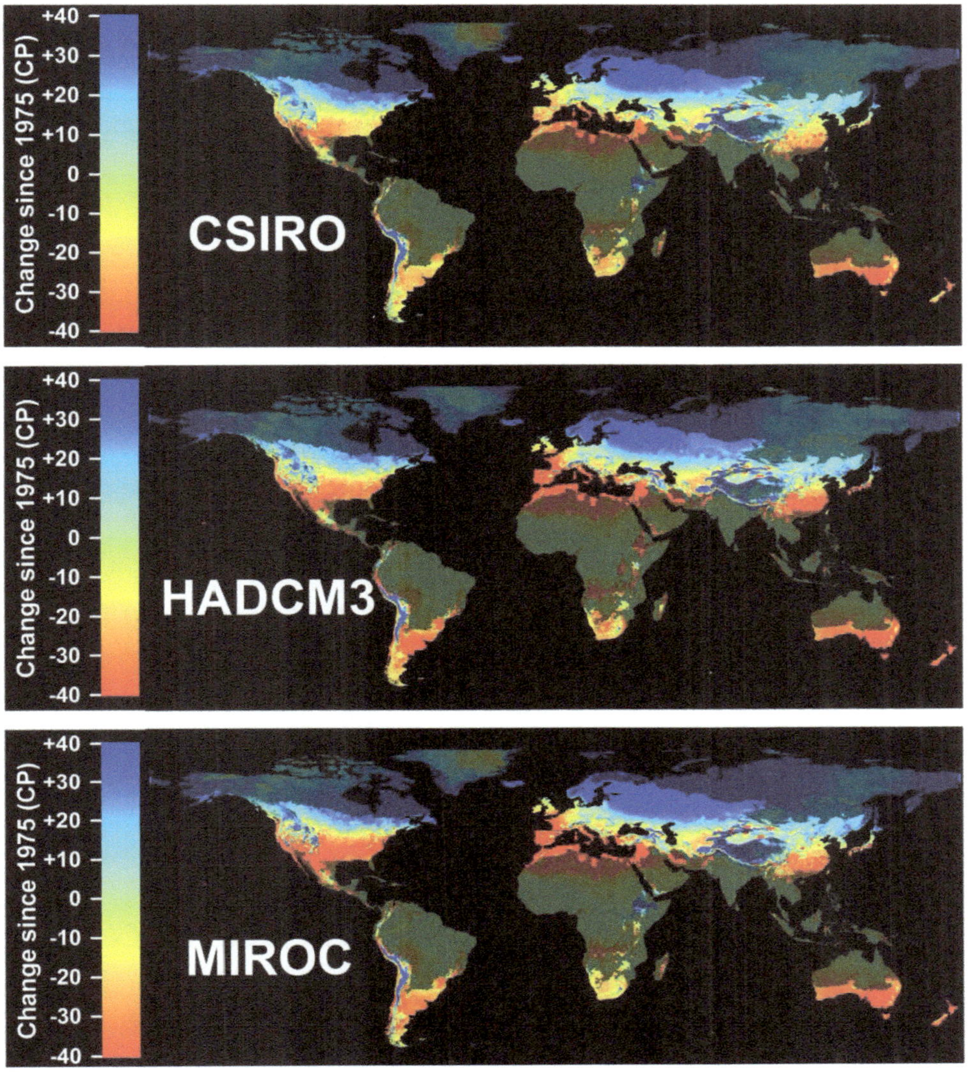

Figure 10. Projected losses in Safe Winter Chill at the end of the 21st century compared to 1975, for three Global Climate Models: CSIRO (top), HADCM3 (middle) and MIROC (bottom). For each scenario, results are averaged over projections for three greenhouse gas emissions scenarios. Areas that are more than 5° away from the closest weather station are shaded, because interpolated results are unreliable.

mean winter chill, because economic success of an orchard operation relies on fulfillment of chilling requirements in most years (e.g. 90% of all years), rather than in an average year. For all trees with lower chilling requirements than available SWC, the dormancy season should be sufficiently long and cold to allow fulfillment of tree-specific temperature needs. All data processing was implemented in JSL, the scripting language of JMP 8 (SAS Institute, Cary, NC, USA).

Spatial analysis

Based on estimates of SWC for every weather station, SWC was spatially interpolated for each climate scenario, using 12-neighbor Kriging with a spherical semivariogram at 0.1 degree spatial resolution (ArcGIS 9.3, ESRI, Redlands, CA, USA). A simple interpolation based on only the available weather stations would not be a very accurate representation of winter chill around the world, because temperatures at many locations may differ substantially from those of the closest weather station, which is often far away. We therefore used a high-resolution temperature dataset obtained from the WorldClim database [48] to correct for temperature variation that was not accounted for by the simple interpolation procedure. We calculated mean annual temperatures at a spatial resolution of 1/24 degree from monthly mean temperatures for the year 2000 given in the database. Because mean annual temperature explains much of the variation in winter chill [21], this dataset was useful for correcting the chilling estimates. To do this, we used the same Kriging procedure as for the winter chill estimates to interpolate a temperature surface from mean annual temperatures at all weather stations. The resulting grid represents the temperatures that correspond to the interpolated chill portion surface (SWC_{int}). Subtracting this grid from the original dataset of mean annual temperatures produced an estimate of the temperature variation that was not accounted for in the original interpolation of chill portion values (T_{diff}). For the correction, the effect of temperature on chill portion numbers was then estimated by a 5th order polynomial regression between mean annual temperatures at all weather stations and the amount of safe winter chill calculated for

Table 1. Mean absolute differences (in Chill Portions) between different climate scenarios [combination of time, greenhouse gas emissions scenario (GHG) and Global Climate Model (GCM)], over all relevant 0.1°×0.1° pixels of the global Safe Winter Chill projections.

Time			1975	2000	mid 21st century									end 21st century								
	GHG				B1			A1B			A2			B1			A1B			A2		
		GCM			C	H	M	C	H	M	C	H	M	C	H	M	C	H	M	C	H	M
1975				6.7	10.0	10.5	15.5	11.1	14.5	17.5	12.8	15.7	17.5	12.8	18.1	20.9	17.8	22.5	27.3	23.3	26.6	29.0
2000			6.7		7.5	7.2	11.9	6.2	9.5	13.9	7.7	10.6	13.7	9.2	14.0	17.1	13.8	18.1	23.5	18.6	22.6	24.8
mid 21st century	B1	C	10.0	7.5		4.0	6.4	4.4	6.1	8.3	4.9	6.5	9.1	3.4	8.4	11.4	8.1	13.4	18.0	13.9	17.3	20.0
		H	10.5	7.2	4.0		6.0	4.9	5.2	7.9	5.3	6.5	8.9	3.8	7.8	11.0	8.1	12.8	17.5	13.6	16.8	19.5
		M	15.5	11.9	6.4	6.0		7.7	5.2	3.1	7.0	6.0	5.5	4.5	3.6	5.9	5.5	8.4	12.4	9.1	12.1	14.6
	A1B	C	11.1	6.2	4.4	4.9	7.7		4.7	9.3	2.5	5.4	9.7	4.7	9.2	12.1	8.8	12.9	18.3	13.2	17.3	19.5
		H	14.5	9.5	6.1	5.2	5.2	4.7		6.5	4.4	2.8	7.4	4.8	5.8	8.9	7.0	9.0	14.8	9.6	14.0	15.9
		M	17.5	13.9	8.3	7.9	3.1	9.3	6.5		7.8	6.5	5.5	5.9	4.0	3.7	4.6	6.7	10.2	7.3	10.0	12.7
	A2	C	12.8	7.7	4.9	5.3	7.0	2.5	4.4	7.8		4.4	8.9	4.3	7.8	10.4	7.0	11.2	16.5	11.3	15.5	17.7
		H	15.7	10.6	6.5	6.5	6.0	5.4	2.8	6.5	4.4		7.4	5.1	5.7	8.5	6.0	8.1	14.0	8.5	12.9	15.0
		M	17.5	13.7	9.1	8.9	5.5	9.7	7.4	5.5	8.9	7.4		7.4	6.0	7.2	7.0	8.8	12.4	9.4	11.9	14.2
end 21st century	B1	C	12.8	9.2	3.4	3.8	4.5	4.7	4.8	5.9	4.3	5.1	7.4		5.8	8.8	5.4	11.0	15.4	11.3	14.6	17.5
		H	18.1	14.0	8.4	7.8	3.6	9.2	5.8	4.0	7.8	5.7	6.0	5.8		5.2	4.9	6.9	10.6	7.9	10.3	13.2
		M	20.9	17.1	11.4	11.0	5.9	12.1	8.9	3.7	10.4	8.5	7.2	8.8	5.2		5.3	5.9	6.8	5.2	8.0	9.5
	A1B	C	17.8	13.8	8.1	8.1	5.5	8.8	7.0	4.6	7.0	6.0	7.0	5.4	4.9	5.3		7.5	10.8	6.8	10.1	13.2
		H	22.5	18.1	13.4	12.8	8.4	12.9	9.0	6.7	11.2	8.1	8.8	11.0	6.9	5.9	7.5		7.7	4.9	6.8	8.0
		M	27.3	23.5	18.0	17.5	12.4	18.3	14.8	10.2	16.5	14.0	12.4	15.4	10.6	6.8	10.8	7.7		7.5	6.2	4.2
	A2	C	23.3	18.6	13.9	13.6	9.1	13.2	9.6	7.3	11.3	8.5	9.4	11.3	7.9	5.2	6.8	4.9	7.5		8.0	8.0
		H	26.6	22.6	17.3	16.8	12.1	17.3	14.0	10.0	15.5	12.9	11.9	14.6	10.3	8.0	10.1	6.8	6.2	8.0		7.1
		M	29.0	24.8	20.0	19.5	14.6	19.5	15.9	12.7	17.7	15.0	14.2	17.5	13.2	9.5	13.2	8.0	4.2	8.0	7.1	

GCMs: C - CSIRO; H - HADCM3; M - MIROC.

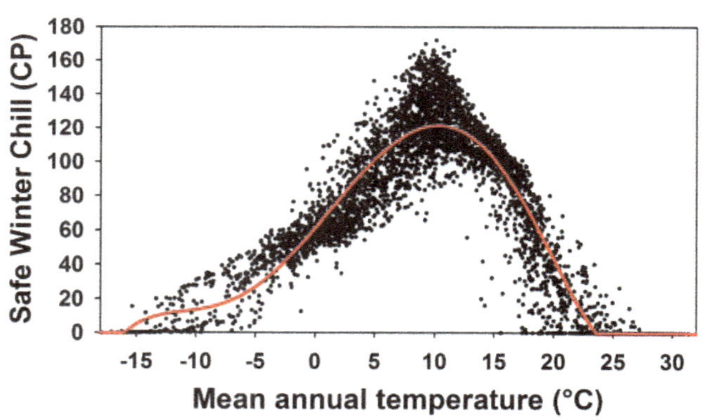

Figure 11. Correlation between mean annual temperature and modeled Safe Winter Chill for the year-2000 scenario. The red line indicates the equation used to correct for unaccounted for variation in temperature during spatial interpolation of site-specific Safe Winter Chill estimates.

them (Fig. 11). The resulting regression equation was:

$$CP(T) = 116.71 + 0.00005 \cdot (T - 10.005)^5 + 0.0015 \cdot (T - 10.005)^4 - 0.0180 \cdot (T - 10.005)^3 - 0.8395 \cdot (T - 10.005)^2 + 0.4690 \cdot T$$

for $-15.996 < T < 23.594$, and $CP(T) = 0$ for temperatures outside this range. In this equation, T is the temperature and CP(T) is the corresponding number of chill portions. The final correction equation was then:

$$SWC_{corr}(x,y) = SWC_{\text{int}}(x,y) + CP(T_{mean}(x,y) + T_{diff}(x,y)) - CP(T_{mean}(x,y)),$$

with T_{mean} being a Kriging surface calculated from mean station temperatures of the respective climate scenario, and x and y the longitude and latitude of each grid cell. SWC_{corr} is the temperature corrected estimate of safe winter chill. On all maps, areas where mean annual temperatures were below 0°C or above 20°C were shaded, because such regions are not suitable for the production of fruits and nuts with chilling requirements. Likewise, all areas that were further than 5° away from the closest useable weather station were shaded, because interpolation results for such grid cells were deemed unreliable. All gridded SWC layers are available from http://treephenology.ucdavis.edu/.

Scenario evaluation

From the surfaces of safe winter chill, site-specific results were extracted for 24 point locations representing important growing regions around the world. Comparing safe winter chill estimates for different combinations of GCM, GHG emissions scenario and point in time provides an impression of the agreement between scenarios, on a case-study basis. We also evaluated differences between scenarios based on the entire distribution over all relevant grid cells (excluding all that were shaded in the maps). Because differences between models in site specific estimates can be both negative and positive, we evaluated the mean absolute difference among all grid cells. The results are indicative of the agreement between models.

Acknowledgments

We are grateful to Heiner Lieth and to the CSTARS lab at UC Davis, who contributed computing capacity, without which this study would not have been possible.

Author Contributions

Conceived and designed the experiments: EL PHB. Performed the experiments: EL. Analyzed the data: EL. Contributed reagents/materials/analysis tools: EL EHG MAS. Wrote the paper: EL EHG MAS PHB.

References

1. Knight TA (1801) Account of Some Experiments on the Ascent of the Sap in Trees. Philosophical Transactions of the Royal Society of London 91: 333–353.
2. Samish RM (1954) Dormancy in Woody Plants. Annual Review of Plant Physiology and Plant Molecular Biology 5: 183–204.
3. Saure MC (1985) Dormancy Release in Deciduous Fruit Trees. Horticultural Reviews 7: 239–300.
4. Vegis A (1961) Samenkeimung und vegetative Entwicklung der Knospen. Handbuch der Pflanzenphysiologie - Encyclopedia of Plant Physiology 16: 168–298.
5. FAOSTAT (2009) Statistical Database of the Food and Agriculture Organization of the United Nations.
6. CIA (2006) CIA World Factbook. Washington D.C., , USA: Central Intelligence Agency.
7. Baldocchi D, Wong S (2008) Accumulated winter chill is decreasing in the fruit growing regions of California. Climatic Change 87: S153–S166.
8. Luedeling E, Zhang M, Girvetz EH (2009) Climatic changes lead to declining winter chill for fruit and nut trees in California during 1950–2099. PLoS ONE 4: e6166.
9. Luedeling E, Blanke M, Gebauer J (2009) Auswirkungen des Klimawandels auf die Verfügbarkeit von Kältewirkung (Chilling) für Obstgehölze in Deutschland. Erwerbs-Obstbau 51: 81–94.
10. Luedeling E, Gebauer J, Buerkert A (2009) Climate change effects on winter chill for tree crops with chilling requirements on the Arabian Peninsula. Climatic Change 96: 219–237.
11. Chandler WH (1942) Deciduous orchards. Philadelphia, USA: Lea & Febiger. 438 p.
12. Richardson EA, Seeley SD, Walker DR (1974) A Model for Estimating the Completion of Rest for Redhaven and Elberta Peach Trees. Hortscience 9: 331–332.
13. Fishman S, Erez A, Couvillon GA (1987) The Temperature-Dependence of Dormancy Breaking in Plants - Computer-Simulation of Processes Studied under Controlled Temperatures. Journal of Theoretical Biology 126: 309–321.
14. Fishman S, Erez A, Couvillon GA (1987) The Temperature Dependence of Dormancy Breaking in Plants: Mathematical Analysis of a Two-Step Model Involving a Cooperative Transition. Journal of Theoretical Biology 124: 473–483.
15. Luedeling E, Zhang M, Luedeling V, Girvetz EH (2009) Sensitivity of winter chill models for fruit and nut trees to climate change. Agriculture, Ecosystems & Environment 133: 23–31.
16. Luedeling E, Zhang M, McGranahan G, Leslie C (2009) Validation of winter chill models using historic records of walnut phenology. Agricultural and Forest Meteorology 149: 1854–1864.
17. Allan P, Linsley-Noakes GC, Holcroft DM, Brunette SA, Burnett MJ, et al. (1997) Kiwifruit research in a subtropical area. Acta Horticulturae 444: 37–42.
18. Erez A (2000) Bud dormancy; phenomenon, problems and solutions in the tropics and subtropics. In: Erez A, ed. Temperate Fruit Crops in Warm Climates. Dordrecht, The Netherlands: Kluwer Academic. pp 17–48.
19. Ruiz D, Campoy JA, Egea J (2007) Chilling and heat requirements of apricot cultivars for flowering. Environmental and Experimental Botany 61: 254–263.
20. Balandier P, Bonhomme M, Rageau R, Capitan F, Parisot E (1993) Leaf Bud Endodormancy Release in Peach-Trees - Evaluation of Temperature Models in Temperate and Tropical Climates. Agricultural and Forest Meteorology 67: 95–113.
21. Luedeling E, Brown PH (2011) A global analysis of the comparability of winter chill models for fruit and nut trees. International Journal of Biometeorology 55: 411–421.
22. NCDC (2008) Database of the National Climatic Data Center National Oceanic and Atmospheric Administration.
23. Semenov MA, Donatelli M, Stratonovitch P, Chatzidaki E, Braruth B (in press) ELPIS: a dataset of local-scale daily climate scenarios for Europe. Climate Research.
24. Semenov MA, Stratonovitch P (2010) The use of multi-model ensembles from global climate models for impact assessments of climate change. Climate Research 41: 1–14.
25. Girvetz EH, Zganjar C, Raber GT, Maurer EP, Kareiva P, et al. (2009) Applied climate-change analysis: the Climate Wizard Tool. PLoS ONE 4: e8320
26. Petri JL, Berenhauser Leite G (2004) Consequences of insufficient winter chilling on apple tree bud-break. Acta Horticulturae 662: 53–60.
27. Blanke M, Kunz A (2009) Einfluss rezenter Klimaveränderungen auf die Phänologie bei Kernobst am Standort Klein-Altendorf – anhand 50 jähriger Aufzeichnungen. Erwerbsobstbau 51: 101–114.
28. Luedeling E, Steinmann KP, Zhang M, Brown PH, Grant J, et al. (2011) Climate change effects on walnut pests in California. Global Change Biology 17: 228–238.
29. Erez A, Samish RM, Lavee S (1966) Role of light in leaf and flower bud break of peach (Prunus persica). Physiologia Plantarum 19: 650–659.
30. Dozier WA, Powell AA, Caylor AW, McDaniel NR, Carden EL, et al. (1990) Hydrogen Cyanamide Induces Budbreak of Peaches and Nectarines Following Inadequate Chilling. Hortscience 25: 1573–1575.
31. George AP, Lloyd J, Nissen RJ (1992) Effects of Hydrogen Cyanamide, Paclobutrazol and Pruning Date on Dormancy Release of the Low Chill Peach Cultivar Flordaprince in Subtropical Australia. Australian Journal of Experimental Agriculture 32: 89–95.
32. Erez A (1995) Means to compensate for insufficient chilling to improve bloom and leafing. Acta Horticulturae 395: 81–95.
33. Griesbach J (2007) Growing temperate fruit trees in Kenya. Nairobi, Kenya: World Agroforestry Center (ICRAF).
34. Vennetier M, Ripert C (2009) Forest flora turnover with climate change in the Mediterranean region: A case study in Southeastern France. Forest Ecology and Management 258: S56–S63.
35. Chmielewski FM, Rötzer T (2001) Response of tree phenology to climate change across Europe. Agricultural and Forest Meteorology 108: 101–112.
36. Parmesan C (2007) Influences of species, latitudes and methodologies on estimates of phenological response to global warming. Global Change Biology 13: 1860–1872.
37. Menzel A, Sparks TH, Estrella N, Koch E, Aasa A, et al. (2006) European phenological response to climate change matches the warming pattern. Global Change Biology 12: 1969–1976.
38. Yu H, Luedeling E, Xu J (2010) Winter and spring warming result in delayed spring phenology on the Tibetan Plateau. Proceedings of the National Academy of Sciences of the United States of America 107: 22151–22156.
39. Wilks DS, Wilby RL (1999) The weather generation game: a review of stochastic weather models. Progress in Physical Geography 23: 329–357.
40. Semenov MA (2009) Impacts of climate change on wheat in England and Wales. Journal of the Royal Society Interface 6: 343–350.
41. Semenov MA, Halford NG (2009) Identifying target traits and molecular mechanisms for wheat breeding under a changing climate. Journal of Experimental Botany 60: 2791–2804.
42. Evans N, Butterworth MH, Baierl A, Semenov MA, West JS, et al. (2010) The impact of climate change on disease constraints on production of oilseed rape. Food Security 2: 143–156.
43. Lazzarotto P, Calanca P, Semenov MA, Fuhrer J (2010) Transient responses to increasing CO2 and climate change in an unfertilized grass-clover sward. Climate Research 41: 221–232.
44. IPCC (2000) Special report on emissions scenarios; Nakicenovic N, Swart R, eds. Cambridge, UK: Cambridge University Press.
45. Linvill DE (1990) Calculating Chilling Hours and Chill Units from Daily Maximum and Minimum Temperature Observations. Hortscience 25: 14–16.
46. Spencer JW (1971) Fourier series representation of the position of the Sun. Search 2: 172.
47. Almorox J, Hontoria C, Benito M (2005) Statistical validation of daylength definitions for estimation of global solar radiation in Toledo, Spain. Energy Conversion and Management 46: 1465–1471.
48. Hijmans RJ, Cameron SE, Parra JL, Jones PG, Jarvis A (2005) Very high resolution interpolated climate surfaces for global land areas. International Journal of Climatology 25: 1965–1978.

17

SERCA2 Regulates Non-CF and CF Airway Epithelial Cell Response to Ozone

Shama Ahmad[1]*, David P. Nichols[1], Matthew Strand[2], Raymond C. Rancourt[1], Scott H. Randell[3], Carl W. White[1], Aftab Ahmad[1]

1 Department of Pediatrics, National Jewish Health, Denver, Colorado, United States of America, **2** Division of Biostatistics and Bioinformatics, National Jewish Health, Denver, Colorado, United States of America, **3** Cystic Fibrosis/Pulmonary Research and Treatment Center, Department of Medicine, The University of North Carolina at Chapel Hill, Chapel Hill, North Carolina, United States of America

Abstract

Calcium mobilization can regulate a wide range of essential functions of respiratory epithelium, including ion transport, ciliary beat frequency, and secretion of mucus, all of which are modified in cystic fibrosis (CF). SERCA2, an important controller of calcium signaling, is deficient in CF epithelium. We conducted this study to determine whether SERCA2 deficiency can modulate airway epithelial responses to environmental oxidants such as ozone. This could contribute to the pathogenesis of pulmonary exacerbations, which are important and frequent clinical events in CF. To address this, we used air-liquid interface (ALI) cultures of non-CF and CF cell lines, as well as differentiated cultures of cells derived from non-CF and CF patients. We found that ozone exposure caused enhanced membrane damage, mitochondrial dysfunction and apoptotic cell death in CF airway epithelial cell lines relative to non-CF. Ozone exposure caused increased proinflammatory cytokine production in CF airway epithelial cell lines. Elevated proinflammatory cytokine production also was observed in shRNA-mediated SERCA2 knockdown cells. Overexpression of SERCA2 reversed ozone-induced proinflammatory cytokine production. Ozone-induced proinflammatory cytokine production was NF-κB- dependent. In a stable NF-κB reporter cell line, SERCA2 inhibition and knockdown both upregulated cytomix-induced NF-κB activity, indicating importance of SERCA2 in modulating NF-κB activity. In this system, increased NF-κB activity was also accompanied by increased IL-8 production. Ozone also induced NF-κB activity and IL-8 release, an effect that was greater in SERCA2-silenced NF-κB-reporter cells. SERCA2 overexpression reversed cytomix-induced increased IL-8 release and total nuclear p65 in CFTR-deficient (16HBE-AS) cells. These studies suggest that SERCA2 is an important regulator of the proinflammatory response of airway epithelial cells and could be a potential therapeutic target.

Editor: Mauricio Rojas, University of Pittsburgh, United States of America

Funding: This work was supported by R01-ES014448 from the National Institute of Environmental Health Sciences (NIEHS), NIH, and by the Max and Yetta Karasik Family Foundation (CWW), R01-HL080322 and CF Foundation Resources Development Program Grant RDP CORE C: R026-R02 (SR). SA is supported by K12 KL2RR025779 from the National Center for Research Resources (NCRR). AA is supported by the Scientist Development Grant award (0830418N) from the American Heart Association. The funders had no role in study design, data collection and analysis, decision to publish, or preparation of the manuscript.

Competing Interests: The authors have declared that no competing interests exist.

* E-mail: ahmads@njhealth.org

Introduction

Atmospheric pollutants such as ozone, particulates and nitrogen oxides continuously challenge airways of urban dwellers. Elevated pollutant levels may contribute to exacerbations, and accelerated decline of lung function, in patients with chronic airway disease like cystic fibrosis (CF) and asthma [1,2,3]. Ozone attacks the lung through oxidative mechanisms, causing disruption of epithelial barrier, increased permeability, influx of neutrophils and generation of cytokines and chemokines [4,5,6,7]. We have previously established the role of ozone-reactive surfactant phospholipids in modulating airway epithelial cell viability in response to ozone [8]. These derivatives not only caused apoptotic cell death but also induced proinflammatory responses that may potentiate airway injury in asthmatics and other susceptible patients like those with cystic fibrosis (CF). Increased oxidative stress and enhanced lipid derived inflammatory mediators are characteristic of CF patient biofluids, during respiratory exacerbations [9]. Acute exacerbations accelerate the clinical progression of CF and hasten decline of lung function. Contribution of environmental ozone to exacerbations in CF symptoms has been suggested [1]. However, identification of mechanisms leading to pulmonary exacerbations in patients with CF is crucial for developing therapies for maintenance of lung function, quality of life and survival.

We have recently established that in CF airway epithelium there is decreased expression of an important ER calcium pump, sarcoendoplasmic reticulum calcium ATPase, SERCA2 [10]. We have also documented a role for SERCA2 in survival of airway epithelium in oxidative stress, including that caused by ozone [10]. SERCA2 is the only pump that actively loads Ca^{2+} back into ER of airway cells for subsequent signaling events once it has been released following IP3 receptor activation on ER. Therefore, abnormalities in this enzyme could have critical consequences. This is evident in SERCA2 gene knockout studies, where mice do not survive to birth [11]. One obvious consequence of decreased activity of this pump would be a sustained increase in cytosolic calcium resulting in passive overloading of cellular calcium stores. Expanded or overloaded calcium stores have been demonstrated by previous studies in cystic fibrosis airway epithelial cells [12,13]. Sustained increase in cytosolic calcium has also been shown to upregulate expression of proinflammatory transcriptional regulators such as NF-κB [14]. Similarly, inhibition of SERCA2 by specific inhibitor thapsigargin and subsequent increase in cytosolic calcium has been shown to potentiate NF-κB activation and IL-8 release response in airway epithelial cells [15,16]. In this study we demonstrate that ozone exposure causes an increased proinflammatory response in CF airway epithelial cells. This enhanced proinflammatory response could be mediated by decreased SERCA2 that induces NF-κB promoter activity. These studies provide insight into the mechanism of noxious inflammatory response in CF, Supporting studies also provide evidence that SERCA2 could regulate NF-κB and that modulation of SERCA2 could be a potential therapeutic strategy for treatment of CF.

Table 1. Demographics of airway tissue donors for epithelial cells.

Donor No.	Category	Age (yr)/Sex	COD*	Genotype
1	NTD	29/M	Head Trauma	Non-CF
2	NTD	30/M	MVA/Head Trauma	Non-CF
3	TD	22/F	Head Trauma	Non-CF
4	CF transplant	22/F		ΔF508/ΔF508
5	CF transplant	27/M		ΔF508/?
6	CF transplant	29/F		ΔF508/ΔF508

*COD, Cause of death, NTD = non transplant donor, TD = excess tissue from transplant donor.
MVA = motor vehicle accident. ? = mutation not identified.

Materials and Methods

Cell culture

Human lung tissue was procured from National Disease Research Interchange (NDRI) under University of North Carolina (UNC) and National Jewish Institutional Review Board (NJIRB)-approved protocols and human bronchial epithelial cell harvest and culture was performed using established procedures previously described in detail [17]. Epithelial cells were removed from lower trachea and bronchi by protease XIV digestion and cells were plated in BEGM medium on collagen-coated dishes as described previously [10,17]. Primary epithelial cells (passage 2; 5×10^5 cells) were seeded onto 12 mm diameter type IV collagen (Sigma) coated Snapwells (0.4 µM pore size, Corning, Corning, NY) and following confluence on day 4-5 were maintained at an air-liquid interface (ALI). Experiments comparing ozone-induced proinflammatory cytokine release in primary non-CF and CF cells (demographics of donors provided in Table 1) were performed with differentiated air-liquid-interface (ALI) cultures grown simultaneously and matched for passage number, number of cells plated, and days in culture. Transformed human CF airway epithelia cell lines, CFBE41o- (CF41o-) and CFBE45o- (CF45o-) and a wild-type, 16HBE14o- (16HBEo-), were provided by Prof. D. Gruenert (California Pacific Medical Center Research Institute, University of California at San Francisco). These cell lines were cultured in Eagle's minimal essential medium (Invitrogen, Carlsbad, CA) supplemented with 10% fetal bovine serum, L-glutamine, and penicillin/streptomycin at 37°C under 5% CO_2. Human bronchial epithelial 16HBEo- cells with stable expression of sense (16HBE-S) and antisense CFTR (16HBE-AS) oligonucleotides were provided by Dr. Pamela Davis, Case Western Reserve University Cleveland, OH and cultured as described previously [10]. Trans-epithelial resistance (TER) was also measured using previously described procedure [18].

Ozone exposures

Exposure of airway epithelial cells to ozone at precise levels was carried out in a computer- controlled *in vitro* exposure chamber [18]. Differentiated primary airway epithelial cells that were exposed to ozone contained a very thin layer of media apically. Cells were rocked so as to expose one side of apical surface at a time to ozone.

Determination of cell survival and protein concentration

To assess cell death Calcein-AM and Propidium iodide were added apically on the inserts after treatment. Caspase 3/7 activity was measured by using Caspase-Glo^R 3/7-assay kit (Promega Corporation, Madison, WI). TUNEL staining on the inserts was performed using ApopTag^R Plus Peroxidase *In Situ* apoptosis kit (Millipore, Billerica, MA). Annexin binding was assessed by using Vybrant apotosis assay kit (Molecular Probes, Eugene, OR), as described before [18]. In this method, apoptotic cells bearing phosphatidylserine in the plasma membrane outer leaflet were identified as those binding Alexa Fluor 488-labeled annexin V. Mitochondrial membrane potential (MMP) determination was performed as described before [19]. Protein concentration in cell lysates was determined using the Bio-Rad DC protein assay kit (Bio-Rad, Hercules, Ca) in a 96-well plate with bovine serum albumin as a standard.

Western blot

Western blots were performed as previously described in detail [20] and the membranes were probed with rabbit polyclonal antibodies against SERCA2 (Affinity Bioreagents, Golden, CO) at 1:1,000 dilution for each, overnight at 4°C. Blots were then washed again with TBS-T and incubated with secondary peroxidase-conjugated IgG (Bio-Rad, Hercules, CA) at 1:2,000 dilution, for 1 h at room temperature. Immunoreactive bands were detected using an ECL detection kit (Pierce, Rockford, IL) followed by exposure to Hyperfilm (Amersham Pharmacia Biotech Inc. UK). Antibody against cytochrome c (Affinity Bioreagents, Golden, CO) was used at a dilution of 1:1000.

Immunofluorescence staining

Cells grown on glass coverslips in 6-well plates were fixed in 4% paraformaldehyde (PFA) for 10 minutes, rinsed in TBS and permeabilized with 0.4% Triton-X-100 in 10 mM sodium citrate for 20 minutes. After blocking in 5% donkey serum for 20 minutes, cells were incubated with rabbit anti-cytochrome c (Affinity Bioreagents, Golden, CO) for 1 hour. Negative controls included normal rabbit IgG at same concentration as the primary antibodies used. A secondary antibody, anti-rabbit or FITC-conjugated donkey anti-mouse was then applied for 60 minutes. Cells were mounted on slides with Prolong Gold-DAPI and allowed to dry overnight. Slides were viewed using a Zeiss, Axiovert 200M fluorescent microscope and digital images recorded using Slidebook software (Intelligent Imaging Innovations, Denver, CO).

Cytokine and total p65 assay

Apical and basolateral media was collected, centrifuged and supernatant was analyzed for cytokines by ELISA at ELISATech (Denver, CO). Nuclear lysates were analyzed for total p65 using

NF-κB p65 (Total) Elisa kit (Invitrogen, Carlsbad, CA). NF-κB inhibitor [6-amino-4- (4-phenoxyphenylethylamino) quinazoline] was obtained from EMD Biosciences La Jolla CA.

Respiratory mucin analysis

Human airway epithelial cell cultures were analyzed for extracellular levels of respiratory mucins by use of a double-sandwich ELISA system (MPY-661P) from Covance (Princeton, NJ) [21]. The cell surface was washed with PBS and incubated with DTT (6 mM) in PBS for 30 min. Supernatants were carefully collected using a 1.0 ml pipette tip and stored at −20°C for subsequent analysis. A delay of 24 h was deliberately used to ensure oxidation of DTT and minimize its effect in assays. For mucin analysis, 96-well plates were coated with purified monoclonal mucin antibody (17B1, Covance, Princeton, NJ) (0.3 μg/well) in 100 μl of coating buffer (0.05 M sodium carbonate buffer, pH 9.6) for 3 h under airtight cover. Wells were then rinsed 3 times using 150 μl of PBS-Tween 20 (0.05%). Samples were diluted in PBS-Tween (0.025%) and added to wells and the plates were sealed and allowed to incubate at 37°C for 4 h. Wells were rinsed 3 times and then reacted with 100 μl of PBS-tween 20 containing 0.4 μg of the alkaline phosphatase-conjugated 17Q2 antibody. Plates were then incubated overnight at 4°C on a rotary shaker set to 50 rpm. Fluids were removed and each well rinsed 3 times and then filled with 100 μl of substrate solution (1 mg/ml ρ-nitrophenyl phosphate disodium in 10% diethanolamine; pH 9.8). Absorbance change was monitored at 405 nm wavelength using a microplate reader. A standard curve was prepared using a stock of respiratory mucins, the protein concentration of which was determined using the optical method of Warburg and Christian [22].

Lentiviral transduction

Lentivirus each carrying 5 different SERCA2 target shRNA (pLKO.1-CMV-tGFP-ATP2A2) and controls (pLKO.1-CMVtGFP-TurboGFP control) were obtained from Sigma (St.Louis MO). Transductions were carried out using standard procedure and according to manufacturer's instructions and SERCA2 knock down was tested using real time RTPCR. The shRNA sequence giving more than 50% SERCA2 knockdown was used.

Adenoviral transduction

Transduction of SERCA2 and GFP-encoding adenoviral vectors was carried out as described before [23]. Ad.SERCA2 was a kind gift of Dr. R. J. Hajjar, Harvard Medical School, Massachusetts General Hospital, Cardiovascular Research Center, Charlestown, MA [24]. Recombinant viruses were added to cell cultures (multiplicity of infection, MOI 10:1) on day 3 of culture, with exposure for 17 hours duration. Transduction efficiency was estimated by observing green fluorescence of adenoviral GFP-transduced cells.

Stable NF-κB reporter cell lines construction

Reporter cell lines were generated by stably transducing 16HBEo- cells with GreenFire1™ pGF1-NF-κB reporter lentiviral vector (Systems Biosciences, Mount View, CA). This vector is a Human Immunodeficiency Virus (HIV) lentiviral vector that expresses destabilized (ds) copGFP reporter and firefly luciferase under the control of four NF-κB response elements and a minimal CMV promoter. The transfected cells were treated with TNFα (10 ng/ml) overnight then sorted using Fluorescence Activated Cell Sorting (FACS) using MoFlo Cell Sorter (Dako Cytomation, Carpinteria, CA) for high GFP expression. Sorted cells were cultured and sorted again after TNF treatment and individual clones were isolated and cultured.

SERCA2 gene silencing using siRNA and luciferase reporter assay

SERCA2 gene silencing was carried out as described previously [10]. Briefly, predesigned human 'SMARTPOOL' SERCA2 siRNAs were purchased from Dharmacon (Lafayette, CO). NF-κB reporter expressing 16HBEo- cell lines were cultured on fibronectin-coated 6-well plates and transfected with 50 nM siRNA using DharmaFECT2 siRNA transfection reagent (Dharmacon; Lafayette, CO) according to manufacturer's instructions as described before [10]. Silencer negative control siRNA was used as a nonspecific siRNA, and mock transfection was used as a negative control. Cells were treated 48 h post-transfection and cell lysates were made using Reporter Lysis Buffer (Promega, Madison WI). Luciferase activities were determined with equal amounts of protein by using a commercially available luciferase assay system (BD Pharmingen, San Jose, CA) and a Monolight 3010 luminometer (Analytical Luminescence Laboratory, San Diego, CA).

Statistical analysis

All statistical calculations were performed with JMP and SAS software (SAS Institute, Cary, NC). Data from experiments using one cell line per disease group were analyzed using two-way analysis of variance (ANOVA), with ozone level and disease (or treatment) category as the two factors, including the interaction term. For experiments involving multiple subjects per disease group (see section entitled '*Ozone-dependent cytokine release in differentiated primary non-CF and CF epithelial cells*' in Results) a linear mixed model was fit for each cytokine and media combination, using ozone level, disease category, and their interaction as predictors; a random intercept term for subjects was also included to account for repeated measures within subjects across factor levels.

Generally, pairwise comparisons were performed using t-tests, comparing the given ozone level to no ozone within a disease or treatment category and between a given disease or treatment category to no disease (or reference category) within a fixed level of ozone. Variables with right-skewed distributions were analyzed on the natural log scale in order to better meet model assumptions. The log transformation was also used for data with multiple subjects in order to stabilize variances between ozone*disease groups. Graphs were constructed based on untransformed sample data.

Results

Exposure to ozone causes acute toxicity to human airways. Using non-polarized 16HBEo- cell lines we have demonstrated previously that exposure to ozone causes enhanced cell death [18]. We have also demonstrated that the airway epithelial cells from CF patients have decreased SERCA2 [10]. SERCA2 inhibition in airway epithelial cells also caused enhanced cell death upon ozone exposure [10]. Therefore, in this study, we systematically investigated ozone toxicity using polarized cultures of non-CF (16HBEo-) and CF (CF41o- and CF45o-) cells. We examined several components of cellular toxicity to assess the effect of ozone exposure.

Exposure to ozone (200 ppb, 8 h) caused increased monolayer disruption in CF cells (CF41o- and CF45o-) when compared to the non-CF controls (16HBEo-). Monolayer disruption in polarized

Table 2. Effect of ozone exposure on transepithelial resistance (TER) in non-CF and CF airway epithelial cells.

Cell line	TER (Mean±SEM) Ohms.cm^2 0 ppb	TER (Mean±SEM) Ohms.cm^2 200 ppb
16HBEo-	522±06	500±45
CF41o-	369±23	190±68* #
CF45o-	394±33	251±41* #

*Indicates significant difference from 0 ppb, $p<0.05$.
# Indicates significant difference from 200 ppb non-CF, $p<0.05$.

cultures was measured by loss of trans-epithelial resistance (TER) as shown in Table 2.

Measurements of uptake of Calcein-AM (live cell) and propidium iodide (dead cell) were also performed in polarized cultures of non-CF and CF cell lines. Propidium iodide uptake by these cells at 8 h of 200 ppb ozone exposure was minimal and could not be quantified using this method. Therefore, we exposed the cells at 500 ppb. Exposure of CF cells (CF41o- and CF45o-) to ozone (500 ppb) caused increased uptake of propidium iodide, further reflecting membrane damage and cell death (Figure 1 A & B). Caspase 3/7 release by non-CF and CF cells exposed to ozone (500 ppb) was also measured. Exposure to ozone caused an increase in caspase release in CF45o- cells at 2 h and was sustained until 6 h of exposure time. CF41o- cells also exhibited an increased caspase release at 4 h of exposure to ozone (Figure 1C). To further, validate and assess mode of cell death, cells were harvested by trypsinization and apoptosis was quantified using the Annexin V-binding method. Exposure to ozone (500 ppb, 8 h) of non-CF (16HBEo-) cells caused an increase in Annexin V-binding, which was significantly greater than that in 0 ppb exposed cells. Ozone exposure of CF cells (CF41o-) caused approximately two-fold increased annexin binding which was also significantly greater than in non-CF (16HBEo-) cells (Figure 1D). Additionally we performed TUNEL staining on the ozone exposed non-CF and CF airway epithelial cell cultures. Ozone exposure caused increased TUNEL positive cells in non-CF and CF airway epithelial cells (Figure S1). CF airway epithelial cells demonstrated enhanced TUNEL positivity as compared to the non-CF cells.

CF airway epithelial cells have decreased SERCA2 expression that could modulate intracellular calcium homeostasis [10,25]. Both these effects could cause enhanced mitochondrial calcium uptake and that, upon further exposure to oxidant stress, could contribute to mitochondrial dysfunction and eventual cell damage [25,26,27]. We determined mitochondrial membrane potential (MMP) and cytochrome c release to assess mitochondrial function upon exposure to ozone in non-CF and CF cells cultured on collagen coated 6-well plates. Since these experiment were performed on cells cultured on collagen coated plastic dishes or glass coverslips, lower 200 ppb ozone exposure was sufficient to demonstrate its effect. Ozone exposure caused a small but nonsignificant decrease in MMP of non-CF (16HBEo-) cells. However, in CF (CF41o-) cells, there was an approximately 40% decrease in MMP after exposure to 200 ppb ozone as compared to 0 ppb ozone exposure. This was also significantly different from non-CF cells exposed to ozone at 200 ppb (Figure 1E). Similarly, we observed enhanced cytochrome c release (immunostaining of cells cultured on collagen-coated coverslips and by Western blots) in CF cells upon ozone exposure (Figure 1F & 1G).

Ozone-induced enhanced cytokine release in CF cell lines

Ambient concentrations of ozone induce airway inflammation through release of proinflammatory mediators from airway epithelial cells [28]. We studied three cytokines viz. IL-8, G-CSF and GM-CSF in both apical and basalolateral compartments of polarized cultures. These cytokines are induced by ozone in bronchial epithelial cells and their levels are further enhanced upon ozone exposure in patients with other airway diseases [28,29]. Exposure to ozone (100 or 200 ppb, 18 h) caused enhanced IL-8, GM-CSF and G-CSF release in the apical media of CF cell line CF41o- as compared to the non-CF 16HBEo- (Figure 2, A-C). We did not observe measurable cytokine content in basolateral media of these cells.

Ozone-dependent cytokine release in differentiated primary non-CF and CF epithelial cells

Air liquid interface (ALI) cultures of human nasal epithelial cells have been shown to release IL-8 upon exposure to ozone [30]. Differentiated primary airway epithelial cells (NHBE cells from a single donor) grown in co-culture with airway smooth muscle cells also release cytokines like IL-6 upon ozone exposure, whereas, the monolayer cultures do not release IL-6 upon exposure to ozone [7]. Studies comparing responses of differentiated primary airway epithelial cells derived from normal and diseased donors are lacking. We extended our study of ozone-mediated toxicity in non-CF and CF airway epithelium by using differentiated cultures of cells obtained from 3 non-CF and 3 CF donors (demographics shown in Table 1) to investigate the proinflammatory cytokine release response to ozone in close to *in vivo* airway epithelial cell models. These studies were performed in 30-day-old fully differentiated airway epithelial cell cultures details of which are described before [10]. We conducted these experiments at a relevant ambient concentration of ozone (200 ppb) and the exposure duration of 18 h. Proinflammatory cytokine production by epithelial cells from individual donors varied considerably. Our initial results with cells from 9 non-CF and 9 CF donors revealed even greater heterogeneity (data not shown). The differentiated cultures of airway epithelial cells secrete large amounts of mucus that may impact the cytokine measurement either by binding them or by limiting ozone access to cell surfaces [31]. Therefore, we compared apical IL-8 harvested by adding media to the apical surface to IL-8 harvested by further dissolution of apical mucus with DTT (Figure S2A). (It is important to note that even with DTT solubilization there was still a considerable amount of mucus gel remaining adherent which was impossible to completely remove without damaging cells.) We observed that the apical yield of solubilized IL-8 released from mucus was ~40% greater when combined with the apical media. We used several controls shown in Figure S2A to ensure the authenticity of the assay including (1) 25 nM IL-8 standard in PBS or DTT (column 1 & 2 of Figure S2A) to demonstrate effect of DTT (2) The reaction was performed with prior incubation with IL-8 antibody or without the primary antibody in the sandwich ELISA to demonstrate the specificity of the reaction. (3) Further, purified porcine mucin was included to rule out the potential contribution of mucins directly to a positive reaction in the ELISA (column 5, Figure S2A). (4) IL-8 was added to porcine mucin to demonstrate potential interference of the mucin in the assay (column 6, Figure S2A). The results demonstrate that considerable IL-8 was retained within mucus. We did not detect mucin interference in the assay. The mucin content was also confirmed for these two fractions (Figure S2B, columns 3 & 4). Ozone exposure can cause enhanced mucin secretion in the upper airways of animal models [32,33]. CF airway epithelial cells might have further enhanced mucin

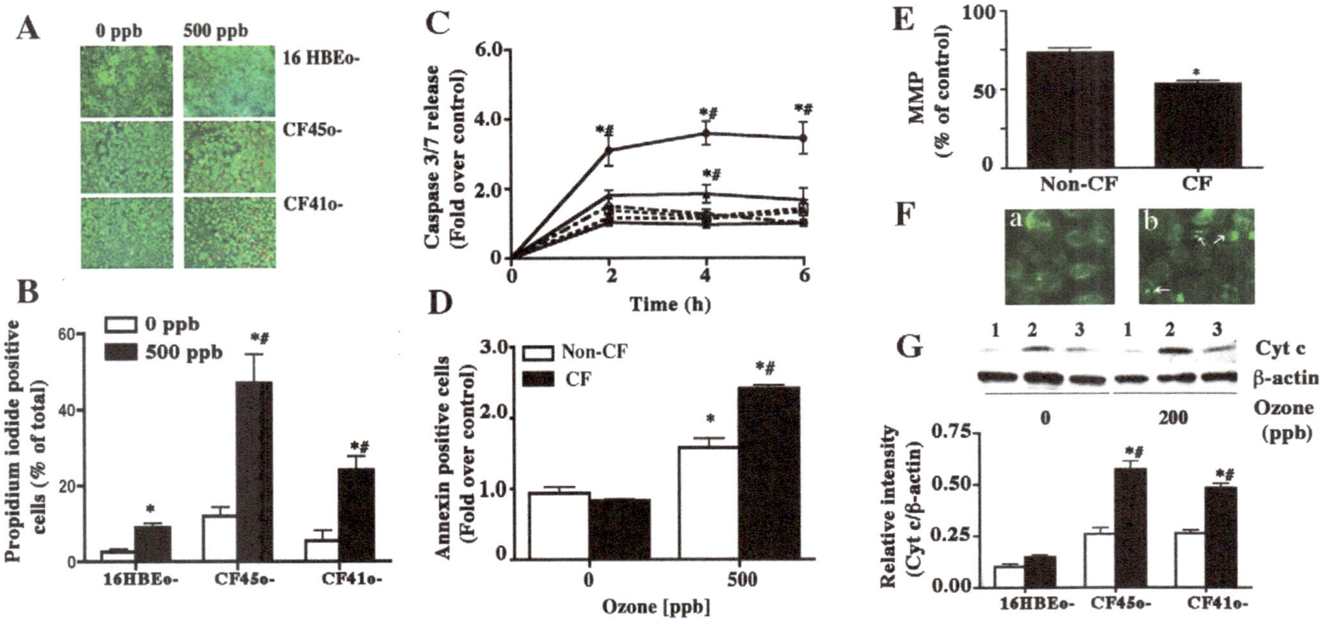

Figure 1. Ozone-induced toxicity in non-CF and CF airway epithelial cells. Polarized cultures of non-CF (16 HBEo-) and CF (CF41o- and CF45o-) cells were exposed to either 0 or 500 ppb ozone for 8 h. At the end of exposure cells were stained with Calcein AM (green, live) and propidium iodide (PI) (red, dead) (A). Quantitation of dead PI positive cells was performed using Image-Pro Plus version 4.0 (Media Cybernetics, Silver Spring, MD) and shown in (B). The percentage of dead cells was determined as the percentage of adherent positive cells relative to total number of adherent cells per ten high-power fields (x20). The assessment was performed blindfolded by two different individuals and the data obtained was analyzed by a statistician. The data shown are mean±SEM (n = 3). The image is representative of three independent experiments. * Indicates significant difference from 0 ppb control and # indicates significant difference from 500 ppb exposed non-CF, $p<0.05$. (C) Polarized cultures of 16HBEo- (-□-) and CF41o- (-△-) and CF45o- (-○-) cells were exposed to either 0 (open symbols) or 500 ppb (closed symbols) ozone for 2, 4 or 6 h. At the end of each exposure supernatant media was collected and analyzed for caspase 3/7 release as described in the Methods. The data shown are mean±SEM (n = 3). * Indicates significant difference from 0 ppb control, and # indicates significant difference from 500 ppb exposed non-CF $p<0.05$. (D) Polarized cultures of non-CF 16HBEo- (white columns) and CF CF41o- (black columns) cells were exposed to either 0 or 500 ppb ozone for 8 h following which cells were collected and Annexin binding assay was performed as described in the Methods. Data is from one representative of 2 experiments. The data shown are mean±SEM (n = 6). The image is representative of two independent experiments. * Indicates significant difference from 0 ppb control and # indicates significant difference from 500 ppb exposed 16HBEo-. Cells $p<0.05$ (E) 16HBEo- and CF41o- cells were cultured on fibronectin-coated 6-well plates and exposed to ozone on the 4th day of plating. Measurement of chloromethyltetramethylrosamine (MitoTracker Orange, Molecular Probes) fluorescence, an indicator of mitochondrial membrane potential (MMP), was performed as described in the Methods. A mean fluorescence intensity (MFI) of cells exposed to 0 ppb ozone was taken as control value. The data shown are mean±SEM (n = 6) The image is representative of two independent experiments. * Indicates significant difference from non-CF cells, $p<0.05$. (F) Ozone-mediated cytochrome c release in non-CF and CF cells. 16HBEo- and CF41o- cells were cultured on glass coverslips in 6-well plates and exposed to ozone (200 ppb) for 4 h. Cells were fixed and processed for cytochrome c releases using immunocytochemistry as described in the text. Representative images of (a) 16HBEo- and (b) CF41o- cells exposed to 200 ppb ozone are shown. The arrows indicate staining of cytochrome c. (G) Western blot of cytochrome c ('Cyt c') in the cytosolic fraction of 0 ppb and 200 ppb (4 h) ozone-exposed 16HBE (1), CF41o- (2) and CF45o- (3) cells. Image is representative of three independent experiments. The lower panel is a bar graph showing quantitation of cytochrome c Western bands. The data shown are mean±SEM (n = 3). * Indicates significant difference from 0 ppb control, and # indicates significant difference from 200 ppb (closed bars) exposed non-CF, $p<0.05$.

secretion. We estimated the apical media mucin content, as well as the solubilized mucin content from the apical surface of ozone-exposed differentiated cultures of non-CF and CF airway epithelial cells (Figure S2C). CF cells had increased mucin secretion relative to non-CF cells. In addition, mucin secretion was further enhanced upon ozone exposure in CF cells (Figure S2C). Therefore, we estimated the total IL-8 content (media+DTT solubilized mucin) harvested from the apical surface of differentiated non-CF and CF airway epithelial cells upon exposure to ozone. The IL-8 content of the basolateral media was also estimated. The overall pattern indicated that IL-8 and G-CSF were present in both apical and basolateral media (Figure 3A, B, C & D), whereas GM-CSF (Figure 3F) was barely detectable in the basolateral compartment. Ozone exposure caused enhanced proinflammatory cytokine (IL-8, G-CSF and GM-CSF) production at the apical surface in non-CF airway epithelial cells. The basolateral IL-8 and G-CSF contents were also significantly increased in the non-CF group upon ozone exposure. There was a significantly increased content of IL-8, G-CSF and GM-CSF on apical surface of 0 ppb exposed cultures of airway epithelial cells from CF donors as compared to the non-CF cultures. However, ozone exposure (200 ppb) did not further alter cytokine levels as compared to 0 ppb in a statistically significant manner in CF cells. The proinflammatory cytokine content of apical surface of 200 ppb ozone-exposed CF cells was still significantly greater than apical cytokine content of 200 ppb ozone-exposed non-CF airway epithelial cells.

SERCA2 modifies ozone-induced cytokine release

A number of studies have implicated intracellular calcium mobilization in the activation of NF-κB and inflammatory signaling responses [15,34]. SERCAs are responsible for (re)

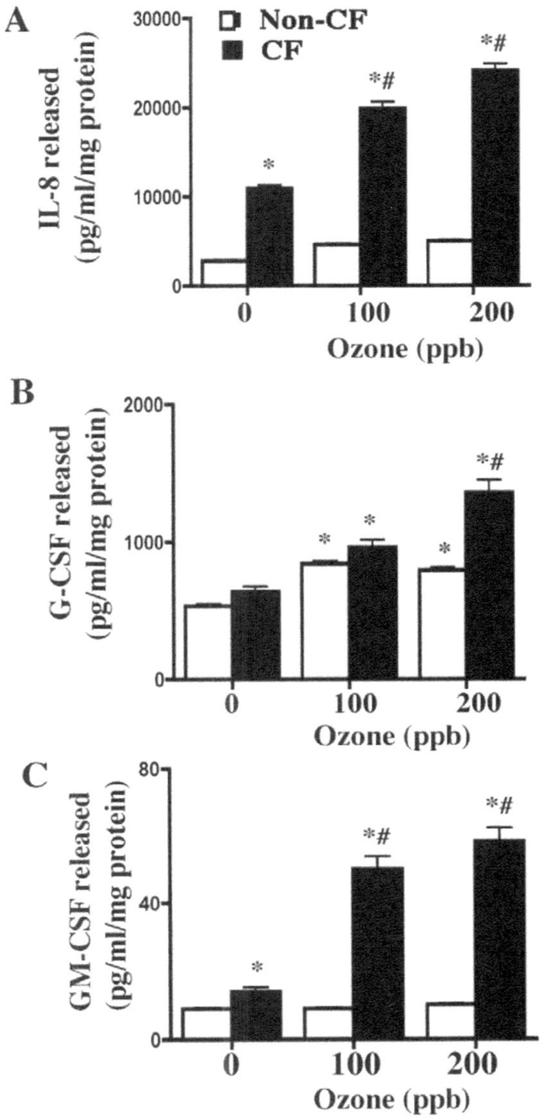

Figure 2. Ozone-mediated cytokine release in polarized cultures of non-CF and CF cell lines. 16HBEo- (white column) and CF41o- (black column) cells were cultured on collagen-coated 30 mm inserts. Once cells were polarized (8-10 days in culture) cells were exposed to 0, 100 or 200 ppb ozone with 200 µl media on the apical surface. At the end of exposure (18 h) an additional 200 µl media was added to the apical surface. Aliquots were collected after 4 h and analyzed for IL-8 (A), G-CSF (B) or GM-CSF (C) as described. The data shown are mean±SEM (n=6). The image is representative of four independent experiments. * Indicates significant difference from 0 ppb control and # indicates significant difference from 100 or 200 ppb exposed non-CF $p<0.05$.

loading of endoplasmic reticulum calcium after such signaling events. Our previous work has established that CF airway epithelial cells have diminished SERCA2 expression [10]. We have also shown that siRNA mediated knockdown of SERCA2 in airway epithelial cells cultured on collagen-coated plastic dishes, causes enhanced cell death due to exposure to ozone, hydrogen peroxide and TNF [10]. Ozone exposure may itself modify SERCA2 expression in primary airway epithelial cells. Therefore, we investigated whether SERCA2 modulation by shRNA-mediated knockdown would affect ozone-induced proinflammatory cytokine release responses of polarized ALI cultures of primary airway epithelial cells. As shown in Figure 4A, using 5 different SERCA2 shRNA targets, we obtained a maximum knockdown using SERCA2 shRNA3 at 1:50 MOI as measured by quantitative real-time RTPCR. Using this shRNA target we also demonstrated a knockdown in SERCA2 protein by western blot (Figure 4B). To observe the effect of SERCA2 knockdown on airway epithelial cell morphology, we cultured control and shRNA-transduced cells on collagen-coated dishes. Cellular morphology was not effected by SERCA2 knockdown as seen in Figure 4C. Based on GFP expression, the SERCA2 shRNA-transduced cells were sorted to purify them from untransduced cells (Figure 4D). Using 7-actinomycin D (7-AAD) dye to identify and exclude dead cells, we also demonstrated that degree of cell death was not increased in SERCA2 shRNA-transduced cells relative to GFP-transduced cells. Thus, SERCA2-shRNA transduction was not toxic (Figure 4D). Ozone exposure of undifferentiated ALI cultures of primary non-CF airway epithelial cells decreased expression of SERCA2 protein (Figure 5A). Short term (10 days) undifferentiated ALI cultures of primary airway epithelial cells expressing lentiviral SERCA2 shRNA were exposed to ozone (200 ppb), and apical media was analyzed for IL-8 release after exposure. Ozone exposure caused increased IL-8 release in these cultures. Specific inhibition of SERCA2 expression caused a further significant increase in IL-8 release upon ozone exposure (Figure 5B). In a complementary approach, we investigated whether SERCA2 overexpression would affect IL-8 release. In primary airway epithelial cells transduced with Ad.SERCA2, IL-8 release due to ozone was significantly decreased as compared to cells expressing Ad.GFP or with untransduced controls (Figure 5C) indicating that SERCA2 overexpression downregulates IL-8 production.

Ozone-induced cytokine release in CF cell lines is NF-κB-mediated

Studies to evaluate the role of NF-κB in ozone-mediated cytokine release were performed on submerged non-confluent, non-polarized cells cultured on fibronectin and collagen-coated 6-well plates. This ensured effective delivery of the inhibitor to the cells. Preincubation with NF-κB inhibitor (10 µM [6-amino-4- (4-phenoxyphenylethylamino) quinazoline] (EMD Biosciences, La Jolla, CA) caused IL-8 levels in ozone-exposed (200 ppb) CF cell line CF41o- to decrease to control values (Figure 6A). In CF cells, there was enhanced ozone-mediated p65 mobilization as well (Figure 6B).

SERCA2 knockdown enhances NF-κB activation and proinflammatory cytokine production

The role of SERCA2 in modulating NF-κB activity was investigated using NF-κB-luciferase reporter gene expressing 16HBEo- cell lines. It was anticipated that inhibition of SERCA2 would induce NF-κB-driven luciferase activity. Inhibition of SERCA2 expression was performed using siRNA approach as described previously [10]. Thapsigargin was used to further decrease SERCA2 activity. Thapsigargin treatment by itself increased NF-κB-driven luciferase activity in both nonsilencing control and SERCA2-silenced cells. Treatment with cytomix (TNFα (20 ng/ml), IL-1β (10 ng/ml), IFγ (10 ng/ml) and LPS (50 ng/ml)) stimulated transcriptional activation of NF-κB that was greatest in SERCA2-silenced cells. Cytomix treatment in the presence of thapsigargin further increased NF-κB-driven luciferase activity in cells transfected with SERCA2-silencing siRNA (Figure 7A). SERCA2 inhibition

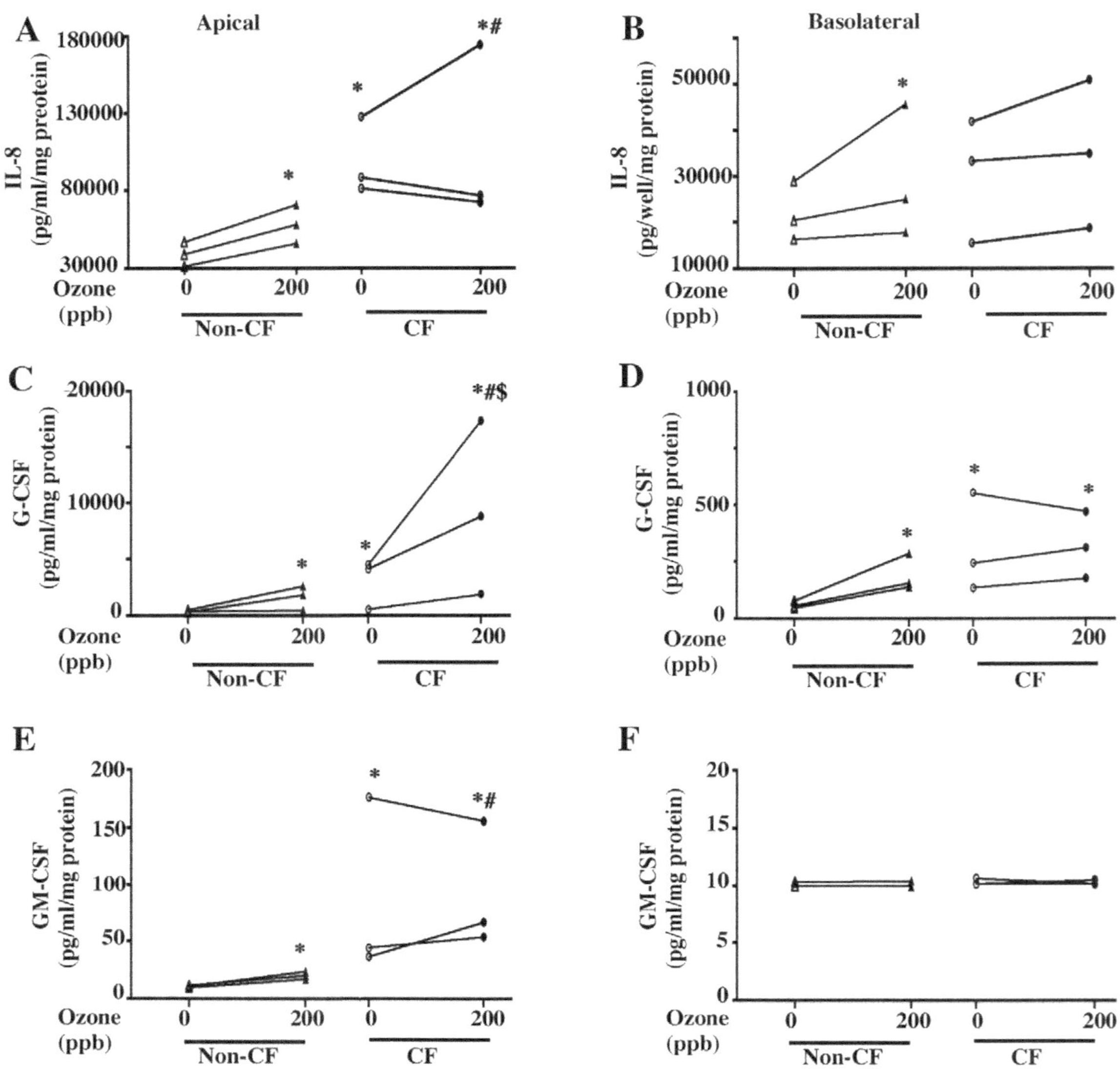

Figure 3. Effect of ozone exposure on the release of proinflammatory cytokines in differentiated cultures of non-CF (△, 0 ppb and ▲, 200 ppb) and CF (○, 0 ppb and ●, 200 ppb) primary airway epithelial cells. Cells were cultured on collagen-coated snapwells and allowed to grow and differentiate for 30 days. Before ozone exposure 100 μl media was added to the apical surface. At the end of exposure (18 h) an additional 200 μl media was added apically. After 4 h the media from apical (A) and basolateral (B) surface was collected and 100 μl of 6 mM DTT in PBS was added on the apical surface. The cells were incubated for 15 min in DTT and the supernatant fluid was carefully collected using a 1.0 ml pipette tip. The media and DTT solubilized mucin were analyzed for cytokines (IL-8 in A & B, G-CSF in C & D and GM-CSF in E & F) as described in the Methods. For apical cytokine content media and DTT solubilized mucin samples were pooled. Each symbol represents the mean of each donor. Three separate experiments were performed each time using one non-CF (donor 1-3) and one CF (4-6) donor. * Indicates significant difference from 0 ppb exposed non-CF, $p<0.05$ and # indicates significant difference from 200 ppb non-CF, $p<0.05$ and $ indicates significant difference from 0 ppb CF, $p<0.05$.

caused increased release of IL-8, as shown by cytomix treatment of NF-κB-luciferase reporter gene expressing 16HBEo- cells (Figure 7B). Interestingly, IL-8 release by thapsigargin treatment alone exceeded IL-8 release by cytomix treatment in both siControl and siSERCA2 groups. Similarly, further increased IL-8 release by cytomix+thapsigargin treatment was observed in both groups, but IL-8 release in the SERCA2-silenced group was greater than in the nonsilenced group (Figure 7B).

SERCA2 inhibition enhances ozone-induced NF-κB activity and IL-8 release

NF-κB-luciferase reporter gene expressing 16HBEo- cells were exposed to 0 or 500 ppb ozone. Ozone (500 ppb) exposure caused an increase in luciferase activity in both siControl and siSERCA2 transfected cells indicating transcriptional activation of NF-κB. However, at 500 ppb, the relative luciferase activity was greater in siSERCA2-silenced cells (Figure 8A). SERCA2 silencing also induced a greater IL-8 release response to ozone (500 ppb) exposure relative to siControl-transfected cells (Figure 8B). These results again indicate that SERCA2 can modulate NF-κB-dependent proinflammatory cytokine production.

SERCA2 overexpression mitigates the enhanced proinflammatory response in CF cells

Previously, Kube et al. have shown that 16HBE-AS, a 16HBE cell line lacking the cystic fibrosis transmembrane conductance regulator

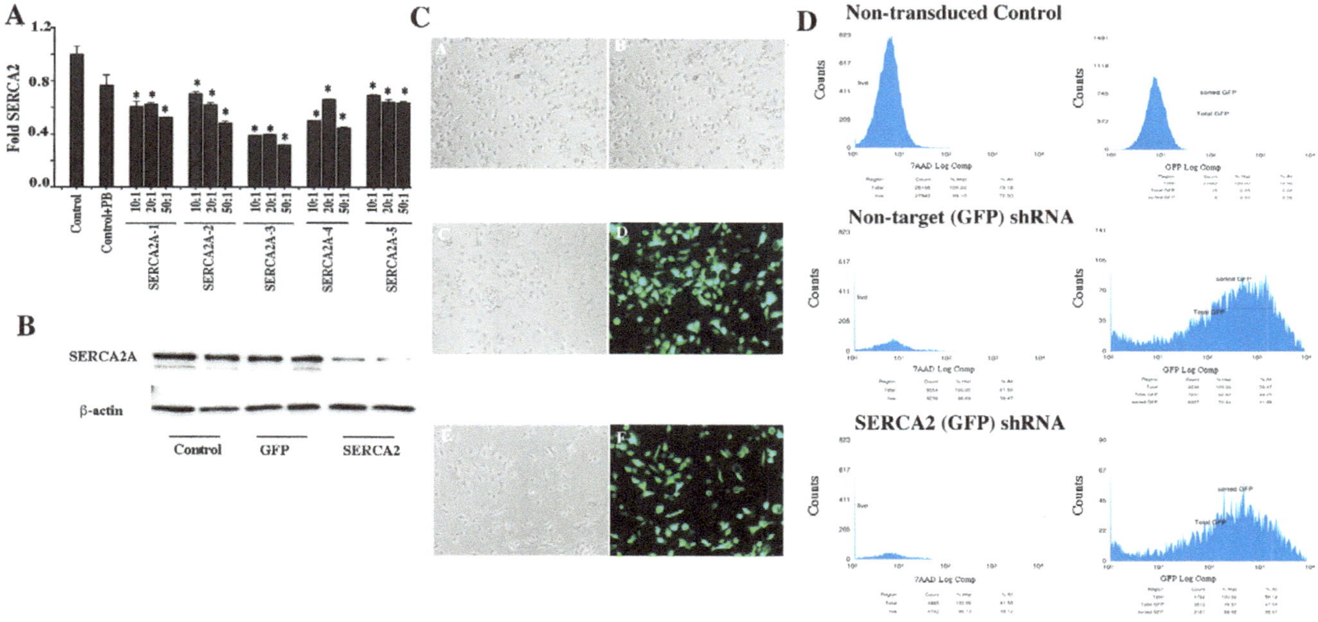

Figure 4. Lentiviral shRNA-mediated SERCA2 knockdown and its effect on cell phenotype and viability. A) Primary airway epithelial cells were cultured on collagen-coated 24-well plates. Replication-deficient lentivirus each carrying 5 different SERCA2 target shRNA (pLKO.1-CMV-tGFP-ATP2A2) (SERCA2A-1-5) were transduced at different MOI according to manufacturer's instructions and SERCA2 knock down was tested using real time RTPCR. The data shown are mean±SEM (n = 3) *$p<0.05$, significant difference between cells treated with SERCA2 shRNA and control. B) Primary airway epithelial cells were cultured on collagen-coated 6-well plates and transduced with SERCA2 shRNA (pLKO.1-CMV-tGFP-ATP2A2) or pLKO.1-CMVtGFP-TurboGFP control. Forty-eight hour post transduction cell lysates were prepared and SERCA2 Western blot was performed. C) Airway epithelial cell phenotype with and GFP expression by control and SERCA2 shRNA transduced cells. Cultured cells retained the morphology of airway epithelial cells (Brightfield light microscopy of untransduced, A & B, transduced C & E) and expressed GFP (Fluorescence microscopy of transduced D & F, matching area to the Brightfield in C & E). D) Flow cytometry of control, SERCA2 shRNA (pLKO.1-CMV-tGFP-ATP2A2) or pLKO.1-CMVtGFP-TurboGFP control cells with 7-AAD a cell death marker.

activity, showed enhanced IL-8 and IL-6 responses compared with the control cell line 16HBE-S upon stimulation with *Pseudomonas aeruginosa* and a cytomix containing TNF-α and IL-1β [35]. This study further suggested that these cell lines might be useful for studying proinflammatory responses in CF airways [35]. We used these matched pair of non-CF and CF cell lines to study the effect of adenoviral-mediated SERCA2 overexpression (Figure S3) on cytomix-induced IL-8 production and NF-κB activity. Untreated control as well as GFP-transduced 16HBE-AS cells exhibited increased basal content of IL-8 that was significantly reduced by SERCA2 overexpression (Figure 9). IL-8 release measured 24 h post-treatment with cytomix was significantly increased in both control and CFTR-deficient cell lines. However, the cytomix-mediated IL-8 release in non-transduced control and GFP-transduced 16HBE-AS cells were significantly enhanced than the control cell line. Ad.SERCA2-transduction significantly abrogated enhanced IL-8 release by CF (16HBE-AS) cells and also diminished the response of control cells to cytomix (Figure 9A). Total p65 was also estimated in nuclear lysates of control and cytomix-treated 16HBE-AS and 16HBE-S cells transduced with Ad.GFP or Ad. SERCA2. Overexpression of SERCA2 also decreased cytomix-induced NF-κB activity, as evidenced by the decrease in total nuclear p65 in both control and CF cell lines (Figure 9B). These results further confirm that SERCA2 can modulate NF-κB activity and proinflammatory responses of airway epithelial cells.

Discussion

Ozone is a highly reactive gas which, when present at higher concentrations, may cause airway damage via complex mechanisms. We have previously established the role of ozone-reactive surfactant phospholipids in modulating airway epithelial cell viability in response to ozone [8]. These derivatives not only caused apoptotic cell death but also induced proinflammatory responses that may potentiate airway injury in asthmatics and other susceptible patients like those with cystic fibrosis (CF). Increased oxidative stress and enhanced lipid-derived inflammatory mediators that are not modulated upon treatment is a characteristic of CF patient biofluids such as sputum during respiratory exacerbations [9]. A role for ozone in contributing to CF respiratory exacerbations has been suggested [1]. However, the type of injury and mechanisms involved in ozone-induced damage to CF airway epithelial cells is not clear. In this study, we investigated responses of CF and non-CF airway epithelium to ozone. We found that exposure of CF airway epithelial cells to ozone caused enhanced membrane damage, cell death and mitochondrial dysfunction. Further, polarized cultures of CF airway epithelial cell lines showed increased proinflammatory cytokine release upon ozone exposure. Similarly, exposure to ozone of differentiated airway epithelial cells from non-CF and CF donors yielded enhanced proinflammatory cytokine production. Ozone exposure decreased SERCA2, and SERCA2 knockdown caused an increased IL-8 release. Ozone-induced IL-8 release was reversed by SERCA2 overexpression. Ozone-induced IL-8 release was found to be NF-κB dependent. Interestingly, in cells with stable expression of the NF-κB luciferase reporter, SERCA inhibition caused increased NF-κB activity upon stimulation by cytomix containing IL-1β, TNFα, IFNγ and LPS. Cytomix and its components have been previously shown to induce NF-κB activity in airway epithelial cells including those of CF patients [36,37,38,39,40]. Increased NF-κB activity was also accompanied

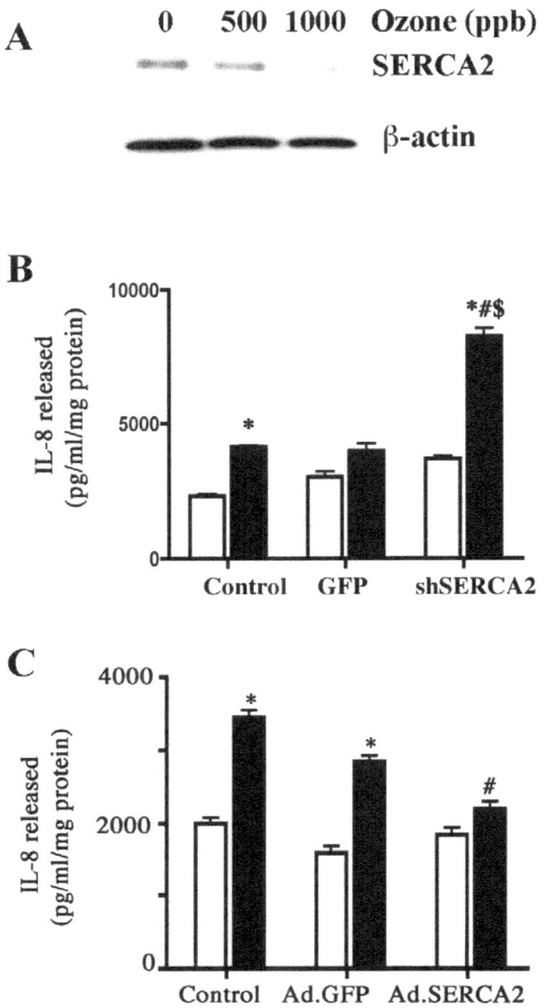

Figure 5. SERCA2 regulates ozone-mediated cytokine release by primary airway epithelial cells. Cells were cultured on collagen-coated snapwells at air-liquid interface (ALI) or on collagen-coated 6-well plates. (A) Primary airway epithelial cells were cultured at ALI and exposed to 500 or 1000 ppb ozone. At the end of exposure (4 h), lysates were prepared and SERCA2 protein was analyzed using Western blot. A representative blot from cells of one donor is shown. (B) Cells transduced with lentiviral GFP control or lentiviral GFP-SERCA2 shRNA were cultured on collagen-coated snapwells at air-liquid interface (ALI) and exposed to 0 (white column) or 200 (black column) ppb ozone as described in legend to Figure 3. Apical media was collected and analyzed. The data shown are mean±SEM (n = 6). The image is representative of two independent experiments. * Indicates significant difference from 0 ppb control and # indicates significant difference from 200 ppb exposed lentiviral GFP transduced cells, and $ indicates significant difference from 0 ppb exposed lentiviral GFP-SERCA2 shRNA transduced cells $p<0.05$. C) Primary airway epithelial cells cultured on 6-well plates were transduced with Ad.GFP or Ad.SERCA2. Exposure to ozone, 0 (white column) or 200 (black column) ppb for 18 h was carried out 48 h post transduction. The data shown are mean±SEM (n = 6). The image is representative of two independent experiments. * Indicates significant difference from 0 ppb and # indicates significant difference from 200 ppb ozone-exposed Ad.GFP transduced cells $p<0.05$.

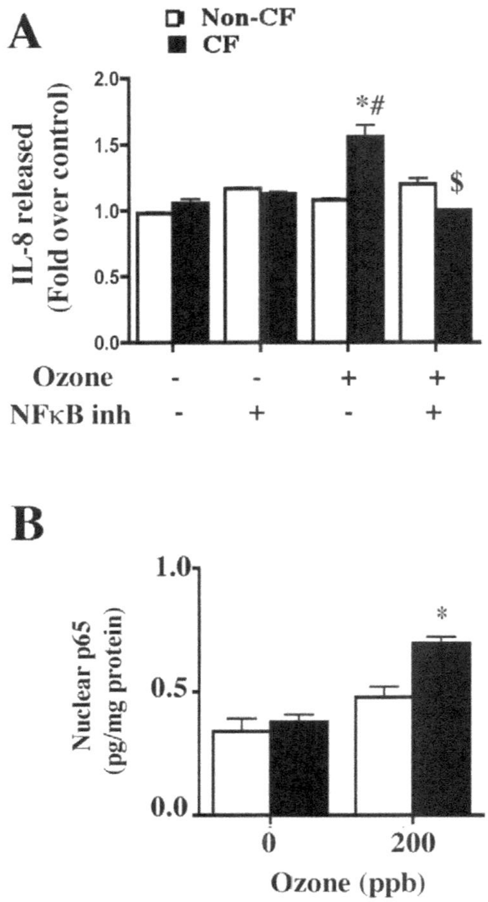

Figure 6. Ozone-induced cytokine release is NF-κB dependent. A) Effect of preincubation of cells cultured on collagen coated 6-well plates for 30 min with 10 μM NF-κB inhibitor [6-amino-4- (4-phenoxyphenylethylamino) quinazoline] or diluent on IL-8 release. The data shown are mean±SEM (n = 3). The image is representative of four independent experiments. * Indicates significant difference from 0 ppb exposed cells, # indicates significant difference from non-CF and $ indicates significant difference from CF cells not treated with NF-κB inhibitor during ozone (200 ppb) exposure $p<0.05$. For analysis of NF-κB activation, nuclear p65 was measured in non-CF and CF cells cultured on collagen-coated 6-well plates as described in the Methods (B). Cells were harvested and nuclear lysates were prepared after exposure to 200 ppb ozone. The nuclear fractions were assessed for p65 using the method described. The data shown are mean±SEM (n = 6). The image is representative of two independent experiments. * Indicates significant difference from 200 ppb ozone-exposed non-CF cells $p<0.05$.

by enhanced IL-8 release in this system. A similar increase in NF-κB activity and IL-8 release was observed upon exposure to higher ozone concentrations in the NF-κB reporter cell lines. Overexpression of SERCA2 by adenovirus-mediated gene delivery abrogated the enhanced proinflammatory response of CFTR deficient CF -like 16HBE-AS cell lines.

Proliferation of mucus (goblet) cells and hypersecretion of airway mucus are important characteristics of CF. Exposure to ozone causes changes in airway fluid secretion similar to CF in animal models of primates and rats [41]. Moreover, exposure to ozone and particulate air pollutants were associated with an increased risk of pulmonary exacerbations and a decline in lung function in CF, suggesting a role of environmental exposures in causing exacerbations [1]. Our findings of decreased SERCA2 expression in CF epithelium and enhanced susceptibility to oxidative stress such as that caused by ozone further suggests that survival responses of CF epithelium are compromised [10]. The

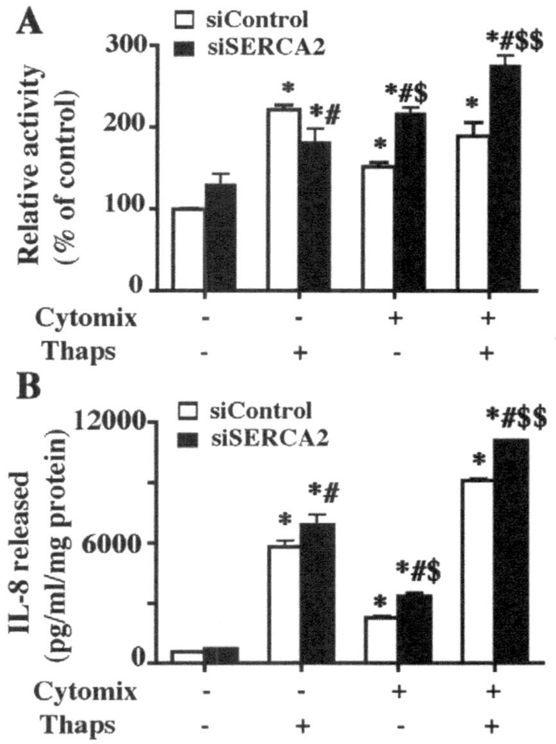

Figure 7. SERCA2 regulates NF-κB promoter activity and IL-8 release. NF-κB luciferase reporter-expressing 16HBEo- cells were cultured on collagen-coated 6-well plates and transfected with either control siRNA or SERCA2 siRNA. Forty-eight hour post transfection cells were treated with 'Cytomix' containing TNFα (20 ng/ml), IL-1β (10 ng/ml), IFNγ (10 ng/ml) and LPS (50 ng/ml) with or without thapsigargin (thaps; 2 μM)). (A) Cell lysates were prepared 24 h after treatment and luciferase activity was measured as described. The data shown are mean±SEM (n = 6). The image is representative of four independent experiments. * Indicates significant difference from untreated siControl transfected cells, # indicates significant difference from untreated siSERCA2 transfected cells, $ indicates significant difference from cytomix-treated siControl transfected cells, and $$ indicates significant difference from cytomix+thaps-treated siControl-transfected cells $p<0.05$. (B) Cells were cultured and treated as described above and supernatant media was collected and analyzed for IL-8. The data shown are mean±SEM (n = 3). The image is representative of two independent experiments. * Indicates significant difference from untreated siControl transfected cells, # indicates significant difference from untreated siSERCA2 transfected cells, $ indicates significant difference from cytomix-treated siControl transfected cells and $$ indicates significant difference from cytomix+thaps-treated siControl-transfected cells $p<0.05$.

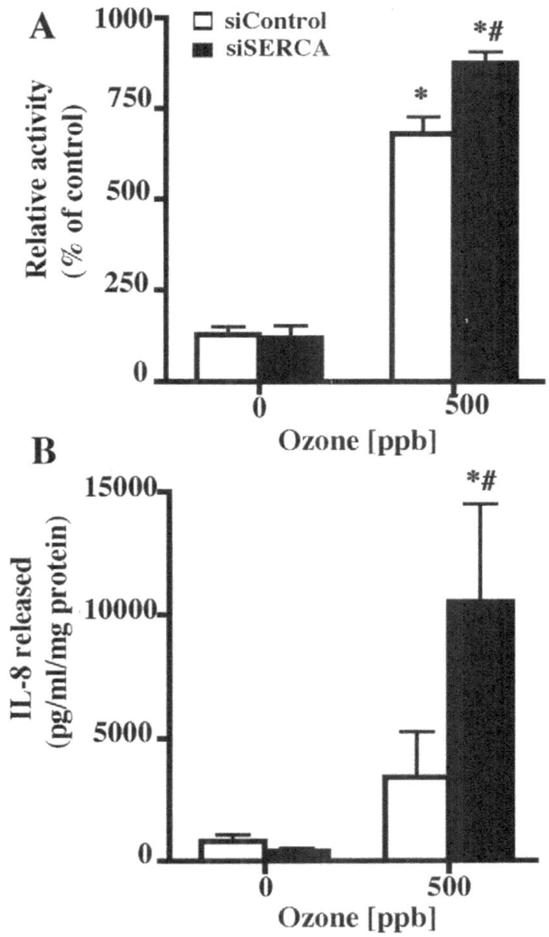

Figure 8. SERCA2 modulates ozone-induced NF-κB activity and IL-8 release. NF-κB luciferase reporter-expressing 16HBEo- cells were cultured on collagen-coated 6-well plates and transfected with either control siRNA or SERCA2 siRNA. Forty-eight hour post transfection cells were exposed to ozone (0 or 500 ppb). (A) Cell lysates were prepared after treatment and luciferase activity was measured as described. The data shown are mean±SEM (n = 6). The image is representative of two independent experiments. * Indicates significant difference from 0 ppb siControl transfected cells, # indicates significant difference from 500 ppb siControl transfected cells $p<0.05$. (B) Cells were cultured and treated as described above and supernatant media was collected and analyzed for IL-8. The data shown are mean±SEM (n = 6). The image is representative of two independent experiments. * Indicates significant difference from 0 ppb siControl transfected cells and # indicates significant difference from 500 ppb siControl transfected cells.

fact that there was increased loss of transepithelial resistance upon exposure to ozone demonstrates greater susceptibility of CF airway epithelial cells to such stress. However, at lower ozone concentrations there was no quantifiable propidium uptake to demonstrate membrane damage and death. The loss of TER without cell death could be attributed to other possible mechanisms such as those that occur upon infection by viral particles [42]. Moreover, lifting of a single cell from a tight monolayer could also dramatically reduce TER. Further, our results of increased membrane damage and apoptosis in CF cell are consistent with our previous findings as well as those of other investigators wherein ozone or ozone-generated oxidized lipids caused apoptosis [8,18,43]. By contrast, acute exposure to higher ozone concentrations caused cell death via non-apoptotic mechanisms in airway epithelium in rats [44]. SERCA2 dysfunction would tend to cause chronically enhanced cytosolic calcium that could be harmful to cellular organelles including mitochondria and even predispose to cell death. Moreover, pre-existing oxidative stress in CF airway epithelium could lead to greater mitochondrial and cellular dysfunction upon exposure to additional oxidizing environmental pollutants such as ozone.

Elevated expression of proinflammatory cytokines appears to contribute to chronic airway disease exacerbations. Accordingly, ozone-induced bronchial epithelial cytokine expression was increased in mild allergic asthmatic subjects as compared to healthy subjects [29]. Those cytokines included, among others, IL-8 and GM-CSF. Decreased SERCA2 expression and increased proinflammatory cytokine production have been recently observed

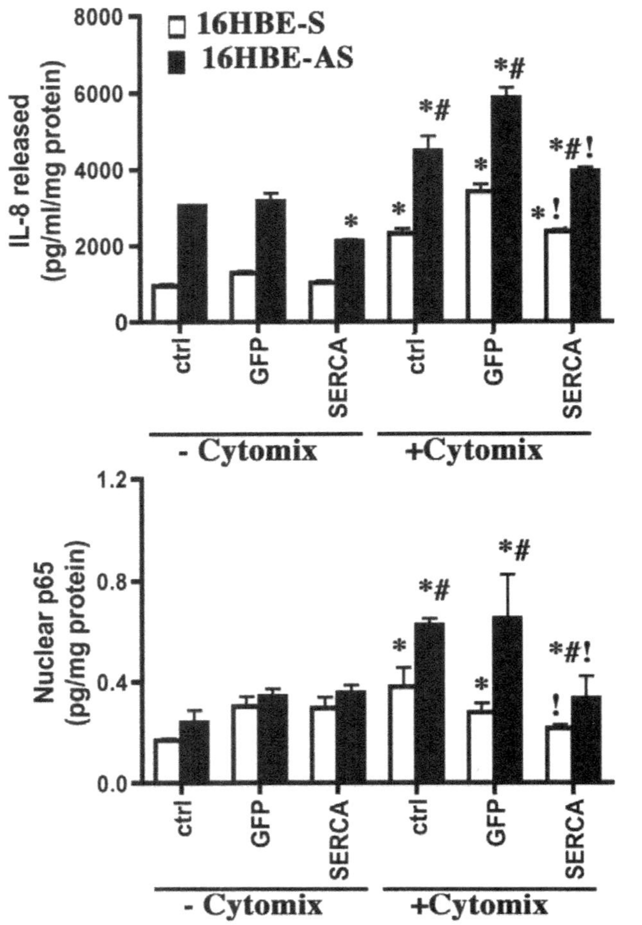

Figure 9. SERCA2 overexpression mitigates cytomix-induced enhanced IL-8 release and NF-κB activity in CF cells. 16HBEo- cell lines stably transfected with sense (16HBE-S, non-CF) and antisense (16HBE-AS, CF) CFTR oligonucleotide were cultured on 6-well plates as described in the Methods. On the 2nd day of plating they were transduced with Ad.GFP or Ad. SERCA2 as described in the Methods. Cells were exposed to cytomix after 24 h incubation. Supernatant media was collected for IL-8 assay (A) and nuclear lysates were prepared for total p65 assay by ELISA. The data shown are mean±SEM (n=6). * Indicates significant difference from untreated control cells, # indicates significant difference from cytomix treated non-CF and ! indicates significant difference from cytomix treated non-CF and CF cells $p<0.05$.

in airways and cells of asthma patients [45]. Our studies with polarized cultures of non-CF and CF airway epithelial cell lines and differentiated cultures of non-CF primary airway epithelial cells demonstrated enhanced stimulation of proinflammtory cytokine at their apical surface when exposed to ozone. The apical proinflammatory cytokine content, with and without ozone exposure, of differentiated primary airway epithelial cells from CF subjects was significantly greater than the cells from airways of non-CF subjects. The fact that significant effect of ozone exposure was not observed in differentiated CF airway epithelial cell cultures could be attributed to several factors, including but not limited to (1) these are differentiated cells, including mucus-secreting cells, and the cell surface is covered with a thick layer of mucus. Ozone is a highly reactive gas, one that likely does not penetrate the cell surface, or the layer covering the cell surface, by more than 0.1 micron. (2) CF cells in culture are covered with greater amounts of cell surface mucus than non-CF cells, and this increases greatly upon ozone exposure. Given that our ability to recover the mucus layer from each cell well was incomplete, we know that our estimate of IL-8 and likely other cytokines produced by each culture was an underestimate of the total cytokines produced, but, among these, this appeared to be proportionately greatest from cultures of CF cells. For this reason, we believe that the degree to which cytokine levels in CF cultures exceeded those present in the non-CF cultures is likely still greater than reported here. (3) We demonstrated that significant amounts of IL-8 are retained in the cell surface mucus layer. Despite the potential interference by cell mucus layer with ozone penetrance, there was an enhancement, albeit inconsistent, of G-CSF and IL-8 by CF primary cells in response to ozone. In addition, CF cells tended to have higher basal levels of all three cytokines assessed, even in absence of ozone exposure. Ozone itself caused decreased SERCA expression in non-CF primary human airway epithelial cells. SERCA expression was already decreased at baseline in the CF cells. This may be another factor contributing to the fact that inflammatory cytokine expression was not consistently increased by ozone in primary CF human airway epithelial cells. Clearly, donor variability also contributed. This may have been related to differences in degree of mucus cell differentiation in different cultures, or other factors. Donor variability in primary human airway epithelial cells has been addressed meticulously by Zabner et al., especially within the context of differentiated epithelial cell cultures from airways of CF subjects [46]. Limitations in cell culture systems may also affect cytokine production. For example, cultures of primary airway epithelial cells from CF patients immersed in culture medium (not at ALI) exhibited higher proinflammatory cytokine production than non-CF cells in response to bacterial pathogens, whereas ALI cultures did not do so in the same study [47,48]. Further the differences from other studies may also be related to technical aspects of handling of the cells; specifically that our rocking exposure system needed for ozone and the potential impact of variable ATP release and cell surface fluid regulation and mediator concentrations in CF and non-CF cultures in the presence and absence of motion may have varied.

Our studies indicated that indeed exposure to ozone of cells with CFTR dysfunction might predispose to proinflammatory cytokine production in an NF-κB dependent manner. CFTR dysfunction-induced enhanced oxidative stress, decreased anti-inflammatory signaling and altered processing of IκB could result in increased cytokine production in CF cells [38,49]. Ozone exposure by itself could cause downregulation of CFTR and decreased CFTR chloride current in human bronchial epithelial cells [50]. Further, a role of CFTR during the initiation of apoptosis in airway epithelial cells also has been proposed [51,52]. We have shown previously that pharmacologic and genetic modulation of CFTR in airway epithelial cells decreases SERCA2. Thus CFTR dependent SERCA2 modulation seems to be an important regulator of airway epithelial cell responses to oxidative stressors such as ozone.

Thapsigargin (a specific SERCA inhibitor) causes enhanced bacterial infection, NF-κB activation and IL-8 release in airway epithelial cells in a cytosolic calcium dependent manner [16]. Although SERCA2 expression itself may be modified by cytokines [53], direct modulation of epithelial production of IL-8 by SERCA2 has not been previously described. Our findings suggest that decreased SERCA2 expression in CF airway epithelial cells contributes to the exaggerated proinflammatory response by modulating NF-κB. The exact mechanisms by which this occurs are potentially important but yet to be investigated. However these studies indicate that pharmacologic modulation of SERCA2 expression and/or activity could prove useful in limiting CF airway inflammation or controlling exacerbations due to environmental pollutants like ozone.

Supporting Information

Figure S1 Effect of ozone exposure on non-CF and CF airway epithelial cells. Polarized cultures of non-CF, 16 HBE and CF, CF45o- and CF41o- cells were exposed to either 0 or 500 ppb ozone for 8 h. At the end of exposure cells were fixed and TUNEL staining was performed *in situ* as described in the Methods section. Panels A-C represents the controls of 16HBE, CF45o- and CF41o- respectively and panels D-F represent the 500 ppb exposed 16HBE, CF45o- and CF41o- cells. Arrows indicate positive TUNEL stained cells. Inset of E shows image focused on positive stained cells.

Figure S2 Retention of cytokines in primary airway epithelial cell surface mucin secretions. Primary airway epithelial cells were cultured on collagen-coated inserts. Cells were maintained at air-liquid-interface (ALI) upon polarization and differentiated for 4 weeks. Cell surface was washed with PBS and 200 µl ALI media was added on the apical surface and cells were incubated for overnight. At the end of incubation apical media was collected. On another set 6 mM DTT containing PBS was added and cells were incubated for 15 min and then apical solution was collected to harvest the mucins. Both type of apical fluid were analyzed for IL-8 (panel A) or respiratory mucins (panel B). Data shown is mean±SEM (n = 6). *Indicates significant difference from "media" content $p<0.05$. Panel C demonstrates the mucin content of non-CF and CF cells upon ozone exposure. Primary non-CF and CF airway epithelial cells from 3 donors each were cultured and exposed to ozone as described in legends to Figure 3. The supernatant media was harvested and DTT (6 mM) containing PBS was added on the apical surface. After 15 min the apical fluid was harvested carefully. Data shown is mean±‴ SEM (n = 9). * Indicates significant difference from 0 ppb non-CF and # indicates significant difference from 200 ppb exposed non-CF cells $p<0.05$.

Figure S3 SERCA2 overexpression by adenovirus-mediated gene transfer. Primary airway epithelial cells were cultured on collagen-coated dishes and recombinant viruses were added to cell cultures (multiplicity of infection, MOI 10:1) on day 3 of culture. Transduction efficiency was estimated by observing green fluorescence of adenoviral GFP-transduced cells. Cell lysates were prepared and analyzed for SERCA2 protein expression by Western. The data shown are mean±SEM (n = 3) * Indicates significant difference between cells treated with Ad.SERCA2 and control (untreated or Ad.GFP treated) $p<0.05$.

Acknowledgments

The authors are grateful to Leslie Fulcher for her expert technical advice on primary CF airway cell culture. Thanks to Joan Loader, Stacy Miller, Tara Jones, Tara Hendry-Hofer, Jenai Kailey and Abhilasha Jain for excellent technical support. The authors are also grateful to Starsa Duskin for cytotoxicity determinations and Glen McConville for maintenance of ozone exposure apparatus.

Author Contributions

Conceived and designed the experiments: SA AA. Performed the experiments: SA AA RCR. Analyzed the data: MS SA RCR. Contributed reagents/materials/analysis tools: SR DN. Wrote the paper: SA CWW.

References

1. Goss CH, Newsom SA, Schildcrout JS, Sheppard L, Kaufman JD (2004) Effect of ambient air pollution on pulmonary exacerbations and lung function in cystic fibrosis. Am J Respir Crit Care Med 169: 816–821.
2. Rabinovitch N, Strand M, Gelfand EW (2006) Particulate levels are associated with early asthma worsening in children with persistent disease. Am J Respir Crit Care Med 173: 1098–1105.
3. Gilliland FD (2009) Outdoor air pollution, genetic susceptibility, and asthma management: opportunities for intervention to reduce the burden of asthma. Pediatrics 123 Suppl 3: S168–173.
4. Harkema JR, Plopper CG, Hyde DM, St George JA, Wilson DW, et al. (1993) Response of macaque bronchiolar epithelium to ambient concentrations of ozone. Am J Pathol 143: 857–866.
5. Bhalla DK (1999) Ozone-induced lung inflammation and mucosal barrier disruption: toxicology, mechanisms, and implications. J Toxicol Environ Health B Crit Rev 2: 31–86.
6. Cho HY, Hotchkiss JA, Bennett CB, Harkema JR (2000) Neutrophil-dependent and neutrophil-independent alterations in the nasal epithelium of ozone-exposed rats. Am J Respir Crit Care Med 162: 629–636.
7. Damera G, Zhao H, Wang M, Smith M, Kirby C, et al. (2009) Ozone modulates IL-6 secretion in human airway epithelial and smooth muscle cells. Am J Physiol Lung Cell Mol Physiol 296: L674–683.
8. Uhlson C, Harrison K, Allen CB, Ahmad S, White CW, et al. (2002) Oxidized phospholipids derived from ozone-treated lung surfactant extract reduce macrophage and epithelial cell viability. Chem Res Toxicol 15: 896–906.
9. Reid DW, Misso N, Aggarwal S, Thompson PJ, Walters EH (2007) Oxidative stress and lipid-derived inflammatory mediators during acute exacerbations of cystic fibrosis. Respirology 12: 63–69.
10. Ahmad S, Ahmad A, Dremina ES, Sharov VS, Guo X, et al. (2009) Bcl-2 suppresses sarcoplasmic/endoplasmic reticulum Ca2+-ATPase expression in cystic fibrosis airways: role in oxidant-mediated cell death. Am J Respir Crit Care Med 179: 816–826.
11. Prasad V, Okunade GW, Miller ML, Shull GE (2004) Phenotypes of SERCA and PMCA knockout mice. Biochem Biophys Res Commun 322: 1192–1203.
12. Ribeiro CM (2006) The role of intracellular calcium signals in inflammatory responses of polarised cystic fibrosis human airway epithelia. Drugs R D 7: 17–31.
13. Ribeiro CM, Paradiso AM, Carew MA, Shears SB, Boucher RC (2005) Cystic fibrosis airway epithelial Ca2+ i signaling: the mechanism for the larger agonist-mediated Ca2+ i signals in human cystic fibrosis airway epithelia. J Biol Chem 280: 10202–10209.
14. Dolmetsch RE, Lewis RS, Goodnow CC, Healy JI (1997) Differential activation of transcription factors induced by Ca2+ response amplitude and duration. Nature 386: 855–858.
15. Fu Z, Bettega K, Carroll S, Buchholz KR, Machen TE (2007) Role of Ca2+ in responses of airway epithelia to Pseudomonas aeruginosa, flagellin, ATP, and thapsigargin. Am J Physiol Lung Cell Mol Physiol 292: L353–364.
16. Ratner AJ, Bryan R, Weber A, Nguyen S, Barnes D, et al. (2001) Cystic fibrosis pathogens activate Ca2+-dependent mitogen-activated protein kinase signaling pathways in airway epithelial cells. J Biol Chem 276: 19267–19275.
17. Fulcher ML, Gabriel S, Burns KA, Yankaskas JR, Randell SH (2005) Well-differentiated human airway epithelial cell cultures. Methods Mol Med 107: 183–206.
18. Ahmad S, Ahmad A, McConville G, Schneider BK, Allen CB, et al. (2005) Lung epithelial cells release ATP during ozone exposure: signaling for cell survival. Free Radic Biol Med 39: 213–226.
19. Ahmad S, White CW, Chang LY, Schneider BK, Allen CB (2001) Glutamine protects mitochondrial structure and function in oxygen toxicity. Am J Physiol Lung Cell Mol Physiol 280: L779–791.
20. Ahmad S, Ahmad A, Ghosh M, Leslie CC, White CW (2004) Extracellular ATP-mediated signaling for survival in hyperoxia-induced oxidative stress. J Biol Chem 279: 16317–16325.
21. Lin H, Carlson DM, St George JA, Plopper CG, Wu R (1989) An ELISA method for the quantitation of tracheal mucins from human and nonhuman primates. Am J Respir Cell Mol Biol 1: 41–48.
22. Warburg O, Christian W (1941) Isolation and crystallisation of fermentation yeast enolase. Biochem Z 310: 384–421.
23. Ahmad A, Ahmad S, Chang LY, Schaack J, White CW (2006) Endothelial Akt activation by hyperoxia: role in cell survival. Free Radic Biol Med 40: 1108–1118.
24. del Monte F, Harding SE, Schmidt U, Matsui T, Kang ZB, et al. (1999) Restoration of contractile function in isolated cardiomyocytes from failing human hearts by gene transfer of SERCA2a. Circulation 100: 2308–2311.

25. Ji Y, Lalli MJ, Babu GJ, Xu Y, Kirkpatrick DL, et al. (2000) Disruption of a single copy of the SERCA2 gene results in altered Ca2+ homeostasis and cardiomyocyte function. J Biol Chem 275: 38073–38080.
26. Aydin J, Andersson DC, Hanninen SL, Wredenberg A, Tavi P, et al. (2009) Increased mitochondrial Ca2+ and decreased sarcoplasmic reticulum Ca2+ in mitochondrial myopathy. Hum Mol Genet 18: 278–288.
27. Lattanzio FA, Jr., Tiangco D, Osgood C, Beebe S, Kerry J, et al. (2005) Cocaine increases intracellular calcium and reactive oxygen species, depolarizes mitochondria, and activates genes associated with heart failure and remodeling. Cardiovasc Toxicol 5: 377–390.
28. Rusznak C, Devalia JL, Sapsford RJ, Davies RJ (1996) Ozone-induced mediator release from human bronchial epithelial cells in vitro and the influence of nedocromil sodium. Eur Respir J 9: 2298–2305.
29. Bosson J, Stenfors N, Bucht A, Helleday R, Pourazar J, et al. (2003) Ozone-induced bronchial epithelial cytokine expression differs between healthy and asthmatic subjects. Clin Exp Allergy 33: 777–782.
30. Spannhake EW, Reddy SP, Jacoby DB, Yu XY, Saatian B, et al. (2002) Synergism between rhinovirus infection and oxidant pollutant exposure enhances airway epithelial cell cytokine production. Environ Health Perspect 110: 665–670.
31. Cebo C, Dambrouck T, Maes E, Laden C, Strecker G, et al. (2001) Recombinant human interleukins IL-1alpha, IL-1beta, IL-4, IL-6, and IL-7 show different and specific calcium-independent carbohydrate-binding properties. J Biol Chem 276: 5685–5691.
32. Larsen ST, Matsubara S, McConville G, Poulsen SS, Gelfand EW (2011) Ozone increases airway hyperreactivity and mucus hyperproduction in mice previously exposed to allergen. J Toxicol Environ Health A 73: 738–747.
33. Wagner JG, Van Dyken SJ, Wierenga JR, Hotchkiss JA, Harkema JR (2003) Ozone exposure enhances endotoxin-induced mucous cell metaplasia in rat pulmonary airways. Toxicol Sci 74: 437–446.
34. Hybiske K, Fu Z, Schwarzer C, Tseng J, Do J, et al. (2007) Effects of cystic fibrosis transmembrane conductance regulator and DeltaF508CFTR on inflammatory response, ER stress, and Ca2+ of airway epithelia. Am J Physiol Lung Cell Mol Physiol 293: L1250–1260.
35. Kube D, Sontich U, Fletcher D, Davis PB (2001) Proinflammatory cytokine responses to P. aeruginosa infection in human airway epithelial cell lines. Am J Physiol Lung Cell Mol Physiol 280: L493–502.
36. Thompson DC, Porter SE, Bauer AK, Das KC, Ou B, et al. (1998) Cytokine-induced nitric oxide formation in normal but not in neoplastic murine lung epithelial cell lines. Am J Physiol 274: L922–932.
37. Stecenko AA, King G, Torii K, Breyer RM, Dworski R, et al. (2001) Dysregulated cytokine production in human cystic fibrosis bronchial epithelial cells. Inflammation 25: 145–155.
38. Venkatakrishnan A, Stecenko AA, King G, Blackwell TR, Brigham KL, et al. (2000) Exaggerated activation of nuclear factor-kappaB and altered IkappaB-beta processing in cystic fibrosis bronchial epithelial cells. Am J Respir Cell Mol Biol 23: 396–403.
39. Perez A, Issler AC, Cotton CU, Kelley TJ, Verkman AS, et al. (2007) CFTR inhibition mimics the cystic fibrosis inflammatory profile. Am J Physiol Lung Cell Mol Physiol 292: L383–395.
40. Nichols DP, Ziady AG, Shank SL, Eastman JF, Davis PB (2009) The triterpenoid CDDO limits inflammation in preclinical models of cystic fibrosis lung disease. Am J Physiol Lung Cell Mol Physiol 297: L828–836.
41. Harkema JR, Hotchkiss JA (1993) Ozone- and endotoxin-induced mucous cell metaplasias in rat airway epithelium: novel animal models to study toxicant-induced epithelial transformation in airways. Toxicol Lett 68: 251–263.
42. Ciarlet M, Crawford SE, Estes MK (2001) Differential infection of polarized epithelial cell lines by sialic acid-dependent and sialic acid-independent rotavirus strains. J Virol 75: 11834–11850.
43. Sathishkumar K, Gao X, Raghavamenon AC, Parinandi N, Pryor WA, et al. (2009) Cholesterol secoaldehyde induces apoptosis in H9c2 cardiomyoblasts through reactive oxygen species involving mitochondrial and death receptor pathways. Free Radic Biol Med 47: 548–558.
44. Oslund KL, Hyde DM, Putney LF, Alfaro MF, Walby WF, et al. (2008) Activation of neurokinin-1 receptors during ozone inhalation contributes to epithelial injury and repair. Am J Respir Cell Mol Biol 39: 279–288.
45. Mahn K, Hirst SJ, Ying S, Holt MR, Lavender P, et al. (2009) Diminished sarco/endoplasmic reticulum Ca2+ ATPase (SERCA) expression contributes to airway remodelling in bronchial asthma. Proc Natl Acad Sci U S A 106: 10775–10780.
46. Zabner J, Scheetz TE, Almabrazi HG, Casavant TL, Huang J, et al. (2005) CFTR DeltaF508 mutation has minimal effect on the gene expression profile of differentiated human airway epithelia. Am J Physiol Lung Cell Mol Physiol 289: L545–553.
47. Aldallal N, McNaughton EE, Manzel LJ, Richards AM, Zabner J, et al. (2002) Inflammatory response in airway epithelial cells isolated from patients with cystic fibrosis. Am J Respir Crit Care Med 166: 1248–1256.
48. Becker MN, Sauer MS, Muhlebach MS, Hirsh AJ, Wu Q, et al. (2004) Cytokine secretion by cystic fibrosis airway epithelial cells. Am J Respir Crit Care Med 169: 645–653.
49. DiMango E, Ratner AJ, Bryan R, Tabibi S, Prince A (1998) Activation of NF-kappaB by adherent Pseudomonas aeruginosa in normal and cystic fibrosis respiratory epithelial cells. J Clin Invest 101: 2598–2605.
50. Qu F, Qin XQ, Cui YR, Xiang Y, Tan YR, et al. (2009) Ozone stress down-regulates the expression of cystic fibrosis transmembrane conductance regulator in human bronchial epithelial cells. Chem Biol Interact 179: 219–226.
51. Barriere H, Poujeol C, Tauc M, Blasi JM, Counillon L, et al. (2001) CFTR modulates programmed cell death by decreasing intracellular pH in Chinese hamster lung fibroblasts. Am J Physiol Cell Physiol 281: C810–824.
52. Jungas T, Motta I, Duffieux F, Fanen P, Stoven V, et al. (2002) Glutathione levels and BAX activation during apoptosis due to oxidative stress in cells expressing wild-type and mutant cystic fibrosis transmembrane conductance regulator. J Biol Chem 277: 27912–27918.
53. Meehan S, Wu AJ, Kang EC, Sakai T, Ambudkar IS (1997) Interferon-gamma induces a decrease in the intracellular calcium pump in a human salivary gland cell line. Am J Physiol 273: C2030–2036.

Comparisons of Three Methods for Organic and Inorganic Carbon in Calcareous Soils of Northwestern China

Xiujun Wang[1,2*], Jiaping Wang[1,3], Juan Zhang[1,3]

1 State Key Laboratory of Desert and Oasis Ecology, Xinjiang Institute of Ecology and Geography, Chinese Academy of Sciences, Urumqi, Xinjiang, China, **2** Earth System Science Interdisciplinary Center, University of Maryland, College Park, Maryland, United States of America, **3** Graduate University of Chinese Academy of Sciences, Beijing, China

Abstract

With increasing interest in the carbon cycle on arid land, there is an urgent need to quantify both soil organic carbon (SOC) and inorganic carbon (SIC) thus to assess various methods. Here, we present a study employing three methods for determinations of SOC and SIC in the Yanqi Basin of northwest China. We use an elemental analyzer for both SOC and SIC, the Walkley-Black method for SOC, a modified pressure calcimeter method for SIC, and a simple loss-on-ignition (LOI) procedure for determinations of SOC and SIC. Our analyses show that all three approaches produce consistently low values for SOC (1–14 g kg^{-1}) and high values for SIC (8–53 g kg^{-1}). The Walkley-Black method provides an accurate estimate of SOC with 100% recovery for most soil samples. The pressure calcimeter method is as accurate as the elemental analysis for measuring SIC. In addition, SOC and SIC can be accurately estimated using a two-step LOI approach, i.e., (1) combustion at 375°C for 17 hours to estimate SOC, and (2) subsequent combustion at 800°C for 12 hours to estimate SIC. There are strong linear relationships for both SOC and SIC between the elemental analysis and LOI method, which demonstrates the capability of the two-step LOI technique for estimating SOC and SIC in this arid region.

Editor: David L. Kirchman, University of Delaware, United States of America

Funding: This study is financially supported by the Program of 100 Talented Young Scientists with the Chinese Academy of Sciences (O972021001)(http://www.cas.cn/ggzy/rcpy/brjh/). The funders had no role in study design, data collection and analysis, decision to publish, or preparation of the manuscript.

Competing Interests: The authors have declared that no competing interests exist.

* E-mail: wwang@essic.umd.edu

Introduction

Soil carbon storage, as the third largest carbon pool in the Earth System, plays an important role in the global carbon cycle and climate change [1]. The majority of carbon, in most soils, is held as soil organic carbon (SOC) whereas in soils of the arid and semiarid regions, the most common form is inorganic carbon, primarily carbonate [2]. Quantifying both SOC and soil inorganic carbon (SIC) is essential to our understanding of the carbon cycle at regional to global scales.

Soil organic carbon is commonly measured by dry combustion with automated analyzers, or a wet chemical oxidation method, i.e., the Walkley and Black method [3]. The automated technique is simple and accurate, but the cost is high and may not be feasible. In addition, this approach often involves pretreatment with acid to remove carbonate for calcareous soils, which may erode the instrument and destroy organic matter in soil samples [4]. On the other hand, the Walkley and Black method has been used as one of the standard methods to determine SOC [5], particularly in China [6]. This technique is less expensive than dry combustion. However, this procedure may lead to variable recovery of SOC [7,8], and hazard Cr due to the use of dichromate.

Soil inorganic carbon can be accurately measured by several methods [9]. Commonly, SIC is determined by measuring CO_2 production after adding HCl acid into soil [10–13]. Sherrod et al. [12] demonstrated that a modified pressure calcimeter method, with addition of $FeCl_2$ to minimize oxidation of organic matter, provided reliable measurement of SIC. Alternatively, dry combustion with automated analyzers can also be utilized to determine SIC. A direct measurement involves dry combustion of soils that are pre-combusted to remove organic matter in an O_2 stream [14]. An indirect measurement is to determine total carbon (without pretreatment with acid to remove carbonate) and SOC separately, then to calculate the difference. Although the automated approaches would provide accurate estimates for SIC, they might not be used widely because of the high costs associated with purchase and maintenance of automated analyzers.

Apart from these traditional methods, loss-on-ignition (LOI) provides an alternative approach, which involves heating soil samples at high temperature to combust soil organic matter (SOM) or carbonate and measuring weight losses [15,16]. The LOI techniques are simple, less expensive and less labor intensive relative to the automated techniques and chemical methods. Thus, the LOI approaches have been used widely to estimate SOM or SOC in agricultural and forest soils and sediments [16–21], and to determine carbonate content in sediments [13,15,22,23]. However, this technique is rarely applied to low fertility soils for SOM or SOC measurement, or used to measure SIC despite some studies showing that the LOI method can give accurate estimates of carbonate for sediments [13,24]. Moreover, there is little utilization of the LOI methods for estimating SOC and SIC in Chinese soils.

The arid and semiarid lands cover approximately 35% of the Earth's land surface, and may play an important role in the global carbon cycle and climate mitigation [1,25,26]. With recent interest in studying the carbon cycle on arid land [27], there is a need to quantify both SOC and SIC of arid soils thus to assess various methods. Here, we select three techniques to determine SOC and SIC contents in the calcareous soils at the Yanqi Basin of northwest China. We employ an elemental analyzer for both SOC and SIC measurements, a wet oxidation for SOC, a modified pressure calcimeter method for SIC, and a simple LOI procedure for determinations of SOC and SIC. The objective of this study is to examine the relationships between these methods, and to test the hypothesis that the LOI technique has capacity of accurately determining SOC and SIC in arid soils.

Materials and Methods

Site Description

There were no specific permits required for the described field studies. We confirm that the location was not privately-owned or protected in any way, and the field studies did not involve endangered or protected species. Our studying area is along the Kaidu River in the Yanqi Basin (Figure 1), which is located on the southeastern flank of the Tianshan Mountain. Brown desert soil is the main soil type, which was developed from limestone parent material and classified as a Haplic Calcisol [28]. The sampling area (approximately 2,200 km) spans both sides of the Kaidu River, which has various land uses, including desert land, shrub land, cropland and grassland. The area has an annual precipitation of less than100 mm and annual evaporation of approximately 2000 mm. Agricultural lands have irrigation systems that extract water from the Kaidu River, whereas the other lands rely on rainfall and underground waters. Annual mean temperature ranges from 7°C to 10°C. On average, the annual maximum temperature is 38°C, and the minimum temperature −35°C.

Soil Sampling and Analyses

We collected 70 soil samples from 15 profiles in November, 2010. While these profiles vary in texture and color because of land use history, we sampled five layers at most profiles, i.e., 0–5 cm, 5–15 cm, 15–30 cm, 30–50 cm, and 50–100 cm, and determined bulk density (BD). For each layer, soils were air-dried, thoroughly mixed, and sieved to pass a 2-mm screen for pH and electric conductivity (EC) measurements. Basic soil properties are given in Table 1. In brief, soils in this region are characterized by high pH (from 8 to 9.4) and high sand/silt contents. Bulk density is slightly lower in the surface soils (0–15 cm: 1.38–1.41 g cm^{-3}) than in below (1.46–1.54 g cm^{-3}). Electric conductivity shows a decreasing trend over depth, from 1.25–1.32 ms cm^{-1} in the 0–15 cm to 0.7 ms cm^{-1} in the 50–100 cm. Representative subsamples were crushed to pass a 0.25 mm screen for SOC and SIC measurements by the following procedures.

Modified Walkley-Black Method for SOC

This method is modified from the traditional Walkley-Black method [3]. A soil sample (~0.4 g) is treated with 5 ml concentrated H_2SO_4 for 4 hours, then mixed with 5 ml 0.5 M $K_2Cr_2O_7$. The mixture is heated at 150–160°C for 5 minutes, and then cooled at room temperature. The solution is transferred into a triangular flask with 100 ml deionized water. Unreacted $K_2Cr_2O_7$ is determined by titrating with 0.25 M $FeSO_4$. Soil organic carbon content is calculated, without a recovery factor, from the difference in $FeSO_4$ used between a blank and a soil solution.

Pressure Calcimeter Method for SIC

We use a procedure similar to that of Sherrod [12]. A subsample of 1.0 g soil and a tiny bottle with 2 ml 6 M HCl containing 3% by weight of $FeCl_2{\cdot}4H_2O$ are placed at the bottom of a 100-ml reaction vessel. After sealing the vessel with a butyl rubber stopper that has an aluminum tearoff seal cap, we mix the acid with soil by turning the vessel sideways. Two hours later, we remove the aluminum tearoff seal cap, and then insert an 18-gauge hypodermic needle that is attached to the pressure transducer and voltage meter. We record the voltage reading to two decimal places after 3 to 5 seconds, and then calculate CO_2 concentration using a calibration curve that is developed by mixing reagent grade $CaCO_3$ with oven-dried laboratory sand. Standards are made based on a final weight of 1.0000 g mixture of sand and $CaCO_3$ to obtain inorganic carbon concentrations of 1.2, 3.6, 6.0, 12.0, 18.0, 24.0, 30.0, 36.0, 42.0, 48.0, 54.0 and 60.0 g C kg^{-1}, respectively.

Elemental Analyzer for SOC and SIC

We used a CNHS-O analyzer (EuroEA3000) for SOC and SIC estimates. For SOC measurement, 20 mg soil is pretreated with 10 drips of H_3PO_4 for 12 hours to remove carbonate. The sample is combusted at 1020°C with constant helium flow carrying pure oxygen to ensure complete oxidation of organic materials. CO_2 production is determined by a thermal conductivity detector. Total soil carbon is measured, using the same procedure without pretreatment with H_3PO_4. Soil inorganic carbon is calculated as the difference between total soil carbon and SOC.

Loss-on-ignition (LOI) Procedure for SOM and SIC

Following Wang et al. [16] and Wang et al. [15], we first place 5.0000 g soil in a quartz glass and heated at 105°C for 12 hours to remove soil moisture. We then combust soil in a programmable muffle furnace (S1849, KOYO LINDBEERE LTD) at 375°C for 17 hours and 800°C for 12 hours. Soil organic matter is calculated as the weight loss between 105°C and 375°C, and SIC as the weight loss between 375°C and 800°C:

$$SOM_{LOI}\,(g\,kg^{-1}) = \left(\frac{Weight_{105^\circ C} - Weight_{375^\circ C}}{Weight_{105^\circ C}}\right) \times 1000 \quad (1)$$

$$SIC_{LOI}\,(g\,kg^{-1}) = \left(\frac{Weight_{375^\circ C} - Weight_{800^\circ C}}{Weight_{105^\circ C}}\right) \times 0.273 \times 1000. \quad (2)$$

The conversion constant of 0.273 in equation (2) is applied to convert mass of CO_2 to mass of carbon.

Results

Soil Organic Carbon

We first compare measured SOC between the automated CNS analysis and the modified Walkley-Black method, and find a strong correlation ($r^2 = 0.93$, $P<0.001$) between the two methods (Table 2). As shown in Figure 2, SOC content is below 15 g kg^{-1} in this region. It appears that the modified Walkley-Black method over-estimates SOC when soil contains extremely low SOC (<5 g

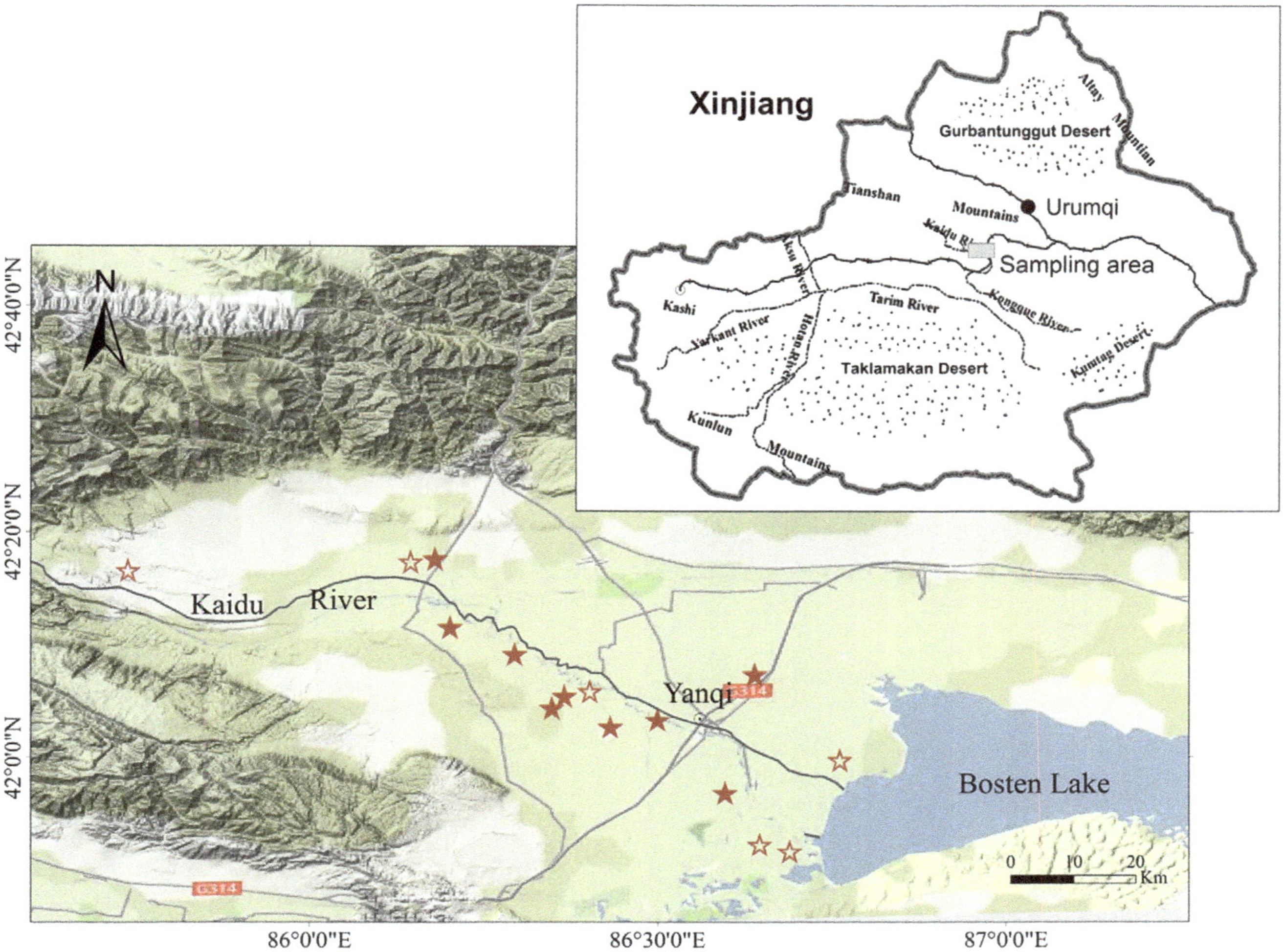

Figure 1. Sampling area with locations of 15 soil profiles. Solid stars denote the sites where topsoil texture was determined.

kg^{-1}). Excluding these data yields an averaged recovery of 0.993. These results indicate that almost 100% SOC in the brown desert soil can be oxidized, which is much higher than previous published recovery rates (less than 82%) for forest soils across Flanders region [7], but in agreement with those (~100%) in most Tasmanian soils [16]. The high recovery of SOC and large correlation coefficient in our study suggest that the modified Walkley-Black method is as accurate as the automated CNS technique except for extremely low SOC soils.

We then compare measured SOM by the LOI method with SOC by the automated CNS technique (Figure 3). As expected, SOM values (0–27 g kg^{-1}) are higher than SOC values

Table 1. Means and standard deviations (in brackets) of soil pH, bulk density (BD, g kg^{-1}) and electrical conductivity (EC, ms cm^{-1}) for each layer, and clay, silt and sand contents (%) of topsoil (0–30 cm) from 8 randomly selected soil profiles.

	pH	BD	EC	Clay	Silt	Sand
0–5	8.92 (0.3)	1.41 (0.1)	1.25 (2.4)	5.6 (2.9)	57.3 (23.0)	37.1 (25.1)
5–15	8.77 (0.3)	1.38 (0.1)	1.32 (1.5)			
15–30	8.78 (0.4)	1.50 (0.2)	1.07 (1.2)			
30–50	8.88 (0.4)	1.54 (0.1)	0.98 (1.2)	–	–	–
50–100	9.00 (0.2)	1.46 (0.1)	0.70 (1.2)	–	–	–

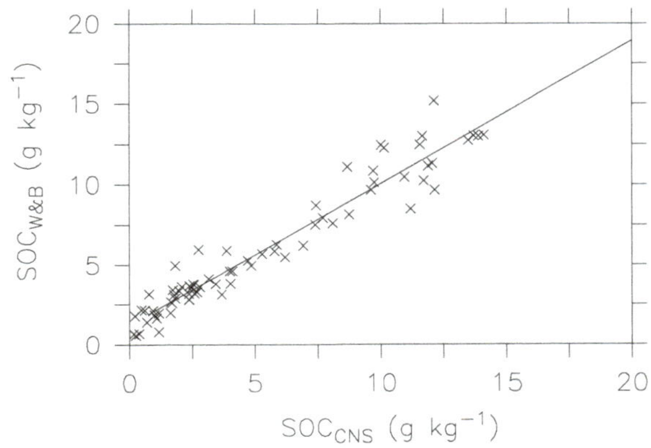

Figure 2. Linear regression of measured SOC by the CNS analyzer and Walkley-Black method.

Table 2. Relationships of soil organic carbon (SOC) and inorganic carbon (SIC) among different methods.

	n	Slope	Intercept	r^2	P *value*
$SOC_{W\&B}$ vs. SOC_{CNS}	70	0.886	1.199	0.92	<0.001
SOM_{LOI} vs. SOC_{CNS}	70	1.792	4.189	0.90	<0.001
SOC_{LOI} vs. $SOC_{W\&B}$	31	0.949	0.443	0.77	<0.001
SIC_{PC} vs. SIC_{CNS}	70	1.025	−1.718	0.99	<0.001
SIC_{LOI} vs. SIC_{CNS}	70	1.043	2.582	0.96	<0.001
SIC_{LOI} vs. SIC_{PC}	70	1.021	4.241	0.98	<0.001

(<15 g kg^{-1}). The large intercept value of 4.189 indicates that there are some non-SOM weight losses (e.g., from structural water and carbonate) during ignition. The conversion factor of 1.792 is slightly larger than the traditional values of 1.724. We find that the conversion factor is 2.71 for those samples with low SOC content (<5 g kg^{-1}) that are mainly subsoils. Based on all 70 soil samples tested, SOC may be calculated from LOI at 375°C for 17 hours, using the following equation:

$$SOC_{LOI} = (SOM_{LOI} - 4.189)/1.792 \quad (3)$$

Figure 4 shows comparison of measured SOC by the Walkley-Black method and calculated SOC by the LOI method using the equation (3). To avoid negative values in calculated SOC, we exclude samples that contain less than 4.189 g kg^{-1} SOM_{LOI} in this comparison. The slope value of 0.949 and significantly high correlation coefficient ($r^2 = 0.77$, $p<0.001$) indicate that the LOI method has potential for estimating SOC in this arid region.

Soil Inorganic Carbon

We assess measured SIC by the pressure calcimeter method and the automated CNS analysis (Figure 5). Clearly, SIC content is much higher than SOC (<15 g kg^{-1}) in this region. The pressure calcimeter method produces slightly lower SIC values (6–51 g kg^{-1}) than the automated CNS technique (9–53 g kg^{-1}). However, there is a high correlation ($r^2 = 0.99$, $p<0.001$) between the two methods. The slope value of 0.996 and large correlation

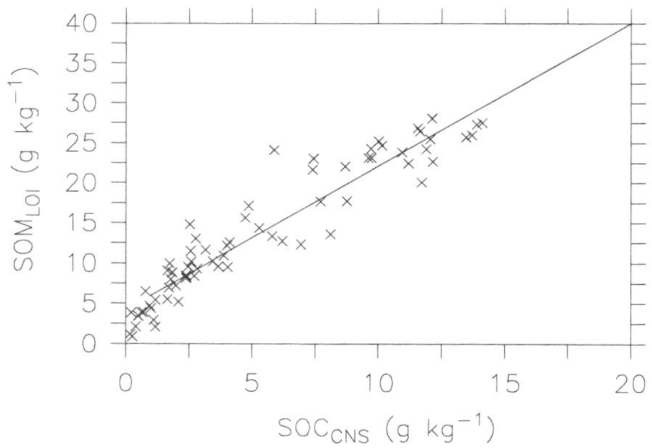

Figure 3. Relationship between SOC measured by the CNS analyzer and SOM determined by the LOI method.

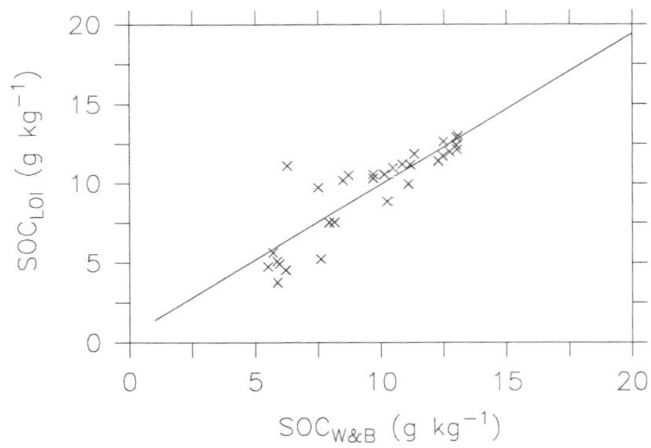

Figure 4. Relationship between SOC calculated by the LOI method and SOC measured by the Walkley-Black method.

coefficient indicate that the pressure calcimeter method is as accurate as the automated CNS technique for the calcareous soils.

Figure 6 presents comparison of measured SIC between the LOI method and the automated CNS analysis. Both methods give a very similar range, and form a linear relationship:

$$SICLOI = 1.043SICCNS + 2.582 \; (r^2 = 0.96, P < 0.001) \quad (4)$$

The intercept value of 2.582 is significantly different from 0, suggesting that there may be some non-SIC weight losses (e.g., dehydration from other compounds) during combustion [22,23,29]. Nevertheless, the factors of slope value close to 1 and significant correlation indicate that the LOI technique can produce accurate estimates for SIC.

As shown in Figure 7, there is also a linear relationship for SIC between the pressure calcimeter method and the LOI technique:

$$SIC_{LOI} = 1.021SIC_{PC} + 4.241 \quad (5)$$

which shows a similar slope to that in the regression between the LOI method and the CNS analysis (i.e., equation 4). The relatively larger intercept in equation (5) reflects the slightly lower SIC values by the pressure calcimeter method, relative to those by the automated CNS analysis (see Table 2).

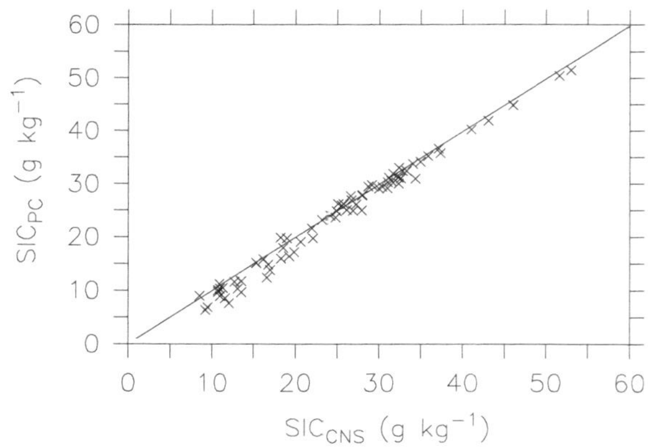

Figure 5. Liner relationship for SIC by the CNS analyzer and pressure calcimeter method.

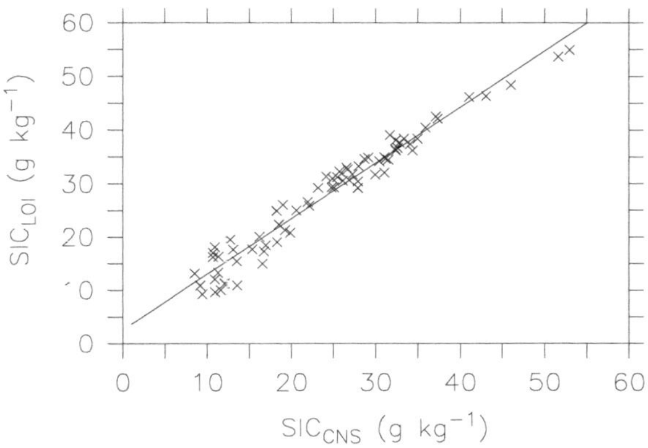

Figure 6. Liner relationship for SIC between the CNS analyzer and LOI method.

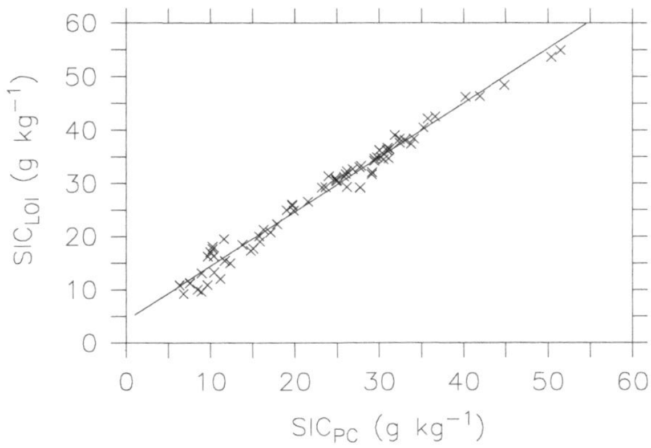

Figure 7. Liner relationship for SIC by the pressure calcimeter and LOI methods.

Discussion

Estimate of SOC

Measurement of SOC has been a common procedure in soil science. Many methods are available, each with advantages and disadvantages in terms of accuracy, expense and convenience [5,30]. The Walkley-Black procedure is considered to be cheap and easy to perform, but may exhibit variable recovery thus a correction factor is often applied to determine the total SOC content for a soil sample. For example, Sleutel et al. [31] applied a correction factor of 1.33 for estimated SOC by the Walkley-Black without external heating. However, an early study [16] showed that without external heating, the Walkley-Black method produced approximately 100% recovery for SOC in most Tasmanian soils. Many modified dichromate oxidation techniques that involve external heating do not require a correction factor [32]. Our study demonstrates that compared with the automated CNS technique, the modified Walkley-Black method with external heating yields almost 100% recovery for SOC in the calcareous soils we tested. However, the Walkley-Black method may over-estimate SOC in soils containing extremely low SOC [8], which is probably attributed to reactions of $K_2Cr_2O_7$ with inorganic soil constituents [33].

While the LOI method has been widely used to estimate SOM and SOC, there is no standard procedure in terms of heating duration and temperature. There have been various combinations with temperature ranging from 150°C to 900°C and duration from 2 hours to 17 hours in the literature, thus a regression between LOI and SOC (measured by an automated analyzer or the Walkley-Black method) varies widely [16,34,35]. In general, with increasing temperature, LOI increases whereas correlation coefficient tends to decrease [35].

One criticism of the LOI technique is that it may result in weight losses from structural water and carbonate during heating [30,34]. These non-SOM weight losses can significantly affect estimates of SOC, particularly in soils containing low SOM and high clay content. As the amount of SOM decreases, dehydroxylation of clays becomes more likely [36]. However, clay content is low in our soils (see Table 1). Thus, weight loss due to dehydroxylation of clays may be small. On the other hand, temperature is the dominant factor affecting LOI, and its relationship with SOC [35]. Therefore, it is critical to choose a temperature that is high enough to completely remove SOM but low enough to prevent dehydroxylation of clay minerals and oxidation of carbonate.

Several studies have shown that a combination of 375°C and 16–17 hours results in a good LOI-SOC relationship, thus it can provide accurate estimates of SOC in many soil groups [16,37,38]. Our study demonstrates that there is a strong relationship between SOC and LOI at 375°C for 17 hours in the calcareous soils tested. While the non-zero intercept may reflect some non-SOM weight losses, the strong relationship indicates the capacity of using the LOI method for SOC estimation.

Estimate of SIC

Soil carbonate is usually quantified by acid dissolution in the reaction below:

$$CaCO_3 + 2H^+ \rightarrow Ca^{2+} + CO_2 + H_2O \quad (6)$$

with the determination of either H^+ consumption or CO_2 production. The methods of H^+ consumption that involve reaction with a strong acid, such as HCl addition and back titration of the unreacted acid, may be problematic owing to H^+ consumption by other soil components [9]. Indeed, our own analyses using the H^+ consumption approach lead to significant underestimates of SIC (by 20%) for the soils we tested (data not shown).

There have been many methods involving determination of CO_2 production following acid dissolution [9]. While these methods are much preferred, precautions must be taken to ensure that there is no interference from organic matter oxidation. Thus, an oxidation inhibitor (e.g., $FeCl_2$) is often used to limit the oxidation of organic matter and the subsequent evolution of CO_2 from this source. Our study shows a strong correlation between the pressure calcimeter method and the automated CNS analysis. Given that measured SIC contents by the pressure calcimeter method are slightly lower than, but not significantly different from those by the automated CNS analysis, we conclude that the pressure calcimeter approach can provide an accurate measurement of SIC for the calcareous soils of this arid region.

In contrast to wide utilization of the LOI method for SOC estimates, there is little in the literature regarding application of the LOI method for SIC estimates. However, the LOI method has been used to measure carbonate content in sediments, following combustion of organic carbon at lower temperature [13,15,22,23]. The simple two-step LOI procedure for estimates of organic carbon and carbonate in sediments [15] is similar to that for determination of organic carbon and carbonate in calcareous soils by dry combustion [14]. Thus, it could be a convenient practice for estimating SOC and SIC.

To the best of our knowledge, this study is the first attempt to evaluate the LOI method for SIC measurement. Our study shows a significantly (r^2>0.96, P<0.001) linear relationship between estimated SIC by the LOI method and determined SIC by direct measurements, i.e., the automated element analyzer and the pressure calcimeter method. Despite the non-zero intercepts and slightly higher values in SIC by the LOI method, the slope values are close to one, which are consistent with those reported for sediment analyses [15]. Our results demonstrate that the LOI method has potential for easily and accurately estimating SIC in calcareous soils of arid regions.

Conclusions

We have compared three methods for determinations of SOC and SIC concentrations in calcareous soils of the Yanqi Basin. In spite of wide ranges in soil properties, SOC is generally low (1–14 g kg^{-1}) whereas SIC is considerably high (9–53 g kg^{-1}). Our study shows that the Walkley-Black method with external heating can provide accurate estimates of SOC (with almost 100% recovery) for most arid soils. The pressure calcimeter method is as accurate as the automated elemental technique for measuring SIC. In addition to these traditional methods, we have also evaluated a two-step LOI approach, i.e., (1) combustion at 375°C for 17 hours to estimate SOC, and (2) subsequent combustion at 800°C for 12 hours to estimate SIC. Our study shows that SOC and SIC can be estimated from LOI for calcareous soils except those with extremely low SOC or SIC contents. This study demonstrates that the two-step LOI technique has potential to be used as a sufficiently accurate technique for estimating SOC and SIC in arid soils. Further testing on a wider range of arid and semi-arid soils is warranted.

Acknowledgments

We are grateful for the reviewer's constructive comments.

Author Contributions

Conceived and designed the experiments: XW JW. Performed the experiments: JW JZ. Analyzed the data: XW. Contributed reagents/materials/analysis tools: XW. Wrote the paper: XW.

References

1. Lal R, Kimble JM (2000) Pedogenic carbonate and the global carbon cycle. In: Lal R, Kimble JM, Eswaran H, Stewart BA, editors. Golbal Climate Change and Pedogenic. Boca Raton, FL, USA.: CRC press. 1–14.
2. Eswaran H, Reich PF, Kimble JM, Beinroth FH, Padmanabhan E, et al. (2000) Global carbon stocks. In: R L, Kimble JM, Eswaran H, Stewart BA, editors. Global climate change and pedogenic carbonates. Boca Raton, FL: Lewis Publishers. 15–27.
3. Walkley A, Black IA (1934) An examination of the Degtjareff method for determining soil organic matter and a proposed modification of the chromic acid titration method. Soil Sci 37: 29–38.
4. Byers SC, Mills EL, Stewart PL (1978) A comparison of methods of determining organic carbon in marine sediments, with suggestions for a standard method. Hydrobiologia 58: 43–47.
5. Nelson DW, Sommers LE, Sparks D, Page A, Helmke P, et al. (1996) Total carbon, organic carbon, and organic matter. Methods of soil analysis Part 3-chemical methods: 961–1010.
6. Zhang WJ, Wang XJ, Xu MG, Huang SM, Liu H, et al. (2010) Soil organic carbon dynamics under long-term fertilizations in arable land of northern China. Biogeosciences 7: 409–425.
7. De Vos B, Lettens S, Muys B, Deckers JA (2007) Walkley–Black analysis of forest soil organic carbon: recovery, limitations and uncertainty. Soil Use and Management 23: 221–229.
8. Conyers MK, Poile GJ, Oates AA, Waters D, Chan KY (2011) Comparison of three carbon determination methods on naturally occurring substrates and the implication for the quantification of 'soil carbon'. Soil Research 49: 27–33.
9. Loeppert R, Suarez D, Sparks D, Page A, Helmke P, et al. (1996) Carbonate and gypsum. Methods of soil analysis Part 3-chemical methods: 437–474.
10. Presley B (1975) A simple method for determining calcium carbonate in sediment samples. J Sediment Petrol 45: 745–C746.
11. Dreimanis A (1962) Quantitative gasometric determination of calcite and dolomite by using Chittick apparatus. Journal of Sedimentary Research 32: 520.
12. Sherrod LA, Dunn G, Peterson GA, Kolberg RL (2002) Inorganic Carbon Analysis by Modified Pressure-Calcimeter Method. Soil Sci Soc Am J 66: 299–305.
13. Heiri O, Lotter AF, Lemcke G (2001) Loss on ignition as a method for estimating organic and carbonate content in sediments: reproducibility and comparability of results. Journal of Paleolimnology 25: 101–110.
14. Rabenhorst MC (1988) Determination of organic and carbonate carbon in calcareous soils using dry combustion. Soil Science Society of America Journal 52: 965–969.
15. Wang QR, Li YC, Wang Y (2011) Optimizing the weight loss-on-ignition methodology to quantify organic and carbonate carbon of sediments from diverse sources. Environmental Monitoring and Assessment 174: 241–257.
16. Wang XJ, Smethurst PJ, Herbert AM (1996) Relationships between three measures of organic matter or carbon in soils of eucalypt plantations in Tasmania. Australian Journal of Soil Research 34: 545–553.
17. David MB (1988) Use of loss-on-ignition to assess soil organic carbon in forest soils Communications in Soil Science and Plant Analysis 19: 1593–1599.
18. De Vos B, Vandecasteele B, Deckers J, Muys B (2005) Capability of loss-on-ignition as a predictor of total organic carbon in non-calcareous forest soils. Communications in soil science and plant analysis 36: 2899–2921.
19. Wright AL, Wang Y, Reddy KR (2008) Loss-on-Ignition Method to Assess Soil Organic Carbon in Calcareous Everglades Wetlands. Communications in soil science and plant analysis 39: 3074–3083.
20. Konen ME, Jacobs PM, Burras CL, Talaga BJ, Mason JA (2002) Equations for predicting soil organic carbon using loss-on-ignition for north central US soils. Soil Science Society of America Journal 66: 1878–1881.
21. Ball DF (1964) Loss-on-ignition as an estimate of organic matter and organic carbon in non-calcareous soils. Journal of Soil Science 15: 84–92.
22. Dean WE (1974) Determination of carbonate and organic matter in calcareous sediments and sedimentary rocks by loss on ignition: comparison with other methods. Journal of Sedimentary Petrology 44: 242–248.
23. Santisteban JI, Mediavilla R, López-Pamo E, Dabrio CJ, Zapata MBR, et al. (2004) Loss on ignition: a qualitative or quantitative method for organic matter and carbonate mineral content in sediments? Journal of Paleolimnology 32: 287–299.
24. Wang Q, Li Y, Wang Y (2011) Optimizing the weight loss-on-ignition methodology to quantify organic and carbonate carbon of sediments from diverse sources. Environmental Monitoring and Assessment 174: 241–257.
25. Scharpenseel HW, Mtimet A, Freytag J (2000) Soil inorganic carbon and global change. In: Lal R, Kimble JM, Eswaran H, Stewart BA, editors. Global Climate Change and Pedogenic Carbonates. Boca Raton, FL: CRC Press. 27–42.
26. Xie JX, Li Y, Zhai CX, Li CH, Lan ZD (2009) CO_2 absorption by alkaline soils and its implication to the global carbon cycle. Environmental Geology 56: 953–961.
27. Stone R (2008) Have desert researchers discovered a hidden loop in the carbon cycle? Science 320: 1409–1410.
28. FAO-UNESCO (1988) soil map of the world: revised legend. World Soil Resources Report No 60 FAO: Rome.
29. Allison L, Black C (1965) Methods of soil analysis. Part 2. Chemical and microbiological properties. Agronomy monograph 9: 1367–1378.
30. Chatterjee A, Lal R, Wielopolski L, Martin MZ, Ebinger MH (2009) Evaluation of Different Soil Carbon Determination Methods. Critical Reviews in Plant Sciences 28: 164–178.
31. Sleutel S, De Neve S, Singier B, Hofman G (2007) Quantification of organic carbon in soils: A comparison of methodologies and assessment of the carbon content of organic matter. Communications in Soil Science and Plant Analysis 38: 2647–2657.
32. Lettens S, De Vos B, Quataert P, van Wesemael B, Muys B, et al. (2007) Variable carbon recovery of Walkley-Black analysis and implications for national soil organic carbon accounting. European Journal of Soil Science 58: 1244–1253.
33. Walkley A (1947) A critical examination of a rapid method for determining organic carbon in soils- effect of variations in digestion conditions and inorganic soil constituents. Soil Science 63: 251–264.
34. Szava-Kovats R (2009) Re-analysis of the Relationship between Organic Carbon and Loss-on-Ignition in Soil. Communications in Soil Science and Plant Analysis 40: 2712–2724.
35. Abella S, Zimmer B (2007) Estimating organic carbon from loss-on-ignition in northern Arizona forest soils. Soil Science Society of America Journal 71: 545–550.
36. Pribyl DW (2010) A critical review of the conventional SOC to SOM conversion factor. Geoderma 156: 75–83.
37. McKeague J (1976) Manual of soil sampling and methods of analysis. Land Resource Research Inst., Res. Br. Agriculture Canada, Ottawa, Canada.
38. Beaudoin A (2003) A comparison of two methods for estimating the organic content of sediments. Journal of Paleolimnology 29: 387–390.

Hyaluronan Signaling during Ozone-Induced Lung Injury Requires TLR4, MyD88, and TIRAP

Zhuowei Li[1], Erin N. Potts-Kant[1], Stavros Garantziotis[3], W. Michael Foster[1], John W. Hollingsworth[1,2]*

1 Department of Medicine, Duke University, Durham, North Carolina, United States of America, 2 Department of Immunology, Duke University, Durham, North Carolina, United States of America, 3 National Institute of Environmental Health Sciences, Research Triangle Park, North Carolina, United States of America

Abstract

Ozone exposure is associated with exacerbation of reactive airways disease. We have previously reported that the damage-associated molecular pattern, hyaluronan, is required for the complete biological response to ambient ozone and that hyaluronan fragments signal through toll-like receptor 4 (TLR4). In this study, we further investigated the role of TLR4 adaptors in ozone-induced airway hyperresponsiveness (AHR) and the direct response to hyaluronan fragments (HA). Using a murine model of AHR, C57BL/6J, TLR4−/−, MyD88−/−, and TIRAP−/− mice were characterized for AHR after exposure to either ozone (1 ppm×3 h) or HA fragments. Animals were characterized for AHR with methacholine challenge, cellular inflammation, lung injury, and production of pro-inflammatory cytokines. Ozone-exposed C57BL/6J mice developed cellular inflammation, lung injury, pro-inflammatory cytokines, and AHR, while mice deficient in TLR4, MyD88 or TIRAP demonstrated both reduced AHR and reduced levels of pro-inflammatory cytokines including TNFα, IL-1β, MCP-1, IL-6 and KC. The level of hyaluronan was increased after inhalation of ozone in each strain of mice. Direct challenge of mice to hyaluronan resulted in AHR in C57BL/6J mice, but not in TLR4−/−, MyD88−/−, or TIRAP−/− mice. HA-induced cytokine production in wild-type mice was significantly reduced in TLR4−/−, MyD88−/−, or TIRAP−/− mice. In conclusion, our findings support that ozone-induced airway hyperresponsiveness is dependent on the HA-TLR4-MyD88-TIRAP signaling pathway.

Editor: Hong Wei Chu, National Jewish Health, United States of America

Funding: This work was supported by the National Institute of Environmental Health Sciences (ES016126, ES020426). The funders had no role in study design, data collection and analysis, decision to publish, or preparation of the manuscript.

Competing Interests: The authors have declared that no competing interests exist.

* E-mail: john.hollingsworth@duke.edu

Introduction

Inhaled ozone significantly contributes to both human morbidity and mortality and is estimated to annually account for approximately 800 premature deaths, 4500 hospital admission, 900000 school absences and more than a million restricted activity days with an estimated $5 billion annual economic burden [1]. There does not appear to be a lower threshold level of ozone associated with detrimental health effects [2,3,4] with each 10 ppb increase in daily ozone being associated with approximately 0.87% increase in total mortality [5]. It is anticipated that levels of ambient ozone will rise with predicted changes in global climate [6]. Understanding the health impacts of ambient ozone is therefore of high clinical interest. Furthermore, understanding the molecular mechanisms that regulate the biological response to inhaled ozone can provide insight into oxidant-related lung injury and reactive airways disease, which could identify novel therapeutic approaches to asthma.

Toll-like receptors (TLRs) are membrane-spanning receptors that recognize structurally conserved molecules that are usually derived from microbes. However, diverse environmental exposures by an inhalational route are also known to activate innate immunity through TLRs, and can thus exacerbate existing allergic airways disease [7,8,9] and contribute to airway hyperresponsiveness [10,11]. For example, toll-like receptor 4 (TLR4) is required for the complete biological and functional responses of airway and lung tissues to inhaled ozone. TLR4, as a candidate gene in response to ozone, was initially identified through forward genetics and supported by the C3H/HeJ (TLR4-mutant) animal, which is protected from sub-chronic exposure to ozone-induced lung injury [12,13]. TLR4-deficient animals were subsequently identified to be protected from ozone-induced airway hyperresponsiveness [11]. TLR4 activates major signal transduction pathways through distinct adaptor proteins capable of contributing to cellular responses specifically myeloid differentiation primary response protein 88 (MyD88) and TIR domain-containing adapter protein for TLR4 (TIRAP). Previous work suggests that MyD88 is required for the complete response to ozone [14] and sub-chronic exposure to ozone results in increased mRNA expression of MyD88 [15]. However, the biologic mechanisms whereby these genes of innate immunity contributed to the response to oxidant lung injury remained elusive.

Previous work supports that the extracellular matrix component, hyaluronan, can function as a signaling molecule when fragmented into low molecular weight [16]. Additionally, hyaluronan can function as an endogenous ligand to TLR4 in context of sterile (non-infectious) tissue injury [17]. Oxidant stress is one mechanism that results in fragmentation of hyaluronan [18,19]. We have recently identified that hyaluronan fragments contribute to the complete biological response to ozone [20]. Furthermore, fragments of hyaluronan function as an endogenous ligand of TLR4 in context of several forms of oxidant lung injury including bleomycin [21], asbestos [22], and ozone [23]. However, the intracellular adaptor molecules required to facilitate hyaluronan-dependent cell signaling remain unknown.

We hypothesized that ozone inhalation results in generation of immunostimmulatory fragments of hyaluronan that signal in a manner dependent on TLR4 and the cytoplasmic adaptor proteins, MyD88 and TIRAP. Similar to prior reports, we identify that both TLR4 and MyD88 are required for development of ozone-induced AHR. We now identify that the adaptor molecule TIRAP is additionally required for development of ozone-induced AHR. The profile of pro-inflammatory cytokines after inhalation of ozone is similar in TLR4−/−, MyD88−/−, and TIRAP−/− mice. Direct challenge of mice to fragments of hyaluronan identifies a role of both MyD88 and TIRAP in development of AHR and release of pro-inflammatory cytokines. These studies provide novel insight into the intracellular adaptor molecules required for the development of airway hyperresponsiveness in response to the endogenous TLR4-ligand hyaluronan.

Materials and Methods

Animal

All experimental protocols were reviewed and approved by the Institutional Animal Care and Use Committee at Duke University Medical Center and performed in accordance with the standards established by the U.S. Animal Welfare Acts. Animals were treated humanely and with regard for alleviation of suffering. Animal work is covered under the following registrations, permits, and accreditations: United States Department of Agriculture (Registrant 55-R-0003 #83); The National Institutes of Health (Assurance 3195-01); and the Association for the Assessment and Accreditation of Laboratory Animal Care, International (File #363). 6–8 weeks old male C57BL/6J mice were purchased from Jackson Laboratory (Bar Harbor, ME). TLR4−/−, MyD88−/−, and TIRAP−/− mice were generously provided by S. Akira [24,25,26] and then backcrossed onto C57BL/6J for at least ten generations. Presented results are representative of at least two experimental repeats utilizing age and gender matched animals.

Ozone Exposure

C57BL/6J, TLR4−/−, MyD88−/− and TIRAP−/− mice were exposed to 1 part per million (ppm) of ozone for 3 h as previously described [27] and phenotyped 20–24 hours after initiation of exposure. The dose and exposure duration were selected based on similar biological response observed in human exposure studies, published deposition fraction data for O_3 [28,29], and our previous observations in the mouse model [27]. Under this exposure dose, wild-type mice maintained comparable airway responsiveness to ozone with less lung inflammation, when compared with higher doses of ozone used in previous publications [20,23]. Briefly, animals in separate caging were placed into a 55-L Hinner-style chamber and exposed to filtered air (FA) or ozone. Ozone was generated by directing 100% oxygen through a UV light ozone generator and mixed with filtered air before supply. Environmental conditions during exposures included air temperatures between 20 to 22°C, relative humidity at 50 to 60% and approximately 20 exchanges of chamber air per hour. Chamber ozone concentration was continuously monitored with a UV light photometer.

HA challenge

Endotoxin-free HA fragments were prepared as previously described [20]. Briefly, high molecular weight hyaluronan (Healon GV, AMO, Santa Ana, CA) was reconstituted to 0.5 mg/ml in 0.02 M acetate/0.15 M sodium chloride, pH 6.0. For the production of low molecular weight hyaluronan, Healon was sonicated 3 seconds for a total of 3 times on ice using a sonic dismembrator (Thermo Fisher Scientific Inc., Pittsburgh, PA). The sizes of HA were confirmed by agarose gel electrophoresis. For *in vivo* experiments, animals were briefly anesthetized by inhalation of isoflurane; 50 µl of 0.5 mg/ml of HA or same amount of vehicle control was then instilled into the lung by oropharyngeal aspiration. This protocol and dose of hyaluronan fragments are used to model ozone-induced reactive airways disease [20]. Animals were allowed to recover for two hours before measurement of AHR.

Whole lung lavage, cell counts and analysis

As previously described [20], mice were euthanized with CO_2 and the lungs were fully inflated three times serially to 25 cmH_2O of 0.9% NaCl. All three lavage returns were pooled together for the following analysis. Cell counts were performed using a hemocytometer and differentials were performed by H&E stained cytospins. Cytokine/chemokines IL-1β, IL-6, KC, MCP-1 and TNFα were determined by Luminex (Bio-Rad, Hercules, CA) using 5-plex reagents from Millipore (Billerica, MA). Assay sensitivities are 1.8 pg/ml for IL-6, 2.0 pg/ml for IL-1β, 1.4 pg/ml for KC, 5.3 pg/ml for MCP-1, 1.0 pg/ml for TNFα. Total protein concentrations in lung lavage fluid were measured by the Lowry Assay (Bio-Rad, Hercules, CA).

Airway Physiology. Airway hyperresponsiveness to methacholine (Mch) challenge was measured 20–24 hours post ozone or air exposure. At this time point, acute ozone inhalation induced significant change in AHR to Mch challenge, which is similar to human response to ozone at 24 hours post exposure [30]. Before measurement, animal was anesthetized with 60 mg/kg of pentobarbital sodium by intraperitoneal injection. Mice were then given neuromuscular blockade (0.8 ml/kg pancuronium bromide) and ventilated with a computer-controlled small animal ventilator (FlexiVent; SCIREQ, Montreal, QC), with a tidal volume of 7.5 ml/kg and a positive end-expiratory pressure of 3 cm H_2O. Lung total resistance was assessed using low frequency oscillation technique whereby airway pressure (via a tracheal cannula) and tidal volume data were generated by the application of a 2 second sine wave volume perturbation (amplitude ~0.2 ml and frequency = 2.5 Hz) [31,32]. Following baseline resistance measurements, mice were challenged in dose related manner with methacholine aerosol (Mch 10–100 mg/mL) and after each Mch dose the lung was hyperinflated to return resistance to baseline levels. After baseline recruitment maneuvers to achieve total lung capacity, baseline total respiratory system resistance (R measurements) were collected and analyzed using the linear single-compartment model with multiple linear regression as described by the manufacturer (SciREQ, Montreal, Canada). The resistance measurements were then averaged at each dose and graphed (R_T, measured in cmH_2O/ml/s) along with the initial baseline measurement.

Statistics

All data are expressed as mean ± SEM. Two-way ANOVA for comparisons among multiple groups was performed using Graphpad Prism 5.0. Student-t test was used for individual comparisons between groups. Significance was defined as two-tailed P value of less than 0.05. Presented results represent at least two replicate studies.

Results

Ozone-induced airway hyperresponsiveness is dependent on TLR4, MyD88 and TIRAP

TLR4, a critical cell surface receptor, mediates the increased airway reactivity to methacholine challenge in mice exposed to both 2 ppm and 3 ppm of ozone [11,14,23]. A lower dose of ozone, 1 ppm, was further reported by our group to be associated with increased AHR [27]. In this study, we investigated whether TLR4 is required for the response to 1 ppm ozone in the development of AHR and whether specific intracellular TLR4 adaptors are required for the response to ozone. After 20–24 hours recovery from inhalation of ozone at 1 ppm for 3 hours, we observed that ozone-exposed wild-type mice demonstrated significantly increased airway responsiveness to 25 and 100 mg/ml of methacholine challenge compared with filtered air exposed-mice. Ozone-exposed TLR4-deficient mice showed significantly reduced AHR when compared with ozone-exposed wild-type animals. Both air-exposed wild-type and TLR4-deficient mice have no difference in airway sensitivity to methacholine challenge (Figure 1A). Therefore, TLR4 contributes to AHR 24 hours after inhalation of 1 ppm ozone for 3 hours. We then investigated the role of TLR4 adaptor molecules in the pathogenesis of ozone-induced AHR. Using gene knockout mice, MyD88 has been reported to be involved in 3 ppm of ozone-induced AHR [14]. We demonstrated that MyD88 contributes to this lower dose ozone exposure (Figure 1B). We additionally investigated the contribution of TIRAP, which is another TLR4 adaptor protein previously reported to function through binding to MyD88 [33]. Similar to both the TLR4-deficient and MyD88-deficient animals, TIRAP-deficient mice failed to demonstrate enhanced AHR in response to methacholine challenge when compared with ozone-exposed wild-type mice (Figure 1C). There were no observed differences in methacholine sensitivity among air-exposed wild-type, TLR4−/−, MyD88−/− and TIRAP−/− animals (Figure 1B and 1C). AHR was also not different between ozone-exposed TLR4−/−, MyD88−/− and TIRAP−/− animals. Together these data support that TLR4, MyD88, and TIRAP contribute to ozone-induced AHR.

TLR4, MyD88 and TIRAP partially contribute to ozone-induced inflammation

Inhalation of ozone is recognized to cause an increase in total cells, neutrophils, and pro-inflammatory cytokine/chemokine levels in BALF. Consistent with previous reports, 1 ppm of ozone exposure increased neutrophil infiltration into the alveolar compartments in wild-type mice. The number of neutrophils in the BALF was also increased in TLR4−/, MyD88−/−, and TIRAP−/− mice exposed to ozone when compared to air (Figure 2A, 2B and 2C). The number of neutrophils was reduced in MyD88−/− mice when compared to ozone-exposed wild-type mice (Figure 2B). This observation suggested that neutrophilic inflammation caused by ozone exposure was partially dependent on a MyD88-mediated pathway. However, we did not observe a role of either TLR4 or TIRAP in ozone-induced neutrophil recruitment into the airspace. We further measured the level of cytokine/chemokine levels in BALF (Figure 3A, 3B, 3C, 3D and 3E). The results demonstrated that the levels of KC, IL-1β, IL-6, MCP-1 and TNFα in BALF from ozone–exposed wild-type animals were each significantly higher than that in BALF from air-exposed control animals. By comparison, the levels of KC, IL-1β, IL-6, MCP-1 and TNFα were also increased in BALF from ozone-exposed TLR4−/− mice, when compared with air exposed controls, but significantly lower than levels in BALF from ozone-exposed wild-type mice. Elevated production of KC, IL-1β, IL-6 and MCP-1 was not detected in BALF from either MyD88−/− or TIRAP−/− mice after either ozone or air exposure. The level of TNFα was reduced in ozone-exposed MyD88−/− mice when compared with ozone-exposed wild-type mice, but not elevated in ozone-exposed TIRAP−/− mice when compared with FA-exposed TIRAP−/− mice. These observations support that TLR4-MyD88-TIRAP signaling pathway partially contributed to ozone-induced enhanced levels of pro-inflammatory cytokines/chemokines. Interestingly, when making comparison between ozone-exposed strains, KC, IL-1β and IL-6 levels in BALF from TIRAP−/− mice were significantly lower than that from either TLR4−/− or MyD88−/− mice; and BALF cytokine levels from MyD88−/− mice were lower than levels from TLR4−/− mice. The observed differences between strains suggest the existence of TLR4-independent signaling pathways that are dependent on these adaptor proteins. In addition, total protein levels were determined in BALF from all experimental animals as a marker of lung injury or permeability. Ozone exposure significantly increased protein levels in BALF from wild-type, TLR4−/−, MyD88−/− and TIRAP−/− mice when compared to air exposed animals. We did not observe significant differences in total protein levels between strains with this low dose of acute exposure to ozone (Figure 4). In conclusion, our observations support that TLR4−/−, MyD88−/− and TIRAP−/− contribute to the generation of pro-inflammatory factors within the airspace after inhalation of ozone.

Inhalation of ozone increased the level of low molecular weight HA in the lung

Previously, we reported that 2 ppm of ozone exposure for 3 hours significantly increased HA level in BALF [20,23] in a manner independent of intact TLR4. Consistent with the previous

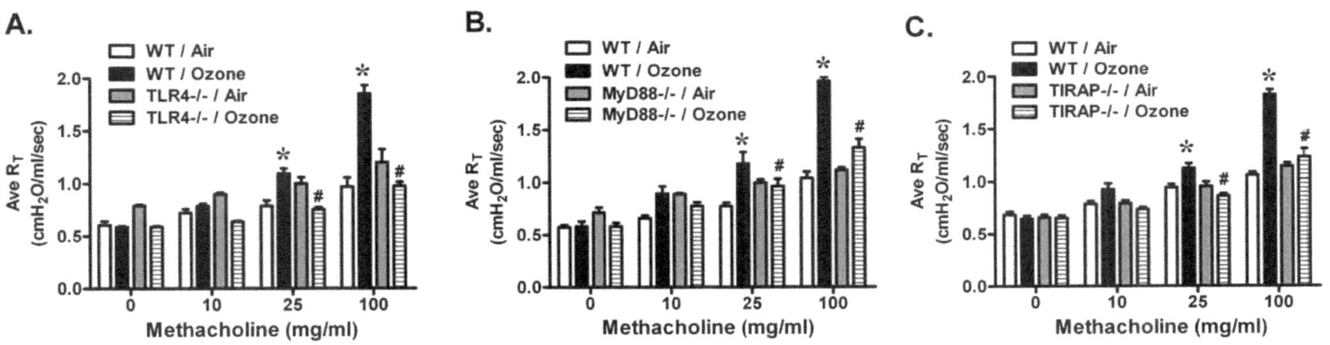

Figure 1. Ozone inhalation increased airway sensitivity to methacholine challenge and was dependent on the TLR4-MyD88-TIRAP signaling pathway. Animals were exposed to filtered air (FA) or 1 ppm of ozone for 3 hours. Airway responsiveness to methacholine challenge was measured 24 h later. **A)** Ozone-induced AHR was increased in WT mice but not in TLR4−/− mice, **B)** MyD88−/−mice, or **C)** TIRAP−/− mice (*$p<0.05$ vs. FA-WT; #$p<0.05$ vs. O3-WT, N = 5–6 per group).

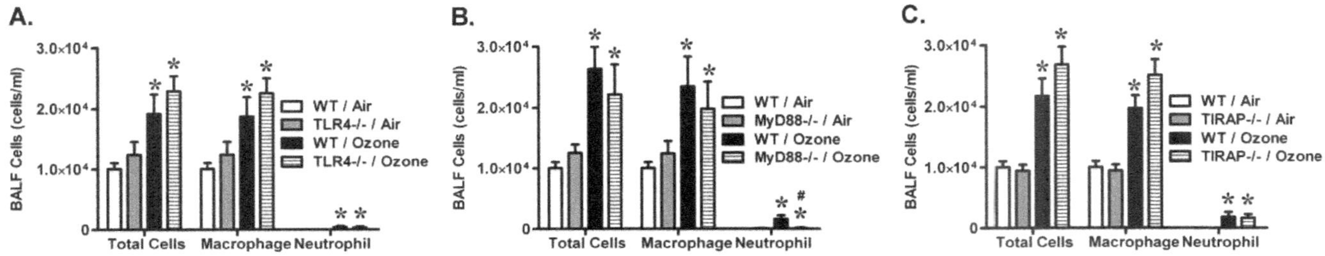

Figure 2. Ozone exposure increased neutrophilic lung inflammation in a manner partially dependent on MyD88, but not TLR4 and TIRAP. Ozone-exposed mice demonstrated increased neutrophil cell counts in BALF when compared to air-exposed animals. Neutrophil recruitment to the airspace was independent of TLR4 (**A**) and TIRAP (**C**), but was partially dependent of MyD88 (**B**) (* $p<0.05$, vs. FA exposed group; # $p<0.05$ vs. O_3-WT, N = 4–6 per group).

observation, 1 ppm of ozone exposure for 3 hours also increased HA levels in BALF in wild-type mice 24-hours after exposure. Similar elevated levels of HA were detected in BALF from ozone-exposed wild-type, TLR4−/−, MyD88−/−, and TIRAP−/− mice, when compared to air exposed mice. The levels of HA in BALF from ozone-exposed TLR4−/− and TIRAP−/− mice were not different from the levels of HA in BALF from ozone-exposed wild-type mice. Surprisingly, ozone-exposed MyD88−/− mice generated a lower level of HA in BALF than ozone-exposed wild-type, TLR4−/−, and TIRAP−/− mice. Baseline HA levels were not different among all air exposed strains (Figure 5). These results demonstrate that inhalation of ozone results in elevated levels of soluble HA and suggest that the level of HA in the BALF after ozone may be partially dependent on MyD88.

HA- induced airway hyperresponsiveness is dependent on TLR4, MyD88 and TIRAP

Previous reports support that hyaluronan can function as an endogenous ligand of TLR4 in the context of non-infectious tissue injury [21,34]. Our previous work demonstrates that TLR4 is required for both the response to ozone [11] and the response to short fragments of HA [23]. To determine whether the TLR4 adaptor proteins MyD88 and TIRAP contribute to the response to short fragments of HA, we directly challenged animals and measured methacholine sensitivity of the airways. As previously described, wild-type, TLR4−/−, MyD88−/−, and TIRAP−/− mice were challenged to either fragments of endotoxin-free HA or vehicle control by oropharyngeal aspiration. Two hours after exposure, the animals were characterized for AHR. We observed that all strains had similar levels of airway responsiveness to vehicle treatment. As previously reported, HA-treated wild-type mice demonstrate elevated AHR when compared to vehicle. However, TLR4−/−, MyD88−/− and TIRAP−/− mice were protected from the effects of HA challenge (Figure 6A, 6B and 6C). Airway sensitivity to methacholine challenge was not different among TLR4−/−, MyD88−/− and TIRAP−/− strains. These findings demonstrate that TLR4, MyD88, and TIRAP contribute to the biological response to short fragments of hyaluronan and contribute to the development of AHR.

HA induced production of pro-inflammatory cytokines is partially dependent on TLR4, MyD88 and TIRAP

Ozone exposure was associated with elevated neutrophils, total protein, pro-inflammatory cytokines/chemokines, hyaluronan, in

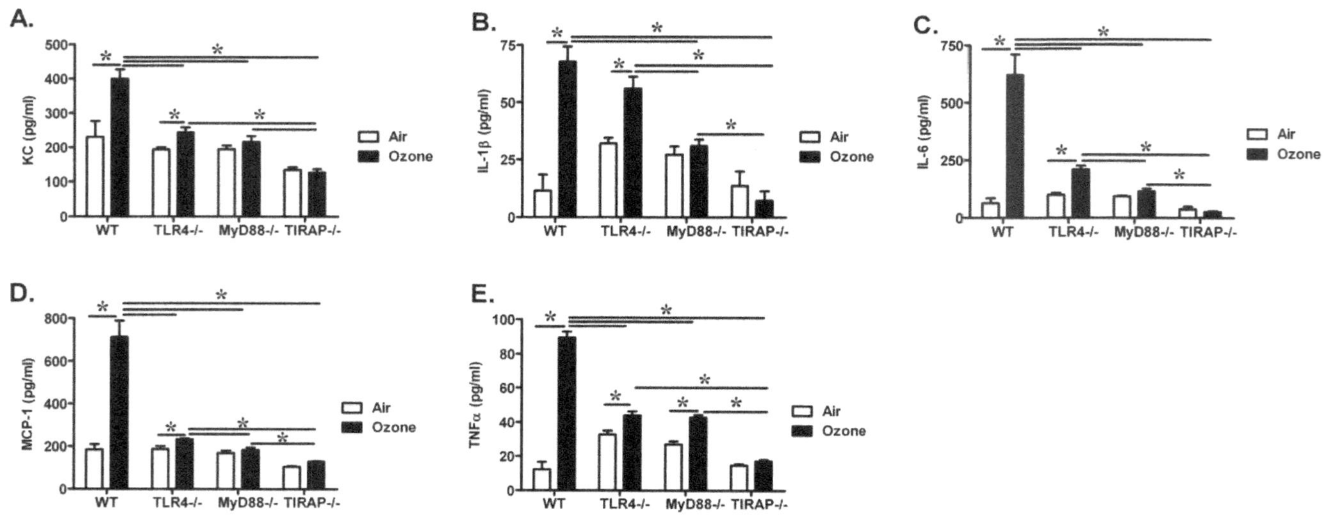

Figure 3. Ozone inhalation increased the level of pro-inflammatory factors in alveolar lavage fluid in a manner partially dependent on TLR4, MyD88, and TIRAP. The level of (**A**) KC, (**B**) IL-1β, (**C**) IL-6, (**D**) MCP-1 and (**E**) TNFα in BALF from air and ozone-exposed WT, TLR4−/−, MyD88−/− and TIRAP−/− mice were measured by luminex beads (*$p<0.05$, vs. FA exposed group; #$p<0.05$, vs. O_3-exposed group comparisons between strains; N = 4–5 per group).

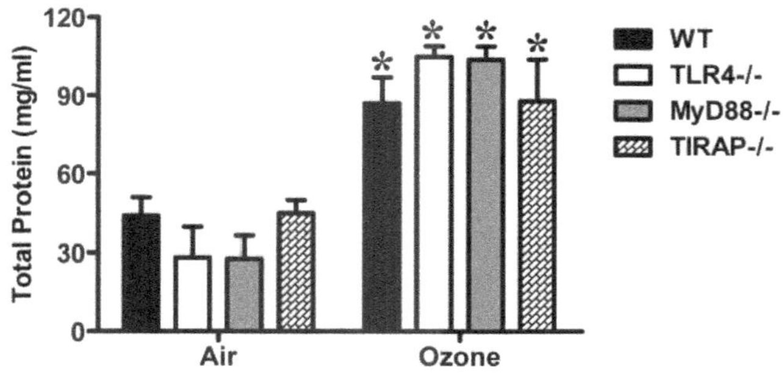

Figure 4. Increased lavage total protein level after ozone exposure was independent of TLR4, MyD88, and TIRAP (*p<0.01, vs. FA exposed group; N = 5–6 per group).

addition to the development of AHR. Direct challenge to fragments of hyaluronan did result in development of AHR in wild-type animals. However, with exposure to hyaluronan fragments, we did not observe elevated levels of either total protein or neutrophils in the BALF (Figure 7A and 7B). After challenge to HA, we did observe elevated levels of pro-inflammatory factors, including KC, IL-1β, IL-6, MCP-1 and TNFα in the BALF of wild-type mice, when compared to vehicle challenge. Similar to the profile of cytokines after inhalation of ozone, the response to HA challenge was partially dependent on TLR4−/−, MyD88−/− and TIRAP−/−. The levels of these cytokines/chemokines in BALF were not different between HA-treated MyD88−/− and TIRAP−/− mice (Figure 8A, 8B, 8C, 8D and 8E). These observations support a central role of TLR4-MyD88-TIRAP in hyaluronan-induced production of pro-inflammatory cytokines in the BALF.

Discussion

In this report, we demonstrate that the response to both ambient ozone and short fragments of hyaluronan are dependent on the TLR4-MyD88-TIRAP signal transduction pathway. We previously reported that the response to ozone is dependent of fragments of hyaluronan [20] and hyaluronan interaction with TLR4 [23]. We build on previous observations by demonstrating that the cytoplasmic adaptor molecules MyD88 and TIRAP are required for ozone-induced AHR. Specifically, we identify that TLR4-MyD88-TIRAP signal transduction pathway is a key determinant for AHR after either ozone or the direct delivery of hyaluronan to airspace. Additionally, we observe a similar profile of pro-inflammatory cytokines/chemokines dependent on TLR4-MyD88-TIRAP after challenge to either ozone or hyaluronan. Taken together, these observations support a striking role for HA-TLR4-MyD88-TIRAP signaling as a central transduction pathway in the development of reactive airways disease.

Our observations in the context of previous work provide additional support that HA, TLR4, MyD88, and TIRAP each contribute to the biological response to ozone and reactive airways disease. However, the dependence of each of these factors is not identical, which suggests that additional factors may contribute to the response to ozone. For example, TLR4-deficient animals are only partially protected from the effects of either ozone or hyaluronan when compared to MyD88-deficient animals. Thus, it appears that additional surface receptors participate to provide a complete innate immune response to ozone and may signal through conserved MyD88-dependent signaling. We and others have previously demonstrated that CD44 is the dominant receptor for hyaluronan and that CD44-TLR4 co-localize in context of non-infectious tissue injury [23,34]. The functional role of the CD44-TLR4 surface complex remains poorly understood. Additionally, toll-like receptor 2 (TLR2) has been previously reported to contribute to both macrophage recognition of hyaluronan (21) and functional response to ozone [14]. After inhalation of 3 ppm ozone, TLR2 partially contributes to AHR, but not neutrophilic inflammation [14]. Characterization of the functional role of additional TLR-surface receptors in divergent models of inhaled oxidative stress could provide a more complete understanding of the role of innate immunity in lung injury. In addition, we observed that the level of TNF-α was significantly lower in the TLR4-deficient mice in response to hyaluronan when compared to wildtype, MyD88−/− and TIRAP−/− mice. This observation suggests the presence of a MyD88-TIRAP independent pathway, which may contribute to the inflammatory response to hyaluronan fragments. We speculate that additional TLR4 adaptor proteins including TIR-domain-containing adapter-inducing interferon-β (TRIF) or tumor necrosis factor receptor-associated factor (TRAF) may modulate the response to ambient ozone. However, the specific role of additional adaptor proteins and additional TLR surface receptors in response to ozone will be the focus of continued investigations.

Given the dominant effects of MyD88-TIRAP in this phenotype when compared to TLR4, consideration has to be given to the potential that additional ligand-receptor interactions may contribute to the response to ozone. For example, it is well recognized that IL-1 signaling utilizes the intracellular adaptor MyD88 [35,36] and that the biological response to ozone is dependent on IL-1 [37,38]. It is plausible that the dominant role of MyD88-TIRAP signaling and cellular response results from both ozone-induced generation of hyaluronan and IL-1. In this paradigm, hyaluronan would function as an endogenous ligand of TLR4 leading to MyD88-dependent signal transduction and additional MyD88-dependent signaling would arise from IL-1 receptor ligation. Thus, each ligand (HA and IL-1) individually may be sufficient to induce cellular responses causing AHR to develop. With respect to this scenario, depletion of both hyaluronan and IL-1 may be required to more effectively ablate ozone-induced airway hyperresponsiveness. Clear understanding of interdependence among TLR receptors, and the specific molecular ligands and cytoplasmic adaptor molecules antecedent to upregulation of signal transduction pathways could provide insight into development of novel therapies to limit over-zealous innate immune responses to microbial and environmental exposures.

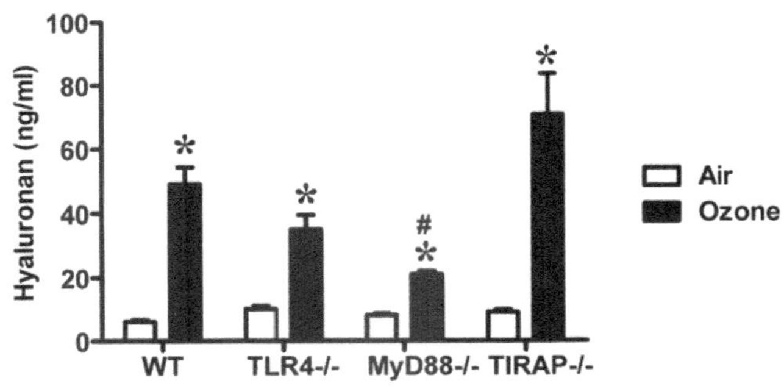

Figure 5. Level of HA in the BALF was increased after exposure to ozone. HA levels were significantly increased in all ozone-exposed groups (*p<0.01, vs. FA exposed group; #p<0.05, vs. O3-exposed WT, group, N = 4–5 per group).

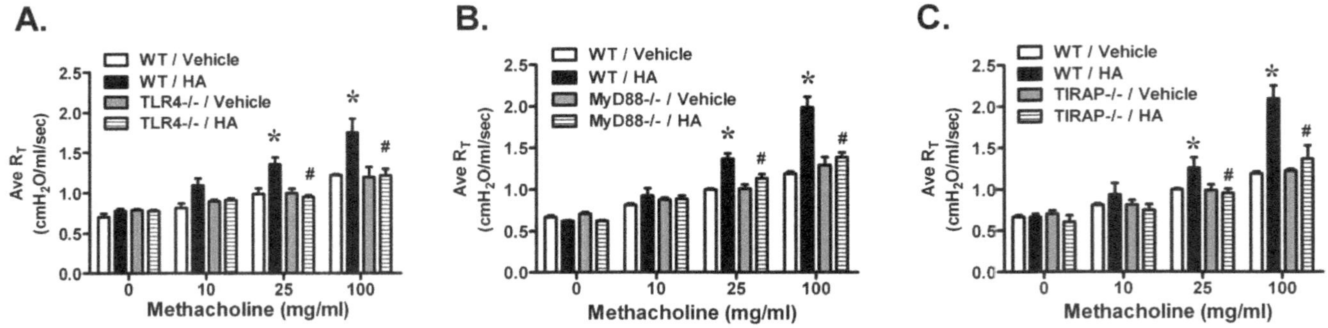

Figure 6. Airway reactivity to HA challenge was dependant on TLR4, MyD88, and TIRAP. HA increased AHR in WT mice, but not in (**A**) TLR4−/−mice, (**B**) MyD88−/−mice, or (**C**) TIRAP−/− mice (*$p<0.05$ vs. vehicle-WT; #$p<0.05$ vs. HA-WT, N = 5 per group).

There are fundamental differences between the two models that we have characterized for airway hyperresponsiveness. With respect to ozone, the animals were challenged to an environmentally relevant concentration level and response was phenotyped 24-hours after challenge. The delay in phenotyping is designed to model observations from studies of human exposure to ambient ozone, as previous epidemiological studies in humans demonstrate a lag of 24–72 hours after high levels of ambient ozone to the development of AHR [39]. In our animal model, we observe the development of ozone-induced AHR at 24-hours after exposure to ozone. After ozone, we do observe increased levels of hyaluronan in the lung even when delayed and investigated at a 24-hour time point. However, the response to hyaluronan is much more immediate [20]. Our previous work supports that ozone induces fragmentation of hyaluronan through oxidant stress [27]. We therefore speculate that ozone results in increased oxidant stress and fragmentation of hyaluronan and release into the airspace and into the sub-epithelial space. These fragments are thus positioned to increase production of pro-inflammatory cytokines/chemokines and recruitment of neutrophils. While we did not observe hyaluronan-induced recruitment of neutrophils into the airspace, we consider that this may be, in part, due to the timing of HA-induced neutrophilia. Interestingly, we observed the development of AHR in the absence of neutrophils in the airspace. While neutrophils are present in the airways after ozone exposure, the role of neutrophils in ozone-induced AHR remains somewhat unclear [40,41]. The specific mechanism of hyaluronan fragments-induced airway hyperresponsiveness remains unknown, but we identify a central role of TLR4-MyD88-TIRAP.

The role of genes of innate immunity in context of non-infectious lung injury appears highly dependent on both the intensity and duration of stimulus and underlying host factors. High levels of ozone (3 ppm×2 hr) or prolonged exposure to low level ozone (0.3×72 hr) can result in considerable lung injury, which is associated with induction of mRNA expression of TLR4 [14,42]. We do not observe induction of TLR4 mRNA with exposure to 1 ppm×3 hours ozone. It is not particularly surprising that the response to various models of lung injury is dependent on divergent host genetics. We recognize that the role of innate immunity in homeostatic maintenance of lung function is dependent on duration and intensity of injury. For example, low-level TLR4 signaling appears protective in some forms of oxidative lung injury [43,44,45], moderate TLR4 signaling facilitates the clearance of pathogens [46,47], and excessive prolonged TLR4 signaling can augment lung injury [48]. Furthermore, in this study, we utilize a low dose exposure to hyaluronan fragments (25 ug/mouse or estimated 1 mg/kg), which activates genes of innate immunity and can induce AHR. Zhao *et al.* provide evidence that a much higher dose of hyaluronan (65 mg/kg) can result in neutrophil recruitment into the lung and induce lung injury. Interestingly, with high dose hyaluronan, TLR4-deficient mice had an enhanced response to hyaluronan suggesting both a TLR4-independent pro-inflammatory signaling pathway and that TLR4-signaling can contribute to induction of a negative regulatory pathway and induction of neutrophil apoptosis [49,50]. Additional differences exist between these experimental protocols including the source and size of hyaluronan and timing of phenotyping. Clear understanding of

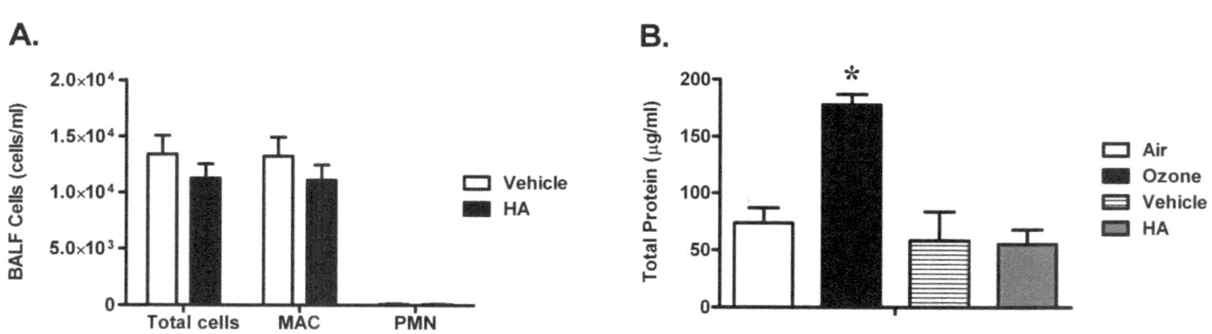

Figure 7. HA challenge was not sufficient for either neutrophilic inflammation or epithelial injury. (**A**) There were no observed differences in cellular inflammation in the airspace 2 hours after direct challenge to HA, (**B**) When compared to ozone challenge, HA challenge had no observed effect on the level of BALF total protein (*$p<0.05$, vs. FA exposed group, N = 5).

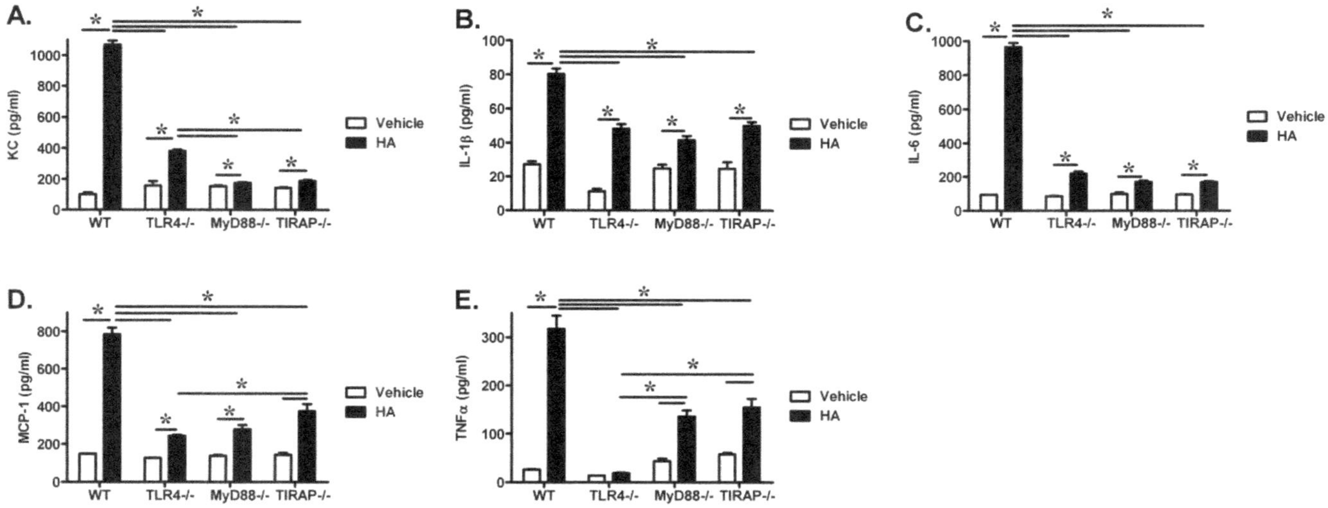

Figure 8. Direct challenge to hyaluronan fragments increased the level of pro-inflammatory factors in alveolar lavage fluid in a manner partially dependent on TLR4, MyD88, and TIRAP. The level of (**A**) KC, (**B**) IL-1β, (**C**) IL-6, (**D**) MCP-1 and (**E**) TNFα in BALF from vehicle or HA-challenged WT, TLR4−/−, MyD88−/− and TIRAP−/− mice were measured by luminex beads (*$p<0.05$, vs. FA exposed group; #: $p<0.05$, vs. O_3-exposed group comparisons between strains; N = 4 per group).

the signaling pathways that contribute to both the induction and resolution of lung inflammation may provide insight into novel therapeutic approaches to lung injury.

Together, our results further highlight the important role of genes of innate immunity in recognition of the damage-associated molecular pattern hyaluronan. Hyaluronan is critical to the development of ozone-induced airway hyperresponsiveness and production of pro-inflammatory cytokines/chemokines. Understanding the mechanisms that regulate the response to ozone has important implications to public health. Recent evidence suggest that given anticipated alterations in the global climate, both the level and biological health impact of ambient ozone will be increased [6]. Furthermore, our mouse model provides fundamental insight into the role of damage-associated molecular patterns in reactive airways disease and novel potential targets for the treatment of asthma. In conclusion, our results support that the HA-TLR4-MyD88-TIRAP signaling axis contributes to the pathogenesis of reactive airways disease.

Author Contributions

Conceived and designed the experiments: JWH ZL WMF SG. Performed the experiments: ZL ENP-K. Analyzed the data: ZL ENP-K. Contributed reagents/materials/analysis tools: JWH ZL WMF SG ENP-K. Wrote the paper: JWH ZL WMF SG ENP-K.

References

1. Katsouyanni K, Zmirou D, Spix C, Sunyer J, Schouten JP, et al. (1995) Short-term effects of air pollution on health: a European approach using epidemiological time-series data. The APHEA project: background, objectives, design. Eur Respir J 8: 1030–1038.
2. Gryparis A, Forsberg B, Katsouyanni K, Analitis A, Touloumi G, et al. (2004) Acute Effects of Ozone on Mortality from the "Air Pollution and Health: A European Approach" Project. Am J Respir Crit Care Med 28: 28.
3. Ito K, De Leon SF, Lippmann M (2005) Associations between ozone and daily mortality: analysis and meta-analysis. Epidemiology 16: 446–457.
4. Levy JI, Chemerynski SM, Sarnat JA (2005) Ozone exposure and mortality: an empiric bayes metaregression analysis. Epidemiology 16: 458–468.
5. Bell M, McDermott A, Zeger S, Samet J, Dominici F (2004) Ozone and short-term mortality in 95 US urban communities, 1987–2000. JAMA 292: 2372–2378.
6. Chang HH, Zhou J, Fuentes M (2010) Impact of climate change on ambient ozone level and mortality in southeastern United States. Int J Environ Res Public Health 7: 2866–2880.
7. Lukacs NW, Smit J, Lindell D, Schaller M (2007) Respiratory syncytial virus-induced pulmonary disease and exacerbation of allergic asthma. Contrib Microbiol 14: 68–82.
8. Wood LG, Simpson JL, Wark PA, Powell H, Gibson PG (2011) Characterization of innate immune signalling receptors in virus-induced acute asthma. Clin Exp Allergy 127: 640–8.
9. Schroder NW, Maurer M (2007) The role of innate immunity in asthma: leads and lessons from mouse models. Allergy 62: 579–590.
10. Williams MA, Cheadle C, Watkins T, Tailor A, Killedar S, et al. (2007) TLR2 and TLR4 as Potential Biomarkers of Environmental Particulate Matter Exposed Human Myeloid Dendritic Cells. Biomark Insights 2: 226–240.
11. Hollingsworth JW, Cook DN, Brass DM, Walker JK, Morgan DL, et al. (2004) The Role of Toll-like Receptor 4 in Environmental Airway Injury in Mice. Am J Respir Crit Care Med 170: 126–132.
12. Kleeberger SR, Levitt RC, Zhang LY, Longphre M, Harkema J, et al. (1997) Linkage analysis of susceptibility to ozone-induced lung inflammation in inbred mice. Nat Genet 17: 475–478.
13. Kleeberger SR, Reddy S, Zhang LY, Jedlicka AE (2000) Genetic susceptibility to ozone-induced lung hyperpermeability: role of toll-like receptor 4. Am J Respir Cell Mol Biol 22: 620–627.
14. Williams AS, Leung SY, Nath P, Khorasani NM, Bhavsar P, et al. (2007) Role of TLR2, TLR4, and MyD88 in murine ozone-induced airway hyperresponsiveness and neutrophilia. J Appl Physiol 103: 1189–1195.
15. Bauer AK, Rondini EA, Hummel KA, Degraff LM, Walker C, et al. (2011) Identification of Candidate Genes Downstream of TLR4 Signaling after Ozone Exposure in Mice: A Role for Heat Shock Protein 70. Environ Health Perspect 119: 1091–7.
16. McKee CM, Penno MB, Cowman M, Burdick MD, Strieter RM, et al. (1996) Hyaluronan (HA) fragments induce chemokine gene expression in alveolar macrophages. The role of HA size and CD44. J Clin Invest 98: 2403–2413.
17. Taylor KR, Trowbridge JM, Rudisill JA, Termeer CC, Simon JC, et al. (2004) Hyaluronan fragments stimulate endothelial recognition of injury through TLR4. J Biol Chem 279: 17079–17084. Epub 12004 Feb 17075.
18. Teder P, Vandivier RW, Jiang D, Liang J, Cohn L, et al. (2002) Resolution of lung inflammation by CD44. Science 296: 155–158.
19. Bai KJ, Spicer AP, Mascarenhas MM, Yu L, Ochoa CD, et al. (2005) The role of hyaluronan synthase 3 in ventilator-induced lung injury. Am J Respir Crit Care Med 172: 92–98.
20. Garantziotis S, Li Z, Potts EN, Kimata K, Zhuo L, et al. (2009) Hyaluronan mediates ozone-induced airway hyperresponsiveness in mice. J Biol Chem 284: 11309–11317.
21. Jiang D, Liang J, Fan J, Yu S, Chen S, et al. (2005) Regulation of lung injury and repair by Toll-like receptors and hyaluronan. Nat Med 11: 1173–1179. Epub 2005 Oct 1123.

22. Kliment CR, Tobolewski JM, Manni ML, Tan RJ, Enghild J, et al. (2008) Extracellular superoxide dismutase protects against matrix degradation of heparan sulfate in the lung. Antioxid Redox Signal 10: 261–268.
23. Garantziotis S, Li Z, Potts EN, Lindsey JY, Stober VP, et al. (2010) TLR4 is Necessary for Hyaluronan-mediated Airway Hyperresponsiveness After Ozone Inhalation. Am J Respir Crit Care Med 181: 666–675.
24. Hoshino K, Takeuchi O, Kawai T, Sanjo H, Ogawa T, et al. (1999) Cutting edge: Toll-like receptor 4 (TLR4)-deficient mice are hyporesponsive to lipopolysaccharide: evidence for TLR4 as the Lps gene product. J Immunol 162: 3749–3752.
25. Adachi O, Kawai T, Takeda K, Matsumoto M, Tsutsui H, et al. (1998) Targeted disruption of the MyD88 gene results in loss of IL-1- and IL-18-mediated function. Immunity 9: 143–150.
26. Yamamoto M, Sato S, Hemmi H, Sanjo H, Uematsu S, et al. (2002) Essential role for TIRAP in activation of the signalling cascade shared by TLR2 and TLR4. Nature 420: 324–329.
27. Li Z, Potts EN, Piantadosi CA, Foster WM, Hollingsworth JW (2010) Hyaluronan fragments contribute to the ozone-primed immune response to lipopolysaccharide. J Immunol 185: 6891–6898.
28. Hatch GE, Slade R, Harris LP, McDonnell WF, Devlin RB, et al. (1994) Ozone dose and effect in humans and rats. A comparison using oxygen-18 labeling and bronchoalveolar lavage. Am J Respir Crit Care Med 150: 676–683.
29. Wiester MJ, Tepper JS, King ME, Menache MG, Costa DL (1988) Comparative study of ozone (O3) uptake in three strains of rats and in the guinea pig. Toxicol Appl Pharmacol 96: 140–146.
30. Que LG, Stiles JV, Sundy JS, Foster WM (2011) Pulmonary Function, Bronchial Reactivity, and Epithelial Permeability are Response Phenotypes to Ozone and Develop Differentially in Healthy Humans. J Appl Physiol.
31. Auten RL, Potts EN, Mason SN, Fischer B, Huang Y, et al. (2009) Maternal exposure to particulate matter increases postnatal ozone-induced airway hyperreactivity in juvenile mice. Am J Respir Crit Care Med 180: 1218–1226.
32. Foster WM, Adler KB, Crews AL, Potts EN, Fischer BM, et al. (2010) MARCKS-related peptide modulates in vivo the secretion of airway Muc5ac. Am J Physiol Lung Cell Mol Physiol 299: L345–352.
33. Horng T, Barton GM, Medzhitov R (2001) TIRAP: an adapter molecule in the Toll signaling pathway. Nat Immunol 2: 835–841.
34. Taylor KR, Yamasaki K, Radek KA, Nardo AD, Goodarzi H, et al. (2007) Recognition of Hyaluronan Released in Sterile Injury Involves a Unique Receptor Complex Dependent on Toll-like Receptor 4, CD44, and MD-2. J Biol Chem 282: 18265–18275.
35. Gasse P, Mary C, Guenon I, Noulin N, Charron S, et al. (2007) IL-1R1/MyD88 signaling and the inflammasome are essential in pulmonary inflammation and fibrosis in mice. J Clin Invest 117: 3786–3799.
36. Fremond CM, Togbe D, Doz E, Rose S, Vasseur V, et al. (2007) IL-1 receptor-mediated signal is an essential component of MyD88-dependent innate response to Mycobacterium tuberculosis infection. J Immunol 179: 1178–1189.
37. Park JW, Taube C, Swasey C, Kodama T, Joetham A, et al. (2004) Interleukin-1 receptor antagonist attenuates airway hyperresponsiveness following exposure to ozone. Am J Respir Cell Mol Biol 30: 830–836.
38. Wu ZX, Barker JS, Batchelor TP, Dey RD (2008) Interleukin (IL)-1 regulates ozone-enhanced tracheal smooth muscle responsiveness by increasing substance P (SP) production in intrinsic airway neurons of ferret. Respir Physiol Neurobiol 164: 300–311.
39. Gent JF, Triche EW, Holford TR, Belanger K, Bracken MB, et al. (2003) Association of low-level ozone and fine particles with respiratory symptoms in children with asthma. Jama 290: 1859–1867.
40. Matsumoto K, Aizawa H, Inoue H, Koto H, Nakano H, et al. (1999) Role of neutrophil elastase in ozone-induced airway responses in guinea-pigs. Eur Respir J 14: 1088–1094.
41. DeLorme MP, Yang H, Elbon-Copp C, Gao X, Barraclough-Mitchell H, et al. (2002) Hyperresponsive airways correlate with lung tissue inflammatory cell changes in ozone-exposed rats. J Toxicol Environ Health A 65: 1453–1470.
42. Kleeberger SR, Reddy SP, Zhang LY, Cho HY, Jedlicka AE (2001) Toll-like receptor 4 mediates ozone-induced murine lung hyperpermeability via inducible nitric oxide synthase. Am J Physiol Lung Cell Mol Physiol 280: L326–333.
43. Qureshi ST, Zhang X, Aberg E, Bousette N, Giaid A, et al. (2006) Inducible activation of TLR4 confers resistance to hyperoxia-induced pulmonary apoptosis. J Immunol 176: 4950–4958.
44. Zhang X, Shan P, Jiang G, Cohn L, Lee PJ (2006) Toll-like receptor 4 deficiency causes pulmonary emphysema. J Clin Invest 116: 3050–3059.
45. Zhang X, Shan P, Qureshi S, Homer R, Medzhitov R, et al. (2005) Cutting edge: TLR4 deficiency confers susceptibility to lethal oxidant lung injury. J Immunol 175: 4834–4838.
46. Chassin C, Picardeau M, Goujon JM, Bourhy P, Quellard N, et al. (2009) TLR4- and TLR2-mediated B cell responses control the clearance of the bacterial pathogen, Leptospira interrogans. J Immunol 183: 2669–2677.
47. Wieland CW, Florquin S, Maris NA, Hoebe K, Beutler B, et al. (2005) The MyD88-dependent, but not the MyD88-independent, pathway of TLR4 signaling is important in clearing nontypeable haemophilus influenzae from the mouse lung. J Immunol 175: 5042–6049.
48. Brass DM, Hollingsworth JW, Cinque M, Li Z, Potts E, et al. (2008) Chronic LPS Inhalation Causes Emphysema-like Changes in Mouse Lung that is Associated with Apoptosis. Am J Respir Cell Mol Biol 39: 584–590.
49. Leu SW, Shi L, Xu C, Zhao Y, Liu B, et al. (2011) TLR4 through IFN-beta promotes low molecular mass hyaluronan-induced neutrophil apoptosis. J Immunol 186: 556–562.
50. Zhao H, Leu SW, Shi L, Dedaj R, Zhao G, et al. (2010) TLR4 is a negative regulator in noninfectious lung inflammation. J Immunol 184: 5308–5314.

High Environmental Ozone Levels Lead to Enhanced Allergenicity of Birch Pollen

Isabelle Beck[1], Susanne Jochner[2], Stefanie Gilles[1,4], Mareike McIntyre[3], Jeroen T. M. Buters[1], Carsten Schmidt-Weber[1], Heidrun Behrendt[1,4], Johannes Ring[3], Annette Menzel[2], Claudia Traidl-Hoffmann[1,3,4]*

1 ZAUM – Center of Allergy & Environment, Member of the German Center for Lung Research (DZL), Technische Universität München/Helmholtz Center, Munich, Germany, **2** Department of Ecology and Ecosystem Management, Ecoclimatology, Technische Universität München, Freising, Germany, **3** Department of Dermatology and Allergy, Technische Universität München, Munich, Germany, **4** Christine-Kühne-Center for Allergy Research and Education (CK Care), Davos, Switzerland

Abstract

Background: Evidence is compelling for a positive correlation between climate change, urbanisation and prevalence of allergic sensitisation and diseases. The reason for this association is not clear to date. Some data point to a pro-allergenic effect of anthropogenic factors on susceptible individuals.

Objectives: To evaluate the impact of urbanisation and climate change on pollen allergenicity.

Methods: Catkins were sampled from birch trees from different sites across the greater area of Munich, pollen were isolated and an urbanisation index, NO_2 and ozone exposure were determined. To estimate pollen allergenicity, allergen content and pollen-associated lipid mediators were measured in aqueous pollen extracts. Immune stimulatory and modulatory capacity of pollen was assessed by neutrophil migration assays and the potential of pollen to inhibit dendritic cell interleukin-12 response. *In vivo* allergenicity was assessed by skin prick tests.

Results: The study revealed ozone as a prominent environmental factor influencing the allergenicity of birch pollen. Enhanced allergenicity, as assessed in skin prick tests, was mirrored by enhanced allergen content. Beyond that, ozone induced changes in lipid composition and chemotactic and immune modulatory potential of the pollen. Higher ozone-exposed pollen was characterised by less immune modulatory but higher immune stimulatory potential.

Conclusion: It is likely that future climate change along with increasing urbanisation will lead to rising ozone concentrations in the next decades. Our study indicates that ozone is a crucial factor leading to clinically relevant enhanced allergenicity of birch pollen. Thus, with increasing temperatures and increasing ozone levels, also symptoms of pollen allergic patients may increase further.

Editor: Simon Patrick Hogan, Cincinnati Children's Hospital Medical Center, University of Cincinnati College of Medicine, United States of America

Funding: The study was supported by a grant of the German Research Council (Deutsche Forschunsgemeinschaft, DFG, TR467/5-1 and ME 179/3-1) (CTH and AM), the HWP grant from of the Technische Universität München (SG), and CK Care, Christine Kühne Center for Allergy Research and Education. The funders had no role in study design, data collection and analysis, decision to publish, or preparation of the manuscript.

Competing Interests: Bet v 1- specific antibodies MAK 2E10G6G7 and 4B10D10F8 were kindly provided by Allergopharma. This does not alter the authors' adherence to all the PLOS ONE policies on sharing data and materials, as detailed online in the guide for authors.

* E-mail: claudia.traidl-hoffmann@lrz.tu-muenchen.de

Introduction

Epidemiological studies show an increasing trend in allergies, leading to a major health problem. Reasons discussed for this trend include a westernized life style with diminished immune stimulation [1] and anthropogenic air pollution [2,3]. Particularly, irritant gases and diesel exhaust particles have been shown to exert adjuvant or aggravating effects on sensitisation and elicitation phases of allergic immune responses [4,5]. As underlying mechanisms, effects on cells of the immune system as well as epithelial barrier disruption are discussed [6]. However, pollutants in ambient air do not only impact humans but also the allergen-carrier itself, i.e. the plant and its pollen. Therefore, the question arises whether the observed increase in allergic diseases in the western world might in part be explained by modified allergenicity of pollen caused by urbanisation and paralleled climate change. These environmental changes – higher temperature, in combination with higher concentrations of specific anthropogenic pollutants – lead to higher tropospheric ozone concentrations. In this scenario UV-radiation delivers the energy for ozone generation, but besides this, higher temperatures can also lead to an increase in ozone formation promoted by emission of highly reactive hydrocarbons from vegetation and evaporation processes [7]. Climate extremes are often observed in cities, which function as heat islands and can be regarded as a mirror of future climate [8]. However, urbanisation is not only characterised by higher temperatures, but also by higher levels of pollutants like particulate matter, carbon dioxide (CO_2) or nitrogen dioxide (NO_2). In this respect, it has to be considered that ozone does not show the same distribution as other pollutants [7,9]. Ozone is a secondary pollutant whose formation underlies complex interactions, depending on the presence of precursors, degrading substances, temperature and UV-radiation. The main precursors are nitrogen oxides (NOx) and volatile organic compounds (VOCs). Especially NOx can be found at high concentrations in urban areas. NO, in turn, rapidly degrades the ozone generated in urban areas, especially during night-time. In rural areas, in contrast, we observe an accumulation of ozone due to lower NO levels and higher biogenic emissions [7]. Some studies already addressed the question of how pollutants affect the allergen carrier, showing that single pollutants strongly differ in their effects [10,11]. The current study expands these observations by analyzing one of the most relevant allergen producers – the birch tree – in its natural environment. Hereby, we work with real exposure conditions to analyse relevant factors under natural circumstances. We take into account the parallel occurrence of different environmental factors as well as mechanisms of plant adaptation. Recent studies showed that pollen do not only release allergens, but also non-allergenic compounds such as pollen-associated lipid mediators (PALMs) (reviewed in [12]), which have been shown to exert immune modulatory and stimulatory effects. Allergenicity, thus, was evaluated in a holistic approach also taking adjuvant factors into account.

This study aimed at understanding how long-term increases in urbanisation and concomitant increases in pollution might influence pollen allergenicity, and how this might translate into immune cell activation and symptoms of pollen-allergic patients.

Materials and Methods

Ethic Statement

The ethical committee of the Technical University of Munich approved the study, and volunteers were enrolled after written informed consent. The study was carried out on private land and the owners of the land gave permission to conduct the study on these sites.

Sampling of Birch Pollen

Catkins were collected from birch trees located in the greater area of Munich (n = 40; see Fig. S1). Sampling of catkins took place during the birch flowering season in spring 2010. Their developmental stages were assessed by an adopted and extended code of the BBCH (Biologische **B**undesanstalt, **B**undessortenamt und **Ch**emische Industrie) [13]. The code included 12 different developmental stages, starting with winter rest (BBCH 50) and ending with end of flowering (BBCH 69). The stages of collection were BBCH 60 (single catkins sporadically emit pollen) and BBCH 61 (10% of the catkins emit pollen). After collection, catkins were air-dried, counted and weighed and pollen was extracted by sieving.

Preparation of Aqueous Pollen Extracts for Bet v 1, LTB_4 and PGE_2 ELISA

Aqueous pollen extracts (APEs) were prepared in 0.1 M NH_4HCO_3, pH 8.1, as previously described [14]. For skin prick tests, APEs were prepared as described in [15]. APE concentrations given refer to the extraction of a given amount of pollen per mL of buffer before centrifugation (e. g. 1 mg pollen was extracted in 1 mL buffer), but do not refer to actual protein concentrations in the extracts.

Bet v 1 ELISA

Bet v 1 levels were determined by sandwich ELISA as described in Buters et al. [14]. Bet v 1-specific antibodies MAK 2E10G6G7 and 4B10D10F8 were kindly provided by Joachim Ganzer, Allergopharma, Hamburg, Germany.

PGE_2 and LTB_4 ELISA

Concentrations of the eicosanoid-like PALMs in APEs were measured by commercially available enzyme immunoassays for prostaglandin E_2 and leukotriene B_4 (GE Healthcare, Germany) according to the supplier's protocol.

Passive Sampling Method for Ozone and Nitrogen Dioxide (NO_2) Determination

Passive samplers for ozone were provided and analysed by PASSAM AG, Männedorf, Switzerland. Passive sampling for ozone and NO_2 were done in parallel. Passive sampling was carried out during the one-week period from May 11th to 18th 2010, i.e. 2 weeks after the last catkins were sampled. Measurements for all 40 birch trees were done in the same time frame and directly at the tree. It is supposed that the traffic volume over the period of 1 week is a representative mean for this location. This data was used to characterize the tree for a relative ozone exposure.

The nitrogen dioxide concentration was measured at the 40 sites according to Palmes' principle [16]. Briefly, stainless steel

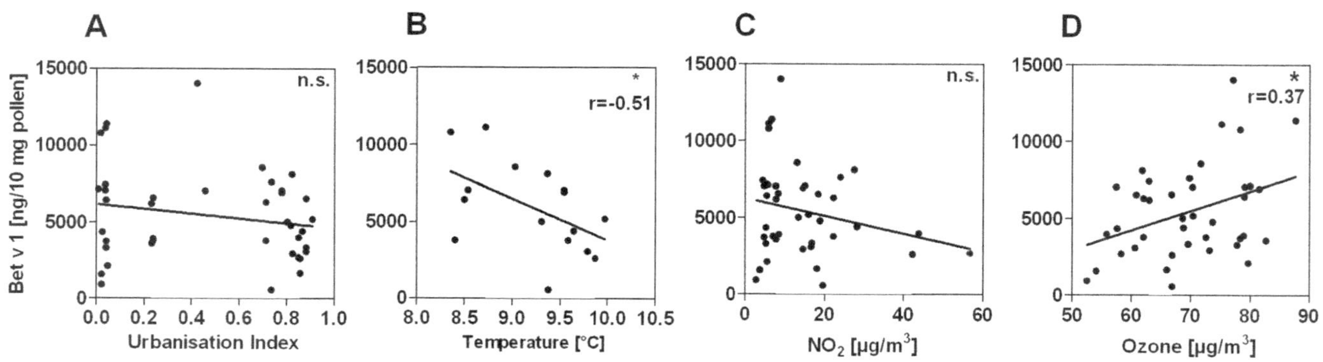

Figure 1. Bet v 1 content of birch pollen in trees against different, urbanisation-related environmental conditions. No significant correlation could be observed between Bet v 1 content and urbanisation index (A; n = 40) or NO_2 concentration (B; n = 40). Bet v 1 showed a significant and negative correlation with temperature (C; n = 16) and was positively correlated with site-specific ozone levels (D; n = 40). *: $p<0.05$.

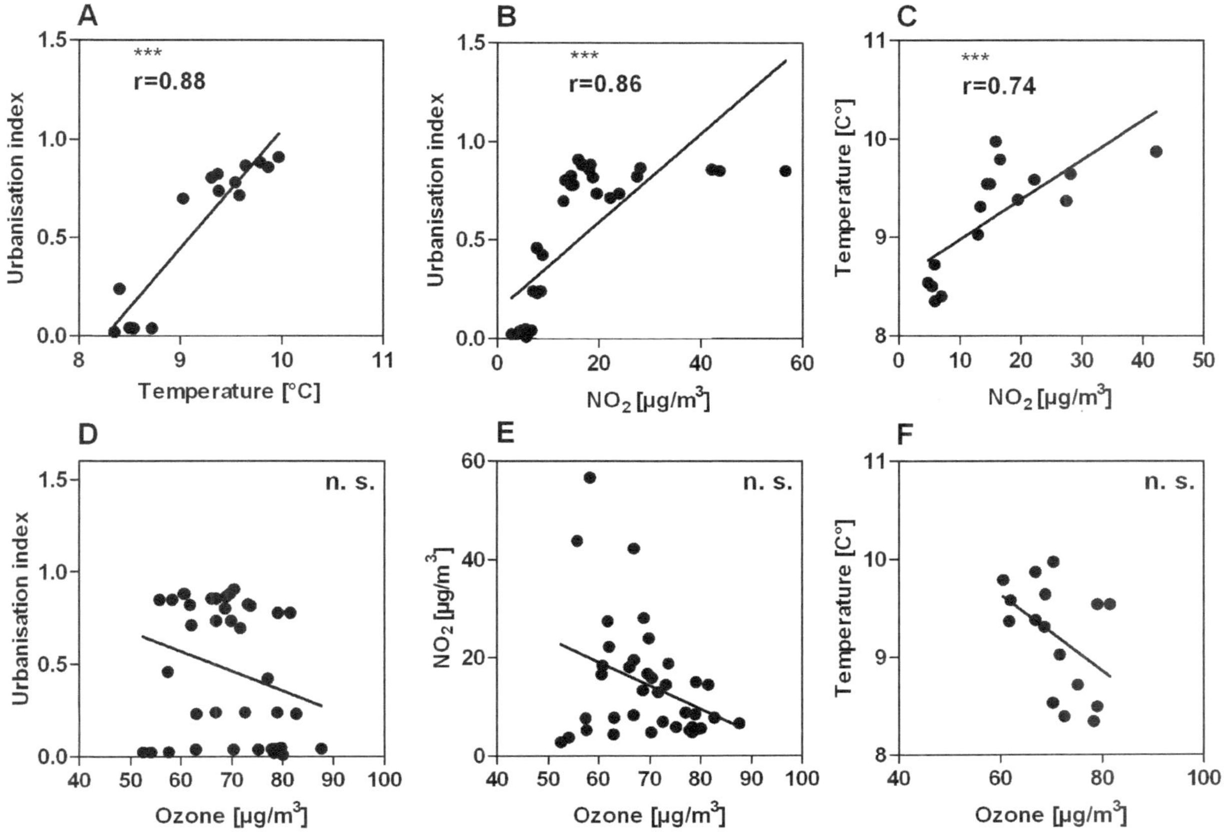

Figure 2. Scatter plots of different environmental parameters. A significant correlation was observed between the parameters urbanisation index (UI) and temperature (A; $n = 16$), (B) UI and NO_2 concentration ($n = 40$) and (C) temperature and NO_2 concentration ($n = 16$). Ozone was not correlated with either (D) UI ($n = 40$), (E) NO_2 ($n = 40$) or (F) temperature ($n = 16$); ***: $p<0.001$.

meshes were immersed in a triethanolamine-aceton mixture and were air-dried for 10 minutes. Three coated meshes were brought into an air-tight tube. NO_2 binds to the coated meshes by forming a triethanolamine-NO_2-complex. NO_2 adsorption was determined photometrically after one week of exposure.

Calculation of Urbanisation Index

An urbanisation index (UI) based on CORINE Land Cover 2000 data (European Environment Agency 2000) was calculated using ArcGIS 9.3. This index reflects the proportion of predefined built up areas (e.g. continuous and discontinuous urban fabric, industrial or commercial units) within a radius of 2 km and thus can vary between 0 and 1; i.e. from a low (UI = 0) to a high (UI = 1) degree of urbanisation [17].

Temperature Measurements

15 birch trees were provided with data recording devices measuring air temperature (HOBO U23-001, Onset Computer Corporation, Southern MA, USA). The devices were fixed in a radiation shield at the northern side of the trees in 3 m height. Air temperatures were recorded every 10 minutes, and daily temperature means were calculated. Temperature data were acquired between 1st July 2009 and 5th May 2010. The mean temperature for this period for each location was calculated.

Blood Donors

Healthy, non-atopic blood donors without a history of allergic diseases were tested by RAST for sensitisation against common allergens including birch allergens. All subjects were tested negative and total IgE was <100 IU/ml. Volunteers did not take any medication for at least 15 days before blood sampling.

Isolation and Culture of Monocyte-derived Dendritic Cells

Monocyte-derived dendritic cells (moDCs) were cultured from human peripheral blood monocytes as described by Gilles et al. [18]. Immature DCs were harvested on day 5 followed by stimulation with LPS (100 ng/ml) plus APEs from the highest (pollen^{O3high}: mean ozone = 85 µg/m^3, n = 2) and lowest (pollen^{O3high}: mean ozone = 54 µg/m^3, n = 2) ozone-exposed birch trees included in the study (mean ozone: 54 µg/m^3 versus 85 µg/m^3). After 24 h, supernatants were collected, and IL-12p70 release was measured by ELISA (BD Pharmingen, Heidelberg, Germany). The IL-12 response to APEs alone was measured exemplary in four APEs and virtually no IL-12 was detected (unstimulated: 18.7 pg/ml; APEs: median (min.-max.): 15.4 (9.9–20.8) pg/ml, n = 4) (data not shown). Viability of the cells after 24 h of culture was tested by propidium iodide staining and subsequent FACS analysis (see Fig. S4, supplementary material). Viability was not decreased by any of the conditions.

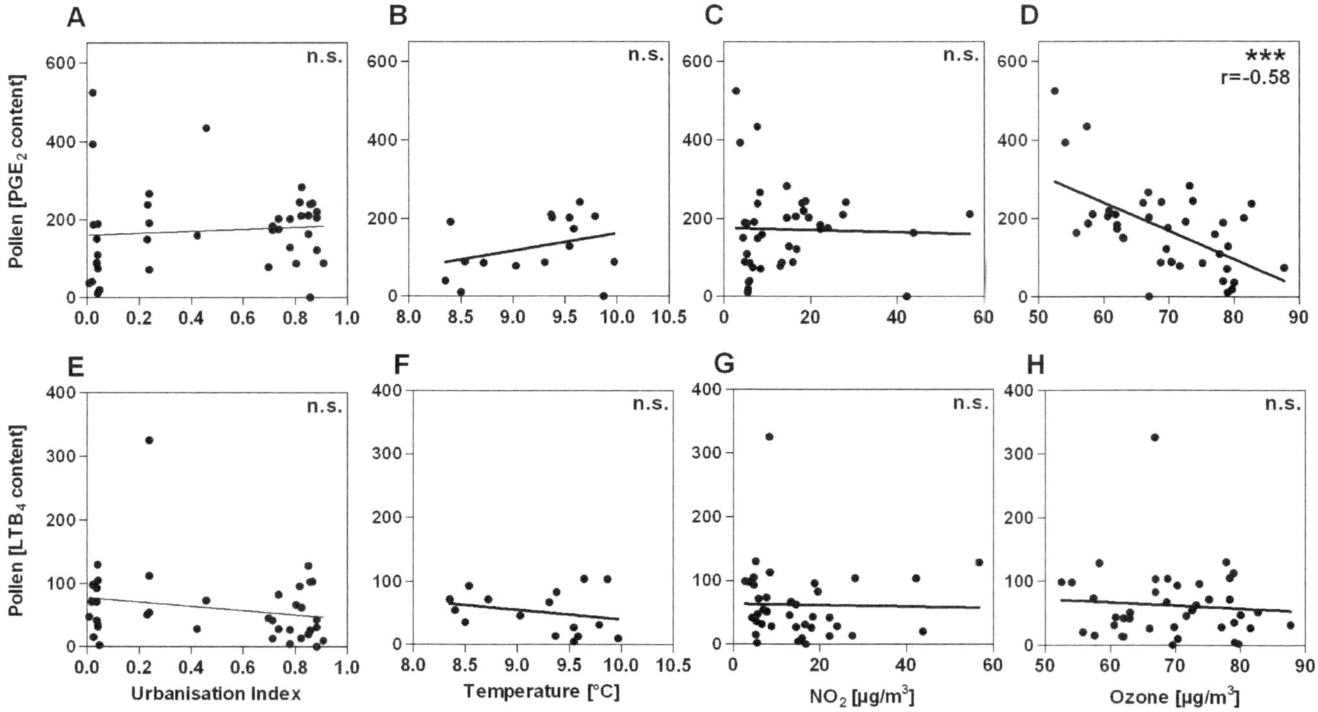

Figure 3. Content of PALMs in pollen samples plotted against different urbanisation-related environmental conditions. No correlation was seen between $PALM_{PGE2}$ and UI (A; n = 40), temperature (B; n = 16) or NO_2 concentration (C; n = 40). A significant association of high ozone concentrations and low $PALM_{PGE2}$ contents was observed (D; n = 40). $PALM_{LTB4}$ did not show any significant correlation. Neither UI (E; n = 40), nor temperature (F; n = 16), NO_2- (G; n = 40) and ozone (H; n = 40) were related to the content of $PALM_{LTB4}$. ***: $p<0.001$.

Neutrophil Migration Assays

The chemotactic activity of APEs from the highest (pollen^{O3high}: mean ozone = 85 µg/m^3) and lowest (pollen^{O3high}: mean ozone = 54 µg/m^3) ozone-exposed birch trees included in the study was evaluated by measuring neutrophil migration through a 5 µm pore polycarbonate membrane (ChemoTx Disposable Chemotaxis System, NeuroProbe). Neutrophils were isolated from peripheral blood as described by Traidl-Hoffmann et al. [19]. APEs were pipetted into the bottom chambers and neutrophils (1 ×10^6 cells/ml) were added to the top of the membrane. After 1 h of incubation the cell suspension was removed and cells that had transmigrated into the lower chamber were recovered and counted with a FACSCalibur (Becton Dickinson, Heidelberg, Germany).

Skin Prick Tests

Birch allergic patients (n = 5) with a specific IgE >0.35 kU/l were pricked on their forearms with APE (10 mg/mL) prepared of pollen samples from the highest (pollen^{O3high}: mean ozone = 85 µg/m^3) and lowest (pollen^{O3high}: mean ozone = 54 µg/m^3) ozone-exposed birch trees of the study. Each sample was pricked in replicates of four (proximally and distally on the same arm, right and left arm), and the mean wheal and flare sizes were calculated out of these four measurements after 15 min.

Statistics

Unpaired t-test was used for Gaussian distributed samples to determine statistically significant differences between groups. For non-Gaussian populations, the non-parametric Mann-Whitney U test was applied. The correlation coefficients (r) and 95% confidence intervals were calculated using the Pearson's approach for Gaussian distributions and the Spearman's approach for non-parametric correlation. Additionally, we applied linear multiple regression analyses based on stepwise variable selection to test which environmental factor is most important for pollen allergenicity. To analyse cell assays, the area under the curve (AUC) was determined and the Wilcoxon matched-pairs signed-ranks test was applied. Prick tests were also analysed by the Wilcoxon matched-pairs signed-ranks test. P values of <0.05 were considered significant (*). **: $p<0.01$, ***: $p<0.001$. Statistics were done with Graph Pad Prism 5, San Diego, CA, USA.

Results

Correlation Analysis of Urbanisation Related Factors and Relationship to Pollen Allergenic Potential

No significant correlation was detected between Bet v 1 content of the pollen specimens and urbanisation index (UI) or NO_2 concentration (Fig. 1A, C: n = 40). However, a negative correlation could be observed for Bet v 1 content and temperature (Fig. 1B; r = -0.51; p = 0.042; n = 16). In contrast, ozone was positively correlated with Bet v 1 content. (r = 0.37; p = 0.017; Fig. 1D; n = 40).

Sites of birch trees were analyzed for environmental parameters. By correlation analyses the relationship of the determined parameters was investigated. A highly significant and positive correlation of the variables UI (n = 40), temperature (n = 16) and NO_2-concentration (n = 40) could be observed (Fig. 2 A, B, C). In contrast, ozone (n = 40) was not statistically associated with any of these parameters (Fig. 2 D, E, F).

To investigate whether ozone or temperature is the factor influencing pollen Bet v 1 content, we applied multiple regression analyses based on stepwise variable selection. Indeed, the role of the independent variable ozone for predicting Bet v 1 content was superior, since temperature has been excluded in the linear model, yielding in an R^2 of 27.6% (data not shown).

To determine immune stimulatory and modulatory potential of the birch pollen specimens, we analysed the content of PALMs in the different pollen extracts. Prostaglandin E_2-like PALMs, termed $PALM_{PGE2}$, harbour the immune modulatory PALMs [20,21]. No statistically significant correlation was observed between $PALM_{PGE2}$ contents and UI (n = 40), temperature (n = 16) or ambient NO_2 (n = 40) (Fig. 3A, B, C). Instead, $PALM_{PGE2}$ content was negatively correlated with ambient ozone (n = 40) levels (r = -0.58; p<0.0001; Fig. 3D). The LTB_4-like PALMs harbour the chemotactic, immune stimulatory PALMs [19]. $PALM_{LTB4}$ contents were not correlated with either UI (n = 40), temperature (n = 16), NO_2 (n = 40) or ozone (n = 40) levels (Fig. 3E, F, G, H).

Immune-stimulatory and Immune-regulatory Capacity of Pollen Samples

To assess whether the altered lipid compositions related to higher and lower ozone concentrations go along with differences in immune stimulatory and –modulatory potential, pollen from differentially ozone-exposed birch trees (mean ozone: 85 µg/m^3 (pollen^{O3high}) versus 54 µg/m^3 (pollen^{O3low})) were chosen and analysed for induction of neutrophil chemotaxis and modulation of dendritic cell (moDC) cytokine secretion. Migration to pollen^{O3high} and pollen^{O3low} was tested in 3 concentrations (Fig. S4) and the AUC was calculated. Pollen^{O3high} induced significantly stronger neutrophil chemotaxis than pollen^{O3low}. Neutrophil migration towards pollen extracts (median AUC (min.-max.)) were 3181 (2214–4429) for pollen^{O3high} versus 2327 (1839–4095) for pollen^{O3low}; p = 0.03 (Fig. 4A). To test the immune modulatory capacity of differentially ozone-exposed pollen, moDCs were stimulated with LPS in the presence and absence of APE, and IL-12 was measured in the supernatants (LPS induced IL-12 response: median (min.-max.): 14000 (8928–14500) pg/ml). APEs were applied in 3 concentrations (Fig. S3) and AUC was calculated. As shown in Fig. 4B, pollen^{O3high} were less potent inhibitors of the moDC's IL-12 response than pollen^{O3low}. IL-12 release of dendritic cells (% of LPS; median AUC (min-max)) after stimulation with APE was 127.7 (67.3–227.0) for pollen^{O3high} versus 71.0 (52.8–187.7) for pollen^{O3low}; p = 0.02.

To test for clinical relevance of enhanced Bet v 1 levels, birch pollen allergic patients were subjected to skin prick tests with APEs prepared from differently exposed pollen. Wheal and flare sizes were significantly larger when patients were pricked with APEs prepared from pollen^{O3high}. Wheal sizes (in mm^2; median (min.-max.)) were 21.3 (9.5–33.8) for pollen^{O3low} and 13.8 (4.7–22.5) for pollen^{O3high}; p = 0.02. Flare sizes (in mm^2; median (min.-max.)) were 256.3 (12.4–343.8) for pollen^{O3low} and 353.1 (43.8–381.3) for pollen^{O3high}; p = 0.005 (Fig. 5A, B). This is in line with higher Bet v 1 content of pollen^{O3high}.

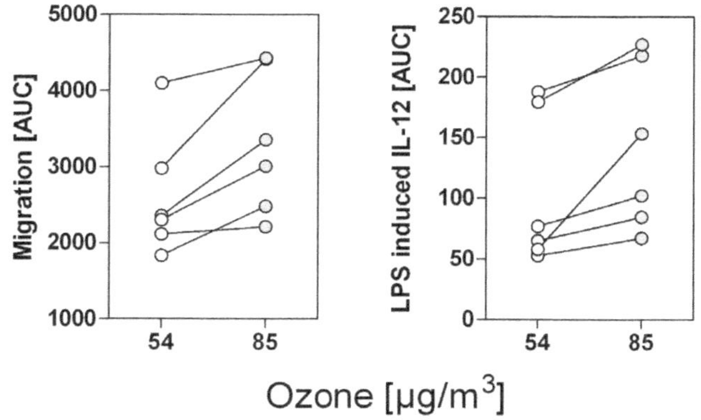

Figure 4. Immune stimulatory versus immune modulatory potential of pollen samples from higher- and lower-ozone-exposed birch trees. Aqueous extracts (APEs) of birch pollen sampled from high and low ozone exposed trees were chosen for neutrophil migration assays and stimulation of monocyte derived dendritic cells. APEs were applied in 3 concentrations and the AUC was calculated. Higher ozone-exposed pollen induced stronger neutrophil chemotaxis compared to pollen samples from lower ozone–exposed trees (**A**). In contrast, birch pollen from lower ozone-exposed trees were more potent in inhibiting the LPS-induced release of IL-12p70 from human monocyte-derived dendritic cells (**B**). APEs were prepared from birch pollen sampled from higher ozone-exposed trees (n = 2; mean ozone: 85 µg/m^3) and from lower ozone-exposed trees (n = 2; mean ozone: 54 µg/m^3). All APEs were tested in n = 3 patients. *: p<0.05 (Wilcoxon matched-pairs signed-ranks test).

Discussion

The present study reveals ozone exposure as important factor enhancing the allergenicity of birch pollen, with clinical relevance for susceptible individuals. Besides a positive relationship of the Bet v 1 content with ozone, we observed a negative association with temperature. Results of studies analysing the influence of temperature on Bet v 1 expression, however, are inconsistent. While Ahlholm et al. [22] reported a positive effect of high temperatures on Bet v 1 content, Helander et al. [23] indicated a negative association. The effect of temperature on the expression of Bet v 1 was also investigated by Tashpulatov et al. [24], using transgenic tobacco plants carrying a Bet v 1a-pomoter-reporter gene fusion. This study showed that warm temperatures positively regulate the activity of the Bet v 1 promoter. Our data indicate the opposite as we observed a negative correlation of temperature with Bet v 1. However, temperature and UI as well as NO_2 concentration were highly and significantly related, while UI and NO_2 did not correlate with Bet v 1. The number of entities for NO_2 and UI were 40, while merely 16 data points for temperature were available. Considering the strong relationship of the parameters UI and NO_2 with temperature it can be assumed that the correlation of temperature and Bet v 1 is observed by chance. Moreover when having a detailed look at the interrelationship of the environmental parameters of this study we also observed a non-significant negative association of ozone and temperature (Fig. 2F; p = 0.07; r = -0,46). This seems to be contradictory as ozone formation is associated with UV radiation and temperature [25]. However, ozone formation also depends on the presence of precursors (e.g. biogenic emissions) and degrading substances (e.g. NO). Rural areas are associated with lower temperatures. At the same time, the composition of ozone forming and -degrading factors in rural areas favor ozone accumulation. This might explain why we observe a negative correlation of ozone and temperature.

It might be hypothesized that a positive influence of ozone on Bet v 1 and the relationship of ozone and temperature implied a negative correlation of temperature and Bet v 1.

To confirm our suggestion, we applied multiple regression analyses based on stepwise variable selection. Indeed, the role of the independent variable ozone for predicting Bet v 1 content was superior, since temperature has been excluded in the linear model (data not shown).

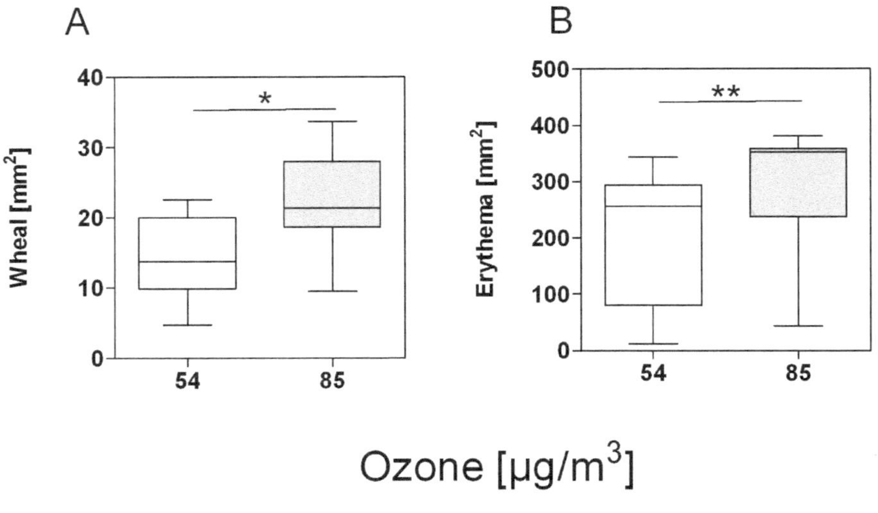

Figure 5. Cutaneous immune response towards pollen from higher and lower ozone-exposed birch trees. APEs were prepared from pollen sampled from higher ozone-exposed trees (n = 2; mean ozone: 85 $\mu g/m^3$) and from lower ozone-exposed trees (n = 2; mean ozone: 54 $\mu g/m^3$). All APEs were tested in n = 5 patients. Higher ozone-exposed pollen induced larger wheals (**A**) and flares (**B**) in skin prick tests compared to lower ozone-exposed pollen. *: $p<0.05$; **: $p<0.01$ (Wilcoxon matched-pairs signed-ranks test).

Besides, Tashpulatov et al. [24] showed that abscisic acid, a stress- and development-related plant hormone, positively regulates the activity of the Bet v 1a promoter. This goes in line with our study showing that ozone, a major stress factor for plants, is associated with a higher Bet v 1 expression. Besides, also former studies in ragweed and grass species gave evidence that ozone impacts on the allergen transcript and content [26,27]. In these studies, plants were exposed to defined ozone concentrations. In our study, trees were subjected to pollen sampling in their natural environment and under natural exposure conditions.

In a holistic approach analysing the allergenicity of birch pollen, we showed that elevated ozone exposure of birch trees was not only associated with increased allergen content in pollen but also with an altered composition of adjuvant PALMs. No relation of $PALM_{LTB4}$ and $PALM_{PGE2}$ content and the degree of urbanisation (UI), temperature or NO_2 concentration was observed. This seems to be in contrast to a recent study [28]. However, this discrepancy is most likely explained by the fact that in the present study, pollen sampling was carried out at defined, distinct maturation stages. As demonstrated in Fig. S2, the content of PALMs in pollen grains differs profoundly depending on the maturation stage of pollen, as does the Bet v 1 content, confirming former results [29].

Notably, $PALM_{PGE2}$ was significantly negatively associated to ozone concentrations. In our in vitro assays, $pollen^{O3high}$ were significantly more chemotactic for neutrophils than extracts of $pollen^{O3low}$. Since $PALM_{LTB4}$ did not significantly differ in differently exposed pollen we hypothesize that other, up to now unknown, substances besides $PALM_{LTB4}$ account for this effect. Moreover, Armstrong [30] could show that PGE_2 is able to inhibit the chemotaxis of neutrophils in a concentration dependent manner. Consequently, lower concentrations of $PALM_{PGE2}$ in $pollen^{O3high}$ could also have contributed to enhanced neutrophil chemotaxis. In contrast, APEs prepared from $pollen^{O3low}$ were significantly more efficient in inhibiting dendritic cell IL-12 secretion, in line with higher levels of immune modulatory $PALM_{PGE2}$. Among the immune modulatory PALMs are plant isoprostanes identified as E_1-phytoprostanes. E_1-phytoprostanes inhibit the IL-12 response in maturing DCs [31,32], finally licensing DCs to differentiate naïve $CD4^+$ T cells into Th2 cells [32]. Transferring our data to allergy mechanisms, we hypothesize that higher $PALM_{PGE2}$ concentrations in low-ozone areas might facilitate *de novo* sensitisation by providing Th2 promoting signals. Higher Bet v 1 concentration and less anti-inflammatory PALMs in $pollen^{O3high}$ might in turn lead to pronounced allergic symptoms in already sensitized individuals.

In summary, urbanisation-related, anthropogenic environmental factors can influence birch trees to produce pollen with altered allergenic potential. Our study emphasizes the correlation of ozone exposure to the pollen content of allergen and non-allergenic, adjuvant factors. It is likely that future climate change with more frequent and intense warm spells [33] as well as increases in urbanisation and anthropogenic air pollutants such as NOx will further enhance the local accumulation of tropospheric ozone. As indicated by this study, ozone might – apart from direct adverse impacts on human health – lead to increased allergic symptoms *via* its impact on the allergen carrier.

Supporting Information

Figure S1 Locations of pollen sampling. Birch pollen were sampled during the birch flowering season of 2010. Red dots represent urban trees, green dots rural trees. Background: CORINE Land Cover 2000 (EEA 2000), major classes: red = urban fabric, green = forest and pastures, yellow = arable land, blue = rivers, lakes (see www.eea.europa.eu/themes/landuse/interactive/clc-download for a complete legend).

Figure S2 Catkin maturation and allergenic potential of pollen. A: Catkins of different maturation stages were collected at different time points from the same trees (BBCH-Code 51–52: n = 1; BBCH-Code 55–65: n = 5) and classified according to a BBCH-Code. Pollen were isolated from the catkins and aqueous pollen extracts were prepared. APEs were then analyzed for the presence of Bet v 1 and PALMs. The content of Bet v 1 peaked at maturation stages 60–61. Inversely to Bet v 1, levels of $PALM_{LTB4}$ and $PALM_{PGE2}$ were high in pollen from immature catkins and decreased during maturation. A concentration minimum of

PALMs corresponded to a maximum in Bet v 1. BBCH-Code: 52: catkins increase in length and show green expansion cracks; 55: enhanced expansion cracks through further increase in length; 60: first catkins emit pollen (sporadically); 61: beginning of flowering: few catkins emit pollen; 65: full flowering: more than 50% of the catkins emit pollen; 67: flowering finishing: just a few catkins still emit pollen.

Figure S3 Immune stimulatory versus immune modulatory potential of high versus low ozone-exposed pollen samples. Aqueous extracts (APEs) of birch pollen sampled from high and low ozone exposed trees were chosen for neutrophil migration assays and stimulation of monocyte derived dendritic cells. APEs were applied in 3 concentrations. Higher ozone-exposed pollen induced stronger neutrophil chemotaxis compared to pollen samples from lower ozone–exposed trees (**A**). In contrast, birch pollen from lower ozone-exposed trees were more potent in inhibiting the LPS-induced release of IL-12p70 from human monocyte-derived dendritic cells (**B**). APEs were prepared from birch pollen sampled from higher ozone-exposed trees (n = 2; mean ozone: 85 $\mu g/m^3$) and from lower ozone-exposed trees (n = 2; mean ozone: 54 $\mu g/m^3$). All APEs were tested in n = 3 patients. *: $p<0.05$ (Wilcoxon matched-pairs signed-ranks test).

Figure S4 Viability of moDCs after stimulation with LPS plus APEs from high- and low ozone-exposed pollen. Viability of monocyte-derived dendritic cells (moDcs) after 24 h of stimulation with LPS (100 ng/ml) and APEs (1, 3, 10 mg/ml) was tested by propidium iodide staining and subsequent FACS analysis. APEs were prepared from birch pollen sampled from higher ozone-exposed trees (n = 2; mean ozone: 85 $\mu g/m^3$) and from lower ozone-exposed trees (n = 2; mean ozone: 54 $\mu g/m^3$). All APEs were tested in n = 3 patients.

Author Contributions

Conceived and designed the experiments: CTH AM. Performed the experiments: IB SG MM. Analyzed the data: IB SJ. Contributed reagents/materials/analysis tools: SJ JB CSW HB JR. Wrote the paper: IB HB SG CTH.

References

1. Strachan DP (1999) Lifestyle and atopy. Lancet 353: 1457–1458.
2. Behrendt H, Friedrichs KH, Kramer U, Hitzfeld B, Becker WM, et al. (1995) The role of indoor and outdoor air pollution in allergic diseases. Progress in Allergy and Clinical Immunology Volume 3, Stockholm 3: 83–89.
3. Kramer U, Koch T, Ranft U, Ring J, Behrendt H (2000) Traffic-related air pollution is associated with atopy in children living in urban areas. Epidemiology 11: 64–70.
4. Diaz-Sanchez D, Riedl M (2005) Diesel effects on human health: a question of stress? American journal of physiology Lung cellular and molecular physiology 289: L722–723.
5. Heinrich J, Wichmann HE (2004) Traffic related pollutants in Europe and their effect on allergic disease. Current opinion in allergy and clinical immunology 4: 341–348.
6. Ring J, Eberlein-Koenig B, Behrendt H (2001) Environmental pollution and allergy. Ann Allergy Asthma Immunol 87: 2–6.
7. Sillman S (1999) The relation between ozone, NOx and hydrocarbons in urban and polluted rural environments. Atmospheric Environment 33.
8. Ziska LH, Gebhard DE, Frenz DA, Faulkner S, Singer BD, et al. (2003) Cities as harbingers of climate change: common ragweed, urbanization, and public health. The Journal of allergy and clinical immunology 111: 290–295.
9. Health Effects Institute B, MA (2010) HEI Panel on the Health Effects of Traffic-Related Air Pollution. 2010. Traffic-Related Air Pollution: A Critical Review of the Literature on Emissions, Exposure, and Health Effects. HEI Special Report 17.
10. Behrendt H, Becker WM, Fritzsche C, Sliwa-Tomczok W, Tomczok J, et al. (1997) Air pollution and allergy: experimental studies on modulation of allergen release from pollen by air pollutants. Int Arch Allergy Immunol 113: 69–74.
11. Darbah JN, Kubiske ME, Nelson N, Oksanen E, Vaapavuori E, et al. (2007) Impacts of elevated atmospheric CO2 and O3 on paper birch (Betula papyrifera): reproductive fitness. ScientificWorldJournal 7 Suppl 1: 240–246.
12. Gilles S, Behrendt H, Ring J, Traidl-Hoffmann C (2012) The Pollen Enigma: Modulation of the Allergic Immune Response by Non-Allergenic, Pollen-Derived Compounds. Current pharmaceutical design.
13. Meier U (2001) Entwicklungsstadien mono- und dikotyler Pflanzen. BBCH-Monography. In: Forstwirtschaft BBfLu, editor. Berlin, Braunschweig. 165.
14. Buters JTM, Kasche A, Weichenmeier I, Schober W, Klaus S, et al. (2008) Year-to-year variation in release of Bet v1 allergen from birch pollen: Evidence for geographical differences between west and south Germany. International Archives of Allergy and Immunology 145: 122–130.
15. Gilles S, Fekete A, Zhang X, Beck I, Blume C, et al. (2011) Pollen metabolome analysis reveals adenosine as a major regulator of dendritic cell-primed T(H) cell responses. J Allergy Clin Immunol 127: 454–461 e451–459.
16. Palmes ED, Gunnison AF, Dimattio J, Tomczyk C (1976) Personal Sampler for Nitrogen-Dioxide. American Industrial Hygiene Association Journal 37: 570–577.
17. Jochner SC, Beck I, Behrendt H, Traidl-Hoffmann C, Menzel A (2011) Effects of extreme spring temperatures on urban phenology and pollen production: a case study in Munich and Ingolstadt. Climate Research 49: 101–112.
18. Gilles S, Mariani V, Bryce M, Mueller MJ, Ring J, et al. (2009) Pollen-derived E1-phytoprostanes signal via PPAR-gamma and NF-kappaB-dependent mechanisms. J Immunol 182: 6653–6658.
19. Traidl-Hoffmann C, Kasche A, Jakob T, Huger M, Plotz S, et al. (2002) Lipid mediators from pollen act as chemoattractants and activators of polymorphonuclear granulocytes. Journal of Allergy and Clinical Immunology 109: 831–838.
20. Traidl-Hoffmann C, Jakob T, Behrendt H (2009) Determinants of allergenicity. J Allergy Clin Immunol 123: 558–566.
21. Behrendt H, Kasche A, von Eschenbach CE, Risse U, Huss-Marp J, et al. (2001) Secretion of proinflammatory eicosanoid-like substances precedes allergen release from pollen grains in the initiation of allergic sensitization. International Archives of Allergy and Immunology 124: 121–125.
22. Hjelmroos M, Schumacher MJ, van Hage-Hamsten M (1995) Heterogeneity of pollen proteins within individual. Betula pendula trees. International Archives of Allergy and Applied Immunology 108.
23. Helander ML, Savolainen J, Ahlholm J (1997) Effects of air pollution and other environmental factors on birch pollen allergens. Allergy 52: 1207–1214.
24. Tashpulatov AS, Clement P, Akimcheva SA, Belogradova KA, Barinova I, et al. (2004) A model system to study the environment-dependent expression of the Bet v 1a gene encoding the major birch pollen allergen. Int Arch Allergy Immunol 134: 1–9.
25. Stathopoulou E, Mihalakakou G, Santamouris M, Bagiorgas HS (2008) On the impact of temperature on tropospheric ozone concentration levels in urban environments. J Earth Syst Sci 117: 227–236.
26. Eckl-Dorna J, Klein B, Reichenauer TG, Niederberger V, Valenta R (2010) Exposure of rye (Secale cereale) cultivars to elevated ozone levels increases the allergen content in pollen. J Allergy Clin Immunol 126: 1315–1317.
27. Kanter U, Heller W, Durner J, Winkler JB, Engel M, et al. Molecular and Immunological Characterization of Ragweed (Ambrosia artemisiifolia L.) Pollen after Exposure of the Plants to Elevated Ozone over a Whole Growing Season. PLoS One 8: e61518.
28. Behrendt H, Kasche A, Traidl C, Plotz S, Huss-Marp J, et al. (2002) Pollen grains contain and release not only allergens, but also eicosanoid-like substances with neutrophil chemotactic activity: A new step in the initiation of allergic sensitization? New Trends in Allergy V: 3–8.
29. Buters JT, Weichenmeier I, Ochs S, Pusch G, Kreyling W, et al. (2010) The allergen Bet v 1 in fractions of ambient air deviates from birch pollen counts. Allergy 65: 850–858.
30. Armstrong RA (1995) Investigation of the inhibitory effects of PGE2 and selective EP agonists on chemotaxis of human neutrophils. Br J Pharmacol 116: 2903–2908.
31. Gilles S, Mariani V, Bryce M, Mueller MJ, Ring J, et al. (2009) Pollen allergens do not come alone: pollen associated lipid mediators (PALMS) shift the human immune systems towards a T(H)2-dominated response. Allergy Asthma Clin Immunol 5: 3.
32. Traidl-Hoffmann C, Mariani V, Hochrein H, Karg K, Wagner H, et al. (2005) Pollen-associated phytoprostanes inhibit dendritic cell interleukin-12 production and augment T helper type 2 cell polarization. J Exp Med 201: 627–636.
33. Schar C, Vidale PL, Luthi D, Frei C, Haberli C, et al. (2004) The role of increasing temperature variability in European summer heatwaves. Nature 427: 332–336.

Dual p38/JNK Mitogen Activated Protein Kinase Inhibitors Prevent Ozone-Induced Airway Hyperreactivity in Guinea Pigs

Kirsten C. Verhein[1*¤a], Francesco G. Salituro[3¤b], Mark W. Ledeboer[3], Allison D. Fryer[2], David B. Jacoby[2]

1 Department of Physiology and Pharmacology, Oregon Health and Science University, Portland, Oregon, United States of America, **2** Division of Pulmonary and Critical Care Medicine, Oregon Health and Science University, Portland, Oregon, United States of America, **3** Vertex Pharmaceuticals, Inc., Cambridge, Massachusetts, United States of America

Abstract

Ozone exposure causes airway hyperreactivity and increases hospitalizations resulting from pulmonary complications. Ozone reacts with the epithelial lining fluid and airway epithelium to produce reactive oxygen species and lipid peroxidation products, which then activate cell signaling pathways, including the mitogen activated protein kinase (MAPK) pathway. Both p38 and c-Jun NH_2 terminal kinase (JNK) are MAPK family members that are activated by cellular stress and inflammation. To test the contribution of both p38 and JNK MAPK to ozone-induced airway hyperreactivity, guinea pigs were pretreated with dual p38 and JNK MAPK inhibitors (30 mg/kg, ip) 60 minutes before exposure to 2 ppm ozone or filtered air for 4 hours. One day later airway reactivity was measured in anesthetized animals. Ozone caused airway hyperreactivity one day post-exposure, and blocking p38 and JNK MAPK completely prevented ozone-induced airway hyperreactivity. Blocking p38 and JNK MAPK also suppressed parasympathetic nerve activity in air exposed animals, suggesting p38 and JNK MAPK contribute to acetylcholine release by airway parasympathetic nerves. Ozone inhibited neuronal M_2 muscarinic receptors and blocking both p38 and JNK prevented M_2 receptor dysfunction. Neutrophil influx into bronchoalveolar lavage was not affected by MAPK inhibitors. Thus p38 and JNK MAPK mediate ozone-induced airway hyperreactivity through multiple mechanisms including prevention of neuronal M_2 receptor dysfunction.

Editor: Heinz Fehrenbach, Research Center Borstel, Germany

Funding: This work was funded by the American Heart Association 0810148Z (to KCV) and the National Institutes of Health HL55543 (to ADF), ES014601 (to ADF), HL54659 (to DBJ), HL071795 (to DBJ), and RR023424 (to DBJ). The funders had no role in study design, data collection and analysis, decision to publish, or preparation of the manuscript. No additional external funding was received for this study.

Competing Interests: FGS and MWL were employed by Vertex Pharmaceuticals at the time of these studies. FGS is currently employed by Sage Therapeutics. This does not alter these authors' adherence to all the PLoS ONE policies on sharing data and materials, as detailed online in the guide for authors. KCV, ADF, and DBJ declare no competing interests exist.

* E-mail: verheinkc@niehs.nih.gov

¤a Current address: National Institute of Environmental Health Sciences, Research Triangle Park, North Carolina, United States of America
¤b Current address: Sage Therapeutics, Cambridge, Massachusetts, United States of America

Introduction

Over half the United States population lives in counties with unhealthy levels of ozone, a major component of smog [1]. Epidemiological studies demonstrate a significant link between exposure to ground level ozone and pulmonary hospitalizations. Exposure to ozone in excess of 0.16 ppm is associated with increased airway reactivity, lung inflammation and exacerbation of asthma in both adults and children [2,3,4].

Ozone induced hyperreactivity is demonstrated by increased reactivity to inhaled methacholine and other agonists, including those causing reflex bronchoconstriction in man [5,6,7]. In animals, ozone induced airway hyperreactivity is demonstrated by increased bronchoconstriction to intravenous methacholine, but this effect is mediated largely via increased acetylcholine release from parasympathetic nerves, since it is blocked by vagal section [8,9]. Direct stimulation of the vagus nerves results in bronchoconstriction that is potentiated in ozone exposed animals and that is associated with loss of function of neural M_2 muscarinic receptors that normally inhibit acetylcholine release [10,11]. Inflammatory cells, especially eosinophils through release of the M_2 inhibitor major basic protein, mediate loss of neuronal M_2 function and airway hyperreactivity in ozone exposed guinea pigs [11].

However, ozone is unlikely to contact inflammatory cells [12]. At the airway epithelial layer, ozone forms reactive oxygen species and lipid peroxides in lungs of humans and animals [13,14]. These end products activate cell signaling pathways, including mitogen activated protein kinase pathways (MAPK) [15]. Activation of the MAPK pathway results in inflammation [16], mucus hypersecretion [17] and airway hyperreactivity [18].

MAPK signaling pathways are important in many cell processes including differentiation, proliferation, activation, degranulation, and migration. Three MAPK subfamilies have been well characterized: ERK, JNK, and p38. The extracellular signal-regulated kinase (ERK) pathway is usually activated by mitogens and growth factors while p38 and c-Jun NH_2 terminal kinase (JNK) pathways are associated with chronic inflammation and are typically activated by inflammatory cytokines, heat shock, and cellular stress [19,20]. Activation of MAPK signaling induces inflammatory cytokine and chemokine production in airway epithelial cells, inflammatory cells, and airway smooth muscle cells [16,21,22]. Humans with severe asthma have increased activated p38 in airway epithelium compared to mild asthmatics or healthy controls, as demonstrated by increased immunostaining of phosphorylated p38 in airway biopsies [23].

Inhibition of MAPKs is protective in allergen challenge models of asthma. Inhibition of p38, either pharmacologically or with antisense oligonucleotides, partially prevents airway hyperreactivity after sensitization and challenge in mice [18,24]. Eosinophil influx into bronchoalveolar lavage is the dominant event in antigen challenged animals, and is prevented by a p38 inhibitor in guinea pigs and mice [25]. Blocking p38 also prevents IL-13 induced mucus metaplasia in human and mouse airway epithelial cells [17,26].

Less is known about the role of the MAP kinases in ozone-induced hyperreactivity. Inhibiting p38 prevents ozone-induced airway hyperreactivity in mice while inhibiting JNK is partially protective [27,28]. Ozone-induced increases in inflammatory cells in bronchoalveolar lavage are significantly inhibited in *Jnk1* knockout mice [29].

The experiments described here use three different MAPK inhibitors to test whether dual inhibition of both p38 and JNK MAPK pathways prevents ozone-induced inflammation and subsequent airway hyperreactivity in guinea pigs.

Methods

Ethics Statement

Guinea pigs were handled in accordance with the standards established by the United States Animal Welfare Act set forth in National Institutes of Health guidelines. All protocols were approved by Oregon Health and Science University Animal Care and Use Committee (protocol A984).

Animals

Specific pathogen-free female Hartley guinea pigs (300–470 g; Elm Hill Breeding Labs, Chelmsford, MA) were shipped in filtered crates, housed in high efficiency particulate filtered air, and fed a normal diet.

Ozone Exposure

Guinea pigs were exposed to 2 ppm ozone or filtered air for 4 hours as described previously [11]. Physiological measurements, airway inflammation, and histological measurements were made one day after a single ozone exposure.

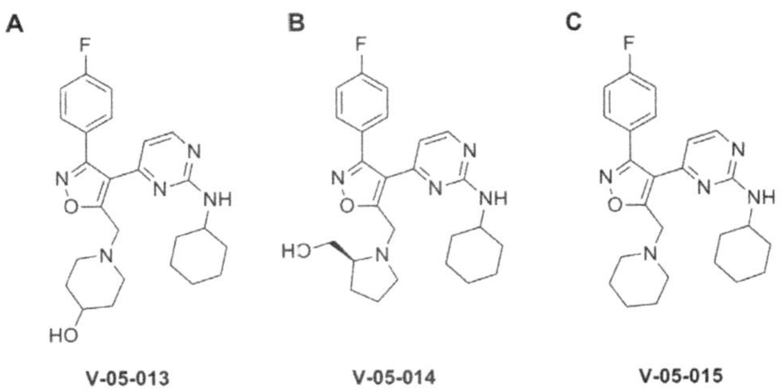

Figure 1. Chemical structures of dual p38 and JNK MAPK inhibitors.

Table 1. Ki values for dual p38 and JNK MAPK inhibitors.

Ki (nM)	V-05-013	V-05-014	V-05-015
p38	13	15	8*
JNK1	170	300	120
JNK2	10	15	4
JNK3	5	10	8

All compounds have a Ki greater than 1 μM for all other kinases tested.
*This one value is an IC50, not a Ki.

Treatment of Guinea Pigs with p38 and JNK MAPK Inhibitors

Animals were given 30 mg/kg intraperitoneally of the dual p38 and JNK MAPK inhibitors V-05-013, V-05-014, or V-05-015 (Vertex Pharmaceuticals, Cambridge, MA) one hour before ozone exposure (Figure 1). These compounds were chosen because of their overall kinase selectivity profile. They are potent and selective dual inhibitors of p38 and JNK (see below and table 1) and they do not show activity against a panel of other kinases at concentrations <1 μM (see characterization data below). Inhibitors were dissolved in 25% DMSO in phosphate buffered saline (PBS). Air exposed control animals were given 25% DMSO in PBS one hour before ozone exposure.

All three drugs have similar kinase inhibition profiles and exhibit potent affinity for both p38 and JNK. Affinity was measured using a kinase inhibition assay. Compounds were assayed for the inhibition of various kinases using a modification of a spectrophotometric coupled-enzyme assay [30]. In this assay, a fixed concentration of activated kinase (10–40 nM) was incubated with various concentrations of a potential inhibitor dissolved in DMSO for 10 minutes at 30°C in a buffer containing 0.1 M HEPES, pH 7.5, containing 10 mM $MgCl_2$, 2.5 mM phosphoenolpyruvate, 200 μM NADH, 2 mM DTT, 30 μg/mL pyruvate kinase, 10 μg/mL lactate dehydrogenase, and 200 μM–500 μM EGF receptor peptide. The EGF receptor peptide has the sequence KRELVEPLTPSGEAPNQALLR. The reaction was initiated by the addition of ATP equal to the ATP Km of the kinase, and the assay plate is inserted into the spectrophotometer's assay plate compartment that was maintained at 30°C. The decrease of absorbance at 340 nm was monitored as a function of time for 10 minutes. The rate data as a function of inhibitor concentration was either fit as an IC50 or to a competitive inhibition kinetic model to determine the compound K_i (see table 1).

Proton NMR spectra for the compounds was recorded on a Bruker Advance instrument with a QNP probe using TMS as the internal standard in the indicated deuterated solvent. LC−MS analyses were performed on a Waters ZQ or ZMD or QuatroII mass spectrometer using the electrospray (ESI) ionization technique. Samples were introduced into the mass spectrometer using chromatography. Methods (LC-MS) consisted of the following: 5–95% Water/Acetontrile (0.1% TFA) over 0.6 min on a Waters Acquity CSH C18, 1.7 μm, 2.1×50 mm column, with a flow rate of 0.6 mL/min. NMR spectra for each compound are below.

V-05-013: 1H NMR (300 MHz, MeOD) 8.20 (d, J = 5.6 Hz, 1H), 7.62 (dd, J = 5.3, 8.6 Hz, 1H), 7.29 (t, J = 8.6 Hz, 1H), 6.64-

Table 2. Baseline cardiovascular and pulmonary parameters.

Group	*n*	Heart Rate (beats/min)	Blood Pressure (mmHg) Systolic	Diastolic	Pulmonary Inflation Pressure (mmH_2O)
Air	7	306±6	40±4	20±3	110±4
Ozone	5	309±10	48±4	23±4	243±21 *
Ozone+V-05-013	5	301±7	40±3	20±2	190±15 * ‡
Ozone+V-05-014	4	286±16	38±4	21±1	225±5 *
Ozone+V-05-015	3	353±7	40±1	18±1	213±7 *
Air+V-05-013	4	291±7	32±2	17±1	102±3
Air+V-05-014	5	306±9	38±2	19±1	102±2
Air+V-05-015	5	327±14	40±3	22±3	94±5

Values are means ± SEM. Baseline pulmonary inflation pressure significantly increased after ozone exposure. Treatment with dual p38/JNK MAPK inhibitor V-05-013 significantly reduced the ozone-induced increase in baseline pulmonary inflation pressure. *$p<0.05$ Significantly different from air exposure. ‡$p<0.05$ Significantly different from ozone exposure.

6.47 (m, 1H), 5.03 (s, H), 4.18-4.00 (m, 1H), 4.18-4.00 (m, 1H), 3.80-3.35 (m, 5H), 1.71 (m, 1H), 1.52-1.23 (m, 5H); LC-MS (method X) $t_R = 0.60$ min., $(M+H^+)$ 452.37. Ki or IC50 was determined to be >1 μM for the following kinases: AKT3, AurA, CDK2, ERK-2, EphA, FLT3, IGF1R, IRAK, ITK, JAK2, JAK3, KDR, MAPKAP2, cMET, MKK4, MKK6, MKK7, PIM1, PKA, PLK1, PRAK, ROCK1, SRC, SYK, TIE2, ZAP70; GSK3β Ki = 0.73 μM.

V-05-014: 1H NMR (400 MHz, DMSO-d6) 11.04 (s, 0.3 H), 10.22 (s, 0.2 H), 8.33 (s, 1 H), 8.08 (s, 1 H), 7.38–7.79 (m, 4 H), 6.68 (s, 1 H), 6.20 (s, 1 H), 5.15–5.23 (m, 0.7 H), 4.86–4.91 (m, 1H), 3.78 (s, 3 H), 3.53 (s, 1 H), 3.33 (s, 1 H), 1.56–2.13 (m, 9 H), 1.09–1.28 (m, 6 H); LC-MS (method Y) $^tR = 2.47$ min., $(M+H^+)$ 452.25; LC-MS (method X) $t_R = 0.63$ min., $(M+H^+)$ 452.37. Ki or IC50 was determined to be >1 μM for the following kinases: AKT3, AurA, COT, CDK2, ERK2, EphA, FAK, GSK3β, IGF1R, IRAK, JAK3, KDR, LCK, MAPKAP2, cMET, MKK4, MKK6, MKK7, NIK, PDK1, PIM1, PKA, PLK1, PRAK, ROCK1, SRC, SYK, TIE2, ZAP70.

V-05-015: 1H NMR (300 MHz, MeOD) 8.20 (d, J = 5.1 Hz, 1H), 7.62 (dd, J = 5.3, 8.6 Hz, 2H), 7.29 (t, J = 8.7 Hz, 2H), 6.68-6.42 (m, 1H), 5.00 (s, 2H), 3.75-3.53 (m, 3H), 3.34-3.15 (m, 2H), 2.10-1.77 (m, 9H), 1.88 (m, 1H), 1.65-1.20 (m, 7H); LC-MS (method X) $t_R = 0.64$ min., $(M+H^+)$ 436.38. Ki or IC50 was determined to be >1 μM for the following kinases: AKT3, AurA, CDK2, ERK2, FLT3, GSK3b, IGF1R, IRAK, JAK2, JAK3, KDR, LCK, MAPKAP2, cMET, MKK4, MKK6, MKK7, PDK1, PIM1, PKA, PLK1, PRAK, ROCK1, SRC, SYK, ZAP70.

Measurement of Pulmonary Inflation Pressure

One day after exposure to ozone, guinea pigs were anesthetized with 1.9 g/kg urethane i.p. (Sigma-Aldrich, St. Louis, MO). This dose produces a deep anesthesia lasting 8–10 hours [31] though no experiments lasted longer than 4 hours.

Physiological measurements were made as previously described [32]. The jugular veins were cannulated for intravenous administration of drugs and the right carotid artery was cannulated to measure heart rate and blood pressure. Both vagus nerves were cut and distal ends placed on platinum electrodes submerged in liquid paraffin. Animals were tracheostomized, ventilated (1 ml/100 g body weight, 100 breaths per minute) and paralyzed with a constant infusion of succinylcholine (10 μg/kg/min iv, Sigma-Aldrich). Pulmonary inflation pressure was measured at the trachea and bronchoconstriction was measured as the increase in pressure over basal inflation pressure produced by the ventilator.

Measurement of Vagally Induced Bronchoconstriction

Electrical stimulation of both vagus nerves (10V, 0.2 ms pulse width, 1–25 Hz, 5 sec duration at 1 minute intervals) produced frequency dependent bronchoconstriction and bradycardia due to release of acetylcholine onto muscarinic receptors. To confirm vagally induced bronchoconstriction was cholinergic, atropine (1 mg/kg iv, Sigma-Aldrich) was given at the end of each experiment.

Measurement of Smooth Muscle M_3 Muscarinic Receptor Function

Recovery from vagal stimulation was confirmed by pulmonary inflation pressure and heart rate returning to baseline before measuring smooth muscle M_3 muscarinic receptor function (5–10 minutes after cessation of vagal stimulation). In vagotomized guinea pigs, M_3 muscarinic receptor function on airway smooth muscle was tested by measuring bronchoconstriction after administration of acetylcholine (1–10 μg/kg iv, Sigma-Aldrich).

Measurement of Neuronal M_2 Muscarinic Receptor Function

Recovery from administration of intravenous acetylcholine was confirmed by pulmonary inflation pressure and heart rate returning to baseline before measuring M_2 muscarinic receptor function (5–10 minutes after the last dose of acetylcholine). To test the function of neuronal M_2 muscarinic receptors, vagally induced bronchoconstriction was measured before and after administration of gallamine (0.1–10 mg/kg iv, Sigma-Aldrich) an M_2 receptor antagonist. Electrical stimulation of both vagus nerves (3–30V, 0.2 ms pulse width, 15 Hz, 5 sec duration at 1 minute intervals) produced reproducible, frequency dependent, bronchoconstrictions. In the presence of normally functioning M_2 receptors, gallamine (0.1–10 mg/kg iv) will block them, resulting in increased vagally induced bronchoconstriction [33]; an effect that is suppressed if M_2 receptors are not responding to endogenous acetylcholine [10].

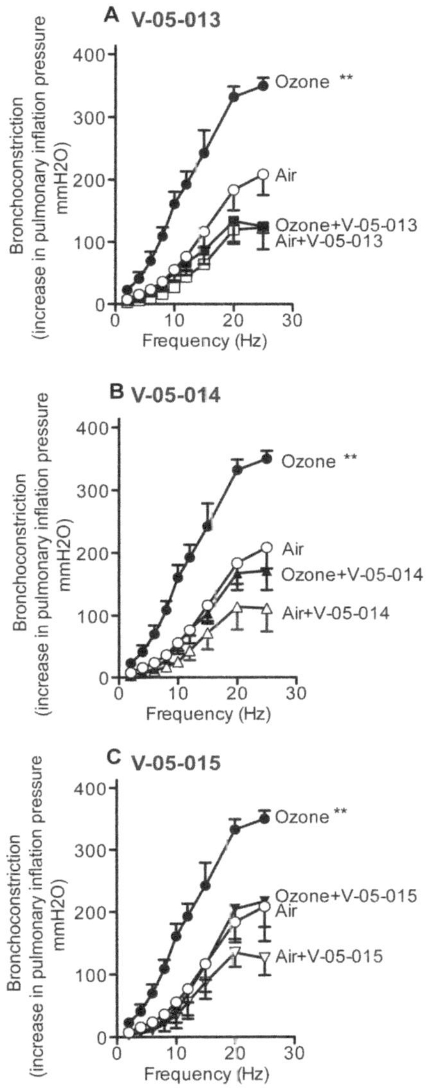

Figure 2. Blocking p38 and JNK MAPK completely prevented ozone-induced airway hyperreactivity mediated by the vagus nerves. In anesthetized and vagotomized guinea pigs, stimulation of the vagus nerves (10V, 0.2 ms pulse width, 1–25 Hz, 5 sec duration at 1 minute intervals) caused frequency dependent bronchoconstriction (A open circles; measured as an increase in inflation pressure in mmH_2O) that is significantly potentiated one day post-ozone exposure (A closed circles). Pretreatment with dual MAPK inhibitors V-05-013 (A closed squares), V-05-014 (B closed triangles), or V-05-015 (C closed inverted triangles) completely prevented ozone-induced airway hyperreactivity. All three dual MAPK inhibitors suppressed parasympathetic nerve activity (A open squares, B open triangles, C open inverted triangles). Ozone and air exposed control data are the same in A-C. *$p<0.05$, **$p<0.01$ Significantly different from air exposed controls. Data are mean ± SEM. n = 4–7.

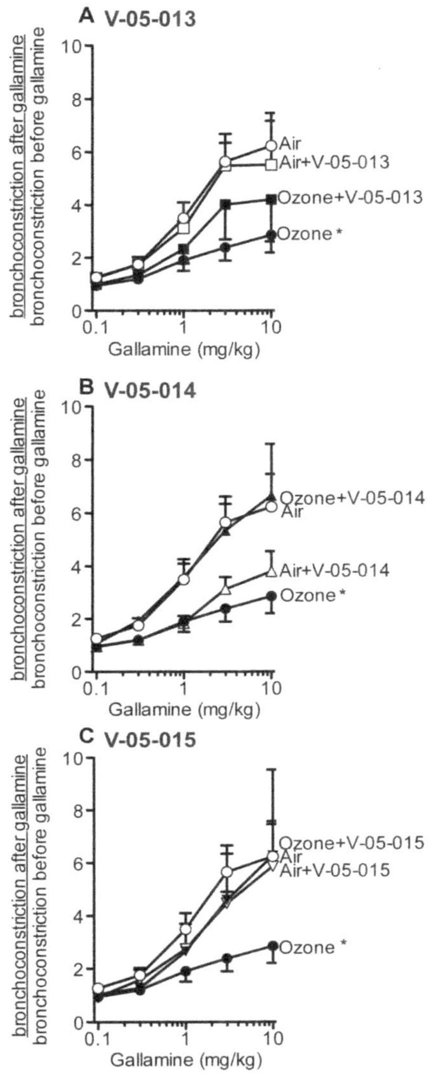

Figure 3. In control (air exposed) guinea pigs electrical stimulation of the vagus nerves (3–30V, 0.2 ms pulse width, 15 Hz, 5 sec duration at 1 minute intervals) resulted in vagally induced bronchoconstriction (measured as an increase in pulmonary inflation pressure; 16±1 mmH_2O). An M_2 receptor antagonist, gallamine, potentiated vagally induced bronchoconstriction up to 6-fold in air exposed animals (open circles) demonstrating that functional M_2 receptors were limiting acetylcholine release. The potentiation by gallamine was decreased in ozone-exposed animals, demonstrating M_2 receptors were dysfunctional after ozone exposure (closed circles). V-05-013 partially prevented M_2 receptor dysfunction (C closed squares), while V-05-014 (B closed triangles) and V-05-015 (C closed inverted triangles) completely protected M_2 receptor function. Vagally induced bronchoconstriction in the absence of gallamine was not different from control among all groups. Ozone and air exposed controls are the same in A-C. *$p<0.05$, **$p<0.01$ Significantly different from air exposed controls. Data are mean ± SEM. n = 4–7.

Bronchoalveolar Lavage (BAL)

At the end of each experiment, the lungs were lavaged with five 10 ml aliquots of phosphate buffered saline (PBS) that contained 100 µg isoproterenol (Sigma-Aldrich). Lavage fluid was centrifuged (400 *g*, 10 min) and the pellets were resuspended in PBS. Cells were counted using a hemocytometer and slides made from centrifuged lavaged cells were stained with Hemacolor (EMD Chemicals, Gibbstown, NJ) and used to determine cell differentials.

Drugs

Acetylcholine, succinylcholine, and urethane were purchased from Sigma (St. Louis, MO) and were dissolved and diluted in PBS.

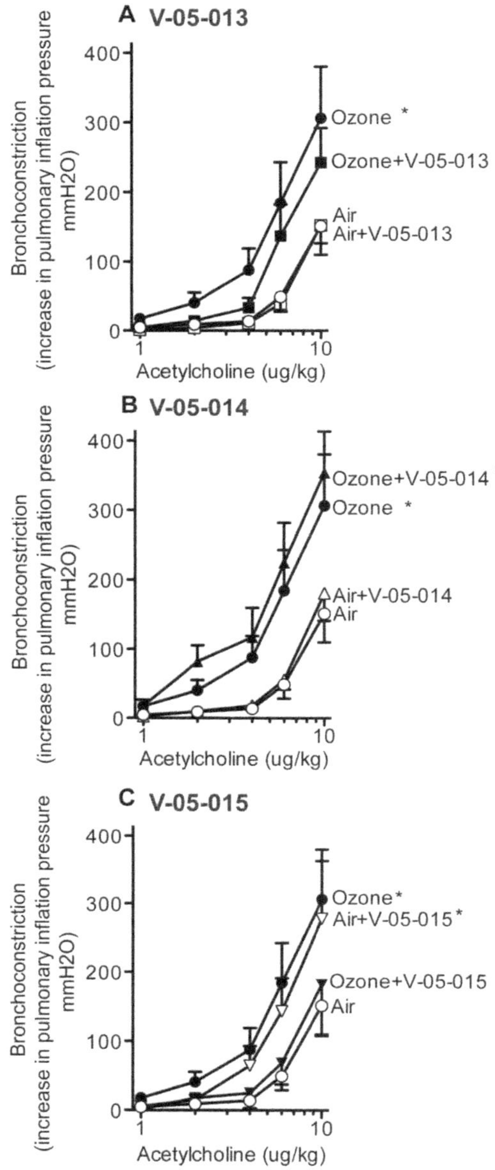

Figure 4. Bronchoconstriction (measured as an increase in inflation pressure in mmH$_2$O) in response to intravenous acetylcholine was significantly potentiated one day post-ozone (closed circles) compared to air exposed controls (open circles), and was not blocked by V-05-013 (A closed squares), or V-05-014 (B closed triangles). V-05-015 (C closed inverted triangles) attenuated ozone-induced smooth muscle hyperreactivity while air-exposed animals pretreated with V-05-015 were hyperreactive to intravenous acetylcholine (C open inverted triangles). Ozone and air exposed controls are the same in A-C. *$p<0.05$, **$p<0.01$ Significantly different from air exposed controls. Data are mean ± SEM. n = 3–7.

Data Analysis and Statistics

All data are expressed as means ± SE. *In vivo* frequency response and dose response curves were compared using two-way ANOVA for repeated measures. Baseline data were analyzed by one-way ANOVA with Bonferroni's correction. A *P* value of less than 0.05 was considered significant. Analyses were made with GraphPad Prism (version 5.0; GraphPad Software, La Jolla, CA).

Results

Baselines

One day after ozone exposure, baseline pulmonary inflation pressure was significantly increased compared to air-exposed controls (Table 2). All the dual p38 and JNK inhibitors partially attenuated the ozone induced increase in baseline airway inflation pressure, although the attenuation only reached statistical significance in the group treated with V-05-013. None of the MAPK inhibitors affected baseline inflation pressure in air-exposed controls. Neither ozone nor the MAPK inhibitors affected baseline heart rate or blood pressure.

Airway Physiology

Ozone significantly potentiated bronchoconstriction in response to electrical stimulation of the vagus nerves compared to air-exposed controls as previously reported (Figure 2). Treatment with any of the dual MAPK inhibitors prevented ozone induced airway hyperreactivity (Figures 2A–C). Vehicle treatment had no effect on vagally mediated bronchoconstriction in either air or ozone exposed animals (data not shown). M_2 muscarinic receptors were dysfunctional in ozone treated animals as gallamine, an M_2 selective inhibitor, potentiated bronchoconstriction in response to vagal stimulation in air-exposed animals but not in ozone-exposed animals (Figure 3); an effect that is consistent with decreased function of neuronal M_2 muscarinic receptors [34]. Ozone induced M_2 receptor dysfunction was prevented by treatment with V-05-014 and V-05-015 (Figure 3B–C), and attenuated by treatment withV-05-013 (Figure 3A). Airway smooth muscle responses to intravenous acetylcholine were potentiated by ozone (Figure 4). This was not prevented by any of the MAPK inhibitors, but was partially attenuated by V-05-015 (Figure 4C). V-05-015 also produced a paradoxical increase in airway response to intravenous acetylcholine in air-exposed animals.

Ozone exposure potentiated falls in heart rate in response to vagal stimulation compared to air-exposed controls (Figure 5A–C). Ozone and air-exposed controls are the same in figure 5A–C. Separate pretreatment with all three dual MAPK inhibitors prevented the ozone-induced potentiation in falls in heart rate and had no effect in air-exposed animals (Figure 5A–C). Falls in heart rate in response to intravenous acetylcholine were not affected by either ozone or the MAPK inhibitor (Figure 5D–F). Ozone and air exposed controls are the same in figure 5D–F.

Bronchoalveolar Lavage and Peripheral Blood

One day after ozone exposure neutrophils were increased in bronchoalveolar lavage (Figure 6D). All the MAPK inhibitors slightly, though not significantly, attenuated the ozone induced increase in neutrophils (Figure 6D). None of the other inflammatory cell types were affected by either ozone or the MAPK inhibitors (Figure 6).

There were no significant differences between inflammatory cells in peripheral blood after either ozone exposure, or treatment with the dual MAPK inhibitors (Figure 7).

Discussion

Ozone induces airway hyperreactivity, measured as potentiation of vagally induced bronchoconstriction, in guinea pigs one day after exposure confirming previous studies [35,36]. Ozone also significantly potentiated bronchoconstriction in response to intravenous acetylcholine; an effect that has also been previously reported [10]. Blocking both p38 and JNK MAPK with three different, but related, inhibitors prevented vagally mediated

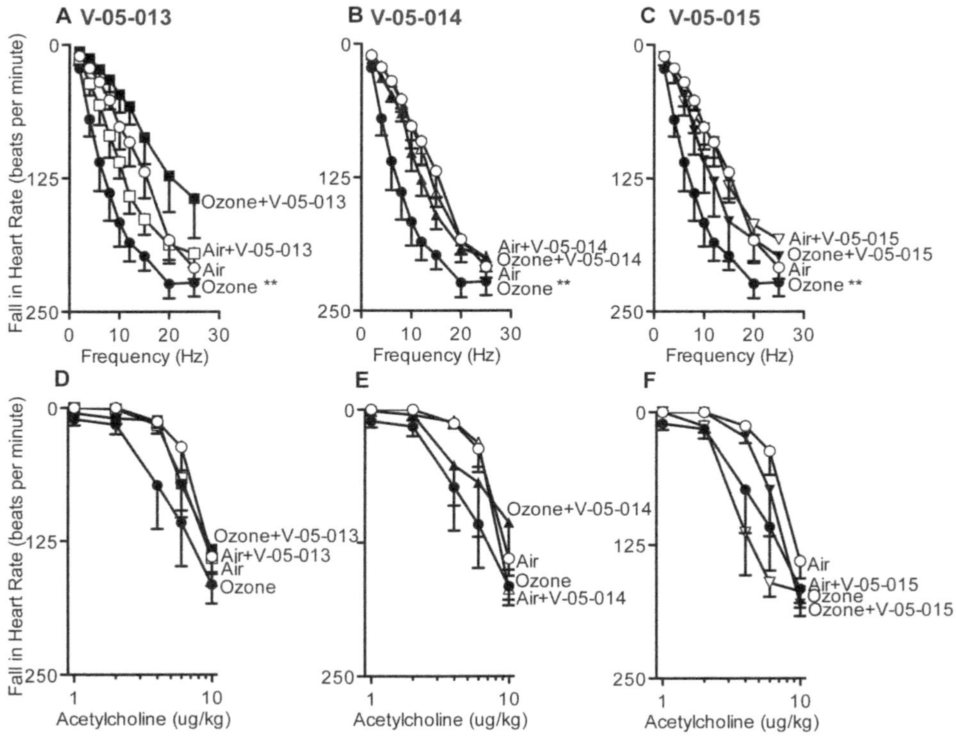

Figure 5. Ozone potentiated vagally mediated falls in heart rate (measured as beats/minute; A–C closed circles) compared to air exposed animals (A–C open circles). Separate pretreatment with all three dual MAPK inhibitors (V-05-013: A closed squares; V-05-014: B closed squares; V-05-015: C closed squares) prevented the ozone-induced potentiation of frequency induced falls in heart rate. Fall in heart rate following intravenous acetylcholine administration was not changed by either ozone, or MAPK inhibitors (D–E). Ozone and air exposed controls are the same for A–C, and are the same for D–F. **$p<0.01$ Significantly different from air exposed controls. Data are mean ± SEM. n = 3–7.

hyperreactivity in ozone-exposed animals but had no effect on inflammatory cell numbers in bronchoalveolar lavage. The prevention of vagally mediated hyperreactivity was associated with prevention of ozone induced M_2 receptor dysfunction that was complete in animals treated with V-05-014 and V-05-015, and partial in animals treated with V-05-013. Ozone induced hyperreactivity to intravenous acetylcholine was partially attenuated by treatment with the MAPK inhibitors.

All three MAPK inhibitors were administered at a dose of 30 mg/kg i.p. one hour before ozone. While the compounds are active with submicromolar potencies, preliminary studies suggested there is a significant shift in *in vivo* potency from the tens of nanomolar to hundreds of nanomolar IC50 s presumably the result of plasma protein binding (unpublished data). Nonetheless, the compounds were chosen because they exhibit adequate pharmacokinetic profiles (table 1) to test our hypothesis *in vivo*. The relatively high *in vivo* clearances and half lives are somewhat limiting, leading to the need for a sufficient dose to demonstrate a role for MAPKs in ozone induced hyperreactivity. However, as with most compounds, different effects could occur at lower doses.

Treatment of air-exposed guinea pigs with any of the three MAPK inhibitors decreased the airway response to vagal stimulation slightly. This effect was most pronounced at high frequency stimulation, but could not be explained by changes in M_2 receptor function, as the effects of gallamine were not potentiated by the MAPK inhibitors in air exposed animals. This effect was also not due to decreased smooth muscle responsiveness, as the effects of intravenous acetylcholine were not decreased by the MAPK inhibitors. Although response to the MAPK inhibitors was variable in air-exposed animals the overall effect with ozone exposure was prevention of ozone-induced airway hyperreactivity. These minor differences may be due to off target effects of the inhibitors, or to the dose of inhibitors used in this study. Thus, in air exposed guinea pigs, p38 and JNK MAPK inhibitors inhibit vagally induced bronchoconstriction by suppressing release of acetylcholine from airway parasympathetic nerves.

The mechanism for this decreased acetylcholine release is unknown. p38 and JNK are involved in nerve regeneration and development [37,38] but whether they inhibit ganglionic transmission, action potentials or transmitter release (by a mechanism separate from M_2 receptors, since there was no change in the response to gallamine) is not well studied. In *Aplysia*, activation of p38 by the peptide neurotransmitter FMRFa leads to long-term depression in sensory neurons in the pleural ganglia [39], although the mechanism is not known. In *Drosophila* motor neurons, expression of constitutively active JNK decreases neurotransmitter release [40] while in primary cultures of rat cortical neurons, IL-1β signaling activates p38, decreasing synaptophysin, a protein involved in synaptic transmission [41]. These varied and sometimes contradictory effects of MAPKs on neural function and transmitter release may be involved in the effects we observed. In neutrophils, activation of p38 MAPK is required for granule exocytosis after stimulation by CXCR1/2 ligands [42]; if neurotransmitter exocytosis were similarly mediated by MAPK, kinase inhibitors would block secretion. Thus, the role of MAPK is cell type dependent and additionally may differ between central

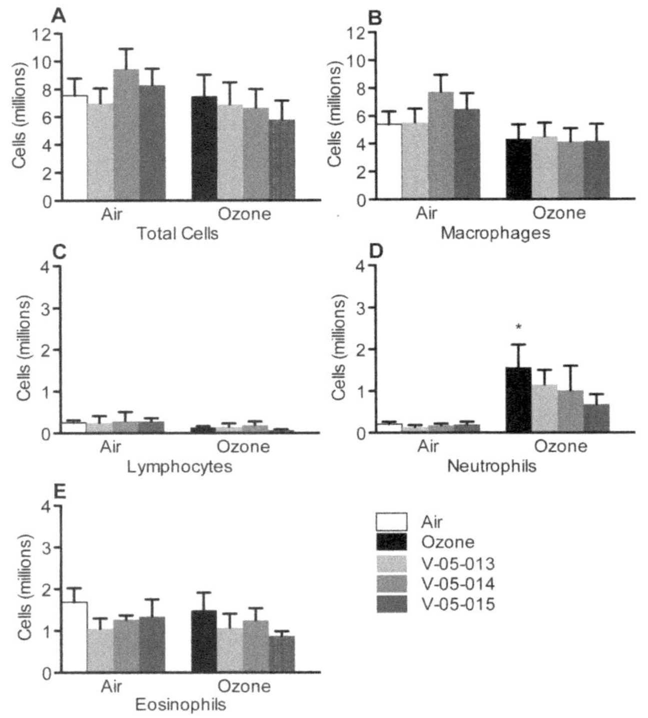

Figure 6. Ozone exposure increased neutrophils in bronchoalveolar lavage (D closed bar). No other inflammatory cell type number was affected by either ozone or the dual p38/JNK MAPK inhibitors. *p<0.05 Significantly different from air exposed controls. Data are mean ± SEM. n = 3–6.

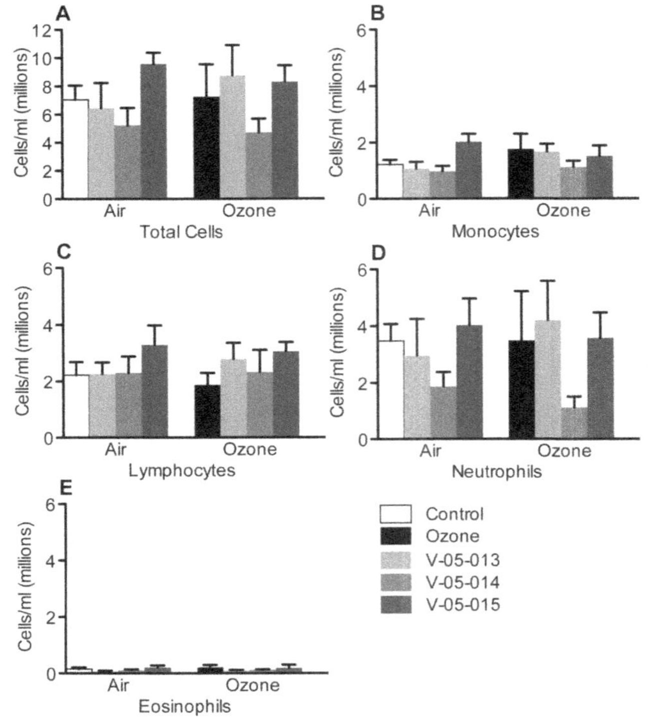

Figure 7. Neither ozone nor the dual p38/JNK MAPK inhibitors affected inflammatory cell numbers in peripheral blood. Data are mean ± SEM. n = 3–6.

neurons where kinases inhibit neurotransmission and peripheral neurons, where they have not been well studied. The data in this paper suggest p38 or JNK MAPK may additionally play a previously unrecognized role in release of acetylcholine from lung parasympathetic nerves.

None of the MAPK inhibitors completely reversed the ozone-induced increase in baseline pulmonary inflation pressure, which is commonly due to airway edema and not increased vagal tone. Ozone also significantly increased the numbers of neutrophils in bronchoalveolar lavage compared to air exposed controls confirming previously published data [35,36]. However, blocking both p38 and JNK MAPK did not prevent the neutrophil influx. No other inflammatory cell population in the lavage was effected by ozone or by the p38 and JNK MAPK inhibitors. Thus, prevention of ozone-induced airway hyperreactivity did not occur via a decrease in airway inflammatory cells.

Previously we have shown major basic protein, released from eosinophils, inhibits neuronal M_2 muscarinic receptor function, thereby increasing acetylcholine release and subsequently leading to increased bronchoconstriction and airway hyperreactivity after ozone exposure [43,44]. Depletion of eosinophils with an antibody to IL-5, or blocking major basic protein with heparin, prevents M_2 receptor dysfunction and ozone-induced airway hyperreactivity one day post-ozone exposure [11]. Thus, although neutrophils are the cells that increase in the bronchoalveolar lavage after ozone, it is tissue eosinophils around airway nerves that mediate ozone-induced hyperreactivity. In eosinophils, eotaxin and IL-5 signal through both ERK and p38 MAPK activation [45,46]. Inhibition of p38 reduces eosinophil degranulation as measured by decreased eosinophil cationic protein release [45]. Major basic protein has also been shown to alter smooth muscle contractility [47]. Thus, while not tested directly in this study, blocking eosinophil degranulation with MAPK inhibitors could also contribute to preventing smooth muscle hyperreactivity.

Thus, p38 and JNK MAPK inhibitors inhibit ozone-induced hyperreactivity by multiple mechanisms. Exposure to high levels of environmental ozone increases hospitalizations from asthma exacerbations. Over 4 million children and 10 million adults with asthma live in counties with unhealthy levels of ozone, and those with asthma are an especially susceptible population to the adverse health effects of ozone [1]. Our data show that treatment with p38 and JNK inhibitors, immediately prior to ozone exposure prevented subsequent development of airway hyperreactivity. Currently there is no specific therapy for ozone related asthma complications and our data suggest both p38 and JNK are potential targets for additional therapeutic candidates; and that inhibitors could be tested as prophylactic treatment for asthma exacerbations on days with anticipated high ozone.

Author Contributions

Conceived and designed the experiments: KCV FGS MWL ADF DBJ. Performed the experiments: KCV. Analyzed the data: KCV. Contributed reagents/materials/analysis tools: FGS MWL. Wrote the paper: KCV ADF DBJ MWL. Designed and characterized the molecules used in this study: FGS MWL.

References

1. Association AL (2009) State of the Air Report.
2. Hiltermann JT, Lapperre TS, van Bree L, Steerenberg PA, Brahim JJ, et al. (1999) Ozone-induced inflammation assessed in sputum and bronchial lavage fluid from asthmatics: a new noninvasive tool in epidemiologic studies on air pollution and asthma. Free Radic Biol Med 27: 1448–1454.
3. Bell ML, McDermott A, Zeger SL, Samet JM, Dominici F (2004) Ozone and short-term mortality in 95 US urban communities, 1987–2000. Jama 292: 2372–2378.
4. Lewis TC, Robins TG, Dvonch JT, Keeler GJ, Yip FY, et al. (2005) Air pollution-associated changes in lung function among asthmatic children in Detroit. Environ Health Perspect 113: 1068–1075.
5. Kreit JW, Gross KB, Moore TB, Lorenzen TJ, D'Arcy J, et al. (1989) Ozone-induced changes in pulmonary function and bronchial responsiveness in asthmatics. J Appl Physiol 66: 217–222.
6. Wagner EM, Jacoby DB (1999) Methacholine causes reflex bronchoconstriction. J Appl Physiol 86: 294–297.
7. Foster WM, Brown RH, Macri K, Mitchell CS (2000) Bronchial reactivity of healthy subjects: 18–20 h postexposure to ozone. J Appl Physiol 89: 1804–1810.
8. Lee LY, Bleecker ER, Nadel JA (1977) Effect of ozone on bronchomotor response to inhaled histamine aerosol in dogs. J Appl Physiol 43: 626–631.
9. Mitchell HW, Adcock J (1988) Vagal mechanisms and the effect of indomethacin on bronchoconstrictor stimuli in the guinea-pig. Br J Pharmacol 94: 522–527.
10. Schultheis AH, Bassett DJ, Fryer AD (1994) Ozone-induced airway hyperresponsiveness and loss of neuronal M2 muscarinic receptor function. J Appl Physiol 76: 1088–1097.
11. Yost BL, Gleich GJ, Fryer AD (1999) Ozone-induced hyperresponsiveness and blockade of M2 muscarinic receptors by eosinophil major basic protein. J Appl Physiol 87: 1272–1278.
12. Pryor WA (1992) How far does ozone penetrate into the pulmonary air/tissue boundary before it reacts? Free Radic Biol Med 12: 83–88.
13. Hamilton RF Jr, Hazbun ME, Jumper CA, Eschenbacher WL, Holian A (1996) 4-Hydroxynonenal mimics ozone-induced modulation of macrophage function ex vivo. Am J Respir Cell Mol Biol 15: 275–282.
14. Kirichenko A, Li L, Morandi MT, Holian A (1996) 4-hydroxy-2-nonenal-protein adducts and apoptosis in murine lung cells after acute ozone exposure. Toxicol Appl Pharmacol 141: 416–424.
15. Kumagai T, Nakamura Y, Osawa T, Uchida K (2002) Role of p38 mitogen-activated protein kinase in the 4-hydroxy-2-nonenal-induced cyclooxygenase-2 expression. Arch Biochem Biophys 397: 240–245.
16. Cui CH, Adachi T, Oyamada H, Kamada Y, Kuwasaki T, et al. (2002) The role of mitogen-activated protein kinases in eotaxin-induced cytokine production from bronchial epithelial cells. Am J Respir Cell Mol Biol 27: 329–335.
17. Atherton HC, Jones G, Danahay H (2003) IL-13-induced changes in the goblet cell density of human bronchial epithelial cell cultures: MAP kinase and phosphatidylinositol 3-kinase regulation. Am J Physiol Lung Cell Mol Physiol 285: L730–739.
18. Nath P, Leung SY, Williams A, Noble A, Chakravarty SD, et al. (2006) Importance of p38 mitogen-activated protein kinase pathway in allergic airway remodelling and bronchial hyperresponsiveness. Eur J Pharmacol 544: 160–167.
19. Denhardt DT (1996) Signal-transducing protein phosphorylation cascades mediated by Ras/Rho proteins in the mammalian cell: the potential for multiplex signalling. Biochem J 318 (Pt 3): 729–747.
20. Kyriakis JM, Avruch J (1996) Protein kinase cascades activated by stress and inflammatory cytokines. Bioessays 18: 567–577.
21. Kalesnikoff J, Huber M, Lam V, Damen JE, Zhang J, et al. (2001) Monomeric IgE stimulates signaling pathways in mast cells that lead to cytokine production and cell survival. Immunity 14: 801–811.
22. Peng Q, Matsuda T, Hirst SJ (2004) Signaling pathways regulating interleukin-13-stimulated chemokine release from airway smooth muscle. Am J Respir Crit Care Med 169: 596–603.
23. Liu W, Liang Q, Balzar S, Wenzel S, Gorska M, et al. (2008) Cell-specific activation profile of extracellular signal-regulated kinase 1/2, Jun N-terminal kinase, and p38 mitogen-activated protein kinases in asthmatic airways. J Allergy Clin Immunol 121: 893–902 e892.
24. Duan W, Chan JH, McKay K, Crosby JR, Choo HH, et al. (2005) Inhaled p38alpha mitogen-activated protein kinase antisense oligonucleotide attenuates asthma in mice. Am J Respir Crit Care Med 171: 571–578.
25. Underwood DC, Osborn RR, Kotzer CJ, Adams JL, Lee JC, et al. (2000) SB 239063, a potent p38 MAP kinase inhibitor, reduces inflammatory cytokine production, airways eosinophil infiltration, and persistence. J Pharmacol Exp Ther 293: 281–288.
26. Fujisawa T, Ide K, Holtzman MJ, Suda T, Suzuki K, et al. (2008) Involvement of the p38 MAPK pathway in IL-13-induced mucous cell metaplasia in mouse tracheal epithelial cells. Respirology 13: 191–202.
27. Williams AS, Issa R, Leung SY, Nath P, Ferguson GD, et al. (2007) Attenuation of ozone-induced airway inflammation and hyper-responsiveness by c-Jun NH2 terminal kinase inhibitor SP600125. J Pharmacol Exp Ther 322: 351–359.
28. Williams AS, Issa R, Durham A, Leung SY, Kapoun A, et al. (2008) Role of p38 mitogen-activated protein kinase in ozone-induced airway hyperresponsiveness and inflammation. Eur J Pharmacol 600: 117–122.
29. Cho HY, Morgan DL, Bauer AK, Kleeberger SR (2007) Signal transduction pathways of tumor necrosis factor-mediated lung injury induced by ozone in mice. Am J Respir Crit Care Med 175: 829–839.
30. Fox T, Coll JT, Xie X, Ford PJ, Germann UA, et al. (1998) A single amino acid substitution makes ERK2 susceptible to pyridinyl imidazole inhibitors of p38 MAP kinase. Protein Sci 7: 2249–2255.
31. Green C (1982) Animal Anesthesia. London: Elsevier.
32. Fryer AD, Stein LH, Nie Z, Curtis DE, Evans CM, et al. (2006) Neuronal eotaxin and the effects of CCR3 antagonist on airway hyperreactivity and M2 receptor dysfunction. J Clin Invest 116: 228–236.
33. Fryer AD, Maclagan J (1984) Muscarinic inhibitory receptors in pulmonary parasympathetic nerves in the guinea-pig. Br J Pharmacol 83: 973–978.
34. Fryer AD, Wills-Karp M (1991) Dysfunction of M2-muscarinic receptors in pulmonary parasympathetic nerves after antigen challenge. J Appl Physiol 71: 2255–2261.
35. Yost BL, Gleich GJ, Jacoby DB, Fryer AD (2005) The changing role of eosinophils in long-term hyperreactivity following a single ozone exposure. Am J Physiol Lung Cell Mol Physiol 289: L627–635.
36. Verhein KC, Jacoby DB, Fryer AD (2008) IL-1 receptors mediate persistent, but not acute, airway hyperreactivity to ozone in guinea pigs. Am J Respir Cell Mol Biol 39: 730–738.
37. Hirai S, Kawaguchi A, Suenaga J, Ono M, Cui DF, et al. (2005) Expression of MUK/DLK/ZPK, an activator of the JNK pathway, in the nervous systems of the developing mouse embryo. Gene Expr Patterns 5: 517–523.
38. Agthong S, Koonam J, Kaewsema A, Chentanez V (2009) Inhibition of MAPK ERK impairs axonal regeneration without an effect on neuronal loss after nerve injury. Neurol Res.
39. Guan Z, Kim JH, Lomvardas S, Holick K, Xu S, et al. (2003) p38 MAP kinase mediates both short-term and long-term synaptic depression in aplysia. J Neurosci 23: 7317–7325.
40. Etter PD, Narayanan R, Navratilova Z, Patel C, Bohmann D, et al. (2005) Synaptic and genomic responses to JNK and AP-1 signaling in Drosophila neurons. BMC Neurosci 6: 39.
41. Li Y, Liu L, Barger SW, Griffin WS (2003) Interleukin-1 mediates pathological effects of microglia on tau phosphorylation and on synaptophysin synthesis in cortical neurons through a p38-MAPK pathway. J Neurosci 23: 1605–1611.
42. Rittner HL, Labuz D, Richter JF, Brack A, Schafer M, et al. (2007) CXCR1/2 ligands induce p38 MAPK-dependent translocation and release of opioid peptides from primary granules in vitro and in vivo. Brain Behav Immun 21: 1021–1032.
43. Fryer AD, Jacoby DB (1992) Function of pulmonary M2 muscarinic receptors in antigen-challenged guinea pigs is restored by heparin and poly-L-glutamate. J Clin Invest 90: 2292–2298.
44. Evans CM, Fryer AD, Jacoby DB, Gleich GJ, Costello RW (1997) Pretreatment with antibody to eosinophil major basic protein prevents hyperresponsiveness by protecting neuronal M2 muscarinic receptors in antigen-challenged guinea pigs. J Clin Invest 100: 2254–2262.
45. Kampen GT, Stafford S, Adachi T, Jinquan T, Quan S, et al. (2000) Eotaxin induces degranulation and chemotaxis of eosinophils through the activation of ERK2 and p38 mitogen-activated protein kinases. Blood 95: 1911–1917.
46. Adachi T, Choudhury BK, Stafford S, Sur S, Alam R (2000) The differential role of extracellular signal-regulated kinases and p38 mitogen-activated protein kinase in eosinophil functions. J Immunol 165: 2198–2204.
47. White SR, Ohno S, Munoz NM, Gleich GJ, Abrahams C, et al. (1990) Epithelium-dependent contraction of airway smooth muscle caused by eosinophil MBP. Am J Physiol 259: L294–303.

The Role of Stream Water Carbon Dynamics and Export in the Carbon Balance of a Tropical Seasonal Rainforest, Southwest China

Wen-Jun Zhou[1,2,3,4], Yi-Ping Zhang[1,2,3]*, Douglas A. Schaefer[1,2], Li-Qing Sha[1,2,3], Yun Deng[1,2,3], Xiao-Bao Deng[1,2,3], Kai-Jie Dai[1,2]

1 Key Laboratory of Tropical Forest Ecology, Xishuangbanna Tropical Botanical Garden, Chinese Academy of Sciences, Mengla, Yunnan, China, **2** Xishuangbanna Tropical Botanical Garden, Chinese Academy of Sciences, Kunming, China, **3** Xishuangbanna Station for Tropical Rain Forest Ecosystem Studies, Chinese Ecosystem Research Net, Mengla, Yunnan, China, **4** University of Chinese Academy of Sciences, Beijing, China

Abstract

A two-year study (2009 ~ 2010) was carried out to investigate the dynamics of different carbon (C) forms, and the role of stream export in the C balance of a 23.4-ha headwater catchment in a tropical seasonal rainforest at Xishuangbanna (XSBN), southwest China. The seasonal volumetric weighted mean (VWM) concentrations of total inorganic C (TIC) and dissolved inorganic C (DIC) were higher, and particulate inorganic C (PIC) and organic C (POC) were lower, in the dry season than the rainy season, while the VWM concentrations of total organic C (TOC) and dissolved organic C (DOC) were similar between seasons. With increased monthly stream discharge and stream water temperature (SWT), only TIC and DIC concentrations decreased significantly. The most important C form in stream export was DIC, accounting for 51.8% of the total C (TC) export; DOC, POC, and PIC accounted for 21.8%, 14.9%, and 11.5% of the TC export, respectively. Dynamics of C flux were closely related to stream discharge, with the greatest export during the rainy season. C export in the headwater stream was 47.1 kg C ha^{-1} yr^{-1}, about 2.85% of the annual net ecosystem exchange. This finding indicates that stream export represented a minor contribution to the C balance in this tropical seasonal rainforest.

Editor: David L. Kirchman, University of Delaware, United States of America

Funding: This study was funded by the National Science Foundation of China (40801035, 40571163), the CAS 135 program (XTBG-T03) and Strategic Priority Research Program-Climate Change: Carbon Budget and Related Issues of the Chinese Academy of Sciences(XDA05020302), the Development Program in Basic Science of China (No. 2010CB833501), and the Yunnan Natural Science Foundation of Yunnan Province, China (2008CD167). The funders had no role in study design, data collection and analysis, decision to publish, or preparation of the manuscript.

Competing Interests: The authors have declared that no competing interests exist.

* E-mail: yipingzh@xtbg.ac.cn

Introduction

Streams and small inland rivers are important links between terrestrial and aquatic ecosystems. Cole et al. [1] suggested that inland waters export 1.9 Pg C yr^{-1}, indicating that regional carbon (C) balances can influence transport into large terrestrial rivers and oceans [2]. Recently, several studies have focused on dissolved organic C (DOC), dissolved inorganic C (DIC), particulate inorganic (PIC) and organic C (POC), and even gaseous C (CO_2, CH_4) in catchment runoff, and on their role in C exports from ecosystems [2–8].

Previous studies showed that the export of dissolved and gaseous C with rivers and streams may vary among forest ecosystems. Shibata et al. [9] found that sum of DIC and DOC export by stream water (7.6 g C m^{-2} yr^{-1}) accounted for only 2% of net ecosystem exchange (NEE) in cool temperate forests of northern Japan, whereas in Canadian boreal forests, C export from surface waters accounted for NEE from 9.5% to 16.4% [10]. In the Amazon, Richey et al. [6] demonstrated that outgassing of CO_2 (1.2±0.3 Mg ha^{-1} yr^{-1}) from rivers and wetlands constituted an important C loss. Also Lloret et al. [8] demonstrated the key role of streams in the C balance of forest catchments in the Amazon Basin. Neu et al. [2] showed that C transported by water comprised about 20% of the total annual C exchange across tropical forest canopies. The roles of surface water in C export vary because of diversity in geographic location, basin-specific soil and vegetation types, catchment topography, climate, and upland-wetland flow paths in forests [4,7,11–16]. As a result, by ignoring the export of CO_2, DOC, DIC, PIC and POC via hydrological pathways, terrestrial C budgets are incomplete and net C sequestration could be overestimated [1].

Surface water and wetland play substantial roles in C balance in the Amazon, the largest tropical forest region in the world [2,6,8]. So far, little is known about the importance of C export by headwater streams on the carbon balance of tropical seasonal rain forests (TSRF) at the northern edge of the tropical zone in southwest China (Figure 1). Despite its relatively high latitude, tropical seasonal rain forest has a moist tropical climate due to the influence of the Himalayas. It is unique in terms of forest type, differing from those in the equatorial region of Southeast Asia and has highly diverse and mixed types of floristic compositions due to its unique geographical location between a tropical zone to the south and a subtropical zone to the north [17]. Consequently, the tropical seasonal rain forest in southwest China is an important biogeographic area in Southeast Asia. Tan et al. [18] and Zhang et al. [19] have reported that TSRF in Xishuangbanna

(XSBN) is a small net C source. Accounting for TC export with stream water may make the loss of C from TSRF at XSBN even larger than earlier anticipated. In order to clarify the role of C export by headwater streams, a study was therefore undertaken in TSRF at XSBN. The objectives of this study were (1) to ascertain the seasonal dynamics of different C components (DIC, DOC, PIC, POC, TIC, TOC, and TC), and (2) to assess the contribution of stream export to the C balance in this tropical seasonal rainforest ecosystem.

Materials and Methods

Ethics Statement

All necessary permits were obtained from Xishuangbanna National Nature Reserve for the described field studies which did not involve endangered or protected species.

Study Site

The study area in XSBN (Dai autonomous prefecture), Yunnan province, China (21.16° N, 101.04° E) (Figure 1), is influenced by the Southwest monsoon and dominated by North Tropical Monsoon weather, with annual average temperature 21.5°C, annual average rainfall 1557 mm and average relative humidity 86%. Based on precipitation data, the rainy season (with 84.1% of the total annual precipitation) [20,21], is between May and October. The dry season is between November and April.

The experimental site is located in the centre of the National Forest Reserve in Menglun, Mengla County, Yunnan province, with relatively little human disturbance. The dominant trees are *Terminalia myriocarpa* and *Pometia tomentosa*, which is typical of tropical forest [17]. The total catchment area is 23.4 ha, the slope is 12°~18°, and the soil type is oxisol formed from Cretaceous yellow sandstone with a pH value of 4.5~5.5 and a clay content of 19.5%~29.5% [22].

Experimental Set-up

Hydrological observations. At the watershed outlet, a 90° V-notch weir instrumented with a water-level recorder was installed. The recorder was set to take averaged discharge measurements at 5-min intervals. Daily and monthly discharges were calculated separately from the stream-height data, as follows:

$$Q = 0.014H^{2.5} \tag{1}$$

$$R = QT/1000F \tag{2}$$

Where Q = discharge (m^3/s); H = water head (m); R = runoff (mm); T = time (s); and F = catchment area (km^2). Stream water temperature (SWT) was recorded at the mid-point of stream depth near the stream outlet. Measurements were made every half hour and stored in a data logger.

Water sample sampling and analysis. Stream water was sampled in the middle of the stream outlet. Stream water samples were collected between 8:00 and 9:00 am local time at the sampling site in high-density polyethylene (HDPE) bottles; sampling bottles were completely filled, allowing no headspace. Bottles were rinsed with distilled water after being washed with 3% HCl solution. Bottles were pre-rinsed three times with the stream water before sample collection. The study was done during two full calendar years, from 1.1.2009 to 31.12.2010. During the dry season, stream water samples were collected once per week, in addition to daily samples during three consecutive days following rain events. Stream water was sampled twice per week during the rainy season in 2009, and once per week in 2010. All water samples were immediately transported to the laboratory in insulated bags.

Following the analysis method of Baker et al. [23], all samples were vacuum-filtered through 0.45-μm GFF (Tianjinshi Dongfang Changtai Environmental Protection Technology Co. Ltd., China) pre-rinsed with deionized water and sample water under vacuum. The filtered and unfiltered water samples were analysed for DOC/DIC and TOC/TIC by TOC/TN analyser (LiquiTOC II, Elementar Analyses System GmbH, Germany) respectively, within 24 hours.

The TOC/TN analyser allows particle size up to 200μm. So in this study, the diameter of particulate matter (PIC and POC) was defined from 0.45 μm to 200 μm. TIC and TOC were defined as less than 200 μm for all the water samples. Concentrations of PIC and POC were calculated by subtracting the DIC and DOC concentrations from the TIC and TOC concentrations, respectively.

Calculations and Statistics

The monthly volume-weighted mean (VWM) concentrations were computed as follows:

$$VWM = \frac{\sum_{i=1}^{n} CiVi}{\sum_{i=1}^{n} Vi} \tag{3}$$

Where VWM is the volume-weighted mean concentration, Ci is the concentration (mg L^{-1}), and Vi is the runoff (m^3 s^{-1}) at sampling time.

The monthly C flux was calculated as the monthly VWM C concentration multiplied by monthly discharge. We calculated TIC, DIC, TOC, and DOC flux directly, and calculated PIC, POC, and TC flux as follows:

$$TC\,flux(F_{TC}) = F_{TIC} + F_{TOC} \tag{4}$$

$$PIC\,flux(F_{PIC}) = F_{TIC} - F_{DIC} \tag{5}$$

$$POC\,flux(F_{POC}) = F_{TOC} - F_{DOC} \tag{6}$$

Where F indicates flux.

The correlations among stream discharge, stream water temperature on the one hand, and concentrations of TIC, TOC, DIC, DOC, PIC, and POC on the other, as well as the correlations among different carbon components, were tested using the Pearson correlation (two tailed), employing the software SPSS 15.0.

Results

Seasonal Variations of Rainfall, Stream Runoff, and Stream Water Temperature

The average annual rainfall and runoff for the two years were 1026.1 mm and 326.9 mm, respectively. These values are less than the past 40 years means [24]. Rainfall and runoff were higher during the rainy season (average 848.9 mm and 279.1 mm, respectively) than during the dry season, confirming earlier reports

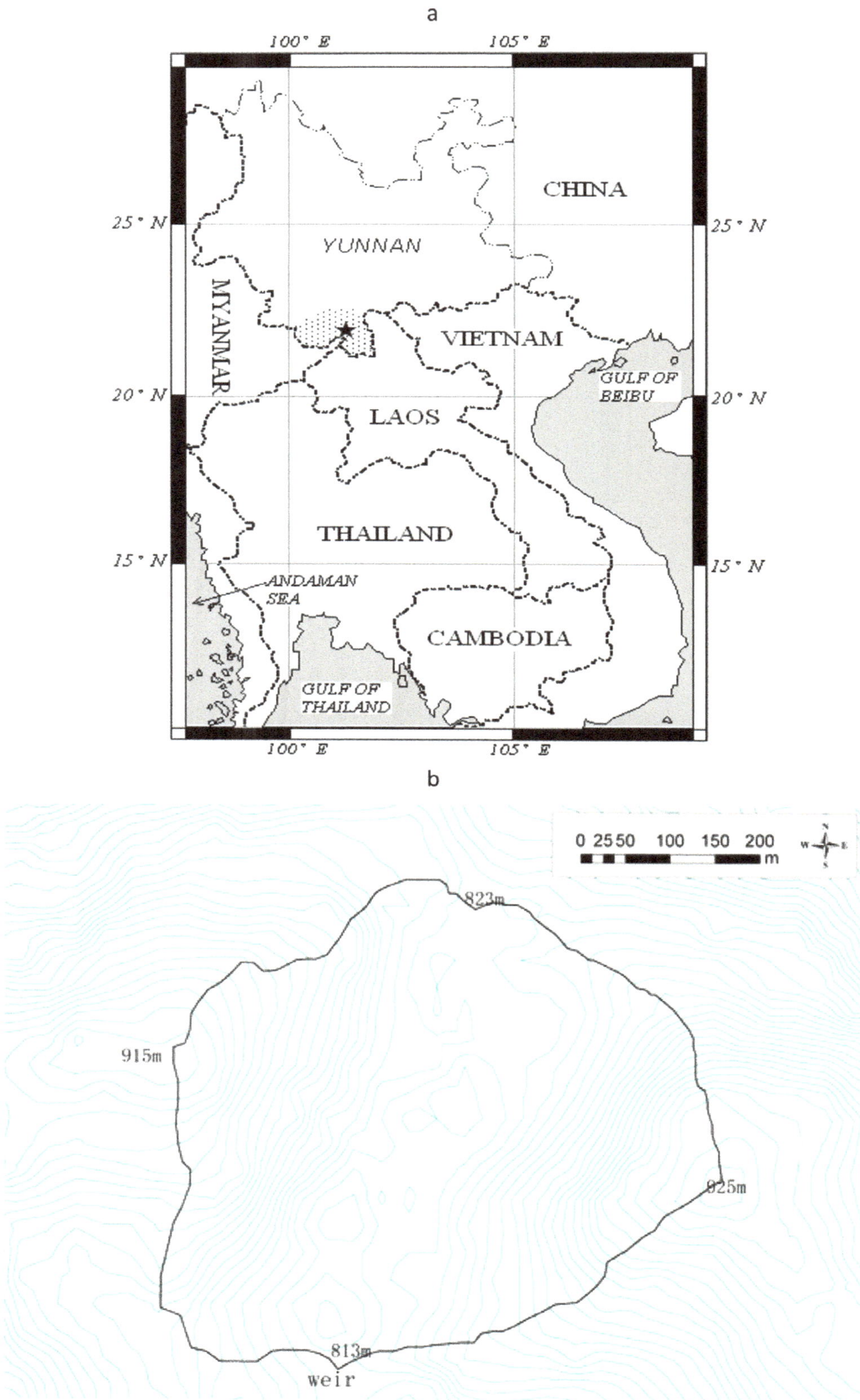

Figure 1. Study site description in Xishuangbanna tropical seasonal rainforest, Southwest China. (a) Location of the study area (indicated by the black star). (b)The catchment site description was from the Advanced Spaceborne Thermal Emission and Reflection Radiometer (ASTER) Global Digital Elevation Model (GDEM) that is a product of METI and NASA.

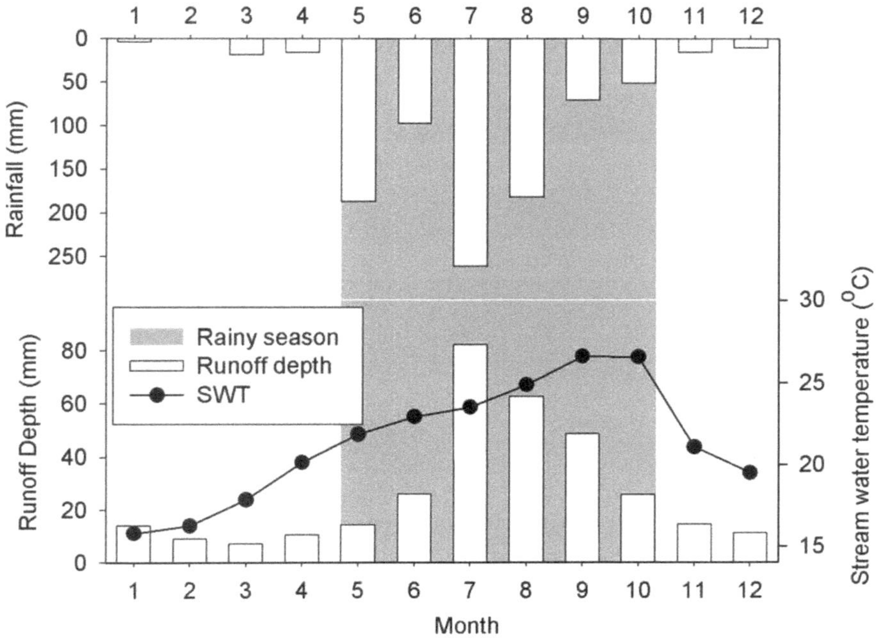

Figure 2. Annual dynamics of rainfall, runoff depth, and stream water temperature in the seasonal tropical rainforest. Values are averages for 2009 and 2010.

on the seasonal dynamics of rainfall and stream discharge [20,25].The seasonal dynamics of rainfall and runoff showed similar patterns and were well correlated ($r = 0.794$, $p < 0.001$, $n = 24$; Figures 2, S1).

Average SWT was 21.4°C, with lowest values in January (15.8°C) and peak values in September (26.6°C; Figures 2b, S1).

C concentration Dynamics

DIC was the largest component of TC. The rank order of the overall contribution of different C forms to TC was as follows: DIC (51.8%)>DOC (21.8%) >POC (14.9%) >PIC (11.5%) (Figure 3). However, these contributions differed seasonally: the contribution of DIC to TC was lowest in July (31.0%) and highest in February (64.9%), DOC had the highest contribution to TC in April and May (35.0%) and the lowest in December (10.0%), and the contribution of POC to TC was greatest in July (Figure 3). The DIC: DOC ratio and its monthly variation (2.9 and 67.1%, respectively) were higher than those for PIC: POC (0.8 and 47.3%, respectively).

Seasonal variations in VWM concentration were different among the various C components (Table 1). The VWM concentrations of TIC and DIC were higher during the dry season than during the rainy season, others were similar.

Monthly VWM concentrations of TIC and DIC were significantly correlated ($r = 0.956$, $p < 0.001$, $n = 24$). The highest monthly VWM concentration was in March as discharge was the lowest; the lowest concentrations were in June and July while stream discharges were relatively high (Figures 2b, 4a). Both TIC and DIC were negatively correlated to discharge and SWT (Table 2). The highest and lowest monthly VWM concentrations of PIC occurred in August (2.0 mg L^{-1}) and September (1.2 mg L^{-1}) while discharge was high. Although floods increased DOC and POC concentrations, the highest VWM concentrations of TOC, DOC and POC occurred during the beginning of the rainy season (Figure 4b) at intermediate values of discharge (Figure 2b).

The lowest values of TOC and DOC were in September during relatively high discharge, but POC was lowest in February when discharge was low. The seasonal dynamic of DOC was different from DIC ($r = 0.157$, $p = 0.464$, $n = 24$), but those of PIC and POC were similar ($r = 0.515$, $p = 0.010$, $n = 24$). The annual variation of PIC (coefficient of variation (CV) = 19.2%) was less than that of TIC (CV = 23.1%) and DIC (CV = 28.4%). The rank order of the coefficients of variation of organic C forms was as follows: POC (CV = 76.4%), DOC (CV = 58.9%), TOC (CV = 51.2%).

Stream C Flux Dynamics and Distribution

Annual TC export was 53.9 kg ha^{-1} and 40.7 kg ha^{-1} in 2009 and 2010, respectively. The dynamics of the fluxes of C differed between the various compounds. With the exception of POC, the

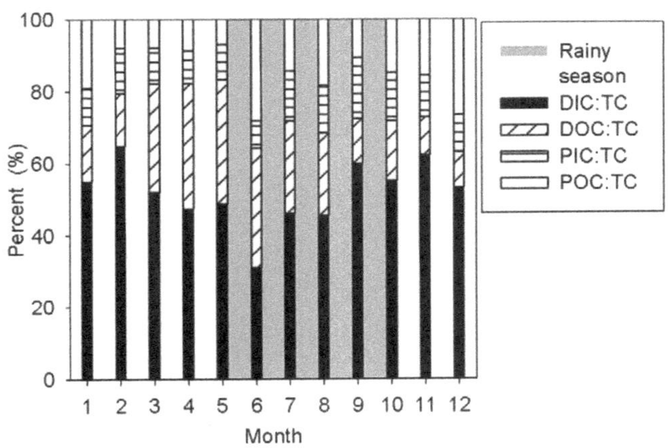

Figure 3. Annual dynamics of percentage difference of various carbon components to TC export, in the headwater stream.

Table 1. Concentration and export of various carbon components in a headwater stream in the tropical seasonal rainforest of Xishuangbanna, Southwest China.

	Carbon concentration (mg C L^{-1})			Carbon export (kg C ha^{-1})				
Season	**Dry season**	**Rainy season**	**Annual**	**Dry season**	**Rainy season**	**Annual average**	**2009**	**2010**
TIC	10.1	6.9	7.5	7.6	21.1	28.7	32.5	24.9
TOC	6.6	6.6	6.6	3.9	14.5	18.4	21.2	15.7
DIC	8.9	5.4	6.0	6.4	16.6	22.9	26.0	19.9
DOC	3.7	3.7	3.7	2.2	8.6	10.8	13.4	8.1
PIC	1.6	1.7	1.7	1.2	4.6	5.8	6.5	5.0
POC	3.1	3.3	3.2	1.7	5.9	7.6	7.7	7.5
TC				11.5	35.6	47.1	53.5	40.6

Carbon concentration is the seasonal volumetric weighted mean (VWM) concentration of 2009 and 2010.

greatest flux of all C components occurred in July when discharge was highest (Figures 2, 5). By contrast, POC export was greatest in June when discharge was intermediate. Due to low discharge also C flux was low in February (DOC, TOC) and in March (inorganic C and POC). Measures of seasonal C export (Table 1) showed that most of the C export for all components occurred during the rainy season.

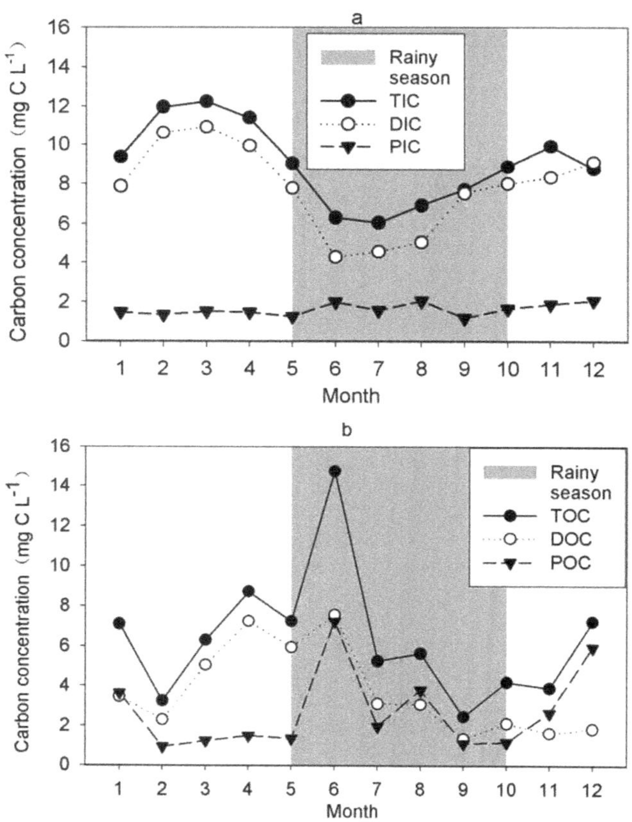

Figure 4. Annual dynamics of monthly volumetric weighted mean concentrations of carbon components in seasonal tropical rainforest. (a) Annual dynamics of monthly volumetric weighted mean concentrations of TIC, DIC, and PIC. (b) Annual dynamics of monthly volumetric weighted mean concentrations of TOC, DOC, and POC.

Contribution Stream Water C Flux in the C Balance of Tropical Seasonal Rainforest

Based on seasonal NEE dynamics (Figure 6a) [19], TC export resulted in an increased net C export ratio (TC: NEE) from 0.54% to 3.30% from April to August, and in a decrease from September to March, ranging from 0.20% to 1.80%. In total, stream C export represented a mean 2.85% (2009, 3.25%; 2010, 2.45%) increase in the annual net carbon export (1660 kg C ha^{-1} yr^{-1}) [19] in the TSRF. The ratios of DIC, DOC, PIC, and POC export to NEE showed the highest fractions of the C components' flux to NEE occurred in August (DIC: NEE = 1.50%, DOC: NEE = 0.75%, PIC: NEE = 0.44%, POC: NEE = 0.60%). The lowest absolute value of ratios of DIC, DOC, and PIC to NEE were observed in November (DIC: NEE = 0.13%, DOC: NEE = 0.024%, PIC: NEE = 0.025%), whereas the lowest value of POC: NEE occurred in February (0.024%; Figure 6b).

Discussion

C Dynamics in a Headwater Stream in the Tropical Seasonal Rainforest of Xishuangbanna

Headwater stream C originates from surface soil, ground water, vegetation (dead and alive), roots and microbial biomass in the forest [8,26,27]. Stream C dynamics depended on rainfall and discharge dynamics in this study and other small catchments [13,16,28]. Accurate calculation of C export with stream water requires representative sampling [12,16]. Our sampling campaigns were throughout 2 years, and included both base flow and storm flow conditions (Figure S1), so that carbon components in stream

Table 2. Pearson correlations for monthly average stream discharge, monthly average SWT, and monthly VWM concentration of carbon components in 2009 and 2010 in a headwater stream in the tropical seasonal rainforest in Xishuangbanna, Southwest China.

Parameters	TIC	TOC	DIC	DOC	PIC	POC
Monthly Q (n = 24)	−0.658**	−0.140	−0.661**	−0.139	0.057	−0.079
SWT (n = 24)	−0.520**	−0.166	−0.482*	−0.103	−0.150	−0.186

*Correlation is significant at the 0.05 level (2-tailed);
**Correlation is significant at the 0.01 level (2-tailed).

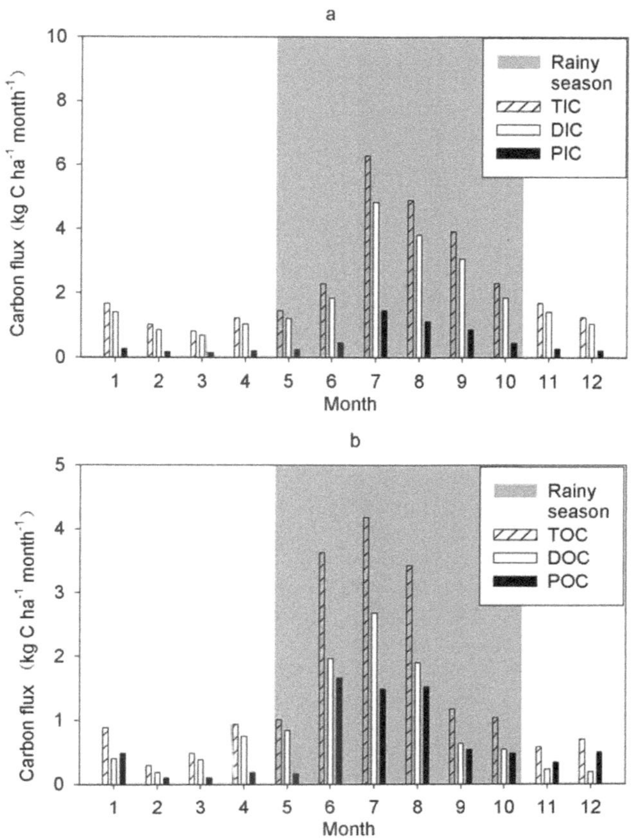

Figure 5. Annual dynamics of flux of carbon components in a headwater stream in tropical seasonal rainforest. (a) Annual dynamics in flux of TIC, DIC, and PIC. (b) Annual dynamics in flux of TOC, DOC, and POC.

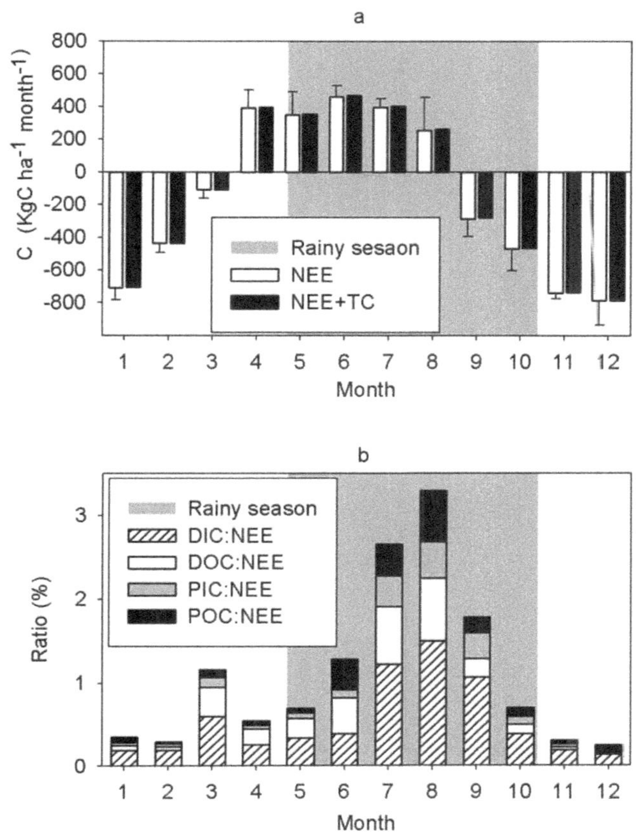

Figure 6. Annual dynamics of ratio of stream carbon export to net ecosystem CO_2 exchange (NEE). (a) Annual dynamics of NEE [19] and the sum of NEE and TC exported by headwater stream. (b) Annual dynamics of ratios of annual flux of DIC (DIC: NEE), DOC (DOC: NEE), PIC (PIC: NEE), and POC (POC: NEE) to NEE.

water were sampled across a wide range of discharge rates (Figure S2). Otherwise, our sampling revealed little particle matter larger than 200 μm only in March when discharge was relatively low (Figure 1). Therefore, this approach ensures the relatively high accuracy of C export calculations, which includes all C matter less than 200 μm in size. Accordingly, our calculations of TC and PC export excluded coarse C particles (>200 μm) in this study.

DIC was the most important C component (Figure 4, Table 1) of C export in the headwater stream of TSRF. This is consistent with the observation that the mineral soil has little organic C (23.88 g kg^{-1}) [22] and that there are few storm events (Figure S1) [2,8], which tend to be higher in DIC than in DOC. Ground water flow, which dominates base flow, is a continuous C conduit from landscape to stream in Amazonian [2,7,8,29,30] and British forests [23]. In streams, groundwater-derived DIC is significantly diluted by surface runoff and interflow [24] and by PIC transferred to the stream by surface runoff [31,32] and lateral movement of soil water during rain events [8,23]. In addition, TIC and DIC concentrations in stream water may decrease with increases in SWT (Table 2) during the rainy season, due to its microbial transformation to DOC or gaseous C [13,31,32]. Similar observations were reported for small streams in the Amazon [32] and in northern California [31].

DOC was the dominated organic carbon form in the headwater stream (Figure 5, Table 1). TOC and DOC dynamics were similar (Figures 4b, S2, Table 1). The annual variance of DOC (CV = 58.9%) concentration was more than that of DIC (CV = 28.4%), and correlations between DOC and stream water discharge differ from those for DIC (Figures 2, S2, Table 2) suggesting different sources for DOC and DIC. Litter fall peaks in late March [33,34], releasing dissolved organic matter (DOM) through decomposition [7]. Lower stream discharge in March and April further increased the DOM concentration. At the beginning of the rainy season, organic C in surface soil and litter leachate is transported to the headwater stream by surface runoff, interflow water, and canopy throughfall [26,29,35] due to persistent rain events [20]. In addition, a large share of DOC from soil is "flushed" during the last rainstorms of the dry season and the first storms in the rainy season [36]. Also, stream DOC concentration peaked in June (Figures 4b, S2), and was lowest in September (Figure 4b) when litter had been decomposed and the store of DOC in the soil had leached gradually during the rainy season [36] (Figure S2). This explains the weak relationship between discharge and DOC concentration (Table 2), but disagrees with the strong positive or negative correlations of streams having varying agricultural land-use intensities in their catchments [28]. The VWM concentrations were similar in the rainy season and the dry season, which contrasts other studies showing DOC concentrations higher during the rainy season [8,16,32].

Stream C export increased as stream discharge increased (Figures 2b, 4). This result supports the notion that organic and inorganic C export in watersheds is always dominated by runoff

Table 3. Ratios of total carbon export to different components of the carbon cycle in a tropical seasonal rainforest stream in Xishuangbanna, Southwest China.

Links of carbon cycle	kg C ha^{-1} yr^{-1}	TC Ratio%	Reference
NEE	1660	2.85	[19]
Soil surface efflux	14564	0.32	[43]
Soil respiration	9491	0.50	
Litter respiration	3245	1.45	
Stem respiration	14~47	100.31~332.39	[42]
Litterfall mass	7180~12850	0.37~0.66	[44]
Fine root mass of 0–20 cm depth	6124	0.77	[45]
Living fine root mass of 0–20 cm depth	5418	0.87	
Dead fine root mass of 0–20 cm depth	707	6.66	

The TC ratio (%) indicates the ratio of TC export by stream to the amount of carbon in different components of the carbon cycle.

amount despite sometimes even smaller concentrations in stream water [3,27]. TOC flux in this study (18.4 kg C ha^{-1} yr^{-1}) was much less than that in Amazonian rivers and streams (100 kg ha^{-1} yr^{-1}) [3], reflecting higher rainfall and stream discharge in the Amazon tropical region [3,20]. In addition to stream discharge, C export is influenced by vegetation, soil type, and soil inorganic and organic carbon content [3,8,26,28]. DIC flux was higher in this study than in the Southern Amazon (11.3 kg C ha^{-1} yr^{-1}) [2], and was within the range of fluxes reported for Basse-Terre Island catchments, Lesser Antilles, during periods of low flow (1.7±0.9 to 14.8±9.4 kg C ha^{-1} yr^{-1}) and flood (7.3±4.2 to 75.7±36.9 kg C ha^{-1} yr^{-1}) [8] where runoff were higher than XSBN. TOC flux in XSBN was higher than that in subtropical forest in China (7.2 to 9.3 kg C ha^{-1} yr^{-1}) [37], although the subtropical forest catchment had more soil carbon stock (164 t C ha^{-1}) [38] than our site (87.0 t C ha^{-1}) [39]. DOC export (12.0, and 9.5 kg C ha^{-1} yr^{-1} in 2010 and 2009, respectively) at XSBN was less than that measured in Juruena headwater catchments in Brazil (31.5 kg C ha^{-1} yr^{-1}) [7], tropical volcanic islands in Guadeloupe (16.0±9.0 to 57.0±26.0 kg C ha^{-1} yr^{-1}). Also, primary tropical forest (20.7±1.89 kg C ha^{-1} yr^{-1}), secondary forest (18.9±1.4 kg C ha^{-1} yr^{-1}), pine reforestation (17.9±9.0 kg C ha^{-1} yr^{-1}) and cabbage cultivation (14.8±1.0 kg C ha^{-1} yr^{-1}) in tropical highlands in northern Thailand [40], and a Wisconsin stream in a peat land catchment (25.0 kg ha^{-1} yr^{-1}) [41] had higher C export rates than XSBN. The differences in C dynamics among these regions may reflect the soil type, stream chemistry, vegetation, or hydrology. Therefore, future studies should consider the complex mechanisms that underlie regional differences in C dynamics.

Role of Stream Water Export in the C Balance

The contributions of the fluxes of all C components to the net C loss (as determined by the NEE) were determined (Figures 5, 6). Stream export increased C output from April to August but the NEE indicated that TSRF was C source in this period. From September to March carbon accumulation of TSRF was smaller than indicated by the C sink suggested by the NEE, due to organic C export with stream water (Figure 6a). In comparison with NEE, all stream water C components were small (Figure 6b; Table 3), even for TC. TC export in TSRF in XSBN (Table 3) matched or exceeded C emission due to stem respiration [42]. Compared to C emission due to soil respiration (SR) [43], and C sinks represented by litter fall [44], and fine root biomass production [45], the contribution of stream export was even smaller (Table 3). So, stream C export by headwater stream is negligible in the overall C balance (NEE, SR, litterfall, and root biomass production) of TSRF at XSBN.

Our study contrasts many others, who have suggested that surface waters are an important export pathway for C in tropical regions [2,6,8] and boreal forests [10,46]. Studies in the southern Amazon state of Mato Grosso showed TC (sum of DIC, DOC and fine-particle carbon) export by stream was 7.34% of NEE (1.5 Mg C ha^{-1} yr^{-1}) [2], which is higher than the ratio of TC: NEE 2.85% (Table 3) in this study. In contrast, Shibata et al. [9] found that DIC and DOC output by stream water (7.6 g C m^{-2} yr^{-1}) accounted for only 2% of NEE in cool-temperate forests of northern Japan, which is less than this study. Reason for relatively small C export by TSRF in the headwater stream at XSBN is that stream discharge was less than that of Amazon tropical regions [2,6,8], although NEE [19] and litter input [34] were similar in Amazon regions and XSBN. Furthermore, DOC export to the ecosystem C balance is small if adsorption to the soil matrix is strong [47,48]. Clays particular oxides have greater potential to adsorb DOC compared with the clay-poor sandy podzolic soils found in Amazonian forest [26]. Hence, fluvial export of C from XSBN's clay rich soils is likely to be lower in the present study area.

Based on the discussions above, the sources of different C components varied, leading to differences in the relative influence of stream discharge and SWT on C concentration and seasonal patterns. The relationship between stream discharge and C concentrations, and the distribution of C fractions differed in their influence on the C budget. A comparison of TC: NEE values in tropical seasonal rainforest at Xishuangbanna and in Amazon tropical forest and boreal forest indicates that stream export represents only a small component of the overall forest C balance in tropical seasonal rainforest.

Supporting Information

Figure S1 Sampling date of stream water 2009–2010 in tropical seasonal rainforest at Xishuangbanna, Southwest China. (a) Rainfall and stream water temperature dynamic during 2009 and 2010. (b) Sampling date and the runoff dynamic during 2009 and 2010.

Figure S2 Dynamics of different carbon components of sampling time during 2009–2010. (a) Dynamics in concentration of TIC, DIC, and PIC. (b) Dynamics in concentration of TOC, DOC, and POC.

Acknowledgments

We thank the staffs and technicians of the Xishuangbanna Station for Tropical Rain Forest Ecosystem who assist with field measurements. The authors greatly appreciate Wen-Jun Liu for the catchment figure drawing, Zhi-Hua Zhou for the sampling assistance, and Zhi-Ling Chen, Wu-Fei Liu, San-Mei A and Li-Fang Ou for laboratory works. Jan Mulder provides helpful comments and English writing during the preparation of the manuscript.

Author Contributions

Contributed to writing the manuscript: L-QS YD X-BD K-JD. Conceived and designed the experiments: W-JZ Y-PZ DAS L-QS. Performed the experiments: W-JZ YD X-BD. Analyzed the data: W-JZ DAS Y-PZ YD. Wrote the paper: W-JZ Y-PZ DAS.

References

1. Cole JJ, Prairie YT, Caraco NF, McDowell WH, Tranvik LJ, et al. (2007) Plumbing the global carbon cycle: integrating inland waters into the terrestrial carbon budget. Ecosystems 10: 172–185.
2. Neu V, Neill C, Krusche AV (2011) Gaseous and fluvial carbon export from an Amazon forest watershed. Biogeochemistry 105: 133–147.
3. Meybeck M (1982) Carbon, nitrogen, and phosphorus transport by world rivers. American Journal of Science 282: 401–450.
4. McDowell WH, Asbury CE (1994) Export of carbon, nitrogen, and major ions from 3 tropical tropical montane watersheds. Limnology and Oceanography 39: 111–125.
5. Richey JE, Brock JT, Naiman RJ, Wissmar RC, Stallard RF (1980) Organic carbon oxidation and transport in the Amazon river. Science 207: 1348–1351.
6. Richey JE, Melack JM, Aufdenkampe AK, Ballester VM, Hess LL (2002) Outgassing from Amazonian rivers and wetlands as a large tropical source of atmospheric CO_2. Nature 416: 617–620.
7. Johnson MS, Lehmann J, Selva EC, Abdo M, Riha S, et al. (2006) Organic carbon fluxes within and streamwater exports from headwater catchments in the southern Amazon. Hydrological Processes 20: 2599–2614.
8. Lloret E, Dessert C, Gaillardet J, Albéric P, Crispi O, et al. (2010) Comparison of dissolved inorganic and organic carbon yields and fluxes in the watersheds of tropical volcanic islands, examples from Guadeloupe (French West Indies). Chemical Geology 230: 65–78.
9. Shibata H, Hiura T, Tanaka Y, Takagi K, Koike T (2005) Carbon cycling and budget in a forested basin of southwestern Hokkaido, northern Japan. Ecological Research 20: 325–331.
10. Benoy G, Cash K, McCauley E, Wrona F (2007) Carbon dynamics in lakes of the boreal forest under a changing climate. Environmental Reviews 15: 175–189.
11. Aldrian E, Chen CTA, Adi S, Prihartanto SN, Sudiana N, et al. (2008) Spatial and seasonal dynamics of riverine carbon fluxes of the Brantas catchment in East Java. Journal of Geophysical Research 113: 10.1029/2007jg000626.
12. Buffam I, Galloway JN, Blum LK, McGlathery KJ (2001) A stormflow/baseflow comparison of dissolved organic matter concentrations and bioavailability in an appalachian stream. Biogeochemistry 53: 269–306.
13. Butturini A, Sabater F (2000) Seasonal variability of dissolved organic carbon in a Mediterranean stream. Biogeochemistry 51: 303–321.
14. Doctor DH, Kendall C, Sebestyen SD, Shanley JB, Ohte N, et al. (2008) Carbon isotope fractionation of dissolved inorganic carbon (DIC) due to outgassing of carbon dioxide from a headwater stream. Hydrological Processes 22: 2410–2423.
15. Goller R, Wilcke W, Fleischbein K, Valarezo C, Zech W (2006) Dissolved nitrogen, phosphorus, and sulfur forms in the ecosystem fluxes of a montane forest in Ecuador. Biogeochemistry 77: 57–89.
16. Raymond PA, Saiers JE (2010) Event controlled DOC export from forested watersheds. Biogeochemistry 100: 197–209.
17. Cao M, Zhang J, Feng Z, Deng J, Deng X (1996) Tree species composition of a seasonal rain forest in Xishuangbanna, Southwest China. Tropical Ecology 37: 183–192.
18. Tan Z, Zhang Y, Yu G, Sha L, Tang J, et al. (2010) Carbon balance of a primary tropical seasonal rain forest. Journal of Geophysical Research 115, D00H26, doi: 10.1029/2009JD012913.
19. Zhang Y, Tan Z, Song Q, Yu G, Sun X (2010) Respiration controls the unexpected seasonal pattern of carbon flux in an Asian tropical rain forest. Atmospheric Environment 44: 3886–3893.
20. Wang X, Zhang YP. (2005) An analysis of the characteristics of rainfall and linear tread in the Menglun area of Xishuangbanna, SW China. Journal of Tropical Meteorology 21: 658–664.
21. Zhang YP, Wang X, Wang YJ, Liu WJ, Liu YH (2003) Comparison research on hydrological effect of the canopy of the tropical seasonal rainforest and rubber forest in Xishuangbanna, Yunnan. Acta Ecologica Sinica 23: 2653–2665.
22. Tang YL, Deng XB, Li YW, Zhang SB (2007) Research on the difference of soil fertility in the different forest types in Xishuangbanna. Journal of Anhui Agricultural Sciences 35: 779–781.
23. Baker A, Cumberland S, Hudson N (2008) Dissolved and total organic and inorganic carbon in some British rivers. Area 40: 117–127.
24. Liu WJ, Liu WY, Lu HJ, Duan WP, Li HM (2011) Runoff generation in small catchments under a native rain forest and a rubber plantation in Xishuangbanna, southwestern China. Water and Environment Journal 25: 138–147.
25. Tan ZH, Zhang YP, Song QH, Liu WJ, Deng XB, et al. (2011) Rubber plantations act as water pumps in tropical China. Geophysical Research Letters 38: L24406.
26. McClain ME, Richey JE, Brandes JA, Pimentel TP (1997) Dissolved organic matter and terrestrial-lotic linkages in the central Amazon basin of Brazil. Global Biogeochemical Cycles 11: 295–311.
27. Rantakari M, Mattsson T, Kortelainen P, Piirainen S, Finér L, et al. (2010) Organic and inorganic carbon concentrations and fluxes from managed and unmanaged boreal first-order catchments. Science of the Total Environment 408: 1649–1658.
28. Wilson HF, Xenopoulos MA (2008) Ecosystem and seasonal control of stream dissolved organic carbon along a gradient of land use. Ecosystems 11: 555–568.
29. Palmer SM, Hope D, Billett MF, Dawson JJC, Bryant C (2001) Sources of organic and inorganic carbon in a headwater stream evidence from carbon isotope studies. Biogeochemistry 52: 321–338.
30. Saunders TJ, McClain ME, Llerena CA (2006) The biogeochemistry of dissolved nitrogen, phosphorus, and organic carbon along terrestrial-aquatic flowpaths of a montane headwater catchment in the Peruvian Amazon. Hydrological Processes 20: 2549–2562.
31. Finlay JC (2003) Controls of streamwater dissolved inorganic carbon dynamics in a forested watershed. Biogeochemistry 62: 231–252.
32. dos Santos Sousa E, Salimon CI, de Oliveira Figueiredo R, Krusche AV (2011) Dissolved carbon in an urban area of a river in the Brazilian Amazon. Biogeochemistry 105: 159–170.
33. Ren YH, Cao M (1999) Comparative study on litterfall dynamics in a seasonal rain forest and a rubber plantation in Xishuangbanna, SW China. Acta Phytoecologica Sinica 23: 418–425.
34. Tang JW, Cao M, Zhang JH, Li MH (2010) Litterfall production, decomposition and nutrient use efficiency varies with tropical forest types in Xishuangbanna, SW China: a 10-year study. Plant and Soil 335: 271–288.
35. Meyer JL, Wallace JB, Eggert SL (1998) Leaf litter as a source of dissolved organic carbon in streams. Ecosystems 1: 240–249.
36. Haaland S, Mulder J (2010) Dissolved organic carbon concentrations in runoff from shallow heathland catchments: effects of frequent excessive leaching in summer and autumn. Biogeochemistry 97: 45–53.
37. Qiao Y, Yin G, Luo Y, Liu Y (2009) Dynamics of total organic carbon (TOC) in hydrological processes and its contributions to soil organic carbon pools of three successional forest ecosystems in southern China. Ecology and Environmental Sciences 18: 2300–2307.
38. Zhou GY, Zhou CY, Liu SG, Tang XL, Ouyang XJ, et al. (2006) Belowground carbon balance and carbon accumulation rate in the successional series of monsoon evergreen broad-leaved forest. Science in China Series D-Earth Sciences 49: 311–321.
39. Lue X-T, Yin J-X, Jepsen MR, Tang J-W (2010) Ecosystem carbon storage and partitioning in a tropical seasonal forest in Southwestern China. Forest Ecology and Management 260: 1798–1803.
40. Möller A, Kaiser K, Guggenberger G (2005) Dissolved organic carbon and nitrogen in precipitation, throughfall, soil solution, and stream water of the tropical highlands in northern Thailand. Journal of Plant Nutrition and Soil Science 168: 649–659.
41. Elder JF, Rybick NB, Carter V, Victoria W (2000) Sources and yields of dissolved carbon in northern Wisconsin stream catchments with differing amounts of peatland. Wetlands 20: 113–125.
42. Yan YP, Sha LQ, Cao M (2008) Stem respiration rates of dominant tree species in a tropical seasonal rain forest in Xishuangbanna, Yunnan, southwest China. Journal of Plant Ecology 32: 23–30.
43. Sha LQ, Zheng Z, Tang JW, Wang YH, Zhang YP, et al. (2005) Soil respiration in tropical seasonal rain forest in Xishuangbanna, SW China. Science in China Series D-Earth Sciences 48: 189–197.
44. Zheng Z, Shanmughavel P, Sha L, Cao M, Warren M (2006) Litter decomposition and nutrient release in a tropical seasonal rain forest of Xishuangbanna, southwest China. Biotropica 38: 342–347.

45. Fang QL, Sha LQ (2005) Study of fine roots biomass and turnover in the rubber plantation of Xishuangbanna. Journal of Central South Forestry University 23: 488–494.
46. Butman D, Raymond PA (2011) Significant efflux of carbon dioxide from streams and rivers in the United States. Nature Geoscience 4: 839–842.
47. Fujii K, Uemura M, Hayakawa C, Funakawa S, Sukartiningsih, etal. (2009) Fluxes of dissolved organic carbon in two tropical forest ecosystems of East Kalimantan, Indonesia. Geoderma 152: 127–136.
48. Fujii K, Hartono A, Funakawa S, Uemura M, Kosaki T (2011) Fluxes of dissolved organic carbon in three tropical secondary forests developed on serpentine and mudstone. Geoderma 163: 119–126.

23

Fine Particle, Ozone Exposure, and Asthma/Wheezing: Effect Modification by Glutathione S-transferase P1 Polymorphisms

Bing-Fang Hwang[1], Li-Hao Young[1], Ching-Hui Tsai[2], Kuan-Yen Tung[2], Pei-Chuan Wang[2], Ming-Wei Su[2], Yungling Leo Lee[2]*

1 Department of Occupational Safety and Heath and Graduate Program, College of Public Health, China Medical University, Taichung, Taiwan, **2** Institute of Epidemiology and Preventive Medicine, College of Pubic Health, National Taiwan University, Taipei, Taiwan

Abstract

Background: There are limited studies on the role of interaction between exposure to ambient air pollution and glutathione-S-transferase (GST) P1 on the risk of asthma/wheezing among children, which provided suggestive, but inconclusive results.

Methods: To assess the joint effect of air pollutants and GSTP1 on asthma/wheezing, we conducted a nationwide cross-sectional study of 3,825 children in Taiwan Children Health Study. The studied determinants were three GSTP1 Ile105Val (rs 1695) genotypes (Ile-Ile; Ile-Val and Val-Val) and expoure to ambient air pollutants. We used routine air-pollution monitoring data for ozone (O_3) and particles with an aerodynamic diameter of 2.5 μm or less ($PM_{2.5}$). The effect estimates were presented as odds ratios (ORs) per interquartile changes for $PM_{2.5}$ and O_3.

Findings: In a two-stage hierarchical model adjusting for confounding, the risk of asthma was negatively associated with $PM_{2.5}$ (adjusted odds ratio (OR) 0.60; 95% confidence interval (CI) 0.45, 0.82) and O_3 (OR 0.74; 95% CI 0.60, 0.90) among Ile105 homozygotes, but positively associated with $PM_{2.5}$ (OR 1.52; 95% CI 1.01, 2.27) and O_3 (OR 1.19; 95% CI 0.91, 1.57) among those with at least one val105 allele (interaction p value = 0.001 and 0.03, respectively). A similar tendency of effect modification between $PM_{2.5}$ and O_3 and GSTP1 on wheezing was found.

Conclusion: Children who carried Ile105 variant allele and exposed to $PM_{2.5}$ and O_3 may be less likely to occurrence of asthma/wheezing.

Editor: James West, Vanderbilt University Medical Center, United States of America

Funding: This study was supported by grant #96-2314-B-039-019, #96-2314-B-006-053, #98-2314-B-002-138-MY3, #99-2314-B-039-021-MY3 and #101-2314-B-002-113-MY3 from the Taiwan National Science Council. BFH was supported by grant CMU# 100-AWARD-07 from China Medical University. The funders had no role in study design, data collection and analysis, decision to publish, or preparation of the manuscript.

Competing Interests: The authors have declared that no competing interests exist.

* E-mail: leolee@ntu.edu.tw

Introduction

Both genetic and environmental factors play important roles in the aetiology of asthma and wheezing. There is probably also genetic susceptibility to the effects of air pollution [1]. Identification of indicators of genetic susceptibility to environmental exposures could be useful from preventive point of view [2]. Recent studies of genes-environment interaction or effect modification for asthma mainly focused on members of the glutathione-S-transferase (GST) superfamily because several members, particular in glutathione-S-transferase p1 (GSTP1), glutathione-S-transferase m1 (GSTM1) and glutathione-S-transferase t1 (GSTT1), are expressed in the respiratory tract and function in processes implicated in asthma pathogenesis, including oxidant defenses, xenobiotic metabolism, and detoxification of hydroperoxides. DNA sequence variants in the GSTP1 Ile105Val locus, GSTM1 null, and GSTT1 null may contribute to susceptibility to oxidative stress and airway inflammation, which are key processes in asthma pathogenesis [3].

In our systematic Medline search, only four studies concerning effect modification betwen exposure to ambient air pollution and GSTP1 on childhood asthma/wheezing provided inconclusive results [4–7]. Melen and colleagues conducted a birth cohort study of 4,089 children in Stockholm, where they reported reduced risks of asthma and persistent wheezing among children with Ile105 homozygotes for NOx exposure during the first two years of life compared with at least one Val105 allele, but the interaction effect was not significant (p-value for interaction p>0.05) [4]. Islam and colleagues conducted a cohort study of 1610 school children in Southern Caifornia indicated negative but not significant association between the risk of asthma and high ozone communities (range from 46.5 to 64.9 ppb) among at least Val105 homozygotes with an adjusted hazard ratio of 0.98 (95% CI 0.4–2.3) [5]. Schroer and colleagues conducted a birth cohort study in

Cincinnati, Ohio found children carrying the Ile105 allele may confer protection from wheezing, but exposure to high level of diesel exhaust particle (DEP≥0.5 ug/m3) converging on a similar pathway may overwhelm the genetic effect [6].

Our previous study suggested an interaction between GSTP1 and outdoor air pollution on childhood asthma [7]. This study did not assess the role of air pollutants such as particles with an aerodynamic diameter of 2.5 μm or less ($PM_{2.5}$), and ozone (O_3). Taiwan Children Health Study (TCHS) offers an opportunity to investigate the joint effects of ambient air pollution exposure and GSTP1 on childhood asthma and wheezing. At the cohort entry, we collected information of allergic symptoms during the past 12 months and also on those important potential determinants of allergic disease in children. In the present study, we elaborated the effect modification relations between exposure to air pollution and GSTP1 on the risk of asthma and wheezing in school children, focusing on $PM_{2.5}$, and ozone (O_3). Further we applied a two-stage hierarchical model to adjust for confounding and to elaborate effect modification on individual-level and to assess the effects of air pollution between communities [8–9].

Methods

Data collection and study population

Taiwan Children Health Study (TCHS) was based on a multi-purpose nationwide design that focused on outdoor air pollutants as primary interest. Communities of Taiwan were selected with the aim of maximizing the variability and minimizing the correlations in criteria outdoor pollutants based on historic routine air monitoring data. A total of 4,102 seventh grade children were recruited from public schools in 14 communities covering diverse parts of Taiwan, which was representative of Taiwanese middle-school children in 2007. The study detail has been described previously [10]. We finally approached 3,816 seventh-grade school-aged children who completed the standard questionnaire and sufficient genetic information (response rate 93.0%).

The study protocol was approved by the Institutional Review Board (National Taiwan University Hospital Research Ethics Committee; number: 200902042R), and complied with the principles outlined in the Helsinki Declaration [11]. We have informed all the participating children/parents and obtained their written consent forms in this study.

Health outcomes

The outcomes of interest were the onsets of asthma and wheezing before the ambient air pollution in recent three years. The definition of asthma subjects was determined by a positive response to the question "Has a physician ever diagnosed your child as having asthma?" Five questions related to current asthmatic symptoms were also asked:

1. In the past 12 months, has your child had dyspnoea with wheezing in the chest? (wheezing)
2. In the past 12 months, has your child's sleep been disturbed because of wheezing? (night wheezing)
3. In the past 12 months, has wheezing ever been severe enough to limit your child's speech to only one or two words at a time between breaths? (dyspnoea at rest)
4. In the past 12 months, has your child's sounded wheezy during or after exercise? (exercise wheeze)
5. In the past 12 months, has your child had a dry cough at night, apart from a cough associated with a cold or chest infection? (night cough)

The criteria of non-asthmatic subjects was (i) no physician diagnosed asthma or dyspnoea with wheezing in the past; (ii) no positive response to any of the five questions concerning current asthmatic symptoms. The definition of wheezing was determined by a positive response to the question "Has your child ever had wheezing or whistling in the chest at any time in the post when he/she did not have a cold or flu?" Non-wheezing subjects were defined on the basis of these reporting not ever having dyspnoea with wheezing, no nocturnal dyspnoea associated with wheezing and without physician diagnosed asthma.

Genetic polymorphisms analysis

Cotton swabs containing oral mucosa were collected and were stored at −80°C before analysis. Genomic DNA was isolated using phenol/chloroform extraction method. The polymorphism of GSTP1Ile105Val, GSTM1 (null/presnt) and GSTT1 (null/present) were detected by real-time polymerase chain reaction using the TaqMan Allelic Discrimination assay on an ABI PRISMTM 7900 Sequence Detector (Applied Biosystems, Foster City, CA) [12]. The sequences of primers and MGB probes used are listed in Table S1. Each subject's clinical status remained anonymous to laboratory staff and genotype assignments were based on two consistent experimental results. About 15% of the randomly selected samples were sequenced, and all of them were consistent with the initial genotyping results.

Exposure assessment

Complete monitoring data for the air pollutants including sulphur dioxide (SO_2), nitrogen dioxides (NO_2), ozone (O_3), carbon monoxide (CO), and particles with an aerodynamic diameter of 2.5 μm or less ($PM_{2.5}$) are available for 14 EPA monitoring stations in TCHS communities. Concentrations of each pollutant are measured continuously and reported hourly—CO by non-dispersive infrared absorption, NO_2 by chemiluminescence, O_3 by ultraviolet absorption, SO_2 by ultraviolet fluorescence, and $PM_{2.5}$ by beta-gauge. Exposure parameters in the present study were 3-year average (2005–2007) and the yearly deviations from the 3-year average concentrations in each municipality, calculated from the 24-hour NO_2, CO, SO_2, $PM_{2.5}$, and 10:00 AM to 6:00 PM 8-hour O_3.

Covariates

Information on potential confounders was obtained from the parental-administered questionnaire. The covariates in the present analyses included age, gender, parental education, family annul income, duration of breast feeding, maternal smoking history during pregnancy, environmental tobacco smoke (ETS), cockroaches note, carpet used, home dampness and mould, and parental atopy. Parental atopy was a measure of genetic predisposition to asthma and it was defined as the father or mother of the index child ever having been diagnosed as having asthma, allergic rhinitis, or atopic eczema.

Statistical methods

The Hardy-Weinberg equilibrium was calculated for the all polymorphisms in our study population using Pearson χ^2 statistics. We estimated adjusted odds ratios in a two-stage hierarchical model using logistic and ecologic model analyses. The models assume two sources of variation: the variation among subjects in the first stage, part of which could be explained by the individual confounders, and the variation of air pollution between communities in the second stage, part of which could be explained by variables measured at municipal level. In the analyses we assumed that 1) the outcome variable follows Bernoulli distribution; 2) intercept terms are random at the municipal level; 3) all the explanatory variables are fixed effects. A logistic regression model was fitted in the first stage for the risk of bronchitic symptoms as a function of site-specific intercepts, j, where $\alpha j = 1, \ldots, 14$, and personal covariates. The adjusted site-specific intercepts and prevalence rates are related by $Pj = e^{\alpha j}/(1+e^{\alpha j})$. In the second stage, these intercept terms representing the logit of the site-specific prevalence rates (Pj; $j = 1, \ldots, 14$), adjusted for personal covariates, were regressed on each site-specific ambient pollutant level by using a linear "ecologic" regression, i.e., logit $\alpha j = \alpha + Uj + \beta Zj$, where Uj denotes the random departure from the general prevalence αj on the logit scale for site j; Zj denotes the ambient pollution level for site j. Thus, β can be interpreted as the log odds ratio (per interquartile changes) for each pollutant, adjusted for personal characteristics. The results from the models are presented as odds ratios (ORs), along with their 95% confidence intervals (CIs). The goodness of fit was assessed with likelihood ratio tests (LR) to determine whether a variable contributed significantly to the model. First, we fitted a full model with a complete set of covariates. To elaborate sources of confounding, we fitted models with different combinations of covariates and compared the effect from models with and without the covariate of interest If the adjusted odds ratio differed from the crude odds ratio by more than 10%, that covariate was included in the final model. The two-stage hierarchical model was used not only to derive more precise estimates of site-specific parameters and site-level effects, but also to adjust for multiple comparisons [13].

We considered the effect of multiple pollutants on the risk of asthma/wheezing. We first fitted one-pollutant models, and then considered two-pollutant models by fitting one (NO_2 or CO) and the other (SO_2 or $PM_{2.5}$) pollutant. Then, we fitted two-pollutant models with O_3 or $PM_{2.5}$ and another pollutant. It was not appropriate fit two-pollutant models with O_3 and $PM_{2.5}$ because of high collinearity (correlation coefficient $r = 0.73$). The two-pollutant models provide estimates of the independent effects of $PM_{2.5}$, and O_3 on the asthma/wheezing controlling for the second pollutant in the model. The effect of each pollutant on the risk of asthma/wheezing was presented as odds ratios (ORs) per interquartile changes for O_3, and $PM_{2.5}$.

We considered the joint effect between air pollutants and GSTP1, GSTM1, or GSTT1 on the risk of asthma and wheezing introducing interaction terms in the model. The community specific average air plloutnats levels were fitted as continuous variables along with the appropriate interaction terms for GSTP1, GSTM1, or GSTT1 variant and air pollution levels. All tests assumed a two-sided alternative hypothesis and a 0.05 significance level.

Results

Characteristics of study population

Hardy-Weinberg equilibrium tests for GSTP1 showed non-significance ($p > 0.05$) in both asthmatic and non-asthmatic groups as well as wheezing and non-wheezing groups. The characteristics of the study population according to the baseline covariates are shown in Table 1. The higher proportion of asthmatic subjects was male and had maternal smoking during pregnancy, any home dampness and mould, gestational age (<37 weeks) and parental atopy. Besides above covariates, the larger proportion of wheezing subjects was also related to the presences of cockroaches and duration of breastfeeding. We adjusted for these factors in the multivariate analysis.

Air pollution

The distributions of the annual mean air pollutant concentrations in the 14 monitoring stations in the year 2005–2007 are presented in Table 2, and the correlations between different pollutants between communities in Table 3. The correlation between NO_2 and CO concentrations was high (0.86), which reflects the common source of motor vehicles. The concentrations of $PM_{2.5}$ and SO_2 were also highly correlated (0.68) indicating common source of stationary fuel combustion, although SO_2 concentrations were also correlated with both traffic-related pollutants. The concentration of O_3 was negatively correlated with the mainly traffic-related pollutants, but positively with $PM_{2.5}$ (0.73) and SO_2 (0.37), and it was only weakly correlated with that of traffic-related and stationary fossil fuel combustion-related air pollutants.

Joint effect of air pollution and GSTP1 variants on asthma

The risk of asthma was negatively associated with $PM_{2.5}$ (OR 0.60; 95% CI 0.45, 0.82 per 16.84 $\mu g/m^3$), and O_3 (OR 0.74; 95% CI 0.60, 0.90 per 8.77 ppb) among Ile-105 homozygotes, but postively associated with $PM_{2.5}$ (OR 1.52; 95% CI 1.01, 2.27 per 16.84 $\mu g/m^3$), and O_3 (OR 1.19; 95% CI 0.91, 1.57 per 8.77 ppb) among those at least one val105 allele (interaction p value = 0.001 and 0.03, respectively) in single pollutant model (Table 4). In the two-pollutant models, the effects estimates for O_3 exposure among Ile-105 homozygotes were stable for the three different combinations of pollutants, varying between 0.72 and 0.76. Stable effect estimates (OR = 0.60) was found for $PM_{2.5}$ exposure between single pollutant models and two-pollutant models (Table 5).

Joint effect of air pollution and GSTP1 variants on wheezing

The effect estimates of $PM_{2.5}$ (OR 0.75; 95% CI 0.59, 0.96 per 16.84 $\mu g/m^3$) and O_3 (OR 0.82; 95% CI 0.69, 0.96 per 8.77 ppb) exposure on the risk of wheezing was substantially reduced among Ile-105 homozygotes compared with among those at least one val105 allele for $PM_{2.5}$ (OR 1.24; 95% CI 0.89, 1.71 per 16.84 $\mu g/m^3$) and O_3 (OR 1.09; 95% CI 0.88, 1.82 per 8.77 ppb) in single pollutant models (Table 4). The effect estimtes for $PM_{2.5}$ and O_3 remain constant as we considered different combination of air pollutants in two-pollutant model (Table 5). Our results shows a signifcant joint effect of GSTP1 varant and air pollution particular in $PM_{2.5}$ (interaction p value = 0.027) and O_3 (interaction p value = 0.049) on wheezing. We did not identify any joint effect between GSTM1 or GSTT1 and air pollutants on asthma/wheezing (Table S2 and Table S3).

Discussion

There were significantly negative associations between asthma/wheezing and increased the 3-year average concentrations of $PM_{2.5}$ and O_3 per interquartile changes among Ile-105 homozygotes children. The results provide evidence that children carrying Ile-105 homozygotes are less likely to development of asthma/wheezing as exposure to ambient air pollution, such as $PM_{2.5}$ and O_3.

Validity of results

The strength of this study was a large and socio-demographically diverse population of children in Taiwan were included. Unbiased observations of the association between genetic polymorphisms and outcomes were expected. We used routine air-pollution monitoring data as the basis for exposure assessment. These data represented reasonably well exposures both in the

Table 1. Distribution of demographic and other characteristic of study subjects.

Characteristic	Asthmatic subjects (n = 295)	Non-asthmatic subjects (n = 3517)	P-value	Wheezing subjects (n = 452)	Non-wheezing subjects (n = 3373)	P-value
	No. (%)	No (%)		No.	%	
Age (years)						
≤12	218 (74.4)	2611 (74.3)	0.80	337 (75.1)	2503 (74.3)	0.28
13–14	68 (22.9)	825 (23.5)		99 (21.8)	796 (23.6)	
≥14	9 (2.7)	76 (2.2)		15 (3.1)	70 (2.1)	
Gender						
Male	164 (55.6)	1708 (48.6)	0.02	249 (55.1)	1631 (48.4)	0.007
Female	131 (44.4)	1809 (51.4)		203 (44.9)	1742 (51.6)	
Parental education (years)*						
<8	48 (16.4)	564 (16.2)	0.78	68 (15.1)	549 (16.4)	0.57
8–11	195 (66.6)	2383 (63.0)		305 (67.8)	2281 (68.2)	
≥12	50 (17.1)	544 (15.6)		77 (17.1)	517 (15.4)	
Family annual incomes*						
Low	85 (31.1)	1180 (36.3)	0.24	150 (35.5)	1121 (35.9)	0.60
Medium	152 (55.7)	1678 (51.6)		215 (50.8)	1624 (52.1)	
High	36 (13.2)	397 (12.2)		58 (13.7)	374 (12.0)	
Environmental tobacco smoke*						
Yes	138 (47.1)	1655 (47.3)	0.95	224 (49.9)	1575 (46.9)	0.24
No	155 (52.9)	1844 (52.7)		225 (50.1)	1781 (53.1)	
Maternal smoking during pregnancy*						
Yes	18 (6.1)	133 (3.8)	0.05	31 (6.9)	122 (3.6)	0.001
No	277 (93.9)	3384 (96.2)		421 (93.1)	3251 (96.4)	
Cockroaches						
Yes	256 (88.0)	3121 (89.5)	0.43	416 (92.2)	2971 (88.9)	0.03
No	35 (12.0)	368 (10.5)		35 (7.8)	371 (11.1)	
Any home dampness and mould						
Yes	179 (60.7)	1924 (54.7)	0.05	293 (64.8)	1817 (53.9)	0.001
No	116 (39.3)	1593 (45.3)		159 (35.2)	1556 (46.1)	
Duration of breastfeeding (month)*						
0	140 (48.3)	1797 (51.9)	0.55	204 (46.0)	1741 (52.4)	0.04
1–3	120 (41.4)	1327 (38.3)		195 (44.0)	1257 (37.8)	
3–6	18 (6.2)	177 (5.1)		27 (6.1)	169 (5.1)	
≥6	12 (4.1)	163 (4.7)		17 (3.9)	157 (4.7)	
Gestational age (weeks)						
≥37	240 (83.3)	3139 (91.4)	<0.001	56 (12.7)	283 (8.6)	0.005
<37	48 (16.7)	294 (8.6)		385 (87.3)	3010 (91.4)	
Carpet used*						
Yes	26 (8.9)	368 (10.5)	0.39	50 (46.9)	346 (10.3)	0.61
No	266 (91.1)	3135 (89.5)		401 (88.9)	3010 (89.7)	
Pet						
Yes	171 (58.0)	2052 (58.3)	0.90	276 (61.1)	1959 (58.1)	0.23
No	124 (42.0)	1465 (41.7)		176 (38.9)	1414 (41.9)	
Parental atopy						
Yes	135 (45.8)	886 (25.2)	<0.001	195 (43.1)	827 (24.5)	<0.001
No	160 (54.2)	2631 (74.8)		257 (56.9)	2546 (75.5)	

*Number of subjects does not add up to total number because data were missing.

Table 2. The mean and distribution of 3-year average air pollutant concentrations between communities, Taiwan 2005–2007.

Pollutant	Mean±SD	Min	Max	Interquartile Range
NO_2 (ppb)	17.68±0.25	10.06	26.83	8.79
CO (100 ppb)	5.24±0.06	3.04	7.78	1.05
SO_2 (ppb)	4.33±0.10	2.16	10.09	1.31
$PM_{2.5}$ (μg/m^3)	33.38±0.50	19.83	51.34	16.84
O_3 (ppb)	44.64±0.39	30.34	59.12	8.77

Abbreviations: NO_2, nitrogen dioxide; $PM_{2.5}$, particles with aerodynamic diameter 2.5 μm or less; SO_2, sulphur dioxide; O_3, ozone; CO, carbon monoxide.

Table 3. Correlations of air pollutants across 14 communities.

	NO_2	CO	SO_2	$PM_{2.5}$	O_3
NO_2	1.00	0.86*	0.55*	0.37	−0.07
CO		1.00	0.16	0.09	−0.33
SO_2			1.00	0.68*	0.37
$PM_{2.5}$				1.00	0.73*
O_3					1.00

Abbreviations: NO_2, nitrogen dioxide; PM_{10}, particles with aerodynamic diameter 10 μm or less; SO_2, sulphur dioxide; O_3, ozone; CO, carbon monoxide.
*Correlation is significant at the 0.05 level.

school and in the home for two reasons. The schools were chosen to be in the vicinity of the monitoring stations. Almost all the children attended schools within one kilometre of their homes, because the density of middle schools in Taiwan is very high. Finally, the two-stage hierarchical modelling took into account the fact that between communities exposure information and multi-pollutant modes were used. The analytical method is better than previous four studies [4–7].

We excluded 286 subjects because of insufficient gene data. This exclusion was unlikely to introduce selection bias, because the characteristic of the excluded individuals did not differ substantially from the included (χ^2 test; p>0.05), as shown in Table S4. Any known or unknown factors such as air exchange, penetration, deposition as well as emission strengths for indoor pollutants could be responsible for the observed association between personal exposure and municipal level exposure. This was a common limitation in all the previous studies assessing the joint effects of air pollution and genetic variant on the risk of astham or wheezing. A continuous variable of air pollution in this study increases the power to detect interaction, but the number of children with different genotypes who are asthma/wheezing or non-asthma/non-wheezing is still limited. We cannot rule out the possibility of false-negaitve results was obtained.

The outcomes of interstes was based on paretnal report, we cannot rule out the possibility of recall bias. We compared recall of asthma status in a subset of the study population and found that the concordance of parental reports of asthma and medical records' documentation of asthma was good.

Assessment of the independent effects of different pollutants is difficult, because urban air pollution constitutes a complex mixture of several compounds. Although all the measured pollutants have several sources, NO_2 and CO are predominantly from vehicle emissions whereas the main sources of SO_2 and $PM_{2.5}$ are stationary fossil combustion processes. In the present study, NO_2 and CO concentrations were highly correlated indicating the common source of motor vehicle traffic emissions. SO_2 and $PM_{2.5}$ concentrations were also correlated, their common sources being stationary fossil fuel combustion. In the multi-pollutant modelling, we were able to control for one stationary fossil fuel pollutant at a time as a potential confounder when assessing the effect of one of the traffic-related pollutants and vice versa.

Synthesis with previous knowledge

The results of the present study, we found negative associations between asthma/wheezing among Ile-105 homozygotes children and exposure to air pollutants particular in $PM_{2.5}$ and O_3 as continuous scale. Our findings support the present the gene-pollution interaction or effect modification between exposure to air pollutants ($PM_{2.5}$ and O_3) and GSTP1 variants, especially the isoleucine (Ile)/valine (Val) polymorphism at amino acid position 105 with respect to childhood asthma/wheezing.

Three previous studies from Southern California [5], Cincinnati [6], and Stockholm [4] have elaborated the effect of exposure to outdoor air pollutants on childhood asthma/wheezing appears to be modified by GSTP1. Our present study and Cincinnati study reported a protective effect among children with Ile105 allele of particles ($PM_{2.5}$ or DEP) on the development of wheezing [6]. Stockholm study showed children with at least Val105 allele appear more susceptible to childhood asthma related to traffic-related air pollutant (NOx) [4], but the findings are inconsistent for Southern California study, where Ile105 homozygotes seems to have higher IgE response after O_3 exposure [5]. Another study in Mexico City reported that asthmatic children with at least Val105 allele are more likely to develop respiratory symptoms as exposure to O_3 than Ile105 homozygotes [14]. Different gene-pollution interaction in different sittings may occur. It is possible that the pattern of exposure can differ substantially between countries or areas and the genotype frequencies also show large global differences [15]. Whether there are different antioxidant mechanisms of GSTP1 in response to specific air pollutants (CO, CO_2, NO_2, SO_2, DEP, and $PM_{2.5}$) still need to be explored as well as possible differences of effect modification between healthy and asthmatic children.

The biological mechanisms by which the toxic effects of air pollutants ($PM_{2.5}$ and O_3) seemed to modify the effect of asthma/wheezing are not well understood. After stratification of the GSTP1 genotypes, the protective effect was only in Ile 105 homozygotes. We speculated that GSTP1, might be involved in detoxification of reactive metabolic intermediates and reactive oxygen species (ROS) from air pollution which might be able to protect against oxidative stress [16–17].

Exposure of bronchial epithelium to particules may generate reactive oxygen species and increase expression of a marker of oxidative stress, haeme oxygenase 1 [18]. Oxdative stress could be induce by particles promotes inflammation via activation of the transcription factors NF-kB and activator protein-1 and other processes implicated in inflammation, including histone acetylation and the MAPK pathway [19]. In vitro studies have shown that particules increase the production of inflammatory mediators (IL8, GM-CSF and ICAM-1) from human airway epithelia cells [20], which suggest that oxidative stress stimulates the proinflammatory response to particle matter. To our best knowledge, our results first showed that GSTP1 Ile 105 genetically determined antioxidant defense modify the adverse effects of fine "respirable" paticules $PM_{2.5}$ on asthma/wheezing.

Table 4. Association between air pollution and asthma and wheezing stratified by GSTP1 genotypes in single pollutant model.

	GSTP1 Ile-Ile		GSTP1 Ile-Val or Val-Val		Interaction p-value
	OR	**95% CI**	**OR**	**95% CI**	
Asthma	(n = 167/2164)		(n = 84/1138)		
NO_2 (8.79 ppb)	0.88	0.64–1.22	1.30	0.83–2.03	0.24
CO (105 ppb)	1.03	0.89–1.20	1.11	0.90–1.36	0.84
SO_2 (1.31 ppb)	0.88	0.77–1.01	1.06	0.89–1.26	0.08
$PM_{2.5}$ (16.84 μg/m3)	0.60	0.45–0.82	1.52	1.01–2.27	0.001
O_3 (8.77 ppb)	0.74	0.60–0.90	1.19	0.91–1.57	0.03
Wheezing	(n = 260/2174)		(n = 137/1141)		
NO_2 (8.79 ppb)	1.02	0.78–1.32	1.15	0.80–1.65	0.81
CO (105 ppb)	1.08	0.96–1.22	1.07	0.90–1.27	0.67
SO_2 (1.31 ppb)	0.91	0.81–1.02	1.01	0.87–1.16	0.26
$PM_{2.5}$ (16.84 μg/m3)	0.75	0.59–0.96	1.24	0.89–1.71	0.027
O_3 (8.77 ppb)	0.82	0.69–0.96	1.09	0.88–1.82	0.049

Two-stage hierarchical analysis adjusting for age, sex, parental education, yearly income, during of breast feeding, gestational age, maternal smoking during pregnancy, environmental tobacco smoke, cockroaches note monthly, carpet, pets, home dampness and mold, parental atopy.

Ozone is a strong oxidant and reacts with the epithelial lining fluid to generate free radicals [21]. O_3 exposure may depletes levles of protective antioxidants (ascorbate and glutathione peroxide) in bronchalveolar lavage fluid [22]. In vitro exposure bronchial epithelia cells to O_3 indicate increases the production of inflammatiory mediators (IL6, IL8, ICAM-1, granulocyte marcophage-colony stimulating factor (GM-CSF), RANTES and tumour necrosis factor α (TNFα). We speculated that GSTP1 105 Ile, leading to repair detoxification of reactive oxygen species, protected from increased the risk of oxidative stress or inflammation from O_3.

The possible explanation for our finding was that GST enzyme, largely expressed in human lung cells, act as detoxifying enzymes and serve as a marker of putative oxidative stress [23]. Furthermore, altered antioxidant defenses, lipid peroxidation, and anti-inflammation pathways were important in asthma/wheezing pathogenesis [1]. Overall, our study suggested that genetically predetermined deplete of antioxident enyzme function as a predisposing factor for susceptibililty to air pollutnats ($PM_{2.5}$ and O_3) and differences in detoxification might contribute to the risk of asthma/wheezing. The differences in direction of effects with GSTP1 genotypes may be due to chance, insufficient power, different populations or ethnic origins, variations of study design and different phenotypes studies. The GSTP1 gene polymorphisms

Table 5. Association between air pollution and asthma and persistent wheezing stratified by GSTP1 genotypes in two-pollutant models.

Asthma	model 1 (O_3+CO)	model 2 (O_3+NO_2)	model 3 (O_3+SO_2)	model 4 (PM_{10}+CO)	model 5 (PM_{10}+NO_2)
GSTP1: Ile-Ile					
O_3 (8.77 ppb)	0.72 (0.58–0.89)	0.72 (0.59–0.88)	0.76 (0.62–0.94)		
$PM_{2.5}$ (16.84 μg/m3)				0.60 (0.45–0.82)	0.60 (0.44–0.83)
GSTP1: Ile-Val or Val-Val					
O_3 (8.77 ppb)	1.30 (0.96–1.77)	1.24 (0.92–1.66)	1.18 (0.89–1.57)		
$PM_{2.5}$ (16.84 μg/m3)				1.52 (1.01–2.27)	1.47 (0.97–2.23)
Wheezing					
GSTP1: Ile-Ile					
O_3 (8.77 ppb)	0.83 (0.69–0.98)	0.81 (0.69–0.96)	0.84 (0.71–0.99)		
$PM_{2.5}$ (16.84 μg/m3)				0.75 (0.59–0.96)	0.74 (0.57–0.95)
GSTP1: Ile-Val or Val-Val					
O_3 (8.77 ppb)	1.14 (0.90–1.45)	1.10 (0.88–1.38)	1.11 (0.88–1.40)		
$PM_{2.5}$ (16.84 μg/m3)				1.23 (0.89–1.71)	1.21 (0.86–1.70)

Two-stage hierarchical analysis adjusting for age, sex, parental education, yearly income, during of breast feeding, gestational age, maternal smoking during pregnancy, environmental tobacco smoke, cockroaches note monthly, carpet, pets, home dampness and mold, parental atopy.

could be a suitable biomarker identifying genetically susceptible children exposed to air pollutants.

Summary

The present study provides additional evidence that children carrying Ile105 variant allele and exposed to ambient air pollutants, such as $PM_{2.5}$ and O_3 may be less likely to occurrence of asthma/wheezing. The observed gene-pollution interaction between GSTP1 gene polymorphisms and exposure to $PM_{2.5}$ and O_3 could help us to understand the aetiology of asthma or wheezing, so allowing earlier prediction and diagnosis of childhood asthma/wheezing, and provide an efficient means of prevention. These results support the roles of genes controlling the antioxidative system and inflammatory response in childhood asthma/wheezing. Diverse detoxification ability of air pollution depended on variations of the GSTP1 polymorphisms. Since air pollution is a complex mixture of compounds and other metabolic genes may be also involved, additional long term prospective study is warranted to explore the relevant role of other genes in determining genetic susceptibility to adverse respiratory outcomes.

Supporting Information

Table S1 Primer and MGB probe sequences for *GSTT1*, *GSTM1*, and *GSTP1* genes variants.

Table S2 Association between air pollution and asthma and wheezing stratified by GSTM1 genotypes in single pollutant model.

Table S3 Association between air pollution and asthma and wheezing stratified by GSTT1 genotypes in single pollutant model.

Table S4 Characteristics of included population, excluded population, and total population.

Author Contributions

Conceived and designed the experiments: BFH YLL. Performed the experiments: CHT KYT PCW. Analyzed the data: BFH. Contributed reagents/materials/analysis tools: LHY MWS YLL. Wrote the paper: BFH.

References

1. Yang IA, Fong KM, Zimmerman PV, Holgate ST, Holloway JW (2008) Genetic susceptibility to the respiratory effects of air pollution. Thorax 63: 555–63.
2. Braback L, Forsberg B (2009) Dose traffic exhaust contribute to the development of asthma and allergic sensitization in children: finding from recent cohort studies. Environ Health 8: 17.
3. Korytina GF, Yanbaeva DG, Babenkova LI, Etkina EI, Victorova TV (2005) Genetic polymorphisms in the cytochromes P-450 (1A1, 2E1), microsomal epoxide hydrolase and glutathione S-transferase M1, T1, and P1 genes, and their relationship with chronic bronchitis and relapsing pneumonia in children. J Mol Med 83: 700–10.
4. Melén E, Nyberg F, Lindgren CM, Berglind N, Zucchelli M, et al. (2008) Interactions between glutathione s-transferase p1, tumor necrosis factors and traffic-related air pollution for development of childhood allergic disease. Environ Health Perspect 116: 1077–84.
5. Islam T, Berhane K, McConnell R, Gauderman WJ, Avol E, et al. (2009) Glutathione s-transferase (GST) p1, GSTM1, exercise, ozone and asthma incidence in school children. Thorax 64: 197–202.
6. Schroer KT, Myers JMB, Ryan PH, Ryan PH, LeMasters GK, et al. (2009) Associations between multiple environmental exposures and glutathione s-transferase p1 on persistent wheezing in a birth cohort. J Pediatr 154: 401–408.
7. Lee YL, LinYC, Lee YC, Wang JY, Hsiue TR, et al. (2004) Glutathione s-transferase p1 gene polymorphisms and air pollution as interactive risk factors for childhood asthma. Clin Exp Allergy 34: 1707–13.
8. Dominici F (2002) Invited commentary: Air pollution and health-what can we learn from a hierarchical approach? Am J Epidemiol 155: 11–15.
9. Greenland S (2000) Principles of multilevel modelling. Int J Epidemiol 29: 158–67.
10. Hwang BF, Lee YL (2010) Air pollution and prevalence of bronchitic symptoms among children in Taiwan. Chest 138: 956–964.
11. 41st World Medical Assembly (1990) Declaration of Helsinki: Recommendations guiding physicians in biomedical research involving human subjects. Bulletin of the Pan American Health Organization 24: 606–09.
12. Tung KY, Tsai CH, Lee YL (2011) Microsomal Epoxide Hydroxylase Genotypes/Diplotypes, Traffic Air Pollution and Childhood Asthma. Chest 139: 839–848.
13. Witte JS, Greenland S, Kim LL, Arab L (2000) Multilevel modeling in epidemiology with GLIMMIX. Epidemiology 2000; 11:684–8.
14. Romieu I, Ramirez-Aguilar M, Sienra-Monge JJ, Moreno-Macías H, del Rio-Navarro BE, et al. (2006) GSTM1 and GSTP1 and respiratory health in asthmatic children exposed to ozone. Eur Respir J 28: 953–959.
15. International HapMap Consortium (2005) A haplotype map of the human genome. Nature 437: 1299–1320.
16. See SW, Wang YH, Balasubramanian R (2007) Contrasting reactive oxygen species and transition metal concentrations in combustion aerosols. Environ Res 103: 317–324.
17. Dellinger B, Pryor WA, Cueto R, Squadrito GL, Hegde V, et al. (2001) Role of free radicals in the toxicity of airborne fine particulate matter. Chem Res Toxicol 14: 1371–1377.
18. Donaldson K, Stone V, Borm PJ, Jimenez LA, Gilmour PS, et al. (2003) Oxidative stress and calcium signaling in the adverse effects of environmental particles (PM10). Free Radic Biol Med 34: 1369–82.
19. Nel AE, Diaz-Sanchez D, Li N (2001) The role of particulate pollutants in pulmonary inflammation and asthma: evidence for the involvement of organic chemicals and oxidative stress. Curr Opin Pulm Med 7: 20–26.
20. Ohtoshi T, Takizawa H, Okazaki H, Kawasaki S, Takeuchi N, et al. (1998) Diesel exhaust particles stimulate human airway epithelial cell to produce cytokines relevant to airway inflammation in vitro. J Allergy Clin Immunol 101 (6 Pt 1): 778–785.
21. Kelly FJ (2003) Oxidative stress: Its role in air pollution and adverse health effects. Occup Environ Med 60: 612–616.
22. Avissar NE, Reed CK, Cox C, Frampton MW, Finkelstein JN (2000) Ozone, but not nitrogen dioxide, exposure decrease glutathione peroxidases in epithelial lining fluid of human lung. Am J Respir Crit Care Med 162 (4 Pt 1): 1342–7.
23. Ercan H, Birben E, Dizdar EA, Keskin O, Karaaslan C, et al. (2006) Oxidative stress and genetic and epidemiologic determinants of oxidant injury in childhood asthma. J Allergy Clin Immunol 118: 1097–1104.

24

Biomass Partitioning and Its Relationship with the Environmental Factors at the Alpine Steppe in Northern Tibet

Jianbo Wu[1], Jiangtao Hong[1,2], Xiaodan Wang[1]*, Jian Sun[1,2], Xuyang Lu[1], Jihui Fan[1], Yanjiang Cai[1]

1 The Key Laboratory of Mountain Environment Evolution and Its Regulation, Institute of Mountain Hazard and Environment, CAS, Chengdu, China, **2** University of Chinese Academy of Sciences, Beijing, China

Abstract

Alpine steppe is considered to be the largest grassland type on the Tibetan Plateau. This grassland contributes to the global carbon cycle and is sensitive to climate changes. The allocation of biomass in an ecosystem affects plant growth and the overall functioning of the ecosystem. However, the mechanism by which plant biomass is allocated on the alpine steppe remains unclear. In this study, biomass allocation and its relationship to environmental factors on the alpine grassland were studied by a meta-analysis of 32 field sites across the alpine steppe of the northern Tibetan Plateau. We found that there is less above-ground biomass (M_A) and below-ground biomass (M_B) in the alpine steppe than there is in alpine meadows and temperate grasslands. By contrast, the root-to-shoot ratio (*R:S*) in the alpine steppe is higher than it is in alpine meadows and temperate grasslands. Although temperature maintained the biomass in the alpine steppe, precipitation was found to considerably influence M_A, M_B, and *R:S*, as shown by ordination space partitioning. After standardized major axis (SMA) analysis, we found that allocation of biomass on the alpine steppe is supported by the allometric biomass partitioning hypothesis rather than the isometric allocation hypothesis. Based on these results, we believe that M_A and M_B will decrease as a result of the increased aridity expected to occur in the future, which will reduce the landscape's capacity for carbon storage.

Editor: Fei-Hai Yu, Beijing Forestry University, China

Funding: This study was financially supported by the Strategic leading science and technology projects, CAS (XDB03030505), the National science and technology support project (2011BAC09B03), Program of the IMDE, CAS (SDS-135-1203-01), and the Science Foundation for Young Scientists of IMDE, CAS. The funders had no role in study design, data collection and analysis, decision to publish, or preparation of the manuscript.

Competing Interests: The authors have declared that no competing interests exist.

* E-mail: wxd@imde.ac.cn

Introduction

Biomass allocation was an important character for the process of characterization of plant physiological ecology [1], moreover, it also was the result of the plant long-term adapted to different environmental conditions [2].The Biomass allocation also reflect show photosynthates are allocated between above-ground and below-ground biomass [3]. Biomass allocation above-ground and below-ground affects plant growth as well as the overall function of the ecosystem and biogeochemical cycles [4,5]. Therefore, the mechanism by which plants respond to variations in the availability of resources in their environment is a major question in plant ecology [6]. Two important hypotheses regarding biomass allocation of plants have been proposed: (i) optimal partitioning and (ii) isometric allocation [2,7,8]. The optimal partitioning hypothesis suggests that plants respond to variations in the environment by partitioning biomass among various plant organs to maximize the plants' growth rate [9,10]. For example, plants in arid regions are rooted deeper than those in humid environments [11,12]. On the contrary, the isometric allocation hypothesis predicts the net primary productivity of the roots *vs* the net primary productivity of the shoots (BNPP:ANPP) isometrically without considering the differences in plant species or community types [13–15]. Thus far, biomass allocation has been widely examined: investigations have focused on individual organisms as well as whole ecosystems. However, no conclusion about biomass allocation has yet been presented.

Optimal partitioning theory might explain the effect of environmental factors on the allocation of plants' photosynthetic products, but this theory does not consider the size of the individual plants [8,16]. The allometric biomass partitioning theory, on the other hand, may resolve biomass allocation patterns in terms of plant size by using standardized major axis (SMA) regression [8,17]. However, this theory does not provide quantitative descriptions about how environmental factors affect biomass allocation. It also cannot explain the mechanism behind how photosynthates are allocated to different organs [18]. Furthermore, it is still hotly debated whether a uniform biomass allocation pattern is applicable to different ecosystems [19].

The alpine steppe is the largest grassland type in the Tibetan Plateau, which contributes significantly to the global carbon cycle [1]. In the alpine grassland ecosystem, few soil nutrients, aridity, and low temperatures limit plant growth [20,21]. According to the optimal partitioning hypothesis, environmental factors likely affect how plant biomass is allocated. At the individual plant level, fewer soil nutrients (particularly nitrogen and water) results in an increase in root biomass. On the contrary, root biomass decreases and shoot biomass increases as soil nutrients increase. This

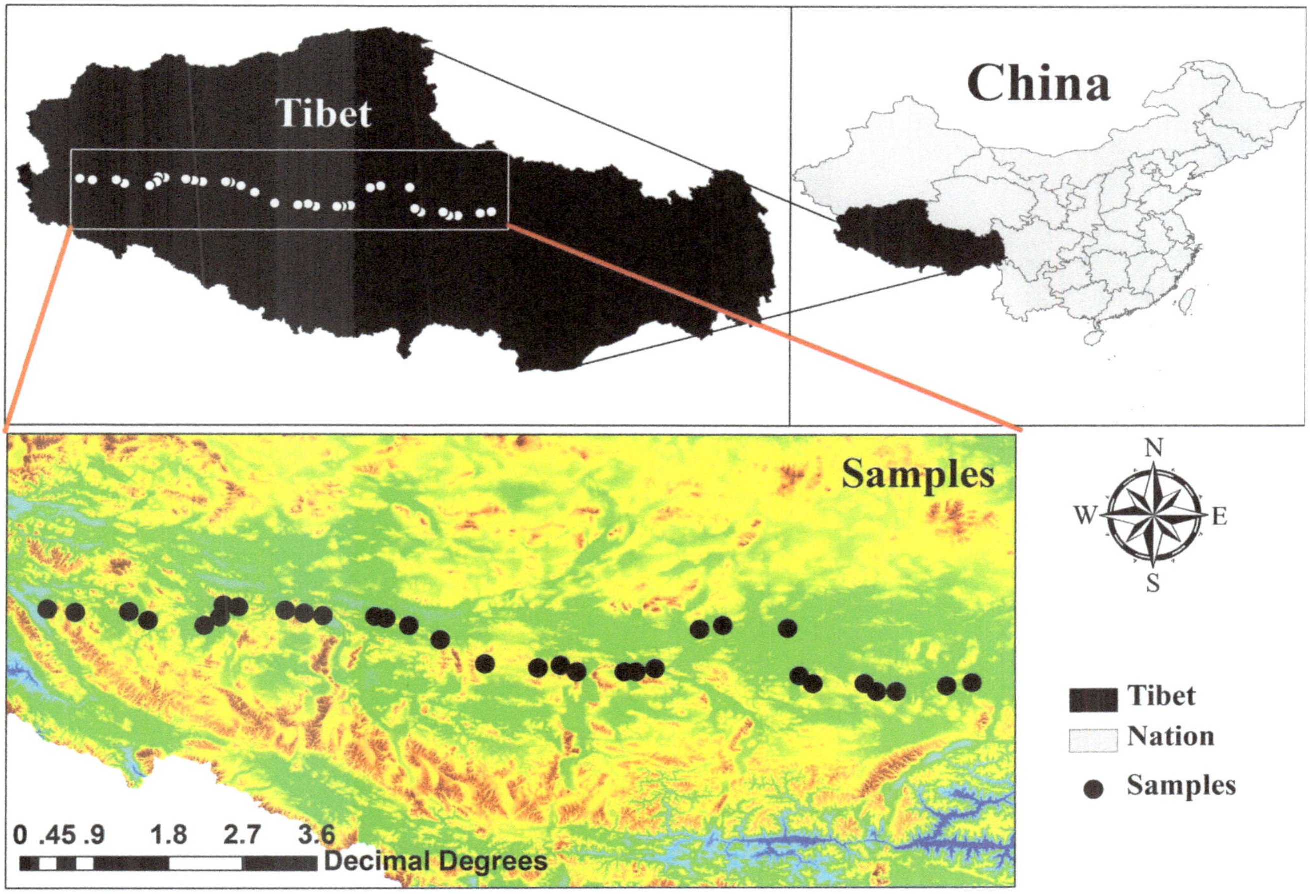

Figure 1. Spatial distribution of the sampling sites in *S. Purpurea* alpine steppe in northern Tibet.

partition model is appropriate for different types of vegetation and life forms of plants [22–26]. Studies have shown that plants allocate more biomass to their roots when water and nutrients in grassland ecosystems are limited [27,28]. Moreover, studies have also suggested that plants allocate photosynthates to root in low-temperature environments, which may increase the rate of nutrient absorption and help the plants adapt to environmental conditions [29–31]. However, Yang et al. (2009a) reported that on the Tibetan alpine grasslands, the relationship between roots and shoots supports the isometric allocation hypothesis [32]. They also found that this isometric relationship is independent of soil nitrogen and moisture [32]. These results indicate that the mechanism of biomass allocation in the alpine steppe is still misunderstood and unverified in alpine and arid environments. Therefore, this subject requires further investigation. In the present study, we investigated (i) the mechanism behind allocating root and shoot biomass in the Tibetan alpine grassland and (ii) the main factors that affect biomass allocation in the alpine steppe of northern Tibet.

Table 1. Site description of *S. purpurea* alpine steppe.

Site	County	Dominant species	Mean annual precipitation (MAP,mm)	Mean annual temperature (MAT,°C)
S1	Nakchu	*S. purpurea Kobre siahumilis*	428.1	-1.5
S2–S7	Baingoin	*S. purpurea Carex moorcroftii*	321.7	-0.8
S8	Xainza	*S. purpurea C.moorcroftii*	304.5	-0.4
S9–S17	Nyima	*S. purpurea C.m oorcroftii*	200	-0.4
S18–S24	Gêrzê	*S. purpurea*	170.1	0.10
S25–S30	Gêgyai	*S. purpurea*	120	0.45
S31–S32	Gar	*S. purpurea*	72.1	0.7

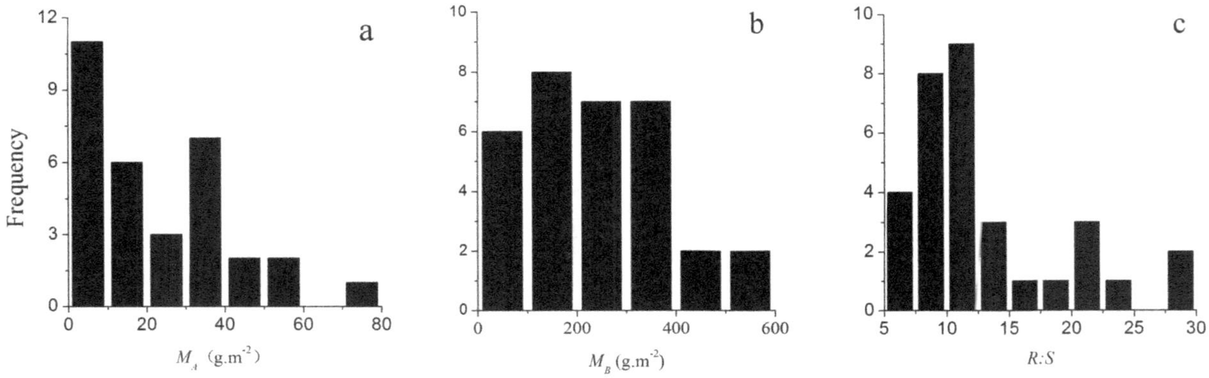

Figure 2. Frequency distributions of (a) above-ground biomass (M_A), (b) below-ground biomass (M_B), and (c) root-to-shoot ratio (*R:S*) in *S. purpurea* alpine steppe.

Materials and Methods

Collecting Biomass and Soil Samples

In August 2012, 32 sites were selected on *Stipa purpurea* alpine steppe from Nagqu County to Gar County in northern Tibet. Sampling sites were established at intervals of 30 km (Fig. 1, Table 1). In each site, no specific permits were required for collecting samples and the field studies did not involve endangered or protected species. We selected flat sites with well-protected vegetation. We harvested the aboveground biomass (M_A) and the belowground biomass (M_B) from three blocks of 0.5 m×0.5 m in each site. We collected M_B from soil depths of 0 cm to 15 cm, where most of belowground biomass is located [33,34].The root samples obtained from the blocks were immediately placed in a cloth bag and then soaked in water to remove the residual soil using a 0.5 mm sieve. Biomass was oven-dried at 65°C until a constant weight was reached, and then it was weighed to the nearest 0.01 g.

Soil samples were collected from two different depths (0–15 cm and 15–30 cm), air-dried, and sieved (2 mm mesh). The fine roots were extracted by hand picking for physical and chemical analyses. The total nitrogen content (TN; TN1:0–15 cm, TN2:15–30 cm) of the soil was determined using the micro-Kjeldahl digestion method. The available nitrogen content (AN; AN1:0–15 cm, AN2:15–30 cm) of the soil was determined using the alkaline hydrolysis diffusion method. All of the element concentrations were expressed as mg·g^{-1} on a dry weight basis.

Data Analysis

M_A in grasslands can be considered as annual aboveground net primary productivity (ANPP). Blow-ground net primary productivity (BNPP) was calculated using Gill's method:

$$\text{BNPP} = M_B \times \left(\frac{\text{live}M_B}{M_B}\right) \times \text{turnover} \qquad (1)$$

where (live M_B/M_B) = 0.6 and turnover = 0.0009(g·m^{-2})×M_A +0.25 [35,36]. In the present study, the value for (live M_B/M_B) was 0.79, which was measured by Zhou (2001) in the Qinghai region [37]. The relationship between log ANPP and log BNPP was constructed using Model II regression [14,15]. The slope (α) and y-intercept (log b) of the allocation function were determined by standardized major axis (SMA) tests [38]. The heterogeneity between slopes was determined by performing a permutation test and was rejected if P>0.05 [15]. We analyzed the correlations between environmental factors and the measured M_A, M_B, and root-to-shoot ratios (*R:S*) using the Pearson correlation. We also examined relationships between M_A, M_B, *R:S*, and environmental factors using regression and ordination space partitioning to find the main environmental factors that affected M_A, M_B, and *R:S*. Analyses were performed using SPSS software version 16.0 (IBM; Armonk, NY).

Results

Variations in the Chemical Properties of the Soil as well as MA, MB, and R:S

Small variations in the chemical properties of the soil along the sampled transect were found. There also was not significance in available nitrogen and total nitrogen between the two soil layers (Table 2). We found large variations in M_A, M_B, and *R:S* along the sampled transects (Fig. 2). M_A ranged from 2.32 g·m^{-2} to 73.6 g·m^{-2}, while M_B ranged from 22.40 g·m^{-2} to 587.32 g·m^{-2}. *R:S* ranged from 6.19 to 29.15 (Table 3). The median values of M_A, M_B, and *R:S* in *S. purpurea* alpine steppe were 17.16 g·m^{-2}, 233 g·m^{-2}, and 11.83, respectively (Table 3).

Biomass Allocation for *S.purpurea* Alpine Steppe

The slope (α) of the plotted relationship between log ANPP and log BNPP of *S. purpurea* alpine grasslands was 0.87 with 95% confidence intervals of 0.75 and 1.01 (Fig. 3). The slope (α) was significantly different from the slope obtained from SMA analysis when the isometric hypothesis was used.

Table 2. Chemical properties of soils in *S. purpurea* alpine steppe.

	Min	Max	Mean	Std. Error	Std. Deviation
AN1 mg·g^{-1}	0.013	0.110	0.057a	0.004	0.025
AN2 mg·g^{-1}	0.008	0.095	0.051a	0.004	0.023
TN1 mg·g^{-1}	0.386	1.630	1.008a	0.061	0.342
TN2 mg·g^{-1}	0.292	1.921	0.980a	0.064	0.362

Table 3. Descriptive statistics of above-ground biomass (M_A), below-ground biomass (M_B), and root-to-shoot (*R:S*) ratio in *S. purpurea* alpine steppe.

	M_A (g·m^{-2})			M_B (g·m^{-2})			*R:S* ratio		
	Min	Max	Median	Min	Max	Median	Min	Max	Median
Present study	2.32	73.6	17.16	22.4	587.32	233	6.19	29.15	11.83
Yang et al. (2009)	9.8	267.4	42.8	44.6	1934.8	206	0.8	13	5.2

Effects of Soil Nitrogen and Environmental Factors on Biomass and R:S

Using the Pearson correlation analysis, we found that M_A and M_B exhibited a significantly positive correlation with available nitrogen in the soil. However, M_A and M_B did not exhibit a significant correlation with total nitrogen (Table 4). The *R:S* ratio also did not exhibit a significant correlation with soil nitrogen (total or available). M_A, M_B, and *R:S* did correlate with the MAP of the sampling sites, while these correlations differed from the ones found with MAT (Table 3). In this study, we found that the regression analysis showed the same results as the Pearson correlation analysis (Fig. 4). Using the ordination space partitioning method, we found that MAP was the main factor that affected M_A, M_B, and *R:S* (Fig. 5).

Discussion

M_A, M_B, and R:S in the Alpine Steppe

In the present study, amounts of M_A and M_B in the alpine steppe (mean = 23.20 g·m^{-2}) were found to be lower than those in the alpine meadows [32] and in the temperate grasslands of China [39]. By contrast, *R:S* in the alpine steppe was found to be higher than it is in China's alpine meadows [32] and temperate grasslands [39] as well as in temperate grasslands of other regions [1]. These results show that precipitation and temperature affect plant growth and biomass allocation [1,40]. Slower root turnover in colder environments might also results in higher R:S ratios [41–43]. M_A, M_B, and *R:S* values found in the present study are not consistent with results reported by Yang et al. (2009a), who performed a field investigation from 2001 to 2004 [32]. *R:S* values have the potential to vary greatly as a result of climate change and anthropogenic activities [44–48].

Mechanism of Biomass Allocation in the Alpine Steppe

Based on the results of our SMA analysis, we found that biomass allocation on the alpine steppe does not fit the isometric hypothesis. By contrast, Yang et al. (2009a) previously reported that biomass allocation on the alpine steppe is supported by the isometric allocation hypothesis [32]. In the harsh alpine ecosystem, scarce precipitation and low temperatures allow plants to allocate more biomass to the roots, which helps plants survive [29–31]. Moreover, roots have also been found to store carbohydrates in alpine grasslands [49,50]. Therefore, biomass allocation in the alpine steppe may reflect the allometric biomass partitioning hypothesis rather than the isometric allocation hypothesis.

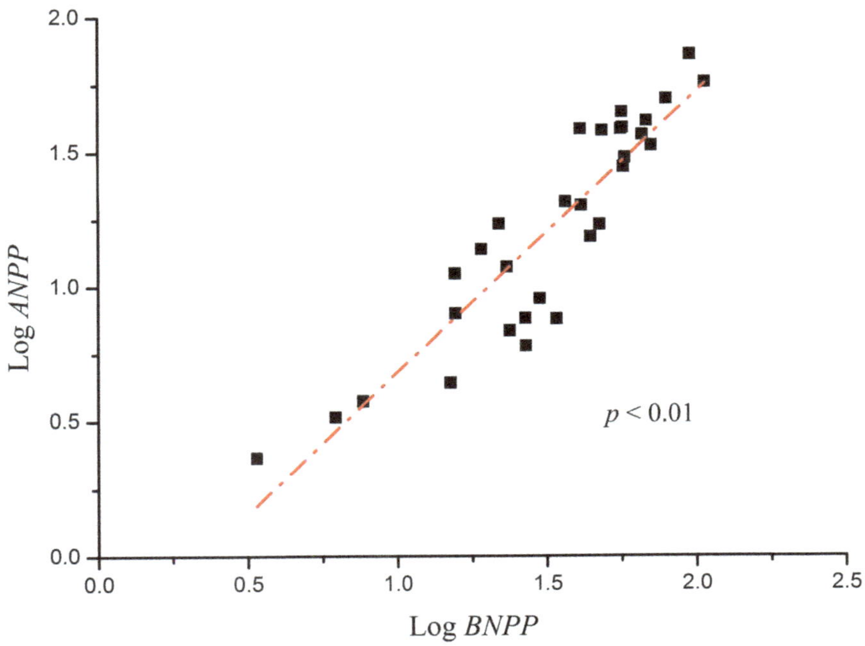

Figure 3. Relationships between above-ground net primary production (ANPP) and below-ground net primary production (BNPP) in alpine steppe by SMA analysis.

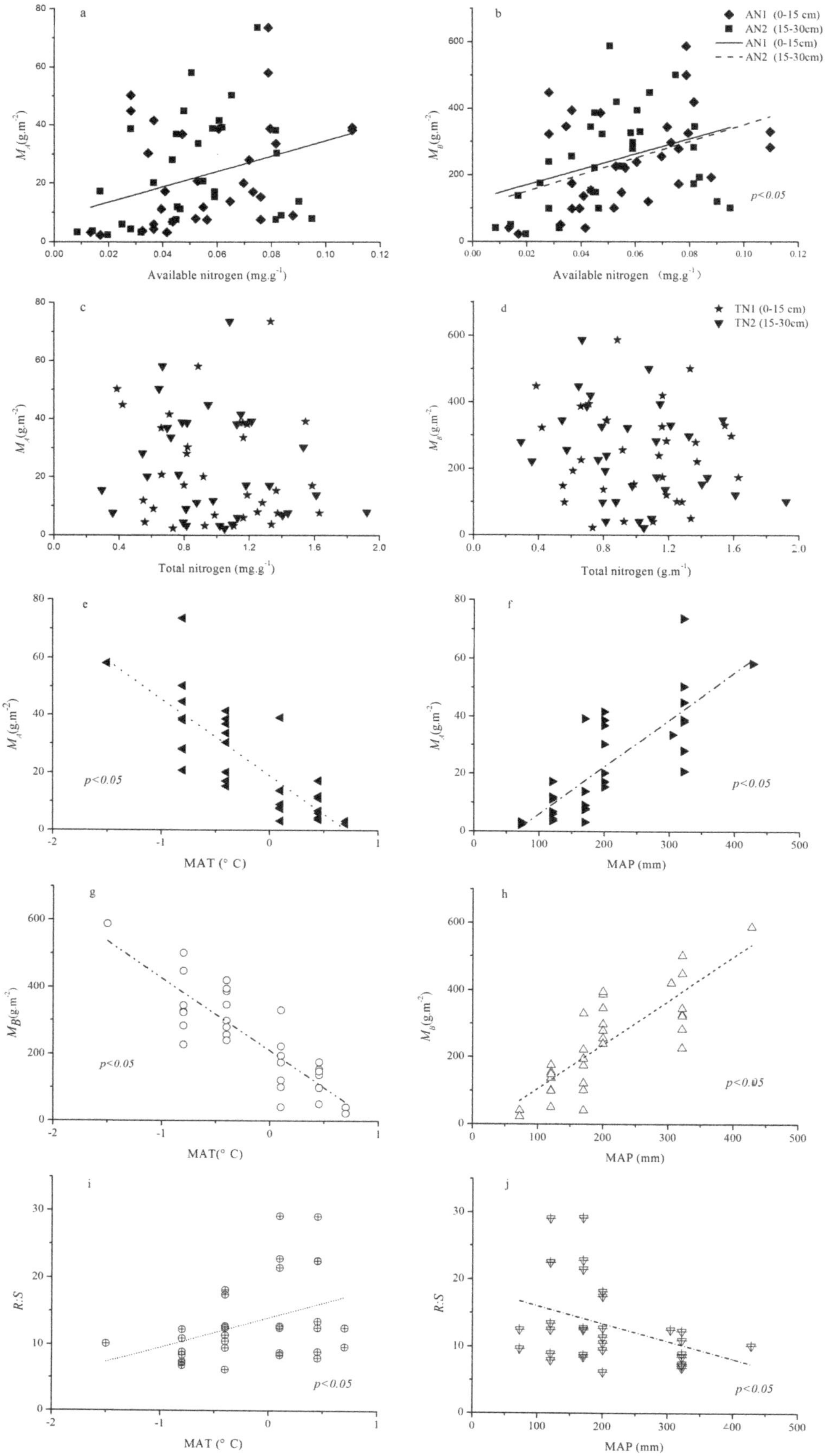

a
b
AN1 (0-15 cm)
AN2 (15-30cm)
AN1 (0-15cm)
AN2 (15-30cm)
p<0.05
Available nitrogen (mg.g-1)
c
d
TN1 (0-15 cm)
TN2 (15-30cm)
Total nitrogen (mg.g-1)
Total nitrogen (g.m-1)
e
f
p<0.05
MAT (° C)
MAP (mm)
g
h
i
j
R:S

Figure 4. Relationships between biomass allocation (M_A, M_B, and R:S) and environmental factors in alpine steppe. Regressions are shown: (a) M_A versus available nitrogen, (b) M_B versus available nitrogen,(c) M_A versus total nitrogen, (d) M_B versus total nitrogen,(e) M_A versus MAT,(f) M_A versus MAP,(g) M_B versus MAT, (h) M_B versus MAP, (i) *R:S* ratio versus MAT, and (j) *R:S* ratio versus MAP.

Relationships between Environmental Factors and MA, MB, and R:S

Precipitation and temperature are considered to be the limiting factors for the growth and distribution of vegetation over the long term [51,52]. In the present study, M_A, M_B, and *R:S* were mainly affected by the environmental factor of precipitation (MAP), as revealed by ordination space partitioning analysis. These results are consistent with those of other reports about the alpine steppe [46,53–56]. The low temperature in the growing season did not limit the growth of alpine plants because these plants shave evolved to survive in the cold alpine climate [57]. The amounts of aboveground and belowground biomass are higher in sites with higher humidity, but the MAT is also relatively low on the alpine steppe. Precipitation is an essential factor that controls the functions of ecosystems in terrestrial biomes, particularly in arid and semiarid ecosystems [58]. Therefore, precipitation is the main factor that influences amounts of biomass in the alpine steppe.

Moreover, in the present study, we found that amounts of M_A and M_B on the alpine steppe were affected by the available

Table 4. Pearson's correlation between M_A, M_B, and *R:S* with the environmental factors.

	AN1(0 to 15 cm)	AN2(15 to 30 cm)	TN1(0 to 15 cm)	TN2(15 to 30 cm)	MAT(°C)	MAP(mm)
M_A	0.351*	0.317	−0.126	−0.165	−0.809**	0.791**
M_B	0.429*	0.372*	−0.088	−0.285	−0.853**	0.817**
R:S	−0.047	−0.083	0.207	−0.082	0.392*	−0.378*

**Correlation is significant at the 0.01 level (two-tailed).
*Correlation is significant at the 0.05 level.

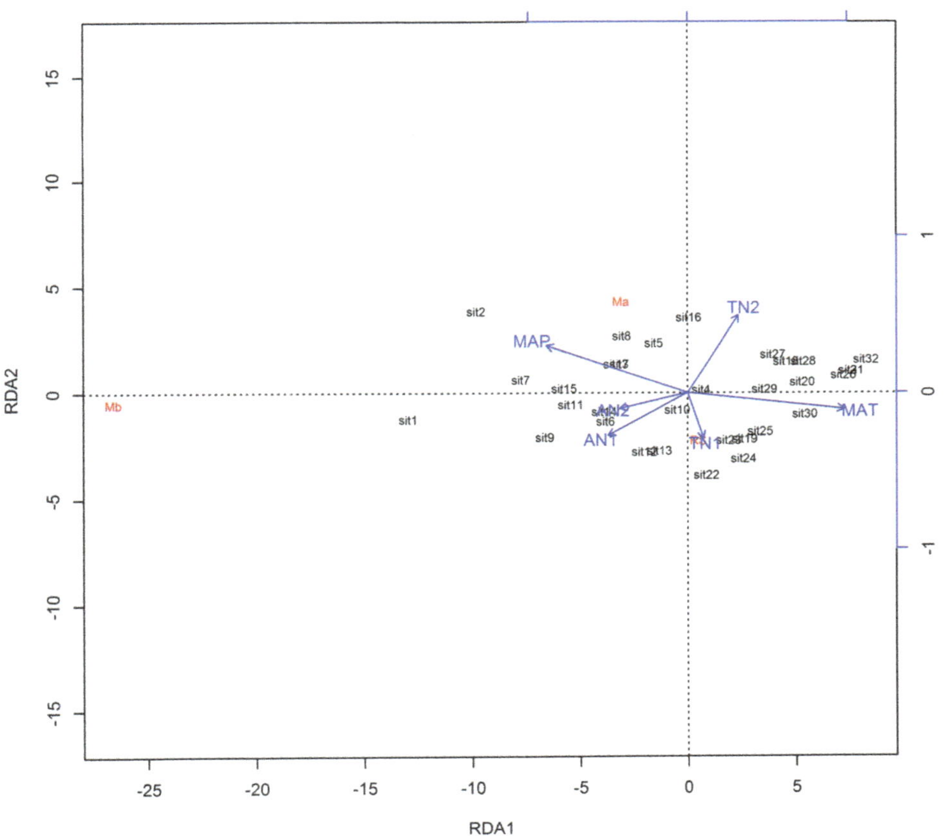

Figure 5. Analysis of the relationship of above-ground biomass (M_A), below-ground biomass (M_B), and root-to-shoot ratio (*R:S*) with the environmental factors by ordination space partitioning method.

nitrogen content in the soil but not by the total nitrogen content of the soil. These results are inconsistent with those from previous studies, which have showed that M_A and M_B are positively related to total nitrogen content [32,59]. Because available nitrogen can be used to approximate the relative supply of nutrients, nitrogen may be another factor that controls ecosystem processes in regions with abundant water resources [60].

Conclusion

As the climate changes, the degree of aridity has been consistently increasing in northern Tibet [61]. Changes in biomass allocation on the alpine steppe are likely to affect the carbon cycle and the general functioning of the alpine ecosystem. In the present study, we found that the *R:S* ratio in the alpine steppe was higher than that of other grassland systems. The amounts of aboveground and belowground biomass as well as the *R:S* ratio were primarily affected by precipitation. The observed biomass allocation was found to follow the allometric biomass partitioning theory rather than the isometric allocation hypothesis. These results suggest that the landscape's capacity to store carbon will potentially decrease as the degree of aridity in northern Tibet increases.

Acknowledgments

This study was financially supported by the Strategic leading science and technology projects, CAS (XDB03030505), the National science and technology support project (2011BAC09B03), Program of the IMDE, CAS (SDS-135-1203-01), and the Science Foundation for Young Scientists of IMDE, CAS.

Author Contributions

Conceived and designed the experiments: JBW XDW. Performed the experiments: JBW JTH XYL JHF YJC. Analyzed the data: JBW JS. Contributed reagents/materials/analysis tools: JBW JS. Wrote the paper: JBW XDW. Figure: JS.

References

1. Mokany K, Raison RJ, Prokushkin AS (2006) Critical analysis of root: shoot ratios in terrestrial biomes. Global Change Biol 11: 1–13.
2. Bazzaz FA (1997) Allocation of resources in plants: state of the sciences and critical questions. In: Bazzaz FA, Grace J, editors. Plant Resource Allocation. San Diego: Academic Press. 1–37.
3. Shipley B, Meziane D(2002)The balanced-growth hypothesis and the allometry of leaf and root biomass allocation. Funct Ecol 16: 326–331.
4. Kuzyakov Y, Domanski G (2000) Carbon input by plants into the soil. J Plant Nutr Soil Sci 163: 421–431.
5. Litton CM, Raich JW, Ryan MG (2007) Carbon allocation in forest ecosystems. Global Change Biol 13: 2089–2109.
6. Lacointe A (2000) Carbon allocation among tree organs: a review of basic processes and representation in functional-structural tree models. Ann Forest Sci 57: 521–533.
7. Müller I, Schmid B, Weiner J (2000) The effect of nutrient availability on biomass allocation patterns in 27 species of herbaceous plants. Perspect Plant Ecol 3: 115–127.
8. McCarthy MC, Enquist BJ (2007) Consistency between an allometric approach and optimal partitioning theory in global patterns of plant biomass allocation. Funct Ecol 21: 713–720.
9. Bloom AJ, Chapin FS, Mooney HA (1985) Resource limitation in plants an economic analogy. Ann Rev Ecol Syst 16: 363–392.
10. Chapin FS III, Bloom AJ, Field B, Waring RH (1987) Plant responses to multiple environmental factors. BioScience 37: 49–57.
11. Scheck HJ, Jackson RB (2005) Mapping the global distribution of deep roots in relation to climate and soil characteristics. Geoderma 126(1–2): 129–140.
12. Mony C, Koschnick TJ, Haller WT, Muller S (2007) Competition between two invasive Hydrocharitaceae (*Hydrilla verticillata* (L. f.) (Royle) and *Egeria densa* (Planch)) as influenced by sediment fertility and season. Aquat Bot 86: 236–242.
13. Enquist BJ, Niklas KJ (2002) Global allocation rules for patterns of biomass partitioning in seed plants. Science 295: 1517–1520.
14. Niklas KJ (2005) Modelling below- and aboveground biomass for non-woody and woody plants. Ann Bot 95: 315–321.
15. Cheng DL, Niklas KJ (2007) Above- and belowground biomass relationships across 1534 forested communities. Ann Bot 99: 95–102.
16. Marcelis LFM, Heuvelink E (2007) Concepts of modelling carbon allocation among plant organs. In: Vos J, Marcelis LFM, Visser PHBd, Struik PC, Evers JB, editors. Functional structural plant modelling in crop production. Dordrecht: Springer. 103–111.
17. Solow AR (2005) Power laws without complexity. Ecol Lett 8: 361–363.
18. Génard M, Dauzat J, Franck N, Lescourret F, Moitrier N, et al.(2008) Carbon allocation in fruit trees: from theory to modelling. Trees 22: 269–282.
19. Han WX, Fang JY(2003) Allometry and its application in ecological scaling. Acta Sci Natur Univ Pekin 39: 583–593. (in Chinese with English abstract).
20. Zhou XM (2001) Kobresia Meadow in China. Beijing: Science Press. 136–139 p.
21. Gugerli F, Bauert MR (2001) Growth and reproduction of *Polygonum viviparum* show weak responses to experimentally increased temperature at a Swiss alpine site. Bot Helv 111: 169–180.
22. Warembourg FR, Estelrich HD (2001) Plant phenology and soil fertility effects on below-ground carbon allocation for an annual (*Bromus madritensis*) and a perennial (*Bromus erectus*) grass species. Soil Biol Biochem 33(10): 1291–1303.
23. Cronin G, Lodge DM. (2003) Effects of light and nutrient availability on the growth, allocation, carbon/nitrogen balance, phenolic chemistry, and resistance to herbivory of two freshwater macrophytes. Oecologia 137: 32–41.
24. Glynn C, Herms DA, Egawa M, Hansen R, Mattson WJ (2003) Effects of nutrient availability on biomass allocation as well as constitutive and rapid induced herbivore resistance in poplar. Oikos 101: 385–397.
25. Vanninen P, Makela A (2005) Carbon budget for Scots pine trees: effects of size, competition and site fertility on growth allocation and production. Tree Physiol 25: 17–30.
26. Grechi I, Vivin P, Hilbert G, Milin S, Robert T, et al. (2007) Effect of light and nitrogen supply on internal C:N balance and control of root-to-shoot biomass allocation in grapevine. Environ Exp Bot 59: 139–149.
27. Dukes JS, Chiariello NR, Cleland EE, Moore LA, Shaw MR, et al. (2005) Responses of grassland production to single and multiple global environmental changes. PLoS Biol 3: 1829–1837.
28. Gao YZ, Chen Q, Lin S, Giese M, Brueck H (2011) Resource manipulation effects on net primary production, biomass allocation and rain-use efficiency of two semiarid grassland sites in Inner Mongolia, China. Oecologia 65: 855–864.
29. Körner CH, Renhardt U (1987) Dry matter partitioning and root length/leaf area ratios in herbaceous perennial plants with diverse altitudinal distribution. Oecologia, 74: 411–418.
30. Wang CT, Cao GM, Wang QL, Jing ZC, Ding LM, et al. (2008) Changes in plant biomass and species composition of alpine Kobresia meadows along altitudinal gradient on the Qinghai-Tibetan Plateau. Sci China Life Sci 51: 86–94.
31. Fan JW, Wang K, Harris W, Zhong HP, Hu ZM, et al. (2009) Allocation of vegetation biomass across a climate-related gradient in the grassland of Inner Mongolia. J Arid Environ 73: 521–528.
32. Yang YH, Fang JY, Ji C, W Han (2009a) Above-and belowground biomass allocation in Tibetan grasslands. J Veg Sci 20: 177–184.
33. Yan Y, Zhang JG, Zhang JH, Fan JR, Li HX (2005) The belowground biomass in alpine grassland in Nakchu Prefecture of Tibet. Acta Ecol Sinica 11: 2818–2823. (in Chinese with English abstract).
34. Li X, Zhang X, Wu J, Shen Z, Zhang Y, et al. (2011) Root biomass distribution in alpine ecosystems of the northern Tibetan Plateau. Environ Earth Sci 64(7): 1911–1919.
35. Gill R A, Kelly RH, Parton WJ, Day KA, Jackson RB, et al. (2002) Using simple environmental variables to estimate below2ground productivity in grasslands. Global Ecol Biogeogr 11: 79–86.
36. Bradford JB, Lauenroth WK, Burke I C (2005) The impact of cropping on primary production in the US Great Plains. Ecol 86: 1863–1872.
37. Zhou XM (2001) Chinese Kobresia Meadow. Beijing: Science Press. 158 p.
38. Falster DS, Warton DI, Wright IJ (2003) (S)MATR: standardized major axis tests and routines. Version 1.0. http://www.bio. mq.edu.au/ecology/SMATR.
39. Ma W, Yang Y, He J, Zeng H, Fang J (2008) Above-and belowground biomass in relation to environmental factors in temperate grasslands, Inner Mongolia. Sci China Life Sci 51(3): 263–270.
40. Hui DF, Jackson RB (2005) Geographical and inter-annual variability in biomass partitioning in grassland ecosystems: a synthesis of field data. New Phytol 169: 85–93.
41. Gill RA, Jackson RB (2001) Global patterns of root turnover for terrestrial ecosystems. New Phytol 147: 13–31.
42. Gao YZ, Giese M, Lin S, Sattelmacher B, Zhao Y, et al. (2008) Belowground net primary productivity and biomass allocation of a grassland in Inner Mongolia is affected by grazing intensity. Plant Soil 307: 41–50.
43. Giese M, Gao YG, Zhao Y, Pan QM, Lin S, et al. (2009) Effects of grazing and rainfall variability on root and shoot decomposition in a semi-arid grassland. Appl Soil Ecol 41: 8–18.

44. Wang GX, Hu HC, Wang YB, Chen L (2007) Response of alpine cold ecosystem biomass to climate changes in permafrost regions of the Tibetan Plateau. J Glaciol Geocryol 29: 671–679. (in Chinese with English abstract).
45. Yu H, Luedeling E, Xu J (2010) Winter and spring warming result in delayed spring phenology on the Tibetan Plateau. Proc Natl Acad Sci U S A 107: 22151–22156.
46. Shi FS, Chen H, Wu Y, Wu N (2010) Effects of livestock exclusion on vegetation and soil properties under two topographic habitats in an alpine meadow on the Eastern Qinghai-Tibetan Plateau. Polish J Ecol 58: 125–133.
47. Li W, Huang HZ, Zhang ZN, Wu GL (2011) Effects of grazing on the soil properties and C and N storage in relation to biomass allocation in an alpine meadow. J Soil Sci Plant Nutr 11: 27–39.
48. Wu JZ, Zhang XZ, Shen ZX, Shi PL, Yu CQ, et al. (2012) Species richness and diversity of alpine grasslands on the Northern Tibetan Plateau: effects of grazing exclusion and growing season precipitation. J Resour Ecol 3: 236–242.
49. Wang JS, Zhang XZ, Zhao YP, Shen ZX, Shi PL, et al. (2008) Spatio- temporal pattern of climate changes in Northern Tibet's Qiangtang Plateau. Resour Sci 12: 1852–1859. (in Chinese with English abstract).
50. Wang L, Niu KC, Yang YH, Zhou P (2010) Patterns of above- and belowground biomass allocation in China's grasslands: evidence from individual-level observations. Sci China Life Sci 53: 851–857.
51. Woodward FI, Williams BG (1987) Climate and plant distribution at global and local scales. Plant Ecol 69: 189–197.
52. Stephenson NL (1990) Climatic control of vegetation distribution: the role of the water balance. Am Nat 5: 649–670.
53. O'Connor TG, Haines LM, Snyman HA (2001) Influence of precipitation and species composition on phytomass of a semiarid African grassland. J Ecol 89: 850–860.
54. Huxman TE, Smith MD, Fay PA, Knapp AK, Shaw MR, et al. (2004) Convergence across biomes to a common rain-use efficiency. Nature 429: 651–654.
55. Yang YH, Fang JY, Pan YD, Ji CJ (2009b) Above-ground biomass in Tibetan grasslands. J Arid Environ 73(1): 91–95.
56. Zhang BC, Cao JJ, Bai YF, Zhou XH, Ning ZG, et al. (2013) Effects of rainfall amount and frequency on vegetation growth in a Tibetan alpine meadow. Clim Chang 118: 197–212.
57. Körner C (2003) Alpine Plant Life: Functional Plant Ecology of High Mountain Ecosystems. Berlin:Springer. 101–111 p.
58. Reynolds JF, Stafford SDM (2002) Do humans cause deserts? In Reynolds JF, Stafford SDM, editors. Berlin: Dahlem University Press. 1–21.
59. Sun J, Cheng GW, Li WP (2013) Meta-analysis of relationships between environmental factors and aboveground biomass in the alpine grassland on the Tibetan Plateau. Biogeosciences 10: 1707–1715.
60. Burke IC, Lauenroth WK, Parton WJ (1997) Regional and temporal variation in net primary production and nitrogen mineralization in grasslands. Ecol 78(5): 1330–1340.
61. Liu X, Chen B (2000) Climatic warming in the Tibetan Plateau during recent decades. Int J Climatol 20: 1729–1742.

Permissions

All chapters in this book were first published in PLOS ONE, by The Public Library of Science; hereby published with permission under the Creative Commons Attribution License or equivalent. Every chapter published in this book has been scrutinized by our experts. Their significance has been extensively debated. The topics covered herein carry significant findings which will fuel the growth of the discipline. They may even be implemented as practical applications or may be referred to as a beginning point for another development.

The contributors of this book come from diverse backgrounds, making this book a truly international effort. This book will bring forth new frontiers with its revolutionizing research information and detailed analysis of the nascent developments around the world.

We would like to thank all the contributing authors for lending their expertise to make the book truly unique. They have played a crucial role in the development of this book. Without their invaluable contributions this book wouldn't have been possible. They have made vital efforts to compile up to date information on the varied aspects of this subject to make this book a valuable addition to the collection of many professionals and students.

This book was conceptualized with the vision of imparting up-to-date information and advanced data in this field. To ensure the same, a matchless editorial board was set up. Every individual on the board went through rigorous rounds of assessment to prove their worth. After which they invested a large part of their time researching and compiling the most relevant data for our readers.

The editorial board has been involved in producing this book since its inception. They have spent rigorous hours researching and exploring the diverse topics which have resulted in the successful publishing of this book. They have passed on their knowledge of decades through this book. To expedite this challenging task, the publisher supported the team at every step. A small team of assistant editors was also appointed to further simplify the editing procedure and attain best results for the readers.

Apart from the editorial board, the designing team has also invested a significant amount of their time in understanding the subject and creating the most relevant covers. They scrutinized every image to scout for the most suitable representation of the subject and create an appropriate cover for the book.

The publishing team has been an ardent support to the editorial, designing and production team. Their endless efforts to recruit the best for this project, has resulted in the accomplishment of this book. They are a veteran in the field of academics and their pool of knowledge is as vast as their experience in printing. Their expertise and guidance has proved useful at every step. Their uncompromising quality standards have made this book an exceptional effort. Their encouragement from time to time has been an inspiration for everyone.

The publisher and the editorial board hope that this book will prove to be a valuable piece of knowledge for researchers, students, practitioners and scholars across the globe.

List of Contributors

Junjun Zhi, Shengpan Lin, Cao Zhang and Qiankun Liu
College of Environmental and Resource Sciences, Zhejiang University, Hangzhou, China

Changwei Jing and Jiaping Wu
Ocean College, Zhejiang University, Hangzhou, China

Stephen D. DeGloria
Department of Crop and Soil Sciences, Cornell University, Ithaca, New York, United States of America

Jens Tank, Karsten Heusser and Jens Jordan
Institute of Clinical Pharmacology, Hannover Medical School, Hannover, Germany

Heike Biller, Olaf Holz, Wolfgang Koch, Norbert Krug and Jens M. Hohlfeld
Fraunhofer Institute for Toxicology and Experimental Medicine (ITEM), Hannover, Germany

André Diedrich
Division of Clinical Pharmacology, Department of Medicine, Autonomic Dysfunction Center, Vanderbilt University School of Medicine, Nashville, Tennessee, United States of America
Department of Biomedical Engineering, Vanderbilt University School of Engineering, Nashville, Tennessee, United States of America

Theodor Framke, Armin Koch and Anika Grosshennig
Institute for Biometry, Hannover Medical School, Hannover, Germany

Yang Xiang, Xiao-Qun Qin, Hui-Jun Liu, Yu-Rong Tan, Chi Liu and Cai-Xia Liu
Xiangya School of Medicine, Central South University, Changsha, Hunan, China

Leticia Britos, Harley McAdams and Lucy Shapiro
Department of Developmental Biology, Stanford University School of Medicine, Stanford, California, United States of America

Eduardo Abeliuk
Department of Developmental Biology, Stanford University School of Medicine, Stanford, California, United States of America
Department of Electrical Engineering, Stanford University, Stanford, California, United States of America

Thomas Taverner and Mary Lipton
Environmental Molecular Sciences Laboratory, Pacific Northwest National Laboratory, Richland, Washington, United States of America

Nie Xiaojun
Key Laboratory of Mountain Surface Processes and Ecological Regulation, Chinese Academy of Sciences, Institute of Mountain Hazards and Environment, Chinese Academy of Sciences and Ministry of Water Conservancy, Chengdu, China
School of Surveying and Land Information Engineering, Henan Polytechnic University, Jiaozuo, China

Zhang Jianhui and Su Zhengan
Key Laboratory of Mountain Surface Processes and Ecological Regulation, Chinese Academy of Sciences, Institute of Mountain Hazards and Environment, Chinese Academy of Sciences and Ministry of Water Conservancy, Chengdu, China

Jiri Jablonsky and Olaf Wolkenhauer
Department of Systems Biology and Bioinformatics, University of Rostock, Rostock, Germany

Martin Hagemann and Doreen Schwarz
Department of Plant Physiology, University of Rostock, Rostock, Germany

Nour El Helou
IRMES (bioMedical Research Institute of Sports Epidemiology), INSEP, Paris, France
Université Paris Descartes, Sorbonne Paris Cité, Paris, France
Faculté de Pharmacie, Département de Nutrition, Université Saint Joseph, Beirut, Lebanon

Muriel Tafflet
IRMES (bioMedical Research Institute of Sports Epidemiology), INSEP, Paris, France
INSERM, U970, Paris Cardiovascular Research Center – PARCC, Paris, France

Geoffroy Berthelot and Andy Marc
IRMES (bioMedical Research Institute of Sports Epidemiology), INSEP, Paris, France
Université Paris Descartes, Sorbonne Paris Cité, Paris, France

Julien Tolaini and Marion Guillaume
IRMES (bioMedical Research Institute of Sports Epidemiology), INSEP, Paris, France

Christophe Hausswirth
Research Department, INSEP, Paris, France

Jean-François Toussaint
IRMES (bioMedical Research Institute of Sports Epidemiology), INSEP, Paris, France
Université Paris Descartes, Sorbonne Paris Cité, Paris, France
Hôtel-Dieu Hospital, CIMS, AP-HP, Paris, France

Matthew J. Kesic
Center for Environmental Medicine, Asthma, and Lung Biology, University of North Carolina Chapel Hill, North Carolina, United States of America

Megan Meyer
Center for Environmental Medicine, Asthma, and Lung Biology, University of North Carolina Chapel Hill, North Carolina, United States of America
Department of Microbiology and Immunology University of North Carolina Chapel Hill, North Carolina, United States of America

Rebecca Bauer
Center for Environmental Medicine, Asthma, and Lung Biology, University of North Carolina Chapel Hill, North Carolina, United States of America
Curriculum in Toxicology, University of North Carolina Chapel Hill, North Carolina, United States of America

Ilona Jaspers
Center for Environmental Medicine, Asthma, and Lung Biology, University of North Carolina Chapel Hill, North Carolina, United States of America
Curriculum in Toxicology, University of North Carolina Chapel Hill, North Carolina, United States of America
Department of Pediatrics University of North Carolina Chapel Hill, North Carolina, United States of America
Department of Microbiology and Immunology University of North Carolina Chapel Hill, North Carolina, United States of America

William D. K. Reid, Christopher J. Sweeting and Nicholas V. C. Polunin
School of Marine Science and Technology, Newcastle University, Newcastle upon Tyne, United Kingdom

Ben D. Wigham
Dove Marine Laboratory, School of Marine Science and Technology, Newcastle University, Cullercoats, United Kingdom

Katrin Zwirglmaier
British Antarctic Survey, Natural Environment Research Council, High Cross, Madingley Road, Cambridge, United Kingdom
Technische Universitaet Muenchen, Wissenschaftszentrum Weihenstephan, Limnologische Station Iffeldorf, Iffeldorf, Germany

Jeffrey A. Hawkes
Ocean and Earth Science, University of Southampton, National Oceanography Centre Southampton, Southampton, United Kingdom

Rona A. R. McGill
Natural Environment Research Council Life Sciences Mass Spectrometry Facility, Scottish Universities Environmental Research Centre, East Kilbride, United Kingdom

Katrin Linse
British Antarctic Survey, Natural Environment Research Council, High Cross, Madingley Road, Cambridge, United Kingdom

Mariona Pinart, Min Zhang, Feng Li, Farhana Hussain, Jie Zhu, Coen Wiegman and Kian Fan Chung
Airways Disease, National Heart & Lung Institute, Imperial College, London, United Kingdom

Bernard Ryffel
Université d'Orléans and CNRS, UMR 6218 Molecular Immunology and Embryology, Orléans, France

Carol Arnosti, Kai Ziervogel and Sherif Ghobrial
Department of Marine Sciences, University of North Carolina, Chapel Hill, North Carolina, United States of America

Andrew D. Steen
Department of Marine Sciences, University of North Carolina, Chapel Hill, North Carolina, United States of America
Center for Geomicrobiology, Aarhus University, Herning, Denmark

Wade H. Jeffrey
Center for Environmental Diagnostics and Bioremediation, University of West Florida, Pensacola, Florida, United States of America

Xue-Dong Lou
Chinese Research Academy of Environmental Sciences, Beijing, China
College of Life Sciences, Northwest Agriculture & Forestry University, Yangling, Shaanxi, China,

Sheng-Qiang Zhai1, Li-Le Hu
Chinese Research Academy of Environmental Sciences, Beijing, China

Bing Kang
College of Life Sciences, Northwest Agriculture & Forestry University, Yangling, Shaanxi, China,

Ya-Lin Hu
Institute of Applied Ecology, Chinese Academy of Sciences, Shenyang, China

Åsa Holmner
Department of Radiation Sciences, Umeå University, Umeå, Sweden

Kristie L. Ebi
ClimAdapt, LLC, Seattle, Washington, United States of America
Department of public health and clinical medicine, epidemiology and global health, Umeå University, Umeå, Sweden

Lutfan Lazuardi
Department of Public Health, Faculty of Medicine, Gadjah Mada University, Yogyakarta, Indonesia

Maria Nilsson
Department of public health and clinical medicine, epidemiology and global health, Umeå University, Umeå, Sweden

Min Wang
Linze Inland River Basin Research Station, Chinese Ecosystem Network Research, Cold and Arid Regions Environmental and Engineering Research Institute, Chinese Academy of Sciences, Lanzhou, Gansu, China
University of Chinese Academy of Sciences, Beijing, China

Yongzhong Su and Xiao Yang
Linze Inland River Basin Research Station, Chinese Ecosystem Network Research, Cold and Arid Regions Environmental and Engineering Research Institute, Chinese Academy of Sciences, Lanzhou, Gansu, China

Tianxing Wang, Jiancheng Shi, Yingying Jing, Tianjie Zhao, Dabin Ji, Chuan Xiong
State Key Laboratory of Remote Sensing Science, Institute of Remote Sensing and Digital Earth, Chinese Academy of Sciences. Beijing, China

Lele Hu
Institute of Systems Biology, Shanghai University, Shanghai, China
Department of Chemistry, College of Sciences, Shanghai University, Shanghai, China

Tao Huang
Key Laboratory of Systems Biology, Shanghai Institutes for Biological Sciences, Chinese Academy of Sciences, Shanghai, China
Shanghai Center for Bioinformation Technology, Shanghai, China

Xiao-Jun Liu
College of Animal Science and Technology, Shihezi University, Shihezi City, Xinjiang, China

Yu-Dong Cai
Institute of Systems Biology, Shanghai University, Shanghai, China
Centre for Computational Systems Biology, Fudan University, Shanghai, China

Eike Luedeling
World Agroforestry Centre (ICRAF), Nairobi, Kenya

Evan H. Girvetz
The Nature Conservancy, Seattle, Washington, United States of America

Mikhail A. Semenov
Rothamsted Research, Harpenden, United Kingdom

Patrick H. Brown
Department of Plant Sciences, University of California Davis, Davis, California, United States of America

Shama Ahmad, David P. Nichols, Raymond C. Rancourt, Carl W. White and Aftab Ahmad
Department of Pediatrics, National Jewish Health, Denver, Colorado, United States of America

Matthew Strand
Division of Biostatistics and Bioinformatics, National Jewish Health, Denver, Colorado, United States of America

Scott H. Randell
Cystic Fibrosis/Pulmonary Research and Treatment Center, Department of Medicine, The University of North Carolina at Chapel Hill, Chapel Hill, North Carolina, United States of America

Xiujun Wang
State Key Laboratory of Desert and Oasis Ecology, Xinjiang Institute of Ecology and Geography, Chinese Academy of Sciences, Urumqi, Xinjiang, China
Earth System Science Interdisciplinary Center, University of Maryland, College Park, Maryland, United States of America

Jiaping Wang and Juan Zhang
State Key Laboratory of Desert and Oasis Ecology, Xinjiang Institute of Ecology and Geography, Chinese Academy of Sciences, Urumqi, Xinjiang, China
Graduate University of Chinese Academy of Sciences, Beijing, China

Zhuowei Li, Erin N. Potts-Kant and W. Michael Foster
Department of Medicine, Duke University, Durham, North Carolina, United States of America

Stavros Garantziotis
National Institute of Environmental Health Sciences, Research Triangle Park, North Carolina, United States of America

John W. Hollingsworth
Department of Medicine, Duke University, Durham, North Carolina, United States of America
Department of Immunology, Duke University, Durham, North Carolina, United States of America

Isabelle Beck, Jeroen T. M. Buters and Carsten Schmidt-Weber
ZAUM - Center of Allergy & Environment, Member of the German Center for Lung Research (DZL), Technische Universität München/Helmholtz Center, Munich, Germany

Susanne Jochner and Annette Menzel
Department of Ecology and Ecosystem Management, Ecoclimatology, Technische Universität München, Freising, Germany

Stefanie Gilles and Heidrun Behrendt
ZAUM - Center of Allergy & Environment, Member of the German Center for Lung Research (DZL), Technische Universität München/Helmholtz Center, Munich, Germany
Christine-Kühne-Center for Allergy Research and Education (CK Care), Davos, Switzerland

Mareike McIntyre and Johannes Ring
Department of Dermatology and Allergy, Technische Universität München, Munich, Germany

Claudia Traidl-Hoffmann
ZAUM - Center of Allergy & Environment, Member of the German Center for Lung Research (DZL), Technische Universität München/Helmholtz Center, Munich, Germany
Department of Dermatology and Allergy, Technische Universität München, Munich, Germany
Christine-Kühne-Center for Allergy Research and Education (CK Care), Davos, Switzerland

Kirsten C. Verhein
Department of Physiology and Pharmacology, Oregon Health and Science University, Portland, Oregon, United States of America
National Institute of Environmental Health Sciences, Research Triangle Park, North Carolina, United States of America

Francesco G. Salituro
Vertex Pharmaceuticals, Inc., Cambridge, Massachusetts, United States of America
Sage Therapeutics, Cambridge, Massachusetts, United States of America

Mark W. Ledeboer
Vertex Pharmaceuticals, Inc., Cambridge, Massachusetts, United States of America

Allison D. Fryer and David B. Jacoby
Division of Pulmonary and Critical Care Medicine, Oregon Health and Science University, Portland, Oregon, United States of America

Wen-Jun Zhou
Key Laboratory of Tropical Forest Ecology, Xishuangbanna Tropical Botanical Garden, Chinese Academy of Sciences, Mengla, Yunnan, China
Xishuangbanna Tropical Botanical Garden, Chinese Academy of Sciences, Kunming, China
Xishuangbanna Station for Tropical Rain Forest Ecosystem Studies, Chinese Ecosystem Research Net, Mengla, Yunnan, China
University of Chinese Academy of Sciences, Beijing, China

Yi-Ping Zhang, Li-Qing Sha, Yun Deng and Xiao-Bao Deng
Key Laboratory of Tropical Forest Ecology, Xishuangbanna Tropical Botanical Garden, Chinese Academy of Sciences, Mengla, Yunnan, China
Xishuangbanna Tropical Botanical Garden, Chinese Academy of Sciences, Kunming, China
Xishuangbanna Station for Tropical Rain Forest Ecosystem Studies, Chinese Ecosystem Research Net, Mengla, Yunnan, China

Douglas A. Schaefer and Kai-Jie Dai
Key Laboratory of Tropical Forest Ecology, Xishuangbanna Tropical Botanical Garden, Chinese Academy of Sciences, Mengla, Yunnan, China
Xishuangbanna Tropical Botanical Garden, Chinese Academy of Sciences, Kunming, China

Bing-Fang Hwang and Li-Hao Young
Department of Occupational Safety and Heath and Graduate Program, College of Public Health, China Medical University, Taichung, Taiwan

Ching-Hui Tsai, Kuan-Yen Tung, Pei-Chuan Wang, Ming-Wei Su and Yungling Leo Lee
Institute of Epidemiology and Preventive Medicine, College of Pubic Health, National Taiwan University, Taipei, Taiwan

Jianbo Wu, Xiaodan Wang, Xuyang Lu, Jihui Fan and Yanjiang Cai
The Key Laboratory of Mountain Environment Evolution and Its Regulation, Institute of Mountain Hazard and Environment, CAS, Chengdu, China

Jiangtao Hong and Jian Sun
The Key Laboratory of Mountain Environment Evolution and Its Regulation, Institute of Mountain Hazard and Environment, CAS, Chengdu, China
University of Chinese Academy of Sciences, Beijing, China

Index